Lecture Notes in Networks and Systems

Volume 1549

The series "Lecture Notes in Networks and Systems" publishes the latest developments in Networks and Systems—quickly, informally and with high quality. Original research reported in proceedings and post-proceedings represents the core of LNNS.

Volumes published in LNNS embrace all aspects and subfields of, as well as new challenges in, Networks and Systems.

The series contains proceedings and edited volumes in systems and networks, spanning the areas of Cyber-Physical Systems, Autonomous Systems, Sensor Networks, Control Systems, Energy Systems, Automotive Systems, Biological Systems, Vehicular Networking and Connected Vehicles, Aerospace Systems, Automation, Manufacturing, Smart Grids, Nonlinear Systems, Power Systems, Robotics, Social Systems, Economic Systems and other. Of particular value to both the contributors and the readership are the short publication timeframe and the world-wide distribution and exposure which enable both a wide and rapid dissemination of research output.

The series covers the theory, applications, and perspectives on the state of the art and future developments relevant to systems and networks, decision making, control, complex processes and related areas, as embedded in the fields of interdisciplinary and applied sciences, engineering, computer science, physics, economics, social, and life sciences, as well as the paradigms and methodologies behind them.

Indexed by SCOPUS, EI Compendex, INSPEC, WTI Frankfurt eG, zbMATH, SCImago.

All books published in the series are submitted for consideration in Web of Science.

For proposals from Asia please contact Aninda Bose (aninda.bose@springer.com).

Sanjay Kumar · Rekh Ram Janghel · Badal Soni ·
Ugo Fiore
Editors

Artificial Intelligence for Next Generation Computing, Volume 1

Select Proceedings of the International Conference, AICTA 2024

Springer

Editors
Sanjay Kumar
Department of Information Technology
National Institute of Technology Raipur
Raipur, Chhattisgarh, India

Rekh Ram Janghel
Department of Information Technology
National Institute of Technology Raipur
Raipur, Chhattisgarh, India

Badal Soni
Department of Computer Science
and Engineering
National Institute of Technology Silchar
Silchar, Assam, India

Ugo Fiore
Department of Computer Science
University of Salerno
Fisciano, Salerno, Italy

ISSN 2367-3370 ISSN 2367-3389 (electronic)
Lecture Notes in Networks and Systems
ISBN 978-981-95-0700-9 ISBN 978-981-95-0701-6 (eBook)
https://doi.org/10.1007/978-981-95-0701-6

This Springer imprint is published by the registered company Springer Nature Singapore Pte Ltd.
The registered company address is: 152 Beach Road, #21-01/04 Gateway East, Singapore 189721, Singapore

Preface

It is with great pleasure that we present the proceedings of the **2nd International Conference on Artificial Intelligence, Computing Technologies, Internet of Things (IoT) and Data Analytics (AICTA 2024)**, held at **NIT Raipur, India on 15–17 November 2024**. This conference brought together leading researchers, engineers, industry experts, and practitioners from around the globe to share cutting-edge advancements, discuss emerging challenges, and explore innovative solutions at the intersection of telecommunications and artificial intelligence.

The rapid evolution of AI-driven technologies and next-generation communication systems has opened new frontiers in research and application. AICTA 2024 served as a dynamic platform for fostering interdisciplinary collaboration, featuring high-quality contributions spanning AI for 5G/6G networks, machine learning in telecommunications, IoT and smart systems, cybersecurity, natural language processing, computer vision, and beyond.

This volume compiles rigorously peer-reviewed papers that reflect the latest theoretical and practical developments in the field. Each submission underwent a meticulous evaluation process by our international Technical Program Committee, ensuring the highest standards of originality, technical soundness, and relevance. We extend our deepest gratitude to the authors for their valuable contributions, the reviewers for their insightful feedback, and the organizing committee for their tireless efforts in making this event a success.

We also acknowledge the support of our sponsors, partners, and institutional collaborators, whose contributions were instrumental in shaping AICTA 2024. Special thanks go to our publishing partner Springer for their continuous support in advancing research in telecommunications and AI.

As we move toward an increasingly connected and intelligent future, the insights and innovations shared at AICTA 2024 will undoubtedly inspire further breakthroughs. We hope this proceedings volume serves as a valuable resource for academia and industry alike, fostering continued progress in these transformative fields.

Raipur, India Sanjay Kumar
Raipur, India Rekh Ram Janghel
Silchar, India Badal Soni
Fisciano, Italy Ugo Fiore

Contents

NLP-Based Approach for Efficient Keyword Identification in Unstructured Text

Minu Choudhary, Sourabh Rungta, and Shikha Pandey

Abstract This paper examines the need for efficient keyword extraction methods to improve search accuracy and indexing due to the increasing volume of digital information. Traditional approaches, often based on word occurrence counts, tend to lose the contextual meaning of the text. The current work proposes a four-step NLP-based method for keyword extraction, including tokenization, POS tagging, noun phrase extraction, and chunking. The novel contribution of this research is the design of a grammatical pattern for extracting keywords (in the form of noun phrases) without losing the actual meaning of the text. This approach aims to retain the contextual meaning of the text. When compared to the most popular NLTK tool RAKE, the proposed method delivers better results while preserving the contextual meaning of the input text.

Keywords Extraction · Information retrieval · NLP · Keywords · RAKE

1 Introduction

The amount of unstructured text data generated across multiple domains is surprising in the current digital era. Unstructured text is a wealth of information that may be found in everything from internet reviews and social media posts to court documents

M. Choudhary (✉) · S. Rungta
Department of Computer Science and Engineering, RCET, Bhilai, Chhattisgarh, India
e-mail: minukum@gmail.com

S. Rungta
e-mail: sourabh@rungta.ac.in

M. Choudhary
School of Computer Science and Engineering, LPU, Phagwara, Punjab, India

S. Pandey
Department of Computer Science and Engineering, BIT, Durg, Chhattisgarh, India
e-mail: shikhamtech2008@gmail.com

S. Kumar et al. (eds.), *Artificial Intelligence for Next Generation Computing, Volume 1*, Lecture Notes in Networks and Systems 1549,
https://doi.org/10.1007/978-981-95-0701-6_1

and scientific publications. However, because these data are unstructured, it is difficult to draw useful conclusions from it. Keyword identification is an important phase in this process, as it forms the basis for several applications like topic modelling, text summarization, and information retrieval. According to author Manning et al. [1], conventional approaches to keyword identification frequently depend on human tagging or basic statistical metrics, which can be laborious and insufficient for capturing the complex meanings and contextual relevance of terms in big text corpora. Since the introduction of natural language processing (NLP), increasingly complex and automated methods have been developed, utilizing cutting-edge algorithms and machine learning strategies to improve the precision and effectiveness of keyword extraction [2]. This paper offers an effective method for identifying keywords in unstructured text using natural language processing (NLP). The suggested methodology is made to handle texts of any length and doesn't have any size restrictions. Tokenization of the text, which divides the input into digestible chunks, is the first step in the process. After tokenization, each token is labelled with the appropriate part-of-speech using Part-of-Speech (POS) tagging, which gives the reader a structural comprehension of the text. The extraction of noun phrases (NPs) [3] is a crucial next step. Noun phrases are important because they frequently hold the most pertinent information in a document. Following extraction, chunking—a procedure that organizes the tokens into coherent phrases based on grammatical patterns—is applied to these noun phrases. The recognized sentences are guaranteed to be logical and pertinent to the given context by this chunking method. Eventually, a collection of significant words or phrases of various lengths that function as the detected keywords are the result of this methodology. The essential information from the unstructured text is condensed into these keywords, which improves data accessibility and facilitates analysis. This method provides a methodical and effective way to identify keywords by utilizing sophisticated natural language processing (NLP) techniques. This opens the door to more sophisticated text processing and information retrieval in big and complicated datasets. Our objective is to present a thorough framework that increases the theoretical knowledge of keyword identification while simultaneously providing scholars and practitioners working with large volumes of unstructured text with useful tools and approaches [4]. Robust keyword identification algorithms are becoming more and more important as the need for effective text processing grows. By this work, we add to the continuing endeavours to leverage natural language processing (NLP) to convert unstructured text into insightful, useful information. A method for adapting language models to specific subject domains through semantic and morphological matching was presented in [5]. A recent [6] analysis traced the development of approaches to keyword extraction over the last 50 years, stressing advances from statistical components and focussing on neural networks, as well as evaluating them in various areas, including news, science, and business. The study [7] advanced the BSKT improvement algorithm developed by the combinatory K-Truss–BERT, which had proven to enhance the keyword extraction effectiveness, outperforming previous algorithms. Several studies [8] attempted measuring the performance of the keyword extraction techniques employed, demonstrating that Novel RAKE outperformed the Bag of Words approach. This study [9] performed a text-mining application towards bridge defect description alternatively used in standard rail lines construction and synthesized BiLSTM_ATT model which outperformed other models in terms of precision, recall, and F-measure in keyword extraction tasks.

The research done so far on the topic keyword extraction recalls the progression of algorithms for extracting key phrases ranging from statistical methods to BERT and BiLSTM_ATT approaches. The customization of language models has been done with the aid of semantic and morphological matching, while performance over classical ones such as Bag of Words has been shown by algorithms BSKT and Novel RAKE. The developments further highlighted issues such as need for domain-specific adaptation and efficient data handling. In particular, the studies show the high importance of advanced algorithms for extracting keywords in relation to relevance and precision of information retrieval in unstructured text datasets in different areas. The rest of the paper is organized as follows. Section 2 contains the literature review. Section 3 contains the proposed methodology. Section 4 contains the experiments done, results, and comparison with RAKE. Section 5 contains the conclusions and future work and scope.

2 Literature Review

Keyword extraction is a critical task in natural language processing (NLP) aimed to identify key terms or phrases that represent the main content of a text. Various methods have been proposed in the literature for keyword extraction, each with its strengths and limitations.

Table 1 summarizes the four different kinds of keyword extraction methods along with their pros and cons. The years represent either the publication or introduction dates of the approaches, with some being continuously developed and refined over time.

The various types of methods available for practically extracting keywords are all efficiently effective; however, each method has its strong and weak points. Older methods based on statistics like TF-IDF, TextRank, and LDA are efficient and simple methods in that they have proven successful and applicable in many fields of information retrieval and document classification but then do not consider the relationships and context of the words within the text.

Machine learning methods offer better features such as supervised learning with feature engineering that allows the use of labels and designed features. These features improve the performance but require a large volume of training data and may bring about overfitting problems. Neural networks especially word embeddings and RNNs allow a better understanding of the semantic and contextual aspect, thus word embeddings and RNN have uses in tasks like sentiment analysis and language modelling. The use of transformer-based models like BERT and variants thereof represents a great step forward in keyword extraction offering top-notch accuracy owing to its bidirectional context comprehension. Nevertheless, these models also come with a lot of computational power and tricky fine-tuning circumstances.

Table 1 Keyword extraction approaches

Sl. no	Approach	Method	Description	Advantages	Disadvantages	Application areas
1	Statistical methods	TF-IDF [10]	Evaluates the importance of a word in a document relative to a corpus	Simple and easy to implement, interpretable results	Ignores semantic relationships, sensitive to document length	Information retrieval, document classification, SEO
		TextRank [11]	Unsupervised algorithm based on PageRank; builds a graph where words are nodes, and edges represent co-occurrence within a fixed window, ranking words based on their centrality	Unsupervised, no need for labelled data, captures co-occurrence relationships	Can be computationally intensive, may miss context-specific keywords	Text summarization, keyword extraction, recommendation systems
		LDA [12]	Latent Dirichlet Allocation; identifies topics within a text, representing each document as a mixture of topics and each topic as a mixture of words	Effective for topic modelling, provides a probabilistic framework	Requires setting the number of topics, may not perform well on short texts	Topic modelling, document clustering, content recommendation
2	Machine learning methods	Supervised learning [13]	Uses algorithms like Naive Bayes, support vector machines (SVM), and decision trees for keyword extraction, requiring labelled training data to learn the significance of keywords	Can be highly accurate with good training data, customizable	Requires large labelled datasets, can overfit to training data	Spam detection, sentiment analysis, named entity recognition
		Feature engineering [14]	Utilizes features such as word frequency, part-of-speech tags, and syntactic dependencies to improve keyword extraction models	Enhances model performance, allows incorporation of domain knowledge	Labour-intensive, feature selection can be subjective	Text classification, entity recognition, information extraction

(continued)

Table 1 (continued)

Sl. no	Approach	Method	Description	Advantages	Disadvantages	Application areas
3	Neural networks	Word embeddings [15]	Methods like Word2Vec and GloVe capture semantic relationships between words, improving keyword extraction	Captures semantic meaning, improves performance on various NLP tasks	Requires significant computational resources, may struggle with rare words	Sentiment analysis, text classification, machine translation
		RNNs (LSTM) [16]	Recurrent Neural Networks, particularly Long Short-Term Memory networks, understand context and sequential dependencies in text	Effective for sequential data, captures long-range dependencies	Computationally expensive, training can be slow	Language modelling, speech recognition, text generation
4	Transformers	BERT and variants [17]	Bidirectional Encoder Representations from Transformers (BERT) and its variants like RoBERTa and DistilBERT, fine-tuned for keyword extraction tasks, providing state-of-the-art results	State-of-the-art performance, captures bidirectional context	Requires large amounts of data and computational resources, complex to fine-tune	Text classification, entity recognition, sentiment analysis
		Transformers for summarization [18]	Transformers used in summarization models indirectly enhance keyword extraction by focussing on key phrases	Improves keyword extraction by leveraging summarization capabilities	May be computationally intensive, requires fine-tuning for specific tasks	Text summarization, document analysis, content generation

(continued)

Table 1 (continued)

Sl. no	Approach	Method	Description	Advantages	Disadvantages	Application areas
5	Hybrid approaches	Combining statistical and neural methods [19]	Integrates statistical methods with neural networks to balance precision and computational efficiency	Balances the strengths of both approaches, flexible and robust	Can be complex to implement, may require significant tuning	Information retrieval, text classification, recommendation systems
		Multi-modal keyword extraction [20]	Combines text with other data modalities, such as images and audio, to enhance keyword extraction in diverse contexts	Enhances keyword extraction by incorporating additional data modalities	Increased complexity, requires alignment of different data types	Multimedia search, video analysis, cross-modal retrieval

Combining statistical and neural based methods in the form of hybrid approaches seems to be the best way of improving keyword extraction since both the paradigms are optimally used. Since multi-modal keyword extraction extracts multiple relevant words from different types of data, it improves the efficiency of the keyword extraction methods by extending their use in the areas of multimedia and cross-modal retrieval.

Thus, the bias of the selected approach should relate closely with what is required in the task underscored; whether more precision in output is needed, contextualization or computation is available. As these methodologies continue to develop and evolve, they will progressively push the boundaries of natural language processing, and the difficulties of efficient contextual keyword extraction in different fields will be a relic of the past.

3 Proposed Methodology

The process begins with the input of a text document of any length, followed by breaking the text into sentences, as they appear in the original document, regardless of its length. Our system accepts both.txt and.pdf file formats. The further steps are as follows:

i. Tokenization: The first step is tokenization, which involves breaking down an input sentence into smaller tokens. Here, we keep the documents in their original form, without applying any removal techniques such as stop word removal or stemming, as these can cause the loss of the actual meaning of the sentences.
ii. Part-of-Speech (POS) tagging: In this step, tokens are used as the input and Part-of-Speech (POS) tagging is used to retrieved part-of-speech label. In this step, Stanford parser third party tool, is used [https://nlp.stanford.edu/software/lex-parser.html]. The grammatical function of each token, such as noun, verb, or adjective, is indicated by POS (Part-of-Speech) tags.
iii. Extraction of Noun Phrases: Determining noun phrases and other grammatical structures requires precise POS tagging. Generally, noun phrases consist of a noun along with any adjectives or determiners associated with it. Since noun phrases frequently contain the most informative substance in a sentence, extracting them is essential. These components are identified using POS tags such as NN, NNP, and NNS. The following are the grammar patterns devised in the experiment for identifying the Noun and Verb families of word components (tokens), with examples shown in Table 2. It is important to note that the meta-symbols ‘,’, ‘*’, and ‘+’ are used to denote ‘various types of that pattern’, ‘zero or more occurrences of that pattern’, and ‘one or more occurrences of that pattern’, respectively, representing all possible regular patterns of nouns.
iv. Chunking: This step entails organizing noun phrases according to their POS tags into meaningful words. The extracted noun phrases undergo additional analysis and refinement. Chunking facilitates the discovery of textual structures

Table 2 Regular patterns for noun and verb families of tokens

Regular expressions (noun and verb family of tokens)	Examples of regular expressions
NP:{<VBN > < NN.* > + }	(NP: shared/VBN control/NN code/NN)
NP:{<VBG > < NN.* > + }	(NP: using/VBG classes/NNS)
NP:{<JJ > < CC > < DT > < NN > }	(NP: little/JJ or/CC no/DT modification/NN)
NP:{<NN.* > + < CD.* > + }	(NP: others/NNS One/CD)
NP:{<RB > < JJ > < NN.* > + }	(NP: sufficiently/RB similar/JJ interfaces/NNS)
NP:{<JJ.* > * < NN.* > + }	(NP: many/JJ different/JJ larger/JJR problems/NNS)
NP:{<NN.* > < CD.* > + }	(NP: Figure/NN 2/CD)
NP:{<NN.* > + }	(NP: object-oriented/JJ programming/NN inheritance/NN)

and patterns that are crucial for keyword extraction. These chunked words are the final keywords.

4 Experimental Setup, Results, and Comparison with RAKE [2]

To make the text documents available in natural (here, in English) language, understandable by the machine, like any other text-mining experiment, we also required a Natural Language Tool Kit (NLTK) which is freely downloadable from its website URL (https://www.python.org/downloads). For this work, python 3.12 framework has been used for experiments. Now the following steps are performed as a part of text pre-processing phase that always precedes the involved text-mining task of the experiment:

i. **Tokenization of sentence**: This method decides token limits, such as words and punctuation symbols in sentences. An example of a sentence and its tokenized output representation is illustrated below:

Input sentence: 'This article describes a method used to calculate the similarity between short English texts, specifically of sentence length.' **Tokenized Output Sentence**: [1–'This', 2–'article', 3–'describes',4–'a', 5–6–'method', 7–'used', 8–'to', 9–'calculate', 10–'the', 11–'similarity', 12–'between', 13- 'short', 14–'English', 15–'texts', 16–',', 17–'specifically', 18–'of', 19–'sentence', 20–'length', 21–'.']

ii. **POS Tagging**: This method assigns grammatical tags (Part-Of-Speech tagging) to each word (fetched token), such as noun, pronoun, verb, adverb, adjective (qualifier) category for detecting grammatical categories of the word forms (tokens) in the sentence. For example, a sentence can be converted into its POS tagged representation as explained below:

Input Sentence: 'This article describes a method used to calculate the similarity between short English texts, specifically of sentence length'. **POS Output Sentence:** [('This', 'DT'), ('article', 'NN'), ('describes', 'VBZ'), ('a', 'DT'), ('method', 'NN'), ('used', 'VBN'), ('to', 'TO'), ('calculate', 'VB'), ('the', 'DT'), ('similarity', 'NN'), ('between', 'IN'), ('short', 'JJ'), ('English', 'NNP'), ('texts', 'NN'), (',', ','), ('specifically', 'RB'), ('of', 'IN'), ('sentence', 'NN'), ('length', 'NN'), ('.', '.')

iii. **Extraction of Noun phrases (NPs):** The noun phrases are extracted from the POS tagged word components of both, the 'abstract' portion along with 'keywords' portion inside suspicious document by fetching word components pertaining to noun family.

While performing experiments, it was observed that recognizing of these regular patterns had been done in a specific sequence, which, otherwise, would fetch incomplete noun phrases and verb phrases of maximal lengths. The above-mentioned list of regular patterns responsible for recognition of correct (noun or verb) chunk sequences is tabulated in Table 2. By checking on correct sequences of the enlisted patterns, the experiments were capable of recognizing almost all combinations of sentence structures combining noun phrases reflecting complete meaning or making sense.

iv. **Chunking:** This text processing step is also called shallow parsing. It has been incorporated in the experiment to help identifying major phrasal constituents in a sentence, namely, noun phrase in the pair of sentences. Each phrasal constituent represents one chunk comprising semantically related word components. Chunking step is necessary before we attempt to explore the internal structure of the sentence or the role of word components in the sentence structures. This step is illustrated well by considering two paragraph instances (shown in Table 3) giving rise to chunked list of noun phrases by using RAKE algorithm and by using current algorithm.

From the Table 3, it is observed that the RAKE method identifies more individual keywords, while current method seems to focus on more meaningful phrases. Now applying comparison of current algorithm with the RAKE algorithm based on precision, recall, and F1 score is presented in Table 4. From the Table4 it is noted that the RAKE method has an average precision of approximately **0.4095**, a perfect recall of **1.0**, and an average F1 score of approximately **0.58**. Our method achieves perfect scores in all three metrics with averages of **1.0** for precision, recall, and F1 score.

5 Conclusion and Future Work and Scope

Here, the scholar would like to emphasize on the point that has already been discussed in [21–25]. This is with reference to, not using third party tools like segmentation, stop word removal, stemmed words, delimiters, numerals, and punctuation symbols,

Table 3 Comparison of current method with RAKE tool

Sl. no.	Input text	Noun phrase chunking by RAKE [2]	Noun phrase chunking by our method [2024]
1	'Natural language processing enables computers to understand human language'	Language Understand Processing Natural Human	Natural language processing Computers Human language
2	'Learning a new language is rewarding. Immersion programs offer intensive practice and cultural exposure'	Rewarding Programs Practice Offer New	New language Immersion programmes Intensive practice Cultural exposure

Table 4 Precision, recall, and F-score of both the methods

Metric	NLTK tool (RAKE,2024)	Current NLP-based Tool (2024)
Precision	0.4095	1.0
Recall	1.0	1.0
F1-score	0.58	1.0

as an embedded layer to the above-mentioned four-layered text processing task. This is due to our own subsumption of carrying out pre-processing tasks without bearing any information. In future work, Scholar will improve the algorithms by using deep learning and statistical methods. Some potential future developments and research directions in keyword extraction methods is required like in the field of multilingual, domain specific and to understand the semantic or contextual knowledge of text.

References

1. Manning CD, Raghavan P, Schütze H (2008) Introduction to information retrieval. Cambridge University Press
2. Rose S, Engel D, Cramer N, Cowley W (2010) Automatic keyword extraction from individual documents. In: Text mining: applications and theory, pp 1–20
3. Chen KH, Chen HH (1994) Extracting noun phrases from large-scale texts: a hybrid approach and its automatic evaluation. arXiv preprint cmp-lg/9405034
4. Vaswani A, Shazeer N, Parmar N, Uszkoreit J, Jones L, Gomez AN, Polosukhin I (2017) Attention is all you need. In: Advances in neural information processing systems, pp 5998–6008
5. Gretter R, Matassoni M, Falavigna D (2021) Seed wordsbased data selection for language model adaptation. arXiv preprint arXiv:2107.09433
6. Nomoto T (2022) Keyword extraction: a modern perspective. SN Comput Sci 4(1):92
7. Guo W, Wang Z, Han F (2022) Multifeature fusion keyword extraction algorithm based on TextRank. IEEE Access 10:71805–71813
8. Muppala G, Devi T (2023) Accurate recasting of giant text into charts using rapid automatic keyword extraction algorithm in comparison with bag of words algorithm. In: 2023

6th international conference on contemporary computing and informatics (IC3I), vol 6. IEEE, pp 2548–2552
9. Guo X, Li J, Shen K, Li L (2024) Extraction of keywords for describing defects in standard rail lines bridge equipment. In: 2024 7th International Conference on Advanced Algorithms and Control Engineering (ICAACE). IEEE, pp 292–295
10. Salton G, Buckley C (1988) Term-weighting approaches in automatic text retrieval. Inf Process Manage 24(5):513–523
11. Mihalcea R, Tarau P (2004) TextRank: bringing order into texts. In: Proceedings of the 2004 conference on empirical methods in natural language processing, pp 404–411
12. Blei DM, Ng AY, Jordan MI (2003) Latent Dirichlet allocation. J Mach Learn Res 3:993–1022
13. Joachims T (1998) Text categorization with support vector machines: learning with many relevant features. In: European conference on machine learning. Springer, Berlin, Heidelberg, pp 137–142
14. Sebastiani F (2002) Machine learning in automated text categorization. ACM Comput Surv (CSUR) 34(1):1–47
15. Mikolov T, Chen K, Corrado G, Dean J (2013) Efficient estimation of word representations in vector space. arXiv preprint arXiv:1301.3781
16. Hochreiter S, Schmidhuber J (1997) Long short term memory. Neural Comput 9(8):1735–1780
17. Devlin J, Chang MW, Lee K, Toutanova K (2018) BERT: pre-training of deep bidirectional transformers for language understanding. arXiv preprint arXiv:1810.04805
18. Liu Y, Lapata M (2019) Text summarization with pretrained encoders. In: Proceedings of the 2019 conference on empirical methods in natural language processing and the 9th international joint conference on natural language processing
19. Hsu WH, Lin CJ (2002) A comparison of methods for multiclass support vector machines. IEEE Trans Neural Netw 13(2):415–425
20. Ngiam J, Khosla A, Kim M, Nam J, Lee H, Ng AY (2011) Multimodal deep learning. In: Proceedings of the 28th international conference on machine learning (ICML-11), pp 689–696
21. Pandey S, Pandey V (2024) Recommendations for ensuring equitable and comparable paragraphs in a research paper. In: Ninth international conference on science technology engineering and mathematics (ICONSTEM). IEEE, pp 1–3
22. Pandey S, Rawal A (2019) Mapping ethical writing guidelines to plagiarism detection heuristics. J Adv Res Dynam Contr Syst JARDCS 4(11):1172–1177
23. Pandey S, Rawal A (2018) An NLP based plagiarism detection approach for short sentences. Int J Rec Technol Eng (IJRTE) 7(4):215–219
24. Pandey S, Sharma HR, Rawal A (2018) Considerable issues over intelligent plagiarism detection methods. Int J Pure Appl Math 119(7):85–96
25. Pandey S, Sharma HR, Rawal A (2018) An improved NLP approach for detection of plagiarism in scientific paper. Int J Appl Eng Res 13(9):6733–6737

Enhancing Cybersecurity Log Analysis with LogGPT: A Novel LLM-Based Approach

Raj Laxmi Singh, Vaishnavi Jadhav, Mahak Chamria, and Vikas Kumar Jain

Abstract Ensuring compliance and finding anomalies in large log collections are major issues in modern cybersecurity environments. This study presents LogGPT, a novel approach that revolutionizes the conventional paradigm of compliance detection and monitoring. Large Language Models (LLMs) are a powerful tool that LogGPT uses to analyze logs without the assistance of other programs. The study examines the inherent drawbacks of the current approaches and offers LogGPT as an advanced substitute. By showcasing LLMs' greater flexibility and contextual awareness compared to more conventional deep learning models, LogGPT offers improved compliance monitoring accuracy by 14% over models like GPT4ALL, and overall, it achieves an accuracy of 89.43%. Experimental results show the effectiveness of the proposed approach.

Keywords Cybersecurity · Large Language Models · Natural language processing

1 Introduction

The necessity for strong compliance monitoring and anomaly identification within large log datasets has become critical in the constantly changing field of cybersecurity. Conventional approaches frequently entail intricate procedures and reliance on other programs, resulting in inefficiencies and possible weaknesses. An important field of research with practical implications for failure identification [1, 2] and the detection of security threats [3, 4]. These logs are structured recordings generated by configured systems or applications intended to communicate relevant details about the state of software or applications as they are being executed. The principal objective of log analysis is to facilitate the rapid identification of anomalous activities, hence enabling expeditious remediation to tackle or alleviate the concerns recorded in the logs. This endeavor forms part of an overarching initiative to bolster system resilience against operational faults, performance deterioration, and deliberately

R. L. Singh · V. Jadhav · M. Chamria · V. K. Jain (✉)
SCSET, Bennett University, Greater Noida, India
e-mail: vikasjain.cse@gmail.com

S. Kumar et al. (eds.), *Artificial Intelligence for Next Generation Computing, Volume 1*, Lecture Notes in Networks and Systems 1549,
https://doi.org/10.1007/978-981-95-0701-6_2

instigated cyber-physical assaults. Moreover, log analysis facilitates the inquiry and examination into the causative factors behind such aberrant activities. This research's focal point pertains to identifying these unusual occurrences within logs. However, the inherent attributes of logs, characterized by their extensive volume, diversity, and contextual intricacies, render routine analysis challenging, necessitating the implementation of automated solutions. Although rule-based or signature-driven automation strategies offer assistance, logs' inherent contextual and semantic intricacies constrain their effectiveness.

In this direction, the LogGPT is designed to solve two key problems in cybersecurity log analysis: the difficulty of interpreting unstructured log data and the inefficiency of current anomaly detection methods. Traditional methods often require extensive pre-processing, including log parsing, which can lead to misinterpretation of complex or context-specific data. Additionally, existing models struggle with high false-positive rates and lack the flexibility to handle new or unseen anomalies in real-time. The LogGPT addresses these challenges by directly analyzing raw log data and improving the accuracy and efficiency of anomaly detection without relying on external pre-processing tools.

The paper is organized as follows: Sect. 2 describes the background and related work. Section 3 presents the proposed methodology along with a flowchart and the steps involved in it. Section 4 describes the dataset used experimental setup and results analysis. Finally, it has been concluded in Sect. 5.

2 Background and Literature Review

2.1 Need for Analysis of Logs

Logs are produced by software-operated applications running on devices or systems to furnish information aiding developers and system engineers in their evaluation of the system's operational status and state. They serve as an audit trail, chronologically documenting events for reference. Log analysis serves the purpose of assisting investigations subsequent to incidents related to the generating system, which may include system defects or unauthorized breaches. These analyses aim to utilize logs as a means to reconstruct the sequence of events during such incidents. During investigation, an analyst or system engineer endeavors to pinpoint abnormal events within the logs. However, detecting such anomalies necessitates expertise in discerning irregularities among the extensive entries within log files.

2.2 Challenges with Log Analysis

Log formats vary based on software development and configuration, making them context-specific and challenging to interpret. Each system or software component adopts unique logging formats and lexicon representations, complicating the analysis. The chronological sequencing of log entries adds a temporal dimension, further emphasizing the need for contextual understanding. Challenges include classifying normal and abnormal entries, requiring both semantic comprehension and limited knowledge about potential anomalies. The structure of logs is typically tailored to how the software is developed or configured to input data into these textual records. Additionally, each system or software component may employ its distinct logging structure and terminology, elucidating the operational state of the system at runtime when entries are made. The content within logs is notably contextual, and specific to the system's environment. For instance, this contextual information includes details such as IP addresses, hostnames, or resource identifiers. Moreover, the content of log entries is influenced by the configuration of the associated system and its environmental variables. Beyond contextual settings, log entries are organized based on the chronological sequence of events or system states, thereby encompassing a temporal dimension.

2.3 Related Work

Within this segment, we examine the prevailing algorithmic advancements in log analysis, detailing their respective strengths and limitations. The existing designs of log analysis algorithms commonly encompass a multi-staged approach preceding the application of analysis. Typically, this sequence initiates with log parsing, wherein raw logs transform structured data features. Subsequently, these extracted features undergo further conversion, transitioning from textual representations to numerical forms. Following this, log partitioning ensues, involving the segmentation of contiguous logs into associative partitions to enhance anomaly classification. These partitions may be time-based, organized by windows of similar operations, or divided based on identifiers present in log entries. After these pre-processing stages, the anomaly detection algorithm is employed.

Notably, there is a scarcity of endeavors aimed at constructing an integrated model capable of directly utilizing raw log data for immediate model training and inference. Notable studies, such as those conducted by Hashemi and Mäntylä [5] and Le and Zhang [6], highlight methods that intake logs sans log parsers. Particularly, Le and Zhang observed that log parsers might lead to inaccuracies in log parsing due to misinterpretation of semantic meanings within log analysis and may not proficiently handle Out-of-Vocabulary (OOV) words. Our approach addresses these concerns by eliminating the necessity for log parsing, and facilitating inferences directly on raw log data inputs.

Numerous algorithms in log analysis concentrate on the crucial task of identifying anomalous events within logs. He et al. [7] and Chen et al. [8] conducted surveys elucidating that these algorithms predominantly fall into supervised and unsupervised machine learning algorithms. These methods may rely on classical machine learning or delve into deep learning algorithms. Supervised learning approaches face constraints linked to the availability of labeled anomalous data within training datasets. Furthermore, even when anomalous data is accessible within log datasets, class imbalance poses a substantial challenge during model training. Additionally, these models may necessitate retraining or reconstruction to internalize new knowledge. Unsupervised learning algorithms [9, 10] grapple with the efficiency of detecting the diverse and evolving anomalies reflected in log entries, which may manifest and fluctuate significantly over extended periods. When novel anomalies emerge, these unsupervised models demand retraining. Our methodology diverges by utilizing a vector database to store only small samples of normal log entries. Employing a Large Language Model (LLM), our approach conducts anomaly detection without any samples of anomalous log entries, presenting itself as a zero-shot classifier.

Recent endeavors have explored the utilization of Large Language Models (LLMs) for log analysis. Qi et al. [11] introduced a framework employing ChatGPT for log-based anomaly detection, experimenting with diverse prompt constructions, window sizes, and input sequences. Their findings revealed challenges in identifying an optimal prompt, encountering limitations in window size, and encountering high rates of false positives. Mudgal et al. [12] developed specific prompts using ChatGPT for log parsing, demonstrating excellent performance. However, within other aspects of log analysis such as anomaly detection and log summarization, LLMs displayed limitations that necessitate further investigation. Liu et al. [13] assessed their LogPrompt model under zero-shot scenarios using various prompt formats and varying numbers of provided log samples. While the zero-shot test results exhibited promise compared to other Deep Learning architectures, the precision scores were notably low. This aspect is suboptimal when considering the application of log analysis for operational and maintenance activities to support system resiliency. These preliminary experiments underscore the need for continued research into applying LLMs for log analysis, especially in anomaly detection within logs, serving as the foundational premise for this ongoing research work. Table 1 compares existing approaches with the proposed one on various parameters.

This comprehensive review of related works establishes the context for LogGPT, offering insights into existing methodologies, challenges, and opportunities in log analysis, compliance monitoring, and anomaly detection. The synthesis of findings from these works will inform the development and positioning of LogGPT in the evolving landscape of cybersecurity. Therefore, this research introduces a groundbreaking solution, LogGPT, designed to revolutionize the approach to compliance monitoring through the application of Large Language Models (LLMs). The LogGPT aims to mitigate the challenges associated with conventional techniques by leveraging the advanced capabilities of the LLMs, offering a more adaptive, context-aware,

Table 1 Comparison LogGPT with existing AI-driven log analysis methods

Method	Log parsing	Handling unstructured data	Anomaly detection	False positive rate	Real-time compliance monitoring
Log prompt	Required	Limited	Moderate	Moderate	No
ChatGPT-based	Required	Limited	Moderate	High	No
GPT4ALL	Required	Moderate	Low	Moderate	No
LogGPT	Not required	Excellent	High	Low	Yes

and accurate methodology for log analysis. This introduction sets the stage for an in-depth exploration of the LogGPT framework, its methods, and its transformable impact on cybersecurity practices.

3 Proposed Solution

3.1 Log Analysis

Our approach is based on using state-of-the-art Large Language Models (Llama2, to be exact) for the complex task of log analysis. Through the use of LLMs' sophisticated natural language processing powers, LogGPT aims to improve log entries' contextual comprehension. This method overcomes the difficult problem of various log formats, allowing LogGPT to recognize subtle information such as IP addresses and hostnames. Employing ongoing training and adaptive learning, LogGPT guarantees unmatched compatibility with diverse log topologies, offering a robust and flexible basis for thorough log analysis.

3.2 Compliance Rule Engine

Advanced log analysis is combined with a dynamic Compliance Rule Engine by LogGPT. Security and compliance teams can easily create and modify compliance rules with the help of this engine. The Compliance Rule Engine functions as the key to detecting deviations from defined compliance standards, easily integrating with insights obtained by LLM-based log analysis. This integrated strategy guarantees both contextual awareness and flexibility in response to the constantly changing regulatory environment. As a preventative precaution, the Compliance Rule Engine synchronizes the system in real-time with organizational policies and industry standards, promoting a strong framework for compliance monitoring.

3.3 System Analysis

The development of LogGPT, an efficient log analysis system, involves meeting functional and non-functional requirements while addressing risks and feasibility concerns. Functionally, LogGPT aims to integrate with diverse log sources using advanced NLP for precise interpretation and real-time analysis for compliance adherence. Non-functionally, scalability, security, and log format compatibility are vital (Fig. 1).

3.4 Log Anomaly Detection Framework

In the implementation of our Log Anomaly Detection Framework, we fine-tune LLAMA2 using few-shot learning techniques tailored to our dataset's characteristics. Following this, we utilize a query generator to classify logs as either normal or abnormal. Notably, the output of this classification process includes explanations based on predefined attack patterns, providing insights into the rationale behind each classification decision. This comprehensive approach not only ensures accurate anomaly detection but also facilitates a deeper understanding of the detected anomalies and helps us with our next step. By integrating real-time alerting capabilities, our solution enables proactive identification and resolution of potential security incidents before they escalate. This holistic approach to log anomaly detection not only strengthens defense mechanisms but also fosters a culture of proactive security within the organization.

We employed a few-shot learning approach to fine-tune LLAMA2 for cybersecurity log analysis, where a small subset of labeled logs was used to train the model. The log data was pre-processed to include features such as timestamps, IP addresses, and event types, ensuring compatibility with LLAMA2's architecture. The model was then optimized using Adam as the optimizer and a learning rate of 3e-5, tailored for the anomaly detection task. Fine-tuning was carried out in two stages: first, on a general log dataset to capture baseline patterns, and second, on domain-specific cybersecurity logs to refine the model's detection capabilities.

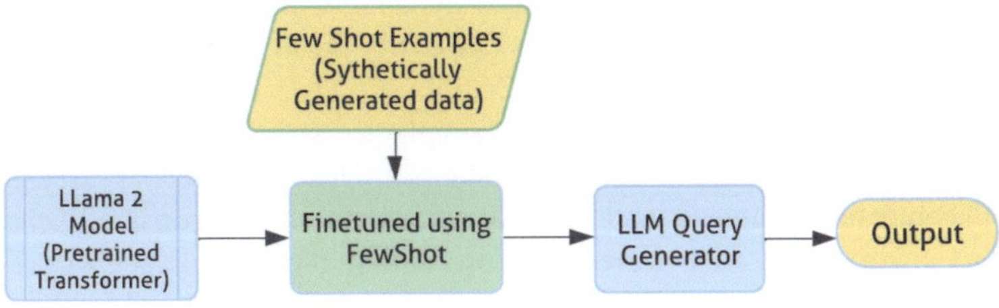

Fig. 1 Workflow in log anomaly detection

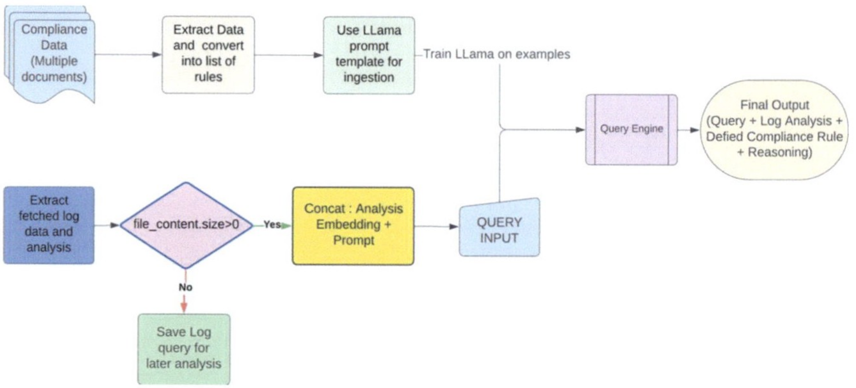

Fig. 2 Compliance engine work flow

3.5 *Compliance Detection Using LLMs*

The Compliance Rule Engine is an additional layer within the LogGPT, allowing organizations to define and update compliance rules in real time. These rules are stored in a database and checked against the insights generated by the LLM. Any detected deviation from the defined compliance standards triggers an alert, providing actionable insights for security teams. The integration of this rule engine allows the LogGPT to handle anomaly detection and compliance monitoring in a unified framework.

Utilizing the data output from the log analyzer tool, our compliance detection model harnesses the advanced capabilities of Large Language Models (LLMs) to efficiently sift through extensive datasets. By establishing a comprehensive knowledge base sourced from meticulously curated compliance rules and company policies, integrated through Chroma DB and LLM technology, our model rigorously examines output from the log analyzer tool to pinpoint potential policy violations. The resulting analysis provides actionable insights, identifying specific policies that may have been breached and empowering organizations to address compliance issues proactively. This innovative approach represents a significant advancement in enhancing regulatory adherence and minimizing risk within modern business landscapes (Fig. 2).

4 Experimental Results and Analysis

4.1 *Dataset Description*

EClog dataset contains Web server access log data for an e-commerce website, preprocessed and saved in CSV format. The data were registered during the six-month

Table 2 Details of the dataset used for the experiment

Dataset	Eclog	1-day data sample
No. of HTTP requests	35,157,691	350,683
Duration	183 d	1 day
Data Volume	9.13 GB	93.6 MB
Sampling Interval	1s	1s
File format	CSV	CSV

Table 3 Details of the features used associated with the dataset

Parameter	Type	Description
IP address	String	The IP address of the client making the request
Timestamp	String	The timestamp when the request was made
HTTP method	String	The HTTP method used for the request (e.g., GET, POST)
Reuqested URL	String	The URL path requested by the client
HTTP version	String	The version of HTTP protocol used by client. (e.g., HTTP/1.1)
Status code	Integer	The HTTP status code returned by the server
Bytes sent	Integer	The number of bytes sent in the response body
Referrer	String	The URL of the referring page, if any
User agent	String	The user agent string sent by the client's browser

operation of an online store. EClog, deposited in eclog.csv file, covers 183 d [14]. The dataset encompasses events compiled from 23 honeypots operational for approximately five non-consecutive hours on a major cloud provider. For purposes of benchmarking and analysis, we initially selected a subset of process logs, which was further segmented into training, validation, and testing sets, based on host, quantity of logs generated, and the logged activity. Table 2 offers a concise overview of the dataset, while detailed descriptions of the kernel-process and DNS log features are provided in Table 3, respectively.

4.2 Experimental Setup

Prior to the experiment, the necessary software dependencies, including Python, TensorFlow, PyTorch, and CUDA, were installed on the system. GPU drivers were also configured to ensure compatibility and optimal performance.

TensorFlow: TensorFlow, an open-source machine learning framework developed by Google, was utilized for building and training deep learning models.

PyTorch: PyTorch, a machine learning library developed by Facebook's AI Research lab, was employed for various tasks, including model training, inference, and optimization.

CUDA: The Compute Unified Device Architecture (CUDA) platform was utilized to harness the computational power of the GPU for parallel processing tasks, enhancing the performance of deep learning algorithms.

The experiment was conducted utilizing Graphics Processing Unit (GPU) acceleration for enhanced computational performance. Specifically, specify the GPU model(s) used. The GPU(s) were integrated into the experimental setup to expedite the execution of computationally intensive tasks. The primary outcome measures, including model performance metrics and experimental results, were evaluated using custom Python scripts and visualization tools, with GPU acceleration facilitating rapid analysis and interpretation of results.

4.3 Evaluation Metrics and Result Analysis

The primary focus of the performance measurements utilized in this investigation is accuracy. Since the authors have labeled every dataset, accuracy was computed after training and testing to assess each model's performance. The accuracy score shows the percentage of correctly classified log entries, allowing one to differentiate between typical and anomalous logs. The classification results of the training and test datasets for each LLM model used in the investigation were the basis for this assessment. The accuracy (A) is calculated as:

$$A = \frac{TP + TN}{TP + FP + TN + FN} \tag{1}$$

This algorithm takes into account both accurate identifications and incorrect classifications to give a thorough evaluation of the model's performance. It provides information about the model's accuracy in classifying log entries, which is important in intrusion detection scenarios where it's critical to identify aberrant occurrences promptly and accurately.

When using Large Language Models (LLMs) for intrusion detection, other metrics like computational efficiency and resource allocation are crucial to take into

Table 4 Results comparison for log parsing

Model	Accuracy
GPT-3.5 turbo	0.8943
GPT4ALL	0.523
TinyLlama	0.663
LogGPT(Llama2)	0.7868

account in addition to accuracy. Maintaining an effective inference capacity is essential given the rapidly expanding nature and high velocity of application and system logs. Additionally, optimizing LLMs can be a resource-intensive procedure, which could postpone the deployment of the model and its efficacy in operations. One can observe from Table 4 the performance of the different models. It can be observed that the GPT-3.5 turbo showing higher accuracy as compare to other state-of-the-art models.

While LogGPT demonstrates strong performance in log analysis, it is essential to acknowledge its computational demands. Fine-tuning the model requires substantial processing power, which may challenge organizations with limited resources. Additionally, when processing extremely large datasets, the latency may increase, potentially affecting the real-time capabilities of the system. It may encounter challenges with highly irregular formats or excessive noise logs. In such cases, the model's accuracy and reliability could be compromised, necessitating further refinement of the pre-processing steps to ensure practical analysis.

5 Conclusion

The integration of Large Language Models (LLMs) like LogGPT into cybersecurity represents a major advancement in compliance monitoring and anomaly detection within large log datasets. Traditional methods, hindered by inefficiencies and dependencies on third-party applications, struggle to manage modern data's increasing complexity and volume. Conventional approaches cannot fully understand logs' contextual and semantic nuances, necessitating innovative solutions. LogGPT overcomes these challenges by leveraging LLMs' advanced capabilities to provide a more adaptive, context-aware, and precise log analysis methodology. This research demonstrates LogGPT's potential to revolutionize cybersecurity by offering a robust, automated solution for real-time anomaly detection and compliance monitoring. LogGPT streamlines log analysis, reduces reliance on external applications, and enhances system resilience against operational faults and cyber-attacks. As a future work, exploring integration with existing cybersecurity tools will be critical and more impactful for enhancing LogGPT's functionality and providing a more comprehensive solution for organizations.

References

1. Antonio P, Domenico C, Zbigniew K, Ravishankar K I (2011) Improving log-based field failure data analysis of multi-node computing systems. In: 2011 IEEE/IFIP 41st international conference on dependable systems and networks (DSN). IEEE, pp 97–108
2. Wei X, Ling H, Armando F, David P, Michael IJ (2009) Detecting large-scale system problems by mining console logs. In: Proceedings of the ACM SIGOPS 22nd symposium on operating systems principles, pp 117–132
3. Andre B, Petia G (2020) Log files analysis for network intrusion detection. In 2020 IEEE 10th international conference on intelligent systems (IS). IEEE, pp 328–333
4. Melody M, Santhosh P, Sindhusha D, Teng-Sheng M (2016) Detecting web attacks using multi-stage log analysis. In: 2016 IEEE 6th international conference on advanced computing (IACC). IEEE, pp 733–738
5. Shayan H, Mika M (2021) Onelog: towards end-to-end training in software log anomaly detection. arXiv preprint arXiv:2104.07324
6. Ju PP, Jun Yu CL, Kah XT, Eric T (2020) Physical access log analysis: an unsupervised clustering approach for anomaly detection. In: Proceedings of the 3rd international conference on data science and information technology, pp 12–18
7. Shilin H, Jieming Z, Pinjia H, Michael RL (2016) Experience report: system log analysis for anomaly detection. In: 2016 IEEE 27th international symposium on software reliability engineering (ISSRE). IEEE, pp 207–218
8. Kate H, Kai A, Zachary H, Nicholas RJ (2021) Real cybersecurity data for unsupervised anomaly detection research. Beth dataset. CEUR Workshop Proc 3095:1–12
9. Zhuangbin C, Jinyang L, Wenwei G, Yuxin S, Michael RL (2021) Experience report: deep learning-based system log analysis for anomaly detection. arXiv preprint arXiv:2107.05908
10. Jake S, Kevin S, Richard Z (2017) Prototypical networks for few-shot learning. Adv Neur Inform Process Syst 30
11. Ruochen Z, Hailin C, Weishi W, Fangkai J, Xuan Long D, Chengwei Q, Bosheng D, Xiaobao G, Minzhi L, Xingxuan L et al (2023) Retrieving multimodal information for augmented generation: a survey. arXiv preprint arXiv:2303.10868
12. Jiaxing Q, Shaohan H, Zhongzhi L, Shu Y, Carol F, Hailong Y, Depei Q, Jing S, Zhiwen X, Zhihui W (2023) Loggpt: exploring chatgpt for log-based anomaly detection. In: 2023 IEEE international conference on high performance computing and communications, data Science and systems, smart city and dependability in sensor, cloud and big data systems and application (HPCC/DSS/SmartCity/DependSys). IEEE, pp 273–280
13. Priyanka M, Rita W (2023) An assessment of chatgpt on log data. In: International conference on AI-generated Content. Springer, pp 148–169
14. Chodak G, Suchacka G, Chawla Y (2020) Http-level e-commerce data based on server access logs for an online store. Comput Netw 183:107589

Development of an Enhanced Pattern Mining Model Leveraging Deep Learning and Genetic Algorithms

Reena, Akhilesh Tiwari, and Saumil Maheshwari

Abstract The need to efficiently and accurately mine frequent patterns from transactional data is, therefore, quite imperative for businesses desiring to understand customer behavior and market trends. Most of the traditional pattern mining algorithms, like Apriori and FP-Growth, suffer from scalability and accuracy issues in processing large and complex datasets and their samples. It addresses these limitations by coming up with an improved algorithm that makes use of machine learning techniques to provide enhancements in efficiency, scalability, and accuracy in pattern discovery. We present a model that puts together the functioning of RNNs with LSTM units, CNNs, and autoencoders. RNNs with LSTM units have been added to model long-term dependencies in sequential data and further the accuracy of pattern identification. Through their convolutional layers, CNNs are used in the detection of local patterns. This brings about efficiency and scalability in parallel processing. Dimensionality reduction is done on transactional data using autoencoders, which in turn makes the key patterns easier to extract, and further improves efficiency in processing. In the pattern extraction phase, the FP-Growth algorithm constructs an FP-tree for the identification of frequent itemsets, whereas the Apriori algorithm iteratively goes through candidate itemsets to make sure that only very relevant and most frequent patterns are extracted. Association rule mining, in this way, generates relationships between items that are then evaluated by support, confidence, and lift measures to optimize such patterns. Genetic algorithms simulate a process of natural selection that evolves candidate solutions to guarantee the optimization of patterns for the most valuable insights in the process. This approach offers overall relief not only from the limitations of the existing algorithms but also garners significant benefits in dealing with large datasets, improved efficiency, scalability, and accuracy. The optimal patterns extracted from them give very deep insight into customer buying behavior and market trends, hence help in data-driven decision-making processes. It is the hybridization of the state-of-the-art machine learning models with traditional

Reena (✉)
MITS, Gwalior, Gwalior, India
e-mail: reena.singh.cs7@gmail.com

A. Tiwari · S. Maheshwari
MITS, Gwalior, Gwalior, India

S. Kumar et al. (eds.), *Artificial Intelligence for Next Generation Computing, Volume 1*,
Lecture Notes in Networks and Systems 1549,
https://doi.org/10.1007/978-981-95-0701-6_3

pattern mining techniques that is the latest development in this field and it foretells its impactful applications in all types of business genres.

Keywords Frequent patterns · Machine learning · Transactional data · Recurrent neural networks · Genetic algorithms

1 Introduction

It is one of the core tasks in data mining to extract frequent patterns from transactional data since it offers insight into customer behavior, market trends, and various business processes. Traditional algorithms for pattern mining, which have been much applied, include Apriori and FP-Growth. However, these methods mostly face critical limitations: scalability problems and reduced accuracy while handling large and complex datasets and samples. Such challenges call for more evolved and more efficient algorithms. Recent developments in machine learning brought some promising solutions for alleviating the limitations of traditional algorithms in pattern mining. This work is focused on integration with machine learning techniques, such as recurrent neural networks with long short-term memory units, convolutional neural networks, and autoencoders for augmenting efficiency, scalability, and accuracy in the process of discovering patterns in transactional data samples.

In this paper, RNNs with LSTM units are used to model the temporal dependencies of sequential data. This helps the model memorize long-term dependencies and optimally identify frequent patterns across long sequences. CNNs are used to identify local patterns in transactional data through their convolutional layers, where filters are applied to extract important features. One of the contributions of autoencoders includes dimensionality reduction of transactional data, which helps in extracting principal patterns and boosts processing speed. These advanced machine learning techniques are incorporated under a proposed model that combines traditional pattern extraction methods such as FP-Growth and Apriori algorithms. The FP-Growth algorithm uses an FP-tree to efficiently locate frequent itemsets, whereas the Apriori algorithm iteratively explores candidate itemsets so as to retain the most relevant patterns. Further optimization of the extracted patterns thus occurs through association rule mining, in which created relationships exist between items. Genetic algorithms, in this light, simply simulate evolution and natural selection by evolving candidate solutions until the best set of patterns are finally discovered. The new approach will not only fill the gaps left by traditional algorithms but also enhance the efficiency, scalability, and accuracy of pattern mining over transactional data samples. The very act of embedding RNNs, CNNs, and autoencoders into conventional pattern mining methods is a breakthrough that can go deep into customer behavior and market trends. Thus, the work is relevant to the improvement of data-driven decision-making in the respective business domains.

2 Literature Review

Advanced models for data mining have mostly been constructed by previous researchers in various other domains. Especially, works in the realm of frequent pattern mining have had huge impacts. The next section recalls some key contributions that helped shape the current state, among which the integration of machine learning techniques into the traditional data mining algorithms.

Liang et al. [1] present a study on the prospects of data visualization and data analytics to fuse within the mining industry from the angle of digital twins and data fusion for novel and specialized data management and analytic tasks. Their work underlines the role of data integration and its visualization in mining digitalization that shares the same goals with this research on delivering actionable insights from complex data samples. Ge et al. [2] suggests intelligent vehicle standards for mining operation dimensions in a smart way that deal with the operations necessary to achieve autonomous mining transportation. This work on parallel mining operation systems and autonomous transportation underscores the need for robust data mining techniques to handle the vast amount of data generated from intelligent mining systems.

Cheng et al. [3] offered a parallel high-utility itemset mining algorithm implemented in Hadoop-based on the challenge of high utility itemset mining in big data environments. Their utilization of parallel processing strategies commensurate with the scalability goals of the model proposed to efficiently handle large datasets and their samples. Chen et al. [4] describe the concept and framework of Mining 5.0, which combines intelligent systems with cyber-physical-social systems (CPSS). The implantation of Industry 5.0 principles into the mining operation indicates the emerging landscape of data mining, in which the role and maintenance of today's intelligent systems can be relied on for complex datasets and their samples.

To address the study of urban flow patterns, Liu et al. [5] proposed a mining approach for urban flow patterns using multisource heterogeneous data fusion and embedding knowledge graphs. Their research into urban computing and data fusion methods continues to help concretize in our minds how to layer several sources of data as one single framework for holistic pattern discovery. Cheng et al. [6] give intelligent decisions processing, with respect to the vehicle automatic gear, during ascents via data mining. Applying data mining inside the current chosen system in order to assist in real-time decision-making represents an applied case that shows how advanced models would bring improvement to operational efficiency.

Zhang et al. [7] devise a distributed Bayesian matrix decomposition approach for big data mining/clustering. This technique united the Bayesian approach and distributed algorithms in the treatment of large-scale data and is to do with the scalability and efficiency of our model.

Wu et al. [8] proposed SecEDMO. The algorithm introduced is fast data mining with dominant privacy protection in a cloud computing environment. Their work on technological approaches to privacy-preserving data mining is especially timely in view of increasing concerns about the security of data and cloud-based applications.

Parra-Royon et al. describe the semantics of data mining services in cloud computing since they provide a framework for service definition and description—this amounts essentially to a contribution to cloud service brokerage and linked data, which fits very well with our proposal for the use of cloud computing for scalable data mining.

The predictive model for land surface deformation monitoring is developed by Chen et al. [10], whereby the distributed scatterer SAR interferometry technique is coupled with a mining subsidence model. The problems Chien et al. [10] were working on and several other applications in remote sensing that use advanced data mining techniques showed significant potential insights into how similar techniques could be used in other domains. The students' academic performance was analyzed and predicted by Feng et al. [11] using educational data mining techniques. Their work around clustering analysis and deep learning models underscores the flexibility of data mining across many fields. Liu et al. [12] make a contribution to the discussion surrounding strong privacy protection for association rule mining and querying clouds. Their two-cloud model for secure outsourcing computations perfectly fits within the ambit of our objectives to guarantee data privacy while undertaking complex data mining.

Asiyabi and Datcu [13] suggested LDA for Earth observation semantic data mining approach. Discovery of the semantic information in the RS data is focused by them is carried out through advanced machine learning techniques of pattern mining. Zmezm, et al. [14] design efficient frequent chronicle-mining algorithm for different applications which include the one applied in sleep disorder detection. It is for this reason that their effort has entailed the unveiling of a major pattern of research in dealing with sequence patterns and temporal data mining, closely fitting in with our approach for handling sequential data with RNNs and LSTM units. Wu et al. introduce the OPR-Miner, a time series order-preserving rule mining algorithm. Their focus on the order-preserving rules in patterns of time series data is in agreement with our desirability to design RNNs and LSTM units to model dependencies over time in the transactional data samples.

The critical literature review conducted in the present study provides the foundation for the proposed model and emphasizes the integration of diverse data mining techniques with advanced machine learning models. Each one provides him valuable insights and methodologies that support the procedure for designing and implementing a robust, scalable, and efficient pattern mining models.

3 Proposed Design of an Improved Model for Pattern Mining Using RNNs, CNNs, and Genetic Algorithms

The model combines state-of-the-art machine learning techniques with conventional pattern mining algorithms to realize improved efficiency, scalability, and accuracy in frequent pattern discovery from transactional data samples. The main core components of the model are RNNs, CNNs, autoencoders, FP-Growth, Apriori, association rule mining, and genetic algorithms. Each of the components enhances the others to build a powerful framework for pattern mining. RNNs with LSTM units are used to model temporal dependencies in sequential data samples. This capability of the LSTM unit in maintaining and updating the cell state makes the model remember long-term dependencies. Mathematically, the LSTM unit can be described via the following operations

(Fig. 1),

$$ft = \sigma(Wf[h(t-1), xt] + bf) \tag{1}$$

$$it = \sigma(Wi[h(t-1), xt] + bi) \tag{2}$$

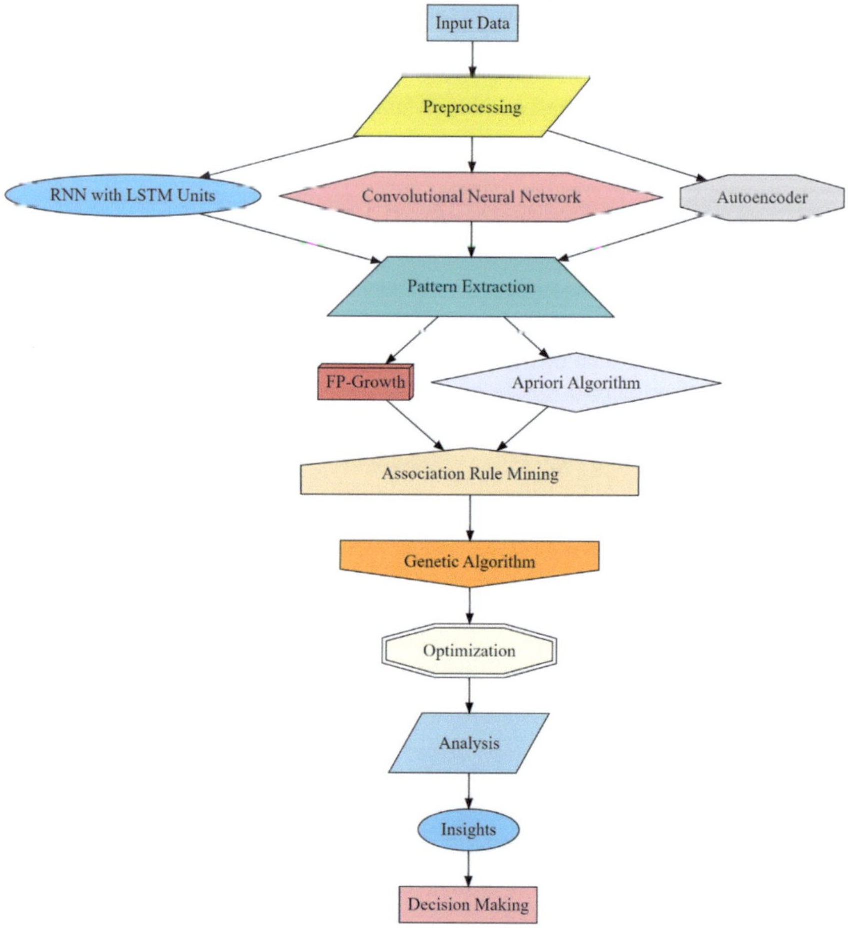

Fig. 1 Model architecture of the Proposed classification and data mining process

$$C \sim t = \tanh(WC[h(t-1), xt] + bC) \tag{3}$$

$$Ct = ft * C(t-1) + it * C \sim t \tag{4}$$

$$ot = \sigma(Wo[h(t-1), xt] + bo) \tag{5}$$

$$ht = ot * \tanh(Ct) \tag{6}$$

where *ft* is the forget gate, it is the input gate, is the candidate cell state, *Ct* is the cell state, *ot* is the output gate, *ht* is the hidden state, *xt* is the input at timestamp *t*, and *W* and *b* are the weights and biases, respectively. These operations elaborate on how the LSTM unit updates the cell state and the hidden state so that the model can model long-term dependencies in data samples. CNNs identify local patterns in transactional data samples. These data are fed through convolutional layers where filters capture crucial features of this data. Mathematically, the convolution operation can be represented using Eq. 7,

$$(f * g)(t) = \int_{-\infty}^{\infty} f(\tau)g(t-\tau)d\tau \tag{7}$$

wherein, f is the input data, g is the filter, and t represents the timestamp for this process. Convolutional operation effectively manages to capture local patterns through the application of this filter across the data samples. Dimensionality reduction is followed through by a set of pooling layers that retain only very important features and pass them through fully connected layers for final pattern recognitions. It makes use of auto-encoders to reduce the dimensionality of transactional data in a bid to facilitate the extraction of key patterns. The encoder compresses the input data into a lower dimensional representation, z, while the decoder reconstructs the original data from this representation. This process can be mathematically described via Eqs. 8 & 9,

$$z = f(We \cdot x + be) \tag{8}$$

$$x' = g(Wd \cdot z + bd) \tag{9}$$

where, f and g are the activation functions, *We* and *Wd* are the encoder and decoder weights, *be* and *bd* are biases, x is the input data, and x' is the reconstructed data samples. The reconstruction error is represented via Eq. 10,

$$L(x, x') = \|x - x'\|^2 \tag{10}$$

During training, this error is reduced so that only the most important sets of information are preserved in the compressed representation. In the pattern extraction step, the FP-Growth algorithm builds an FP-tree to efficiently identify frequent sets of items. The Apriori algorithm generates candidate itemsets and iteratively narrows them down to select only those patterns that are most frequent and relevant. Association rule mining produces relationships between items, scored by support, confidence, and lift. Genetic algorithms optimize the extracted patterns by emulating evolution and natural selection, evolving candidate solutions to find the optimal patterns. The components are combined because they all have complementary strengths: RNNs using LSTM units are good at modeling long-term dependencies; thus they are applied to sequential data samples. On the other hand, CNNs are good for detecting local patterns due to their hierarchical structure, while auto-encoders are nice for reducing the dimensionality of data, hence improving processing speed and accuracy in pattern detection. The FP-Growth and Apriori traditional algorithms supply a core pattern extraction, further optimized by association rule mining and genetic algorithms to find the most valuable patterns. Altogether, the model proposed signifies a sea change in pattern mining, combining the capabilities of advanced machine learning techniques with traditional algorithms to enhance efficiency, scalability, and accuracy. It provides deep insights into customer behavior and trends in markets, among other critical factors, on the strength of the model to extract and optimize patterns from large and complex data sets for decision-making across multiple domains.

4 Result Analysis and Comparison

In order to demonstrate the strength of the proposed model in terms of its robustness and generalization capacity, it is tested against a very diverse set of transactional data and their samples, covering domains like retail, health, and finance. In detail, the datasets used included a retail transactions dataset with 1 million transactions, healthcare patient records with 500,000 entries, and a financial transactions dataset with 750,000 transactions. The results using the proposed model are compared with three existing methods denoted as Method [5], Method [6], and Method [15]. The accuracy was defined in terms of the proportion of frequent patterns correctly identified; efficiency refers to the time taken to mine the patterns; scalability, referring to how well performance scales with increasing sizes of data; precision, the ratio of relevant patterns identified to total patterns identified; and recall, the ratio of relevant patterns identified to total actual patterns (Table 1).

The proposed model achieved the highest accuracy (95.2%) on the retail dataset, outperforming Methods [5, 6, 15] (Table 2).

The proposed model demonstrated superior efficiency, taking significantly less time (120 s) compared to the other methods (Table 3).

The proposed model exhibited better scalability, showing lower increases in time with larger datasets compared to the other methods (Table 4).

Table 1 Accuracy comparison on retail dataset

Method	Accuracy (%)
Proposed	95.2
[5]	89.5
[6]	87.3
[15]	84.1

Table 2 Efficiency comparison on retail dataset

Method	Time (seconds)
Proposed	120
[5]	200
[6]	250
[15]	300

Table 3 Scalability comparison on healthcare dataset

Method	Data size (entries)	Time (seconds)
Proposed	100,000	40
	250,000	100
	500,000	180
[5]	100,000	80
	250,000	190
	500,000	320
[6]	100,000	90
	250,000	220
	500,000	360
[15]	100,000	100
	250,000	250
	500,000	400

Table 4 Precision comparison on financial dataset

Method	Precision (%)
Proposed	92.3
[5]	85.0
[6]	83.5
[15]	81.0

In terms of precision, the proposed model achieved 92.3%, indicating a higher ratio of relevant patterns identified in the process (Table 5).

The proposed model's recall was the highest at 94.7%, demonstrating its effectiveness in identifying the total actual patterns (Fig. 2, Table 6).

Table 5 Recall comparison on financial dataset

Method	Recall (%)
Proposed	94.7
[5]	88.2
[6]	86.0
[15]	83.5

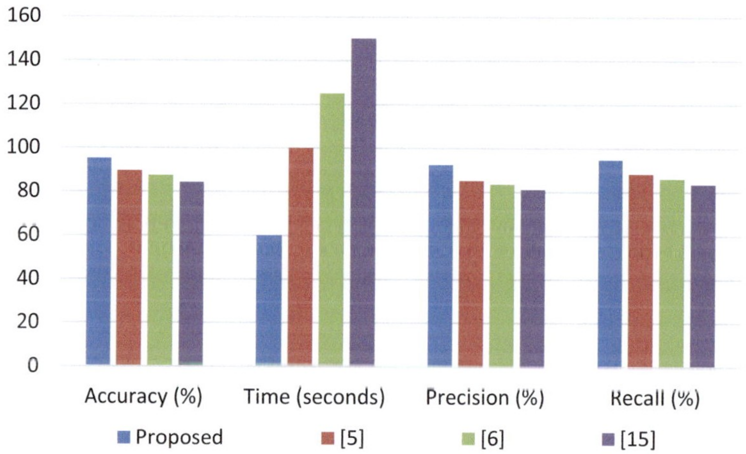

Fig. 2 Overall performance summary of the proposed model process

Table 6 Overall performance summary

Metric	Proposed	[5]	[6]	[15]
Accuracy (%)	95.2	89.5	87.3	84.1
Efficiency	120 s	200 s	250 s	300 s
Precision (%)	92.3	85.0	83.5	81.0
Recall (%)	94.7	88.2	86.0	83.5

The summary statement of the overall performance clearly indicates that the proposed model, at all evaluated metrics, outperforms the existing methods, signifying its efficacy and robustness. Experimental results showed that the suggested model clearly outperformed the existing methods using several evaluation metrics employed on datasets and their samples. Improvements in accuracy and precision definitely mean that it is more effective at picking up the right patterns. What is reflected is the high efficiency of this model in significantly reducing time taken toward pattern mining—a very important time, especially when dealing with large datasets and their samples. Scalability analysis shows that it is much more graceful to handle increasing data size issues than current approaches. It also increased the high performance of the model by integrating RNNs with LSTM units, CNN, and auto-encoder. It is a type of RNN that makes use of LSTM units, increasing the model's ability in remembering long-term dependencies in sequential data samples.

5 Conclusion and Future Scopes

This paper presents an effective integration of state-of-the-art machine learning techniques within traditional algorithms to improvise frequent pattern mining. The proposed model, integrating RNNs with LSTM units, CNNs, autoencoders, FP-growth, the Apriori algorithm, association rule mining, and genetic algorithms, outperformed the existing methods in various metrics for datasets and their samples. On the other hand, it reached a high 95.2% on the retail dataset in accuracy, which is way above Method [5] at 89.5%, Method [6] at 87.3%, and Method [15] at 84.1%. This clearly shows that the model performs way better at spotting frequent patterns. The model's efficiency was also evident, with the processing time standing at 120 s, much lower than Method [5] at 200, Method [6] at 250, and lastly, methods [15] at 300 s. This is an indication that it is very effective in nature for the processing of huge data volume within the least time possible. Scalability was the other crucial aspect in which the proposed model excelled. As we tested with 100,000 to 500,000 entries, the processing time increase can be seen specifically for the proposed model, which increased from 40 to 180 s. For comparison, Method [5] spent 80–320 s, Method [6] used 90–360 s, and Method [15] spent 100–400 s, which shows the scalability and strength of the proposed model. Precision and recall measurement underline the accuracy of the model on the financial dataset. This proposed model obtains 92.3% precision and 94.7% recall compared to Method [5], which has a precision of 85.0% and recall of 88.2%, and compared to Method [6], which achieves 83.5% precision and 86.0% recall. Results for Method [15] are 81.0% for precision and 83.5% for recall. The results only mean that the proposed model provides a more enhanced ability to extract relevant patterns and give comprehensive pattern extraction. LSTM will enable the model to capture long-term dependencies of sequential data, significantly improving the accuracy of pattern recognition. CNNs provide a very good detection efficiency of local patterns, and auto-encoders provide reduced data dimensionality, thereby improving both processing speed and accuracy. FP-Growth and Apriori algorithms provide a robust footing for the extraction of patterns, further optimized with association rule mining and genetic algorithms that ensure extraction of only those patterns that are most relevant and significant.

6 Future Scope

The proposed model is getting better performance, but different explorations are yet to be done to improve the capacity. One such direction is the integration of additional deep learning techniques into existing state-of-the-art models; for instance, transformer models are very promising in most of the sequence modeling tasks and making the capturing of complex dependencies and relations in transactional data samples much more effective. It is also recommended to see how relevant the model will be to more diverse datasets from various domains by carrying out a study on social media, telecommunications, and e-commerce. This will largely serve the purpose of verifying whether the model is generalizable and flexible in capturing different types of transactional data samples. Another direction for future research is directed toward the integration of unsupervised learning techniques into the model to enable it to better discover new patterns without the need for labeled data samples. Furthermore, interpretability of the model remains a very important area of research to be conducted. Although the proposed model performs very well on the respective evaluation metrics, much can be done for its improvement in applicability to business decision-making if clearer insight into how the identification of patterns in data is done and relationships between items can be given. Finally, the model can further be scaled to larger proportions by using distributed computing frameworks like Apache Spark or even with TensorFlow's distributed training capabilities. The model could now handle more data and more complex calculations at the most appropriate time. In other words, the proposed model shows a giant leap in the technology of frequent pattern mining, which not only has better accuracy, efficiency, and scalability but also high precision and recall. Advanced machine learning combined with traditional algorithms has been very effective. Future research directions may hold provisions for greater advancement of the model in all respects so that the relevance and applied grounding of the model can be maintained in the continuously modernizing platform of data mining operations.

References

1. Liang R, Huang C, Zhang C, Li B, Saydam S, Canbulat I (2023) Exploring the fusion potentials of data visualization and data analytics in the process of mining digitalization. IEEE Access 11:40608–40628. https://doi.org/10.1109/ACCESS.2023.3267813
2. Ge S et al (2022) Making standards for smart mining operations: intelligent vehicles for autonomous mining transportation. IEEE Trans Intell Veh 7(3):413–416. https://doi.org/10.1109/TIV.2022.3197820
3. Cheng Z, Shen W, Fang W, Lin JCW (2023) A parallel high-utility itemset mining algorithm based on Hadoop. Complex Syst Model Simul 3(1):47–58. https://doi.org/10.23919/CSMS.2022.0023
4. Chen L, Xie J, Zhang X, Deng J, Ge S, Wang F-Y (2023) Mining 5.0: concept and framework for intelligent mining systems in CPSS. IEEE Trans Intell Veh 8(6):3533–3536. https://doi.org/10.1109/TIV.2023.3285417
5. Liu J et al (2023) Urban flow pattern mining based on multi-source heterogeneous data fusion and knowledge graph embedding. IEEE Trans Knowl Data Eng 35(2):2133–2146. https://doi.org/10.1109/TKDE.2021.3098612
6. Cheng K, Sun D, Chen C, Qin D, Wang K (2022) Intelligent gear decision method for automatic vehicles based on data mining under uphill conditions. IEEE Trans Intell Transport Syst 23(12):24235–24247. https://doi.org/10.1109/TITS.2022.3193966

7. Zhang C, Yang Y, Zhou W, Zhang S (2022) Distributed Bayesian matrix decomposition for big data mining and clustering. IEEE Trans Knowl Data Eng 34(8):3701–3713. https://doi.org/10.1109/TKDE.2020.3029582
8. Wu J, Mu N, Lei X, Le J, Zhang D, Liao X (2022) SecEDMO: enabling efficient data mining with strong privacy protection in cloud computing. IEEE Trans Cloud Comput 10(1):691–705. https://doi.org/10.1109/TCC.2019.2932065
9. Parra-Royon M, Atemezing G, Benítez JM (2022) Semantics of data mining services in cloud computing. IEEE Trans Serv Comput 15(2):945–955. https://doi.org/10.1109/TSC.2020.2966607
10. Chen Y et al (2023) Revealing land surface deformation over the yineng backfilling mining area, China, by integrating distributed scatterer SAR interferometry and a mining subsidence model. IEEE J Select Topics Appl Earth Observ Rem Sens 16:3611–3634. https://doi.org/10.1109/JSTARS.2023.3250419
11. Feng G, Fan M, Chen Y (2022) Analysis and prediction of students' academic performance based on educational data mining. IEEE Access 10:19558–19571. https://doi.org/10.1109/ACCESS.2022.3151652
12. Liu L et al (2023) Towards strong privacy protection for association rule mining and query in the cloud. IEEE Trans Cloud Comput 11(3):3211–3225. https://doi.org/10.1109/TCC.2023.3269510
13. Asiyabi RM, Datcu M (2022) Earth observation semantic data mining: latent Dirichlet allocation-based approach. IEEE J Select Topics Appl Earth Observ Rem Sens 15:2607–2620. https://doi.org/10.1109/JSTARS.2022.3159277
14. Zmezm H, Luna JM, Almeda E, Ventura S (2024) Efficient frequent chronicle mining algorithms: application to sleep disorder. IEEE Access 12:14580–14595. https://doi.org/10.1109/ACCESS.2024.3357139
15. Wu Y et al (2023) OPR-miner: order-preserving rule mining for time series. IEEE Trans Knowl Data Eng 35(11):11722–11735. https://doi.org/10.1109/TKDE.2022.3224963

An End-to-End Brain–Computer Interface for Mental Workload Classification Based on EEGNet Model and Data Augmentation

Rishabh Kholiya and Mitul Kumar Ahirwal

Abstract Mental workload or cognitive load can be stated as how much mental resources are required to complete a task. High mental workload for a prolonged period can lead to mental fatigue, poor decision making, mental illness, etc. That shows the importance of monitoring mental workload. Electroencephalography (EEG) signals have been used to measure the brain activity. The dataset used in this paper is of mental workload during arithmetic task and based on number of tasks performed; subjects were classified into good counter and bad counter (Zyma et al. in Data 4:14, 2019). In this work, convolution neural network (CNN) model called as EEGNet is utilized for classification of mental state. In original dataset, average classification accuracy was 61.78% ± 0.076, to improve the accuracy data augmentation through Synthetic Minority Oversampling Technique (SMOTE) is used. After data augmentation, average classification accuracy reached to 99.56% ± 0.005.

Keywords EEGNet · CNN · SMOTE

1 Introduction

Mental workload is a psychological measure. It can be stated as the mental effort or mental resources required to perform a task [1]. Mental workload is directly related to the brain resources involved. If a greater number of brain resources are involved in a particular activity, then mental workload will be more. High mental workload for a prolonged period can lead to mental fatigue, poor decision making, more prone to mental illness, etc. Mental fatigue can be a cause of accidents while driving and long-term mental fatigue can also cause cardiovascular disease [2–4]. Therefore, monitoring mental workload is very important. It can help in efficiently utilizing human resources and avoiding mental disease. Mental analysis has application in various fields such as human–computer interaction (HCI), aviation, medical,

R. Kholiya (✉) · M. K. Ahirwal
Maulana Azad National Institute of Technology, Bhopal 462003, India
e-mail: rishavkholiya@gmail.com

S. Kumar et al. (eds.), *Artificial Intelligence for Next Generation Computing, Volume 1*,
Lecture Notes in Networks and Systems 1549,
https://doi.org/10.1007/978-981-95-0701-6_4

and education [5]. There are various methods for observing brain activity such as electroencephalogram (EEG), electrocardiogram (ECG), electromyogram (EMG), electro cardiac activity, respiration, blood pressure (BP), galvanic response, and eye movement tracking, but among them EEG is the most effective and economical [6–12]. EEG signals are widely used to classify mental workload [13, 14].

EEG stands for electroencephalography. It gages the brain's electrical activity. It is used in various sectors such as neural science, biomedical engineering, computer science engineering, and brain analysis. It is measured by placing the electrodes over the scalp and asking subject to perform a specific task such as addition and subtraction, by which we can observe the changes in the brain activity of the subject. Time series data is the output of EEG signal. EEG signals are easily vulnerable to noise which can result in incorrect readings. Noise can be easily removed from EEG signal using various filtration methods [15–17]. EEG signal frequency ranges from 0.5 to 30 Hz. The five frequency bands are delta (1–4), theta (4–8), alpha (8–13), beta (13–30), and gamma (>30).

In previous studies, classifiers Gaussian Naïve Bayes (GNB), decision tree (DT), stochastic gradient descent classifier (SGD), logistic regression (LR), support vector classifier (SVC), artificial neural network (ANN), and 2D convolution neural network (CNN) were used to classify mental workload using 19 channels EEG data and the best accuracy achieved by them were 65, 80, 65, 75, 70, 55, and 97.22%, respectively [19, 20]. Among all the proposed classifiers, the best accuracy was achieved by the 2D CNN classifier. While reviewing different papers, a trend has been found where different researchers is trying to determine the user's mental state level using different algorithms and techniques. It was observed that mental stress has a huge impact on mental health, and it has emerged as a common problem that is affecting millions of lives. By detecting high mental workload stress level at early stage, warning can be provided to peoples to avoid mental-related disease. Therefore, this work is an attempt to contribute to betterment of society by classifying the mental states. In this study, EEG signal dataset is used to classify the mental workload [18]. It was observed that previous models were classifying the data into good counter and bad counter with the limited dataset available, and the training parameters were comparatively high which increases the computation and space [19, 20]. To overcome this, SMOTE technique has been used to balance and upscale the dataset [21] and EEGNet(4,2) model has been implemented for classifying the up sampled data into good counter and bad counter. EEGNet is a CNN model [22]. Deep convolutional neural networks (CNNs) have gained prominence due to their success in tackling complex image classification problems [23–27]. CNN is being widely used for processing and analyzing visual data. CNN contains various layer by help of which we try to mimic the human virtual cortex so that model can learn and predict. CNN has various applications in image classification, image segmentation, medical image analysis, object detection, image generation, etc. There are certain challenges of deep learning including lack of training data, imbalanced data, interpretability of data, gradient problem, uncertainty scaling, model compression, overfitting and vanishing gradient problem [28–30].

Novelties of this paper are (i) comparing it with previous state-of-the-art studies, the used model (EEGNet) has a smaller number of parameters and greater accuracy.

(ii) Limited and unbalanced dataset is a major problem in the used dataset. Training CNN model with limited data can result in misclassification and biased model. This issue is resolved by using data augmentation technique (SMOTE).

Paper is arranged in the following sections: Sect. 2, we discuss methodology, it has four subsections which are dataset description, synthetic minority oversampling technique (SMOTE), model architecture, and metrics for performance evaluation. Section 3 contains experiment and results. In Sect. 4, comparison of model with previous models is done and Sect. 5 contains Conclusion.

2 Methodology

2.1 Dataset Description

There are 36 subjects' data used in this study. All the participants were healthy and of matched age. These participants were asked to perform an arithmetic subtraction task, and based on the number of tasks they can perform in one minute, they were categorized into two classes. EEG signals of the subjects were recorded while they were performing the arithmetic task (subtraction). Neurocom EEG 23-channel system is used for recording. Twenty-three electrodes were placed in 10/20 international scheme on human scalp. Electrodes are placed over the scalp on the regions partial, central, frontal and temporal. Recordings are done for the time of 60 s. Each channel's sample rate was 500 Hz. ECG electrodes were not used, and only 19 channels data are used for classification. Data preprocessing is done to eliminate any noise. Independent component analysis (ICA) is done to remove artifacts. Hence, all the recordings don't have any kind of distortion or noise [18].

2.2 Synthetic Minority Oversampling Technique (SMOTE)

SMOTE is used to balance the class distribution in dataset [21]. It is used to handle imbalanced data distribution. It generates new points from minority class using nearest neighbor. Algorithm used to up sample the data is given below:

1. Apply SMOTE on Imbalanced dataset.
2. Remove some samples from classes alternatively (first time from class 1, next time from class 0, then next time from class 1…) and save it somewhere.
3. Again, apply SMOTE on unbalanced dataset.
4. Add the data which is saved in step 2 to the output of step 3. You will again get imbalanced data.
5. Repeat the process.

2.3 EEGNet Architecture

CNN model called as EEGNet specifically designed to work on the EEG dataset [22]. CNN is most popular and widely used method for classification and prediction. This model has achieved better performance compared to previously proposed CNN models. EEGNet(4,2) is used, here 4 and 2 denotes, 4 temporal filters (filter used in CONV2d layer) and 2 spatial filters per temporal filter (Depth multiplier used in Depthwise CONV2D). The architecture EEGNet(4,2) model is shown in Table 1. Convolution 2D layer is used for feature extraction, batch normalization layer for stabilization and faster convergence, DepthwiseConv2D and SeparableConv2D layer to reduce the parameters on feature extraction. Activation layer for nonlinearity, average pooling layer for down sampling, dropout layer to prevent overfitting, flatten layer to convert features to 1D vector, and softmax function for final classification are used.

Training parameters reduced significantly (78.44%) compared to previously proposed models, achieving higher accuracy compared to previous models.

2.4 Performance Measure

To assess the effectiveness of the model, a confusion matrix is created. Confusion matrix uses the following terms:

- True positive (TP)—No of points for which model has predicted positive class and it belongs to positive class only.
- True negative (TN)—No of points for which model has predicted the negative class and it belongs to negative class only.
- False positive (FP)—No of points for which model has predicated positive whereas it belongs to negative class.
- False negative (FN)—No of points for which model has predicated negative class whereas it belongs to positive class.

Based on these terms, five performance measures are calculated. The equations to calculate these performance measures using the confusion matrix parameters are given below in Eqs. (1–5):

$$\text{accuracy} = (\text{TP} + \text{TN}) \div (\text{TP} + \text{TN} + \text{FP} + \text{FN}) \tag{1}$$

$$\text{precision} = (\text{TP}) \div (\text{TP} + \text{FP}) \tag{2}$$

$$\text{recall} = \text{TP} \div (\text{TP} + \text{FN}) \tag{3}$$

Table 1 The architecture of EEGNet(4,2) model

Layer (type)	Input shape	Output shape	Param	Trainable
conv2d_1 (Conv2D)	(None, 19, 31000, 1)	(None, 19, 31000, 4)	256	256
batch_normalization_3 (BatchNormalization)	(None, 19, 31000, 4)	(None, 19, 31000, 4)	16	8
depthwise_conv2d_1 (DepthwiseConv2D)	(None, 19, 31000, 4)	(None, 1, 31000, 8)	152	152
batch_normalization_4 (BatchNormalization)	(None, 1, 31000, 8)	(None, 1, 31000, 8)	32	16
activation_2 (Activation)	(None, 1, 31000, 8)	(None, 1, 31000, 8)	0	0
average_pooling2d_2 (AveragePooling2D)	(None, 1, 31000, 8)	(None, 1, 7750, 8)	0	0
spatial_dropout2d_2 (SpatialDropout2D)	(None, 1, 7750, 8)	(None, 1, 7750, 8)	0	0
separable_conv2d_1 (SeparableConv2D)	(None, 1, 7750, 8)	(None, 1, 7750, 8)	192	192
batch_normalization_5 (BatchNormalization)	(None, 1, 7750, 8)	(None, 1, 7750, 8)	32	16
activation_3 (Activation)	(None, 1, 7750, 8)	(None, 1, 7750, 8)	0	0
average_pooling2d_3 (AveragePooling2D)	(None, 1, 7750, 8)	(None, 1, 968, 8)	0	0
spatial_dropout2d_3 (SpatialDropout2D)	(None, 1, 968, 8)	(None, 1, 968, 8)	0	0
flatten (Flatten)	(None, 1, 968, 8)	(None, 7744)	0	0
dense (Dense)	(None, 7744)	(None, 2)	15490	15490
softmax (Activation)	(None, 2)	(None, 2)	0	0

Note Padding = same, Filters used in conv2d later, i.e., first layer called as temporal filter is 4 and filter used in depthwise conv2d layer is 2 per temporal filter
Total params: 16,170 (63.16 KB)
Trainable params: 16,130 (63.01 KB)
Nontrainable params: 40 (160.00 B)

$$\text{f1} - \text{score} = 2 \times \left(\frac{\text{recall} * \text{precison}}{\text{recall} + \text{precison}} \right) \tag{4}$$

$$\text{Miss} - \text{Classification} = 1 - \text{Accuracy} \tag{5}$$

3 Experiment and Results

Initial dataset contains 36 subject, and it was imbalanced. Ten subjects belong to bad counter and 26 subjects belongs to good counter. The dataset is divided into training and testing in the ratio 70:30 and training dataset is split again into training and validation in the ratio 70:30. Five trails each for epoch 100 were ran and after each trail test accuracy and all performance measures were calculated. Relationship between training accuracy and validation accuracy on original dataset is illustrated in Fig. 1. Relationship between training loss and validation loss on original dataset is illustrated in Fig. 2. EEGNet(4,2) model was tested on original dataset, achieving an average accuracy of 61.78%. The results of EEGNet(4,2) on original dataset is shown in Table 2.

Dataset has been balanced and upscaled to 304 subjects using SMOTE technique. EEGNet(4,2) model is applied on the upscaled dataset. The dataset is divided into

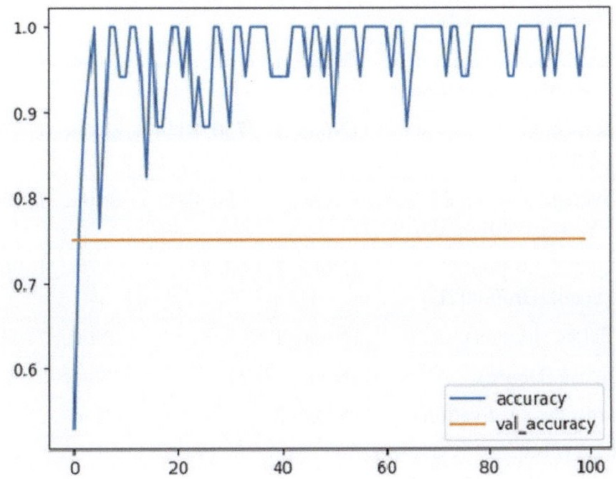

Fig. 1 Training accuracy versus validation accuracy

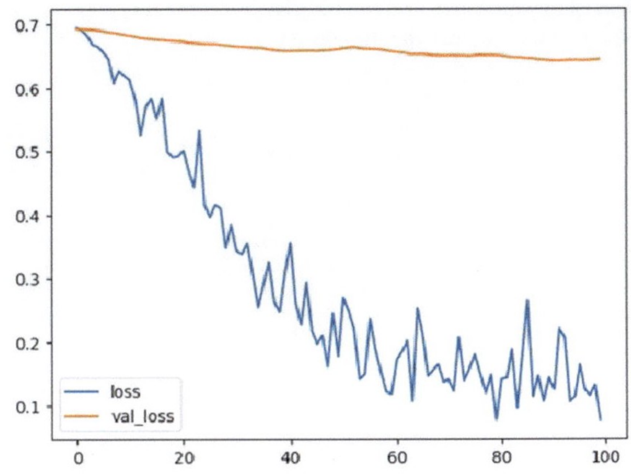

Fig. 2 Training loss versus validation loss

Table 2 Results of EEGNet(4,2) on original dataset

	Trail1	Trail2	Trail3	Trail4	Trail5
Train accuracy	0.941	1.000	1.000	1.000	1.000
Validation accuracy	0.875	0.750	0.625	0.625	0.375
Test accuracy	0.545	0.727	0.636	0.545	0.636
Precision	0.545	0.727	0.636	0.545	0.636
Recall	1.000	1.000	1.000	1.000	1.000
F1-score	0.705	0.842	0.777	0.705	0.777
Miss-classification	0.454	0.272	0.363	0.454	0.363

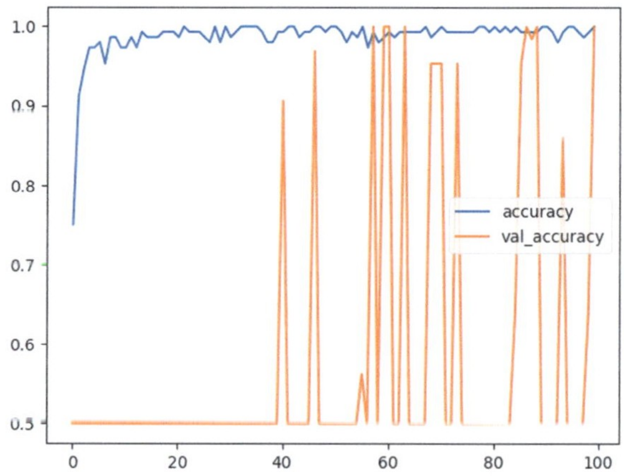

Fig. 3 Training accuracy versus validation accuracy

training and testing in the ratio 70:30 and training dataset is split again into training and validation in the ratio 70:30. Five trails each for epoch 100 were ran and after each trail test accuracy and all performance measures were calculated. Relationship between training accuracy and validation accuracy for upscaled data is illustrated in Fig. 3. Relationship between training loss and validation loss for upscaled data is illustrated in Fig. 4. EEGNet(4,2) model was tested on upscaled dataset, achieving an average accuracy of 99.56%. The results of EEGNet(4,2) on upscaled dataset is shown in Table 3.

Precision, Recall, f1-score, and Miss-Classification are calculated on Test data.

4 Discussion

EEGNet(4,2) model has been successfully implemented. Previously build 2D CNN model on the same dataset has achieved the best accuracy to 97.2% as shown in Table 4 and the number of trainable parameters used is 74,189. The dataset is not balanced

Fig. 4 Training loss versus validation loss

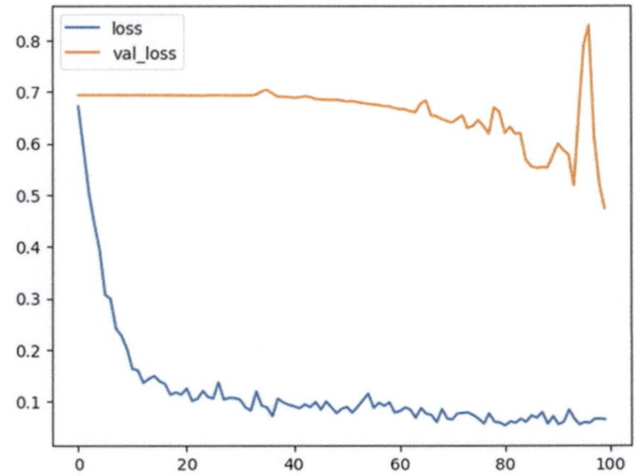

Table 3 Results of EEGNet(4,2) on upscaled dataset

	Trail1	Trail2	Trail3	Trail4	Trail5
Training accuracy	1.000	0.993	0.986	0.993	0.993
Validation accuracy	1.000	0.953	1.000	1.000	0.968
Test accuracy	0.989	1.000	1.000	0.989	1.000
Precision	0.978	1.000	1.000	1.000	1.000
Recall	1.000	1.000	1.000	0.978	1.000
F1-score	0.989	1.000	1.000	0.987	1.000
Miss-classification	0.010	0.000	0.000	0.010	0.000

and upscaled in previous models [21, 22]. Dataset has been upscaled and balanced and by the usage of EEGNet(4,2) model number of parameters significantly reduced to 16,130 which results in lower computation cost and limited memory usage. The EEGNet(4,2) model was tested on the original imbalance dataset without upscaling, achieving an average accuracy of 63.45%. EEGNet(4,2) model was tested on an upscaled and balanced dataset, achieving an average accuracy of 99.56%. Comparing it to all the previous models proposed, the accuracy was highest. Additionally, we did t-test between EEGNet(4,2) model and rest of the other models and got the p-value of 0.01083 for traditional models and 0.07850 for 2D CNN model. Table of comparison for different classifiers and models is illustrated in Table 4. Previously, highest accuracy of 97.22% and average accuracy of 93.74% was achieved by 2D CNN model [22]. The following limitations has been observed in the previously proposed models:

Existing models are trained over limited unbalanced data and have issues with accuracies.

Existing 2D CNN model has a high number of trainable parameters (74,189) and needs more computation and memory usage.

Table 4 Performance measure for different classifiers and models

Classifiers	Accuracy	Precision	Recall	F1 score
GNB [21]	65.00	66.70	70.00	65.30
DT [21]	80.00	83.30	80.00	79.30
SGD [21]	65.00	70.00	70.00	67.30
LR [21]	75.00	70.00	90.00	78.00
SVC [21]	70.00	73.30	70.00	69.30
ANN [21]	55.00	60.00	100.00	66.00
2D CNN [22]	97.22	96.29	100.00	98.11
EEGNet	63.45	63.45	1.00	77.78
EEGNet and SMOTE	99.56	99.57	99.57	99.54

5 Conclusion

In this work, dataset has been upscaled and balanced using SMOTE technique. EEGNet(4,2) a compact CNN model that utilizes the concept of depthwise and separable convolutions has been used to classify the dataset into good counter and bad counter efficiently. Average classification accuracy achieved using EEGNet(4,2) model is 99.56%, standard deviation and F1-score achieved are 0.005 and 99.54%, respectively. Moreover, trainable parameters were significantly reduced by 78.44% as compared to previously developed 2D CNN model. For future scope, EEGNet(4,2) model can be utilized for other EEG datasets and performance of EEGNet(4,2) and EEGNet(8,2) model can be compared.

References

1. Gevins A (1997) High-resolution EEG mapping of cortical activation related to working memory: effects of task difficulty, type of processing, and practice. Cereb Cortex 7(4):374–385
2. Hu J (2017) Automated detection of driver fatigue based on AdaBoost classifier with EEG signals. Front Comput Neurosci 11
3. Goldfarb M et al (2022) Severe mental illness and cardiovascular disease: JACC state-of-the-art review. J Am Coll Cardiol 80(9):918–933
4. Bosom Maarten AS, Theo FM, Monicque M (2005) Effects of mental fatigue on attention: an ERP study. Cogn Brain Res 25(1):107–116
5. Plechawska-Wójcik M, Tokovarov M, Kaczorowska M, Zapała D (2019) A threeclass classification of cognitive workload based on EEG spectral data. Appl Sci 9(24):5340
6. Paas F, Tuovinen J, Tabbers H, Van Gerven PWM (2003) Cognitive load measurement as a means to advance cognitive load theory. Educ Psychol 38:63–72
7. Antonenko P, Paas F, Grabner R et al (2010) Using electroencephalography to measure cognitive load. EducPsychol Rev 22:425–438
8. Hossain D et al (2019) Cognitive load measurement using galvanic skin response for lis-tening tasks. In: 2019 4th International Conference on Electrical Information and Communication Technology (EICT), Khulna, Bangladesh, pp 1–4

9. Fowler A, Nesbitt N, Canossa A (2019) Identifying cognitive load in a computer game: an exploratory study of young children. In: 2019 IEEE Conference on Games (CoG), London, United Kingdom, pp 1–6
10. Park J et al (2024) Assessing workload in using electromyography (EMG)-based prostheses. Ergonomics 67(2):257–273
11. Heine T et al (2017) Electrocardiographic features for the measurement of drivers' mental workload. Appl Ergonom 61:31–43
12. Charles RL, Nixon J (2019) Measuring mental workload using physiological measures: a systematic review. Appl Ergon 74:221–232
13. Ahirwal MK (2021) Analysis and identification of EEG features for mental stress. In: Bhateja V, Peng SL, Satapathy SC, Zhang YD (eds) Evolution in computational intelligence. advances in intelligent systems and computing, vol 1176. Springer, Singapore. https://doi.org/10.1007/978-981-15-5788-0_19
14. Ahirwal MK, Kumar A, Singh GK (2021) Biomedical signals. In: Computational intelligence and biomedical signal processing. SpringerBriefs in electrical and computer engineering. Springer, Cham. https://doi.org/10.1007/978-3-030-67098-6_1
15. Ahirwal MK (2020) Analysis and identification of EEG features for MentalStress, In: 8th International Conference on Frontiers of Intelligent Computing: Theory and Applications (FICTA), NIT Karnataka, Surathkal, India (In Press)
16. Ahirwal MK, Kumar A, Singh GK (2013) EEG/ERP adaptive noise canceller design with controlled search Space (CSS) approach in cuckoo and other optimization algorithms. IEEE/ACM Trans Comput Biol Bioinf 10(6):1491–1504
17. Ahirwal MK, Kumar A, Singh GK (2016) Study of ABC and PSO algorithms as optimized adaptive noise canceller for EEG/ERP. Int J Bio-Inspired Comput 8(3):170–183
18. Zyma I, Tukaev S, Seleznov I, Kiyono K, Popov A, Chernykh M, Shpenkov O (2019) Electroencephalograms during mental arithmetic task performance. Data 4(1):14
19. Rai AA, Ahirwal MK (2022) Electroencephalogram-based cognitive load classification during mental arithmetic task. In: ADCOM 2020: 26th annual international conference on advanced computing and communications, lecture notes in electrical engineering, vol 869. Springer, Singapore. https://doi.org/10.1007/978-981-19-0019-8_36
20. Aman AR, Mitul Kumar A (2021) Cognitive load classification during arithmetic task using single convolution layer based 2D-CNN. In: 3rd international conference on machine learning, image processing, network security and data sciences (MIND-2021) at NIT Raipur
21. Singh U, Mitul Kumar A (2023) Improved classification of mental workload using one dimensional convolutional neural network and SMOTE technique. In: Proceedings of the 2023 7th international conference on graphics and signal processing
22. Lawhern VJ et al (2018) EEGNet: a compact convolutional neural network for EEG-based brain–computer interfaces. J Neural Eng 15(5):056013
23. Krizhevsky A, Sutskever I, Hinton GE (2012) Imagenet classification with deep convolutional neural networks. In: Pereira F, Burges CJC, Bottou L, Weinberger KQ (eds) Advances in Neural information processing systems, pp 1097–1105. [Online]. Available: http://papers.nips.cc/paper/4824-imagenet-classification-with-deep-convolutional-neural-networks.pdf
24. Simonyan K, Zisserman A (2014) Very deep convolutional networks for large-scale image recognition. CoRR, vol. abs/1409.1556, 2014. [Online]. Available: http://arxiv.org/abs/1409.1556
25. Szegedy C, Liu W, Jia Y, Sermanet P, Reed SE, Anguelov D, Erhan D, Vanhoucke V, Rabinovich A (2014) Going deeper with convolutions. CoRR, vol. abs/1409.4842. [Online]. Available: http://arxiv.org/abs/1409.4842
26. He K, Zhang X, Ren S, Sun J (2015) Deep residual learning for image recognition. CoRR, vol. abs/1512.03385, 2015. [Online]. Available: http://arxiv.org/abs/1512.03385
27. Huang G, Liu Z, Weinberger KQ, van der Maaten L (2016) Densely connected convolutional networks. CoRR, vol. abs/1608.06993, 2016. [Online]. Available: http://arxiv.org/abs/1608.06993

28. Alzubaidi L, Zhang J, Humaidi AJ, Al-Dujaili A, Duan Y, Al-Shamma O,... Farhan L (2021) Review of deep learning: concepts, CNN architectures, challenges, applications, future directions. J Big Data 8:1–74
29. Bhatt D, Patel C, Talsania H, Patel J, Vaghela R, Pandya S,... Ghayvat H (2021) CNN variants for computer vision: history, architecture, application, challenges and future scope. Electronics10(20):2470
30. Li Z, Liu F, Yang W, Peng S, Zhou J (2021) A survey of convolutional neural networks: analysis, applications, and prospects. IEEE Trans Neural Netw Learn Syst 33(12):6999–7019

Vaquita Dolphin Detection on Remote Sensing Data Using YOLOV8

Jyothi Priya Inturi, Sunny Nalluri, Sri Lakshmi Dammu, and Hamsa Kumari Kovvasu

Abstract Dolphins belong to the group of marine mammals called cetaceans. They hold a key place at the top of the marine food chain and are essential to preserving the ecosystem's general equilibrium. However, various factors such as gillnet entanglement, climate change, fisheries bycatch, and plastic pollution have led to the endangerment of certain dolphin species, such as the Vaquita dolphins. Therefore, there is an urgent need for detection and classification methods to conserve these species and protect their populations. The application of cutting-edge deep learning models such as You Only Look Once (YOLOv8) for object detection is selected to detect Vaquita dolphins in satellite images. This methodology holds promise for enhancing our understanding of dolphin populations and informing conservation efforts in critical habitats. By leveraging YOLOv8 for detection, we have achieved promising results in identifying Vaquita dolphins with higher accuracy. The results demonstrated promising accuracy and performance metrics, with the YOLOv8 model achieving mAP of 0.927.

Keywords Vaquita dolphin detection · Endangered species · Deep learning (CNNs) · Remote sensing · You Only Look Once (YOLOv8)

1 Introduction

The conservation of vaquita dolphins presents a complex challenges, primarily stemming from human activities. Despite attempts efforts to curb illegal fishing practices, the vaquita, a critically endangered species inhabiting the Gulf of California, continues to face grave threats. Incidental entanglement in illegal gillnets set for totoaba,a fish prized in traditional Chinese medicine, remains the most pressing concern, driving the vaquita population perilously close to extinction. Furthermore, the escalation

J. P. Inturi (✉) · S. Nalluri · S. L. Dammu · H. K. Kovvasu
Department of Computer Science and Engineering, Velagapudi Ramakrishna Siddhartha Engineering College, Vijayawada, India
e-mail: inturijyothipriya@gmail.com

S. Kumar et al. (eds.), *Artificial Intelligence for Next Generation Computing, Volume 1*, Lecture Notes in Networks and Systems 1549,
https://doi.org/10.1007/978-981-95-0701-6_5

of maritime traffic and industrial activities in their habitat increases the risk of vessel strikes and underwater noise pollution, disrupting their natural behaviors and communication.

In response to these challenges, conservationists are exploring innovative approaches, such as utilizing high-resolution satellite imagery, to monitor vaquita populations more effectively. Spearheaded by organizations like the International Committee for the Recovery of the Vaquita (CIRVA), initiatives are underway to leverage satellite technology to complement traditional survey methods. While satellite detection rates may not match those of on-the-ground surveys, they offer a valuable tool, especially in remote and inaccessible regions like the vaquita's habitat. By integrating satellite observations with existing conservation strategies, there is hope for bolstering efforts to safeguard the future of this critically endangered species and deepen our understanding of vaquita ecology.

1.1 Basic Concepts

Space-Borne Remote Sensing refers to the use of satellite or aerial sensors to collect data about the Earth's surface and atmosphere. Many uses for this data exist, such as object and change detection, land cover classification, and more. You Only Look Once (YOLO) is currently the most popular real-time object detector, which can be widely accepted for the following reasons: (a) Lightweight network architecture. (b) Effective feature fusion methods.YOLOv8 is designed to combine the advantages of many real-time object detectors. The key feature of YOLOv8 is that it is extensible. Yolov8 is designed to work with all versions of YOLO and switch between them, making it easy to compare their performance, which is a great benefit for researchers working on YOLO projects. Therefore, YOLOv8 was selected as the baseline.

1.2 Motivation

Dolphins are magnificent creatures that inhabit the ocean, but they are facing numerous threats in their environment. Among these, the Vaquita Dolphins are the most endangered of the world's marine mammals. Urgent action is required to safeguard these majestic creatures, whose conservation is crucial to maintaining the balance of marine ecosystems and also to prevent their extinction. Therefore, it is important to take immediate measures to protect Vaquita Dolphins from the growing threat of extinction.

1.3 Problem Statement

The primary challenge addressed by this study is the lack of comprehensive and efficient methods for monitoring and tracking vaquita dolphins in the Gulf of California. Traditional survey methods are often logistically challenging and costly, especially in remote areas. Additionally, the manual detection and classification of dolphins from satellite imagery are time consuming and prone to errors. Therefore, there is a pressing need to develop automated solutions that can effectively detect, classify, and track vaquita dolphins in the Gulf of California using satellite imagery and deep learning techniques.

1.4 Scope

The present work will be limited to detecting vaquita dolphins in North Pacific Ocean from remote sensing data.

1.5 Objectives

The following are the proposed work's objectives:

1. To collect a dataset of dolphin types for the North Pacific Ocean (Gulf of California).
2. To study how deep learning can be exploited for recognition, of dolphin in remote sensing picture.
3. To propose a system by using deep learning algorithm to detect, that enables extraction of dolphin found in North Pacific Ocean (Gulf of California).

1.6 Advantages

1. Automated detection saves time and resources.
2. Achieves high precision in identifying dolphins populations.
3. Can cover large areas cost-effectively.
4. Provides timely information for decision making.

2 Literature Survey

Chien et al. [1] demonstrated an innovative approach to monitor Taiwanese white dolphins using an autonomous virtual UAV. They employ a virtual Unmanned Aerial Vehicles (UAV) equipped with a customized YOLOv4 detection model. This model is built on images of Taiwanese White Dolphins from Cetacean Lab at NTU. It can identify the side view of these dolphins with good accuracy with the precision of @0.5AP.

Frainer et al. [2] demonstrated the challenges posed by noisy environments and human activities on the detect ability of dolphin sounds, emphasizing the need for automated classifiers in long term monitoring efforts. The research balanced the dataset to address the variability in vocal production among dolphin species and their behavioral contexts, ensuring equal representation for accurate classification. The study utilized Convolutional Neural Networks for detecting and also identifying dolphin vocalizations, achieving high accuracy rates for species identification. The best detection model was designed using a custom built setup and reached an accuracy of 84.4% for all dolphin sounds on the test set, and 89.5% for sounds with clear signal.

Chiruvella et al. [3] states that deep learning framework like Convolutional Neural Networks have limitations in predicting the presence of components without optimizing discriminative features. They simplify the procedure for photographing whales and dolphins, making image recognition 99% faster for researchers. This method uses a Deep Convolutional Neural Network with a loss function known as Additive Angular Margin Loss. To find the most Common class among different features, they use K-Nearest Neighbors, Which calculates the distances to various classes.

Maglietta et al. [4] introduce a novel method know as Neural Network Pool ,which is based on deep learning and specifically designed for the photo identification of Rissos dolphins. RUSBoost algorithm was initially applied for recognizing unknown dolphins in photo identification studies. Convolutional Neural Networks were employed in the research to analyze unbalanced data sets for Risso's dolphin photo identification. A new method called Neural Network Pool was developed for self-operate photo identification of unknown verses known Risso's dolphins using CNNs. NNPool and RUSPool were compared in the study, showcasing the effectiveness of NNPool in recognizing unknown dolphins.

Martino et al. [5] presence data integration is crucial in ecological science, allowing for a comprehensive understanding of ecological processes for research and conservation purposes. Integrated distribution modeling (IDM) is a novel approach that combines different datasets to provide cohesive summaries of species' potential and realized distribution, overcoming challenges posed by heterogeneous data sources. The research paper focuses on modeling the spatial allocation of two dolphin species in the central Mediterranean Sea by integrating data from traditional research surveys, adaptive sampling procedures, distance sampling, and social media platforms. The project used a spatial log Gaussian Cox process model that incorporates different detection functions and thinning for each data source.

Ivanchikova and Tregenza [6] utilize the echolocation click logging device monitors cetaceans passively. The KERNO-Fv1.0 classify into four species categories: "NBHF" for narrow band high frequency clicks, "Other cet" for other odontocete species, "Sonar" for manmade sonars, and "Unclassified" for clicks that do not meet the minimum criteria for the other categories. The error rates for harbor porpoises and dolphins in the Black Sea were found to be acceptable for further study without additional data editing, with boat sonar misclassifications being a main source of errors. The study aimed to estimate the false positive error rate in F-POD data collected in the Black Sea and determine the usability of the data without editing.

Davidesco and Diamant [7] developed a new interceptor adapted to differentiate anthropogenic and biophonic sounds, focused on the dolphin whistle-like signal. Thus, the statistics of the signal phase, which is discussed in the proposed method, can be used for clustering as a metric, compared to the biomimicked signals. This method can be described as the approach which utilizes phase locked loop for defining the phase of a given signal and determination of an approximate entropy for classification. Analytical findings indicate that the interception scheme works as a counter measure to distinguish between real and biomimicked signals depending on different features, which would be beneficial in recognizing disguised communication situations in underwater networks.

Ansari and Zafar [8] introduced a method for integrating Dolphin Swarm Optimisation (DSO) with object identification and classification, which are essential in surveillance video objects look different when they relate to the camera, lighting, and viewing angles, thus it is necessary to improve real-time identification and classification of things. As per them it was suggested that combining Dolphin Swarm Optimization with an improved Sine Cosine Algorithm-based Support Vector Machine classifier can help solve this issue. The foreground segmentation here works by using frame differencing whereas feature extraction is done through Histogram of Orientated Gradients and the DSO-ISCA-SVM classifier ensures accurate classification.

3 Proposed Model

3.1 Architecture

The provided architecture is the working model for our project. The architectural design illustrates how YOLOv8 detects vaquita dolphins. Having remarkable speed and accuracy YOLO is a real-time object recognition system that identifies objects in photos. It handles object detection as an unique regression issue, utilizing entire images to predict class probabilities and bounding boxes in a single evaluation. When compared to more conventional techniques that make use of region proposal networks, this strategy enables YOLO to achieve faster detection times. The most recent version, YOLOv8, has a number of improvements over earlier iterations. It makes use of an upgraded neck design with Feature Pyramid Networks and Path

Aggregation Networks for better multiscale feature representation, as well as an updated backbone network for effective feature extraction. In order to balance speed and accuracy, YOLOv8 also adds sophisticated loss mechanisms. This makes it incredibly useful for a variety of applications, such as driverless cars and animal protection.

3.2 Methodology

The section contains the key modules, which demonstrate how YOLOv8 functions. These modules cover data preparation, augmentation, annotation, and architecture.

3.3 Dataset

The dataset for detecting vaquita dolphins was collected from the National Center for Biotechnology Information (NCBI), leveraging drone images captured for conservation efforts. Vaquita dolphins are critically endangered, with only a few dozen individuals remaining, making the collection of high-quality data imperative for their conservation. To achieve this, the YOLOv8 object detection model was employed due to its effectiveness in identifying objects within images. By utilizing NCBI's repository of drone images, researchers aimed to train the model to accurately detect vaquita dolphins, thereby aiding in their conservation efforts (Fig. 1).

3.4 Data Preprocessing

To ensure successful model training, there are numerous essential steps in the data preprocessing for images. Start by organizing the data by eliminating or fixing any damaged or unnecessary photos. To keep things consistent, resize images to a standard size for consistent brightness and contrast, normalize pixel values to a standard scale, often between 0 and 1. Applying filters will reduce noise, eliminate artifacts, and enhance image quality. Ultimately, in order to guarantee efficiency and compatibility, transform the annotations and photos into the format that the model requires (Fig. 2).

(a) RESIZING: Optimizing object detection models requires resizing satellite photos. Standardizing image size guarantees consistent input, which improves feature extraction and processing speed. We can enhance detection accuracy and maximize computational performance by matching image sizes to model needs, like YOLOv8. Maintaining data consistency and enabling efficient model training depend on proper scaling.

Fig. 1 Architecture of proposed model

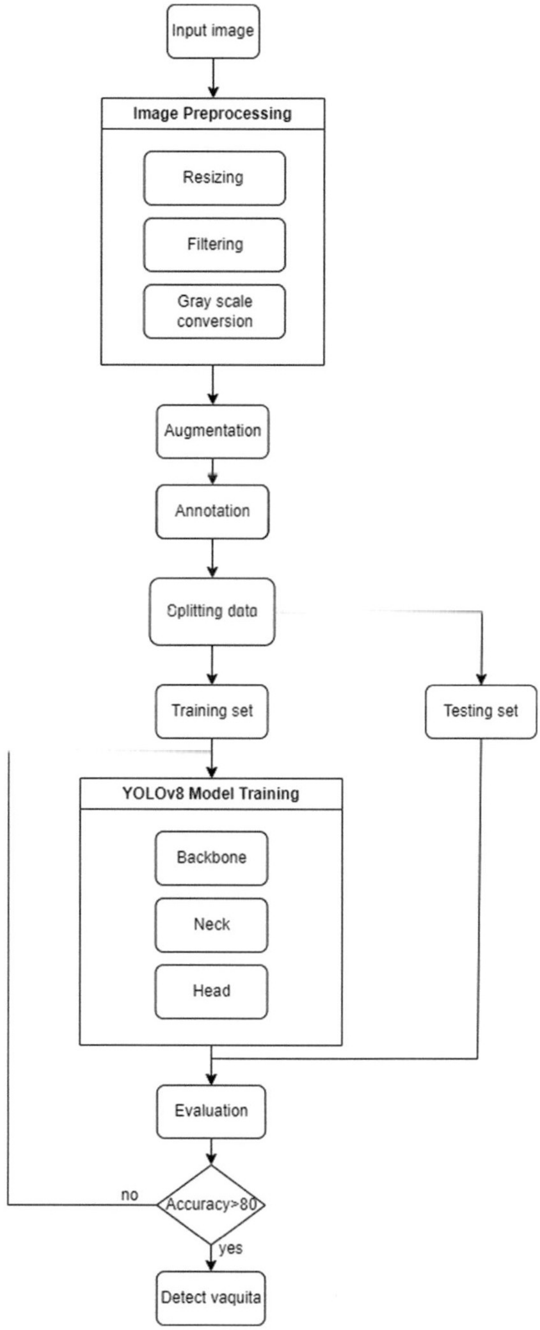

Fig. 2 Drone images from National Center for Biotechnology Information

(b) FILTERING: A basic step in image processing is image filtering, which improves or extracts particular visual properties. Using methods such as high-pass and low-pass filters, we can highlight edges or smooth out images, respectively. Image quality is enhanced via noise reduction, which is accomplished with filters such as median and Gaussian blur. While filtering in satellite images helps to highlight characteristics such as dolphins for further research, sharpening filters improve image quality. In order to ensure optimal model performance through cleaner and more meaningful input data, this preprocessing step is essential for preparing images for tasks like object detection.
(c) AUGMENTATION: Data augmentation is an efficient technique that expand the size and varies of the training dataset to improve the performance of object detection algorithms. We generate new training instances and increase the resilience of the model by implementing transformations like rotations, flips, and scaling. This helps to avoid overfitting. Augments like brightness changes and random cropping work especially well when applied to satellite imagery. In the end, enhancing data is an essential stage in creating dependable and precise object detection models (Fig. 3).

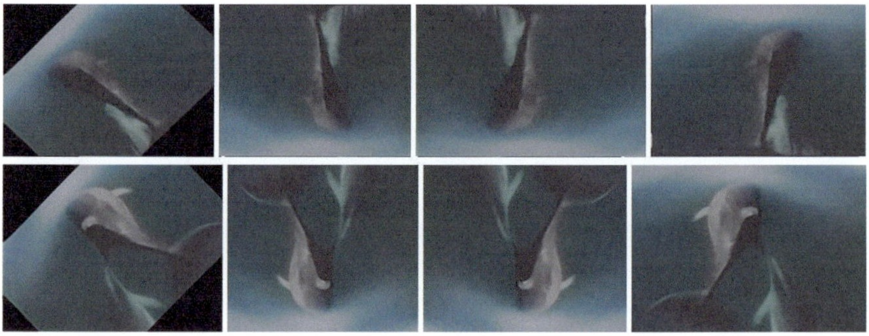

Fig. 3 Augmented images

(d) ANNOTATION: To train accurate object detection algorithms, annotation is essential. Extensive annotation of items of interest in satellite photos will supply vital information from which the model can be trained. Accurate class labels and precise bounding boxes are essential for teaching the model to discriminate between various objects. To further improve the model's comprehension, qualities and characteristics can be annotated. The model's performance is strongly impacted by the quality of the annotations, highlighting the significance of precision and consistency in this process.

(1) YOLO v8:

(a) BACK BONE: The core of a YOLOv8 model is the feature extraction engine, which gradually picks up complex patterns in images. It begins by recognizing simple aspects in previous layers, such as edges and textures, and then iteratively expands on this understanding to identify deeper characteristics in later levels. A complete representation of the input image is generated by this hierarchical method, which is necessary for efficient object detection.
(b) NECK: In a YOLOv8 design, the neck acts as a link between the detecting head and the feature extractor. It combines feature maps from several backbone levels to create a feature pyramid that can hold objects of various sizes. The integration of contextual information across many scales by the neck improves the model's precision in item location in intricate situations. Reducing spatial resolution and dimensionality can have an adverse effect on the overall quality, therefore this fusion process entails a trade-off between computing economy and model performance even though it increases detection precision.
(c) HEAD: The final component of a YOLOv8 network that makes accurate predictions about objects in an image is the head. In order to do this, it creates bounding boxes that specify the position and dimensions of possible objects. The head gives each bounding box a confidence score that represents the probability that an object will be found within that area in order to measure the level of trust in these predictions. In addition, the head classifies the items it

has observed by giving them labels based on a predetermined set of categories. The result of this complex procedure is the creation of comprehensive object detection findings, which offer useful data for further image analysis tasks.

4 Results and Discussions

The results of our project demonstrate the successful deployment of the YOLOv8 object detection system, specifically adapted to identify vaquita dolphins using drone imagery. The system processes high-resolution images captured over North Pacific Ocean, focusing on regions where vaquita dolphins are known to inhabit. Figure 4 illustrates the original images alongside the annotated images, showcasing the bounding boxes and class labels generated by the model. Following the detection phase, Fig. 6 displays the images with the detected dolphins highlighted, confirming the effectiveness of the detection process.

The visual data for this study were sourced from the National Center for Biotechnology Information (NCBI), featuring drone images aimed at conservation efforts for the critically endangered vaquita dolphins. Preprocessing steps included resizing

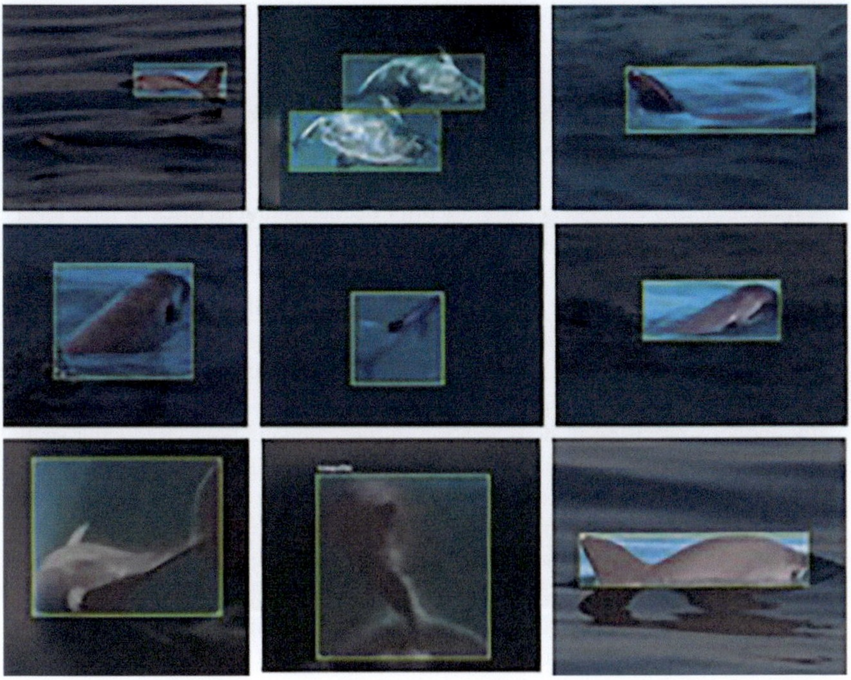

Fig. 4 Annotation images

the images to a standard size, filtering to reduce noise and enhance image quality, and applying data augmentation techniques such as rotations, flips, and brightness adjustments to make the model stronger. The results are presented in a sequential manner, starting with the original images, followed by preprocessing stages, and culminating in the detection and classification of the dolphins.

The processed images exhibit significant clarity and detail, attributable to the preprocessing and augmentation techniques employed. Radiometric correction, noise removal using Gaussian filtering, and enhancement techniques were instrumental in refining the images, ensuring consistent pixel values, and improving the overall quality for better feature detection. The successful implementation of YOLOv8 in this context highlights its ability for real-time object detection in conservation efforts, offering a valuable tool for monitoring and protecting endangered species like the vaquita dolphin (Figs. 5 and 6).

Following preprocessing, a series of analytical steps were conducted to evaluate the effectiveness of the YOLOv8 model in detecting vaquita dolphins. Initially, images underwent resizing, filtering, and augmentation to standardize the dataset and enhance model training. The dataset, derived from drone images provided by the National Center for Biotechnology Information (NCBI), was critical given the endangered status of vaquita dolphins. The YOLOv8 model was used for object detection, and it performed remarkably well, recognizing and localizing vaquita dolphins in photos with an accuracy of 92.8%. The model's efficiency in identifying these dolphins is demonstrated by its high accuracy, which greatly aids conservation efforts. The model's performance was evaluated using metrics like recall, precision, and F1-score, as illustrated in Fig. 7. The aforementioned measures underscore the model's precision and dependability in identifying vaquita dolphins, an essential aspect of their preservation. The effectiveness of the detection process was further evaluated using various metrics. Precision, recall, and mAP were key metrics in this assessment.

Figure 8 displays the evolution of both the training and testing accuracy across 10 epochs. The training accuracy, represented by the blue line, shows a steady improvement over the epochs, mirroring the upward trend of the testing accuracy line depicted

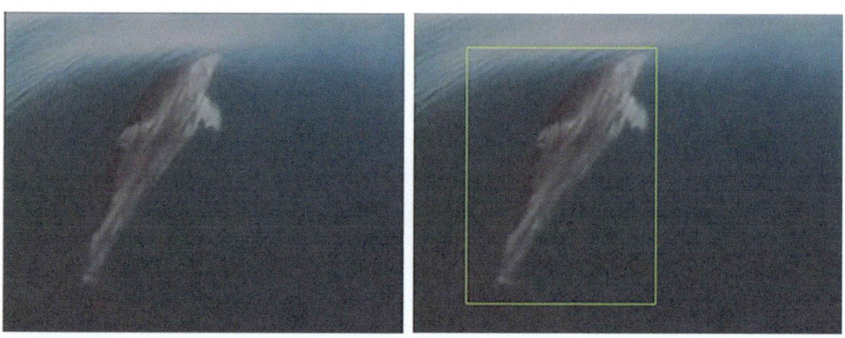

Fig. 5 Presents the orginal image and annotation image

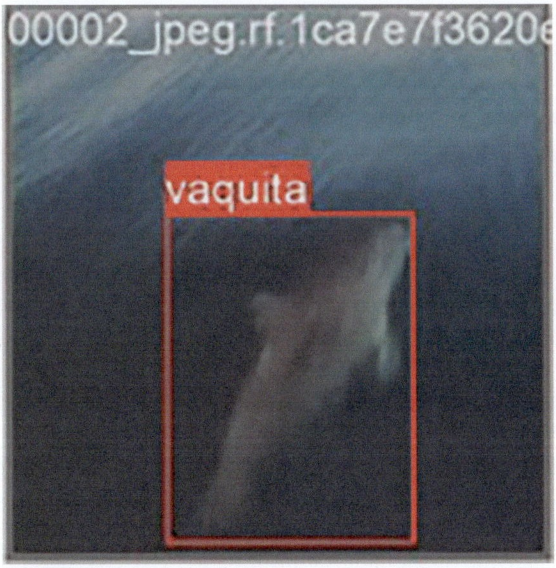

Fig. 6 Represent the image after detection

Model	IoU(or mAP50)	Precision	Recall	mAP50-95
YoloV5	0.861	0.858	0.775	0.393
YoloV8	0.928	0.849	0.866	0.54

Fig. 7 Assessment criteria

in orange. Figure 9 is Precision-Recall curve while the Precision-Recall curve evaluates its precision recall balance across different decision thresholds, vital for tasks like object detection and segmentation with imbalanced class distributions (Figs. 10 and 11).

5 Conclusion and Future Work

In conclusion, the "Vaquita Dolphin detection on Remote Sensing Data Using YOLOv8" project represents a pioneering stride in leveraging advanced technologies for the monitoring and conservation of vaquita dolphins. Through our innovative approach, we have laid the foundation for a robust detection system capable of accurately identifying and tracking these majestic creatures across vast oceanic territories. Moving forward, our focus remains steadfast on refining and expanding our detection

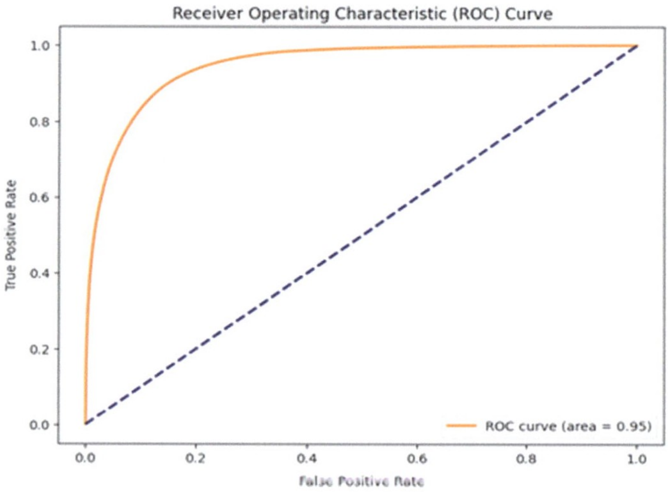

Fig. 8 Receiver operating characteristic curve

Fig. 9 Precision-Recall curve

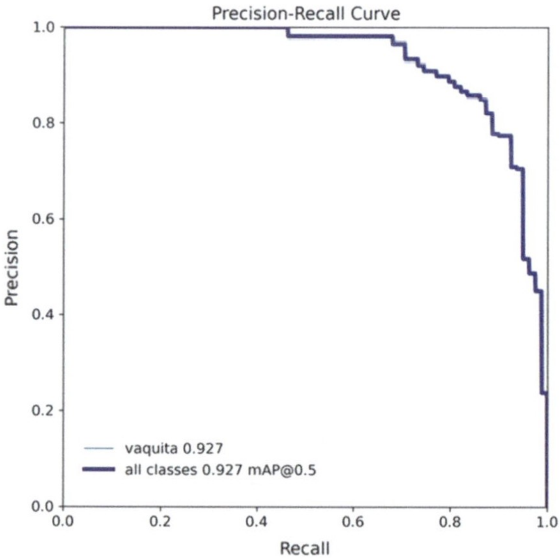

capabilities, with the aim of providing critical data to inform conservation strategies and safeguard the future of vaquita dolphins.

In the future, we aim to expand by incorporating dolphin classification algorithms for species differentiation and enhancing tracking with latitude and longitude data. This will allow precise monitoring of dolphins movements across vast oceanic regions. Additionally, we plan to develop a user friendly interface on edge devices for real-time visualization of whale trajectories in the field. These advancements

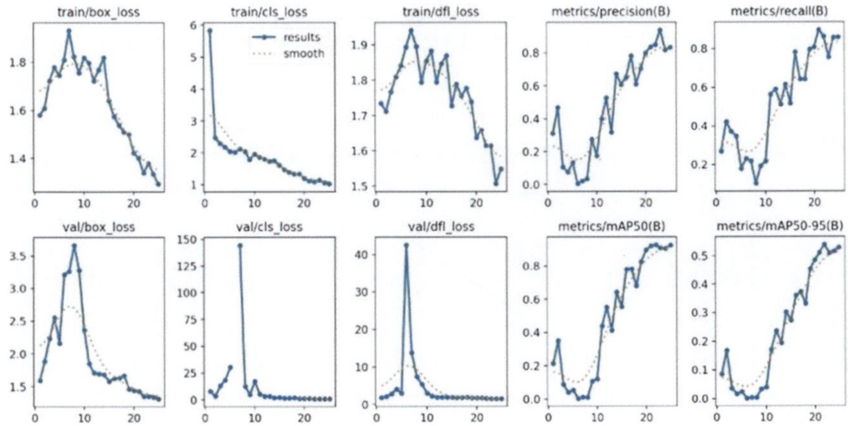

Fig. 10 Training and validation losses during object detection

Work	Methodology	Accuracy Test	Accuracy	Dataset
Guilherme Frainer et al.(2023)	Convolution Neural Networks	Accuracy	89.50%	Aerial Imagery (cetaceans)
Sara Martino et al.(2021)	Deep Learning	Accuracy	88.90%	Aerial Imagery (cetaceans)
Itamar Davidesco et al.(2022)	Convolution Neural Networks	Precesion	90%	Passive Acoustic Dataset
Bergler et al. (2021)	YOlO for detection	mAP	0.94	Satellite imagery (Killer Whales)
	ResNet for classification	Accuracy	95%	
Shaharyar Alam Ansari et al.(2022)	U-Net and FPN	IOU	0.88	Synthetic Dataset
Proposed Work	YOLOV5	mAP	0.86	Aerial Imagery (Vaquita Dolphin)
	YOLOV8	mAP	93%	

Fig. 11 Comparison with other models

will improve the accuracy of our detection system and provide valuable insights into dolphin behavior and distribution, contributing to marine conservation efforts globally.

References

1. Chien T-Y, Chen C-F, Chen Z-Y (2022) Study of Taiwanese white dolphins detection and tracking techniques by utilizing autonomous unmanned aerial vehicles. In: OCEANS 2022, Hampton Roads, VA
2. Frainer G, Dufourq E, Fearey J, Dines S, Probert R, Elwen S, Gridley T (2023) Automatic detection and taxonomic identification of dolphin vocalisations using convolutional neural networks for passive acoustic monitoring. Ecol Inform 78:102291. https://doi.org/10.1016/j.ecoinf.2023.102291
3. Chiruvella H, Sriraj IA, Sreenivasulu Y (2022) Whale and dolphin species detection using hybrid efficient net architecture. SJ Impact Fact 10:7538
4. Maglietta R et al (2020) Convolutional neural networks for Risso's dolphins identification. IEEE Access 8:80195–80206. https://doi.org/10.1109/ACCESS.2020.2990427
5. Martino S, Pace DS, Moro S, Casoli E, Ventura D, Frachea A, Silvestri M, Arcangeli A, Giacomini G, Ardizzone G, Lasinio J (2020) Integration of presence-only data from several sources: a case study on dolphins' spatial distribution. Ecography 44:1533–1543
6. Ivanchikova J, Tregenza N (2023) Validation of the F-POD: a fully automated cetacean monitoring system. PLoS ONE 18(11):e0293402
7. Davidesco I, Diamant R (2022) Detection of Dolphin whistle-like biomimicking signals by phase analysis. IEEE Access 10:36868–36876. https://doi.org/10.1109/ACCESS.2022.3165058
8. Ansari SA, Zafar A (2022) A fusion of dolphin swarm optimization and improved sine cosine algorithm for automatic detection and classification of objects from surveillance videos. Measurement 192:110921
9. Deniz A, Hasan A, Soydemir C, Nuray S, Ergüden D (2019) Report of stranding of striped dolphin *Stenella coeruleoalba* (Meyen, 1833) in Mersin Bay (Northeastern Mediterranean Sea, Turkey). Commagene J Biol 3:114–116
10. Azzellino Arianna, Gaspari Stefania, Airoldi Sabina, Nani B (2008) Habitat use of cetaceans along the continental slope and adjacent waters in the western Ligurian Sea. Deep Sea Res I Oceanogr Res Papers 55:296–323. https://doi.org/10.1016/j.dsr.2007.11.006
11. Chavez-Rosales S, Palka DL, Garrison LP et al (2019) Environmental predictors of habitat suitability and occurrence of cetaceans in the western North Atlantic Ocean. Sci Rep 9:5833
12. Corkeron Peter, Minton Gianna, Collins Timothy, Findlay Ken, Willson Andrew, Baldwin Robert (2011) Spatial models of sparse data to inform cetacean conservation planning: an example from Oman. Endangered Species Res 15:39–52. https://doi.org/10.3354/esr00367
13. Stephen D, Wade P, Slooten E, Barlow J (2008) Design and field methods for sighting surveys of cetaceans in costal and riverine habitats. Mammal Rev 38:19–49
14. Freitas M, Jensen FH, Tyne J, Bejder L, Madsen PT (2014) Echolocation parameters of Australian humpback dolphins (*Sousa sahulensis*) and indo-Pacific bottlenose dolphins (*Tursiops aduncus*) in the wild. J Acoust Soc Am 137(6):3033–3041
15. Karnowski J, Hutchins E, Johnson C (2015) Dolphin detection and tracking. In: Proceedings of the 2015 IEEE winter applications and computer vision workshops, pp 51–56
16. Borowicz A et al (2019) Aerial-trained deep learning networks for surveying cetaceans from satellite imagery. PLoS ONE 14(10):e0212532
17. Lee PQ et al (2021) Beluga whale detection in the cumberland sound bay using convolutional neural networks. Can J Remote Sens 47(2):276–294

18. Patel M et al (2018) The influence of input image scale on deep learning-based beluga whale detection from aerial remote sensing imagery. In: IGARSS 2023–2023 IEEE international geoscience and remote sensing symposium. IEEE, 2023. Moustapha Cisse, Yann N. Dauphin, David Lopez-Paz Mixup: beyond empirical risk minimization, ICLR 2018. arXiv:1710.09412v2
19. Jocher G, Chaurasia A, Qiu J (2023) YOLO by ultralytics. GitHub. https://github.com/ultralytics/ultralytics. Accessed on 27 March 2024
20. Mingxing T, Quoc L (2019) Efficientnet: rethinking model scaling for convolutional neural networks. In: International conference on machine learning, PMLR
21. Al-Wadud A, Mohammad S et al (2021) A dynamic histogram equalization for image contrast enhancement. IEEE Trans Consum Electr

Enhancing Surveillance Systems with Deep Learning-Based Anomaly Detection

M. Evany Anne, M. Brindha, and N. Sivakumaran

Abstract An anomaly in surveillance is defined as any suspicious activity in video footage. An object with an unusual motion pattern or trajectory entering the frame is considered an anomaly. Examples include bike accidents or a person driving at high speed in a crowded place. This paper focuses on detecting anomalies in single-scene surveillance footage, as it is particularly useful for real-world applications where cameras are located at a fixed position. An unsupervised approach is used, employing an autoencoder and decoder model to minimize the reconstruction error between the input and generated output. The model is trained, and the reconstruction error and training loss are observed. After training, regularity scores are calculated for each test case. The proposed model outperforms other models in many scenarios, achieving a precision of 0.87 at a standard threshold of 0.75, which is better compared to other encoder-decoder models. However, the accuracy of the model decreases as the threshold increases.

Keywords Reconstruction error · Anomaly · Auto encoder · Computer vision

1 Introduction

In computer vision, anomaly detection is a significant and difficult problem [1, 2]. Numerous methods for detecting anomalies, including clustering and detection algorithms, have been introduced [3]. Many methods, such as topic modelling and

M. E. Anne (✉) · M. Brindha
Department of Computer Science and Engineering, National Institute of Technology, Tiruchirappalli, Tamil Nadu, India
e-mail: evany@nitt.edu

M. Brindha
e-mail: brindham@nitt.edu

N. Sivakumaran
Department of Instrumentation and Control Engineering, National Institute of Technology, Tiruchirappalli, Tamil Nadu, India
e-mail: nsk@nitt.edu

S. Kumar et al. (eds.), *Artificial Intelligence for Next Generation Computing, Volume 1*,
Lecture Notes in Networks and Systems 1549,
https://doi.org/10.1007/978-981-95-0701-6_6

histogram-based approaches, encode dependencies between variables and establish a relationship between them in the situation of packed scenes. Recently, autoencoders [4] have been used to detect abnormalities by calculating a deviation between the input and the output. They learn patterns of typical behaviour [5]. In general, they require no annotated data and can be trained unsupervised. The study in which the reconstruction error is calculated at test time in order to categorize frames as normal or anomalous served as the inspiration anomaly in video is a matter of study from a long time, this is due to the modelling anomaly events which required learning and understanding of complex visual pattern [6]. Previous methods mainly focus on all novel events which basically works on unsupervised learning pattern, in these model is being trained with both normal frames and validation is being done with both normal and abnormal frames [7]. The goal is to create a collection of bases to represent normal frames and to detect frames with significant reconstruction loss or error as anomalies for inference, for instance, using autoencoders or sparse coding. However, because of the limitations in data and processing, these methods are only used with limited datasets and straightforward scenarios, which are not satisfactory for real-world surveillance applications [8, 9]. Theoretically attractive though it may seem, treating all events as anomalies has drawbacks for real-world surveillance applications [10]. If our model sees only walking people, then riding a bike will comes under an anomaly event. But for security reasons, it shouldn't be viewed as a general abnormality. Recent research starts to use supervision for real-world anomaly identification as some anomalous activities in real-world applications may have clear definitions, such as various criminal events that follow precise patterns. With hundreds of both abnormal and conventional video clips, the UCSD-Crime dataset is currently the largest dataset for anomaly detection with realistic anomalies. In the training set, it contains both normal and anomalous films, and it provides frame-level annotations for the validation set [11]. Weak supervision is provided by video-level annotations, and detection performance has been greatly enhanced by algorithms that take advantage of this weak supervision. Anomaly detection tasks that target certain kinds of abnormalities, like traffic incidents on highways, have also been studied [12]. Given the constancy in camera angles, foreground patterns, and backdrops, geometric prior knowledge and physical notions can be employed to manually design detection pipelines. To find anomaly occurrences, several typical works rely on object detection. The aim of the paper is to find a better and accurate method to find anomaly in CCTV videos existing solutions for anomaly detection have proven effective in many cases, they do have certain limitations like False Positives, False Negatives, Scalability, Lack of Contextual Understanding, Imbalanced Data.

The objective of the paper includes:

- Find better method for single-scene anomaly detection in videos.
- Reduce computational cost and reduce human effort of live monitoring.
- Improve the chances of finding anomaly using variational auto encoder.
- To find model which better accuracy on less training.

2 Literature Review

Early unsupervised Video anomaly detection techniques can be broadly categorized into three categories: unsupervised, weakly supervised, and supervised learning, depending on the experimental context and training data.

2.1 Classic Machine Learning

Probabilistic models and classical machine learning were used in early unsupervised techniques. In order to find patterns, Kim et al. [13] suggested combining probabilistic Principal Component Analysis (PCA) with optical flow characteristics. By utilizing knowledge from research on crowd behaviour, including the social force model, Mehran et al. [14] investigated interaction forces in crowds to gain a better understanding of normal crowd behaviour. They then classified normal and anomalous frames using Latent Dirichlet Allocation (LDA) and Bag of Words (BoW). In order to determine temporal normalcy, Li et al. [13] used several Dynamic Textures(DT) models. A random forest model was then trained to distinguish between usual and anomalous frames. Sparse coding was also utilized for anomaly detection.

2.2 Deep Learning

Recent advances in deep learning have enabled the use of large-scale datasets and powerful computational resources. In the context of unsupervised anomaly detection, several approaches have been developed based on deep autoencoders (AE) [4]. One method involves using a Fully Convolutional Network (FCN)-based autoencoder to learn both motion features and discriminative patterns, with the regularity score determined by the reconstruction error of the AE model. To better capture temporal relationships within videos, another approach integrates FCN with Long Short-Term Memory (LSTM) networks, forming a ConvLSTM-AE, which enhances the performance of the AE framework. Additionally, there has been exploration into combining sparse coding with Recurrent Neural Networks (RNNs).

2.3 Weakly Supervised Methods

The increasing volume of video content on social media platforms like YouTube has made it possible to access and annotate a large number of anomalous videos. Performance can be enhanced by incorporating supervision for specific application scenarios where the anomalous events are clearly defined. Recent research often involves a weakly supervised setting, where training is limited to video-level annotations. In this setup, videos are labelled as either normal or anomalous, but the precise timing of anomaly events within each video is not specified (i.e., weak supervision). Sultani et al. [15] frame this weakly supervised problem as Multiple Instance Learning (MIL). This approach assumes that each typical video contains at least one normal frame in every frame of the video. In a normal video, every frame should be normal, whereas an anomalous video must contain at least one frame with an anomaly. One approach proposes a graph-based MIL framework combined with anchor dictionary learning, with all experiments conducted in a weakly supervised setting using the UCSD dataset. Another study introduces a deep learning-based solution and presents the UCF-Crime dataset, a large collection of surveillance videos featuring real crime-related anomalies. However, in weakly supervised models, dealing with data limitations can sometimes prevent the development of strong predictive models, leading to reduced accuracy.

2.4 Supervised Learning

Recent studies primarily focus on frame-level annotated training videos for specific scenarios where backgrounds and objects are well-defined, such as highways and vehicles in traffic accident detection. A common approach combines object detection with geometric prior knowledge, supplemented by additional supervision from publicly available datasets. For instance, vehicles are first detected using Faster-RCNN, and then an attention-based LSTM module is applied to learn the accident score. In recent works related to the AI City Challenge, frame-level accident annotations are provided in the training set. Additionally, some methods model the background and spatial relationships using semantic segmentation, geometric priors, and object detection, incorporating extra supervision from other public datasets. Shuai Bai and Gaoang Wan's 2019 contributions to the AI City Challenge [16, 17] also utilize frame-level accident annotations in their training process.

3 Proposed Methodology

The proposed model for anomaly detection in videos consists mainly of two parts first is computing the reconstruction error the image passes as in input from encoder and comes out as output as reconstructed image from decoder. In the second step, regularity score is being calculated for each frame. Lower the regularity score chances of anomaly is more in particular frame. Previous model works on normal auto encoder and decoder technique. But in proposed model variational convolutional auto encoder model is used which gives better output in many scenarios compared to traditional approach (Fig. 1).

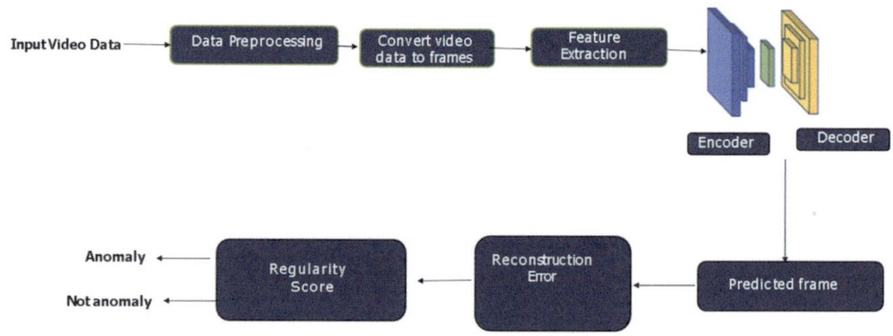

Fig. 1 Proposed methodology

Input: Training data $X = \{x_1, x_2, \ldots, x_n\}$
Initialize the encoder and decoder network parameters θ_e and θ_d
for *each epoch* **do**
    for *each batch of data* $x_b \in X$ **do**
        Encode x_b to latent variables z using $\mu(x_b)$ and $\sigma(x_b)$
        Sample z from $q_\phi(z|x_b) \sim \mathcal{N}(\mu(x_b), \sigma(x_b))$
        Decode z to reconstruct $\hat{x_b}$
        Compute reconstruction loss: $L_{\text{recon}} = \|x_b - \hat{x_b}\|^2$
        Compute KL divergence: $L_{\text{KL}} = D_{\text{KL}}(q_\phi(z|x_b) \| p(z))$
        Compute total loss: $L = L_{\text{recon}} + \beta L_{\text{KL}}$
        Update network parameters θ_e and θ_d using gradient descent
    end
end
Output: Trained VAE model with parameters θ_e and θ_d

Algorithm 1: Variational Autoencoder (VAE) Process

This algorithm describes the training procedure of a Variational Autoencoder (VAE). First, the parameters for the encoder and decoder networks need to be set up. Every epoch, the algorithm goes over batches of the training data. Based on samples obtained from this distribution, the encoder assigns a latent variable with predefined means and variances to each batch. These latent variables are then used by the decoder to recreate the input. The approach computes two types of losses: the reconstruction loss, which measures how much the reconstructed data matches the original, and the KL divergence, which measures how much the learnt latent distribution deviates from a previous distribution. The total loss is obtained by adding these losses together, and it is then modified by applying a weight parameter. The process repeats until the model is trained, resulting in optimized encoder and decoder parameters.

3.1 Data Pre-processing

The training data from video frame is being divided into temporal sequences. Then resize these in number of 10 by using sliding window technique. Resize each frame of the video to size of 256×256 for all input images with the same resolution and pixel and scale between 0 and 1. Data augmentation is being used since the number of parameters is very large.

3.2 LSTM Convolutional Autoencoder

In the first step a convolutional LSTM autoencoder in keras is created. The establishment of encoder and decoder is done for autoencoder. For encoder the ordered sequence of frame is being provided as input as seen from Fig. 2. For motion encoding of model, the model further spitted into two main components one is spatial and other one is temporal encoder, with the output of the former serving as latter's value. It then mirrors encoder, the decoder and aims to reassemble the video sequence.

Input: Training data $X = \{x_1, x_2, \ldots, x_n\}$
Initialize encoder and decoder parameters θ_{enc} and θ_{dec}
for *each epoch* **do**
    for *each sequence* x_s *in* X **do**
        Encode x_s using LSTM to obtain hidden state h_t
        Pass h_t through convolutional layers to get latent z
        Decode z to reconstruct $\hat{x_s}$
        Compute $L_{\text{recon}} = \|x_s - \hat{x_s}\|^2$
        Update θ_{enc} and θ_{dec} using gradient descent
    end
end
Output: Trained LSTM Convolutional Autoencoder

Algorithm 2: LSTM Convolutional Autoencoder Process

This method outlines the training process for the LSTM Convolutional Autoencoder. The process begins by initializing the encoder and decoder parameters. For each epoch, the method processes every sequence in the training data by first encoding it using an LSTM network, which generates a hidden state. Convolutional layers are then applied to this hidden state to create a latent representation. The original sequence is reconstructed by decoding the latent representation. The reconstruction loss is calculated as the squared difference between the original and reconstructed sequences. The network parameters are then updated using gradient descent based on this reconstruction loss. This process is repeated until the autoencoder is fully trained, resulting in optimized encoder and decoder components.

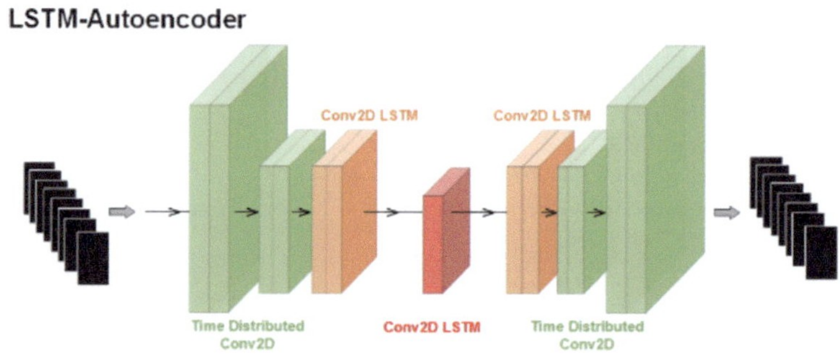

Fig. 2 LSTM autoencoder

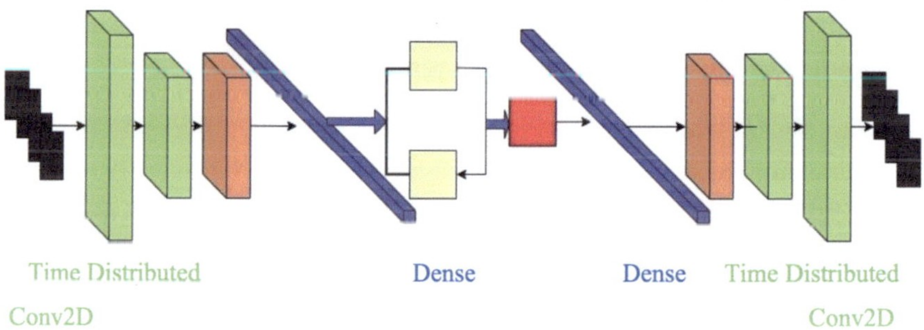

Fig. 3 LSTM Variational Autoencoder

3.3 LSTM-Variational Autoencoder

To enhance the basic autoencoder model which is established on autoencoders, the generation dimensions is being introduced. The latent region is not continuous, and it is quite difficult to explore it. The mean and standard deviation is used for Variational Autoencoder instead of individual points of the vector. By applying this a better rendering performance is observed and it considerably more effective at converting input samples into this rich latent space. As shown in Fig. 3 the AEC model has not very success on how to construct latent valent representing the input space. To overcome this problem instead of using vector, a gaussian distribution with both mean and standard deviation is being used. And finally, a regularization layer is added between input and output which makes reconstruction error more common as it pays less attention to finer details. Lower the reconstruction error sensitivity threshold, the more effectively the model work.

Input: Training data $X = \{x_1, x_2, \ldots, x_n\}$
Initialize encoder and decoder parameters θ_{enc} and θ_{dec}
for *each epoch* **do**
  for *each sequence x_s in X* **do**
    Encode x_s using LSTM to get hidden state h_t
    Compute $\mu(h_t)$ and $\sigma(h_t)$ for latent z
    Sample $z \sim \mathcal{N}(\mu(h_t), \sigma(h_t))$
    Decode z to reconstruct $\hat{x_s}$
    Compute $L_{\text{recon}} = \|x_s - \hat{x_s}\|^2$
    Compute $L_{\text{KL}} = D_{\text{KL}}(q_\phi(z|h_t)\|p(z))$
    Compute total loss $L = L_{\text{recon}} + \beta L_{\text{KL}}$
    Update θ_{enc} and θ_{dec} using gradient descent
  end
end
Output: Trained LSTM Variational Autoencoder

Algorithm 3: LSTM Variational Autoencoder (VAE) Process

The training procedure for an LSTM-based Variational Autoencoder (VAE) is described by this algorithm. First, the encoder and decoder parameters are initialized. The training data's sequences are first encrypted using an LSTM for each epoch in order to produce hidden states. These states are then utilized to calculate the latent variable distribution's mean and standard deviation. By sampling and decoding a latent vector, the original sequence is reconstructed. By combining the reconstruction loss and KL divergence, a total loss is calculated and utilized to update the network parameters via gradient descent. Until the VAE is fully trained, this process is repeated.

4 Architecture Description

The AEC model has not very success on how to construct latent valent representing the input space Layers description and optimization used in the model.

1. Time Distributed Layer: A time distributed layer helps in getting single output by sending multiple input image through the network.
2. Convolutional LSTM2D Layer: A convolutional LSTM is equivalent to Convolutional 1D model layers, main difference is both input and recurrent transformations use convolutional layers. Convolutional model layers are able to perform similar work compared to the LSTM model, the convolutional autoencoder perform convolution operation by maintaining the given image dimensions.
3. Deconvolution Layer: For image reconstruction, autoencoders and GANs frequently employ deconvolution layers, which is also known as transposed convolution used for up sampling. A kernel is used which is like the kernel used in the convolutional, and to increase the output size the back-propagation method is being used.

4. Custom Lambda layer: For representation of mean and standard deviation the custom layer is used by maintaining VAE distributions. Xavier algorithm is used for initialization and training to stop the signal from "exploding" to a high value or decaying to zero for all architectures. The learning rate of 1e-4 and the decay of 1e-5 which uses a two mean squared loss function.

Regularity Score

The reconstruction error of the video at the location (x, y) for the pixel intensity I in the frame t of the video is given as:

$$e(x, y, t) = \|I(x, y, t) - fw(I(x, y, t))\|^2 \tag{1}$$

where Fw is the learned model by the LSTM convolutional autoencoder. By summing the pixel wise errors the reconstruction error of model is given as:

$$e(t) = \sum_{(x,y)} e(x, y, t) \tag{2}$$

Now on reconstructing the reconstruction cost of 10 frames can be calculated using the formula:

$$\text{sequence\_reconstruction\_cost}(t) = \sum_{t'=t}^{t+10} e(t') \tag{3}$$

Abnormality score $Sa(t)$ can be computed by using scale between 0 and 1.

$$sa(t) = \frac{\text{sequence\_reconstruction\_cost}(t) - \min(\text{sequential\_reconstruction\_cost}(t))}{\max(\text{sequential\_reconstruction\_cost}(t)) - \min(\text{sequential\_reconstruction\_cost}(t))} \tag{4}$$

The regularity score $Sr(t)$ is derived by subtracting the value abnormality scores from Eq. 4

$$Sr(t) = 1 - Sa(t) \tag{5}$$

After the regularity score $Sr(t)$ the value of t in range [0, 19] is being computed.

5 Results

This section includes the findings and results of the model like data description in which different datasets which are used are explored, performance evaluation helps in finding accuracy and comparing different models and experimental results shows results given by model.

5.1 Dataset Description

There are various dataset available for detecting anomaly in single scene. Some of the datasets that are used are UCSD ped1 and ped2, Avenue dataset. For the project UCSD dataset is used as UCSD dataset is widely used for checking anomaly in different scenarios. The Ped1 and Ped2 dataset contains 34 training and 36 testing videos. The frames of the videos have low resolution for faster computation which is 158×238. For the testing video at least one frame contain anomaly activity while the training dataset contains only normal activity.

5.2 Performance Evaluation

The accuracy of a model is calculated by dividing the number of correct predictions (true positives and true negatives) by the total number of predictions (true positives, true negatives, false positives, and false negatives), as shown in the following formula:

$$\text{Accuracy} = \frac{\text{Number of Correct Predictions}}{\text{Total Number of Predictions}} \tag{6}$$

A machine learning model's performance during the training phase is assessed using the metric of training accuracy. It measures the percentage of instances in the training dataset that were accurately predicted. The model's training accuracy can be used to gauge how successfully it is picking up new information from the training set. A model that can capture patterns and generalize well within the training dataset is said to have a high training accuracy. It's crucial to remember that strong performance on untrained data is not always guaranteed by a high training accuracy. When the model gets overly specialized to the training data and performs badly on novel, untested samples, overfitting has taken place. A machine learning model's performance during the validation phase is evaluated using a statistic called validation accuracy. It gauges the percentage of cases in the validation dataset that were accurately predicted. By comparing the model's projected labels for the validation set to the actual labels and figuring out the percentage of accurate predictions, vali-

(a) Input Image (b) Reconstructed Image

Fig. 4 Input image versus reconstructed image

dation accuracy is determined. It offers an estimate of the model's performance on hypothetical data. The metrics like precision, recall, F1 score, and accuracy are then used to assess the model's classification performance.

5.3 Experimental Results

Variational Autoencoder takes the input image which is given to encoder and reconstructed image which comes out from decoder. The difference between input image and reconstructed image helps in calculating reconstruction error. More the reconstruction error more the chances of anomaly Fig. 4 represents input versus reconstructed image.

5.4 Training and Validation Accuracy

The first graph shows the accuracy of the model during both training and validation phases over 120 epochs. The blue line represents the training accuracy, while the orange line represents the validation accuracy. Both accuracies increase steadily as the number of epochs progresses, indicating that the model is learning effectively. Around 20 epochs, the validation accuracy begins to plateau, closely following the training accuracy, which suggests that the model is generalizing well to unseen data without overfitting.

5.5 Training and Validation Loss

The second graph illustrates the loss values for both training and validation phases, also over 120 epochs. The blue line represents the training loss, while the orange line represents the validation loss. Both loss values decrease sharply in the initial epochs, indicating rapid learning. As the number of epochs increases, the loss values continue to decrease, albeit at a slower rate, and eventually converge. The close alignment of training and validation losses indicates that the model is not suffering from significant overfitting, as there is no substantial divergence between the two (Fig. 5 and Table 1).

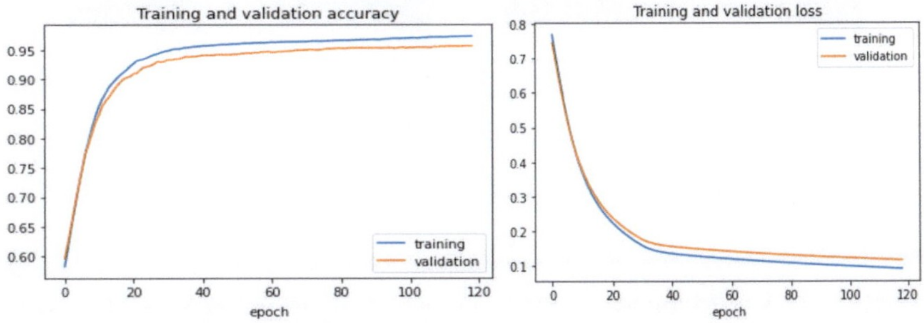

(a) Training Accuracy (b) Training Loss

Fig. 5 Training accuracy and loss

Table 1 Result on training model

Epoch	Loss	Accuracy
1	0.7693	0.5829
2	0.7146	0.6160
3	0.6645	0.6505
4	0.6163	0.6809
5	0.5688	0.7159
6	0.5260	0.7441
7	0.4864	0.7735
145	0.0818	0.9766
146	0.0815	0.9771
147	0.0802	0.9777
148	0.0807	0.9777
149	0.0808	0.9777
150	0.0800	0.9782

5.6 Regularity Score Analysis at Different Thresholds

The regularity score $S_r(t)$ is an important metric used to assess how typical or anomalous a frame is within a given video sequence. A higher regularity score indicates that the frame is more likely to be considered normal, while a lower score suggests potential anomalies. The graphs above depict the regularity score for frames across a video sequence at different threshold levels: 0.75, 0.80, and 0.99 (Fig. 6).

Regularity Score at 0.75 Threshold

The first graph illustrates the regularity score at a threshold of 0.75. The score starts relatively high, indicating that the frames are initially considered normal. As the sequence progresses, there is a slight dip, suggesting the potential onset of an anomaly, followed by a gradual recovery. The overall trend shows fluctuations, reflecting the sensitivity of the model to minor deviations from the norm at this threshold.

Regularity Score at 0.80 Threshold

In the second graph, with the threshold increased to 0.80, the regularity score demonstrates a similar pattern to the 0.75 threshold, but with slightly higher initial values. This suggests that the model is more stringent in classifying frames as normal or abnormal. The increased threshold appears to smooth out some of the smaller fluctuations observed in the first graph, indicating a more stable assessment of the frames.

Regularity Score at 0.99 Threshold

The third graph shows the regularity score at a high threshold of 0.99. Here, the regularity score begins at a very high level, reflecting the model's strict criteria for normalcy. The fluctuations in the score are more pronounced, highlighting the model's heightened sensitivity to even minor deviations. This threshold level is effective in identifying frames that deviate significantly from the norm, as indicated by the sharp dips in the score.

These graphs provide insights into how the model's performance varies with different regularity score thresholds. A lower threshold (0.75) may allow for a broader classification of frames as normal, potentially missing subtle anomalies. Conversely, a higher threshold (0.99) imposes stricter criteria, making it more effective at detecting subtle anomalies but potentially leading to more false positives. The choice of threshold depends on the specific application and the desired balance between sensitivity and specificity in anomaly detection.

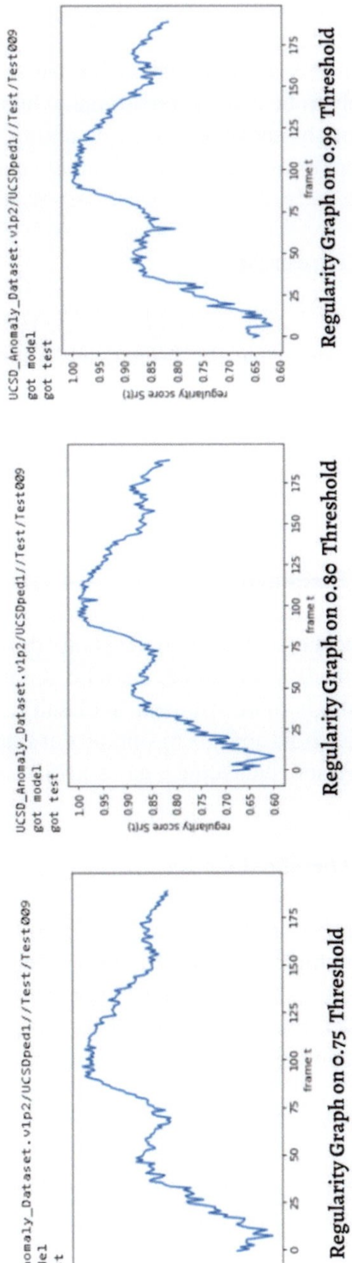

Fig. 6 Regularity score

Table 2 Performance metrics of LSTM-Variational Autoencoder models at different thresholds

Models	Threshold	F1	Precision
LSTM-Variational Autoencoder	0.75	0.304	0.87
LSTM-Variational Autoencoder	0.80	0.452	0.784
LSTM-Variational Autoencoder	0.85	0.587	0.733
LSTM-Variational Autoencoder	0.90	0.69	0.681
LSTM-Variational Autoencoder	0.95	0.750	0.641
LSTM-Variational Autoencoder	0.975	0.744	0.607
LSTM-Variational Autoencoder	0.99	0.740	0.59

5.7 *Result on Training Model*

Table 2 shows F1 score and precision of LSTM-Variational Autoencoder at different threshold 0.85 threshold is considered as standard threshold [18] and model predicts precision of 0.733 and F1 score of 0.58.

6 Conclusions

In this paper, an anomaly detection model is being implemented successfully using convolutional variational auto encoder. By plotting regularity score on different threshold values, we found out that variational auto encoder outperforms other auto encoder slightly. The proposed method gets trained with less loss after each iteration. The purposed model predicts anomaly good accuracy as compared to other auto encoder model. The proposed model can be trained with less computational resources and can give better results in many scenarios compared to other models.

Acknowledgements The research work is supported by the C-DAC, Trivandrum under the project "Next generation Emergency Response Support System"

References

1. Park C (2018) Anomaly pattern detection on data streams. In: 2018 IEEE international conference on big data and smart computing (BigComp), Shanghai, China, pp 689–692. https://doi.org/10.1109/BigComp.2018.00127
2. Zhao Z, Zhang Y, Zhu X, Zuo J (2019) Research on time series anomaly detection algorithm and application. In: IEEE 4th advanced information technology, electronic and automation control conference (IAEAC). Chengdu, China, 16–20. https://doi.org/10.1109/IAEAC47372.2019.8997819

3. Nisha P, Jai Kaur P (2015) A survey of clustering techniques and algorithms. In: 2015 2nd International Conference on Computing for Sustainable Global Development (INDIACom), New Delhi, India, pp 304–307
4. Shah N, Ganatra A (2022) Comparative study of autoencoders-its types and application. In: 2022 6th international conference on electronics, communication and aerospace technology, Coimbatore, pp 175–180. https://doi.org/10.1109/ICECA55336.2022.10009387
5. Rezvanian AR, Imani I, Ghassemian G (2020) Patch-based sparse and convolutional autoencoders for anomaly detection in hyperspectral images. In: 2020 28th Iranian Conference on Electrical Engineering (ICEE), Tabriz, Iran, pp 1–5. https://doi.org/10.1109/ICEE50131.2020.9261057
6. Ajayan AR, Al-Doghman F, Chaczko Z (2018) Visualizing multimodal big data anomaly patterns in higher-order feature spaces. In: 2018 26th International Conference on Systems Engineering (ICSEng), Sydney, NSW, Australia, pp 1–9. https://doi.org/10.1109/ICSENG.2018.8638195
7. Shin J, Kaneko Y, Miah ASM, Hassan N, Nishimura S (2024) Anomaly detection in weakly supervised videos using multistage graphs and general deep learning based spatial-temporal feature enhancement. IEEE Access 12:65213–65227. https://doi.org/10.1109/ACCESS.2024.3395329
8. MR M, Makker M, Ashok A (2019) Anomaly detection in surveillance videos. In: 2019 26th international conference on high performance computing, data and analytics workshop (HiPCW), Hyderabad, India, pp 93–98. https://doi.org/10.1109/HiPCW.2019.00031
9. Nawaratne R, Alahakoon D, Silva D, Yu X (2020) Spatiotemporal anomaly detection using deep learning for real-time video surveillance. IEEE Trans Industr Inform 16(1):393–402. https://doi.org/10.1109/TII.2019.2938527
10. Zhou JT, Zhang L, Fang Z, Du J, Peng X, Xiao Y (2020) Attention-driven loss for anomaly detection in video surveillance. IEEE Trans Circ Syst Video Technol 30(12):4639–4647. https://doi.org/10.1109/TCSVT.2019.2962229
11. Dengxiong X, Bao W, Kong K (2021) Multiple instance relational learning for video anomaly detection. In: 2021 International Joint Conference on Neural Networks (IJCNN), Shenzhen, China, pp 1–8. https://doi.org/10.1109/IJCNN52387.2021.9534124
12. Lv H, Zhou C, Cui Z, Xu C, Li Y, Yang J (2021) Localizing anomalies from weakly-labeled videos. IEEE Trans Image Process 30:4505–4515. https://doi.org/10.1109/TIP.2021.3072863
13. Jaechul K, Kristen G (2009) Observe locally, infer globally: a space-time mrf for detecting abnormal activities with incremental updates. In: 2009 IEEE conference on computer vision and pattern recognition. IEEE, pp 2921–2928
14. Mehran RO, Alexis SM (2009) Abnormal crowd behavior detection using social force model. In: 2009 IEEE Comput Soc Conferen Comput Visi Pattern Recogn Workshops, CVPR Workshops. pp 935–942. 10.1109/CVPR.2009.5206641
15. Waqas S, Chen C, Mubarak S (2018) Real-world anomaly detection in surveillance videos. In: Proceedings of the IEEE conference on computer vision and pattern recognition, pp 6479–6488
16. Bai S, Zhiqun HY, Lei WW, Zhu C, Sun M, Yan J (2019) Traffic anomaly detection via perspective map based on spatial-temporal information matrix. In Procceding CVPR Workshops
17. Gaoang W, Xinyu Y, Aotian Z, Hung-Min H, Jenq-Neng H (2019) Anomaly candidate identification and starting time estimation of vehicles from traffic videos. In: AI city challenge workshop, IEEE/CVF computer vision and pattern recognition (CVPR) Conference, Long Beach, California
18. Oded K, Michal K, Or P (2023) A procedure for anomaly detection and analysis. Eng Appl Artific Intell 117:105503,ISSN 0952-1976. https://doi.org/10.1016/j.engappai.2022.105503 (https://www.sciencedirect.com/science/article/pii/S0952197622004936)
19. Nasaruddin N, Muchtar K, Afdhal A et al (2020) Deep anomaly detection through visual attention in surveillance videos. J Big Data 7:87
20. Zahid Y, Tahir MA, Durrani NM, Bouridane A (2020) IBaggedFCNet: an ensemble framework for anomaly detection in surveillance videos. IEEE Access 8:220620–220630. https://doi.org/10.1109/ACCESS.2020.3042222

21. Duman E, Erdem OA (2019) Anomaly detection in videos using optical flow and convolutional autoencoder. IEEE access 7:183914–183923
22. Mahmudul H, Jonghyun C, Jan N, Amit KR-C, Larry SD (2016) Learning temporal regularity in video sequences. In: Proceedings of the IEEE conference on computer vision and pattern recognition, pp 733–742

Semantic Segmentation of Indian Road Scene Images with a Focus on Small Objects

J. Umamahesh and C. V. Jawahar

Abstract Semantic segmentation is a well-known task in computer vision, with many applications in autonomous navigation. This task aims to partition the image (set of pixels) into multiple labeled subsets of pixels, often referred to as regions. Such a segmentation is key to the scene understanding. This also results in localization of the objects in the image. There exists many research papers that study semantic segmentation in outdoor (often road) scenes. Semantic segmentation of small objects is specially challenging, and critically useful. However, this did not get sufficient attention in the past. The problem is challenging due to (i) availability of only small number of pixels on the small objects, (ii) class imbalance in learning, etc. These make the today's deep learning architectures less effective in performance when it comes to small objects. This can be directly observed from the fact that semantic segmentation performance on small objects in many popular datasets has low accuracy. In this paper, we investigate the challenges associated with and the directions for design of algorithms for segmenting small objects. It is well known that loss functions affect performance in semantic segmentation. The segmentation of small objects also depends heavily on the loss functions used in training the deep learning-based solutions. We also investigate this in this paper. This problem is more severe in Indian situations where many small objects are often seen on the road. The small objects (such as pedestrians) are also very important to segment out for autonomous navigation and driver-assistance systems. In the Indian driving scenario, the Indian Driving Dataset (IDD) provides a class of annotated small objects. These are captured and annotated in an unstructured environment. Thus, we focus primarily on segmentation of small objects in IDD.

Keywords Deep learning architectures · Small objects · Semantic segmentation · Loss functions · Indian driving dataset

---

J. Umamahesh (✉) · C. V. Jawahar
CVIT, KCIS, IIIT Hyderabad, Hyderabad 500032, India
e-mail: umamahesh.j@research.iiit.ac.in

S. Kumar et al. (eds.), *Artificial Intelligence for Next Generation Computing, Volume 1*,
Lecture Notes in Networks and Systems 1549,
https://doi.org/10.1007/978-981-95-0701-6_7

1 Introduction and Related Work

Detection and segmentation of objects in road scenes is crucial for computer vision algorithms involved in tasks such as road surveillance, navigation, intelligent transportation, and autonomous driving. Detection and segmentation helps in localizing the objects of interest in a natural image. In road scenes, often small objects are extremely valuable. They could be missed on a quick scan of the image, but could be costly if you miss to detect such objects (such as pedestrians). Occurrence of small objects in the road scenes needs specific attention. Small objects arise due to small physical objects or low resolution of the objects. This could also be due to other factors such as occlusion, background clutter, and variations in lighting conditions. All these make the detection and segmentation challenging.

A small object is one that appears tiny within the video or image frame, like a distant paraglider or soccer ball, often due to being tracked from afar. It may also refer to objects that are physically smaller in the real world. As per the MS-COCO metric evaluation, small objects are those with an area of 32×32 pixels or less, a threshold generally accepted for datasets involving common objects [1] in the given image frame size.

To localize the object in the image, we need to segment the pixels. Semantic segmentation is a computer vision task that involves partitioning an image into segments, where each pixel is assigned a label corresponding to a particular object class. Unlike object detection, which provides bounding boxes around objects, semantic segmentation offers a pixel-level understanding of visual content, making it crucial for applications requiring precise localization such as autonomous driving, medical imaging, and scene understanding.

Early approaches to semantic segmentation relied on traditional methods like region-based algorithms and graph cuts, but the advent of deep learning has significantly advanced this field. Convolutional neural networks (CNNs) have been particularly useful, with architectures such as fully convolutional networks (FCNs) pioneering the use of end-to-end training for segmentation tasks [2]. Subsequent models like U-Net [3], specifically designed for biomedical image segmentation, and DeepLab [4], which introduced atrous convolution for capturing multi-scale context, have further pushed the boundaries of performance and accuracy in semantic segmentation.

Performance of the deep learning algorithm depends on (a) architecture of the network, (b) data used for training the network, and (c) the loss function used for the training. Researchers have shown that choice of loss functions is an important aspect in semantic segmentation [5, 6]. However, they do not study the effect specifically about small objects. Some research based on deep learning-based small object detection [7] is available. However, similar papers based on deep learning-based small object segmentation are unavailable which shows that small object segmentation is less addressed as compared to small object detection. The segmentation of small objects is not even studied in the Indian setting. To study the segmentation in Indian setting, we use India Driving Dataset [8].

Problem Space: In this paper, we (a) focus on the role and significance of small objects in semantic segmentation, (b) explore the significance of small objects in Indian driving situations, and (c) study the performance of the deep learning architecture with respect to the loss function used for training.

2 Materials and Methods

Now we delve into the details of smaller objects on the road scenes and challenges of segmenting them in an image.

Cars, pedestrians, and traffic signs are usual objects that fall under the ambit of extraction of small objects. Among these, pedestrians fall under the category of vulnerable objects; however, they also fall under the category of small objects in IDD. In the Indian driving scenario, one may find other objects that fall under the category of small objects. A dataset to consider in this scenario is the Indian Driving Dataset (IDD) [8]. This is one of the unique dataset that enables and benchmarks detection and segmentation of objects in the unstructured environments. The baselines on this dataset demonstrated the domain gap and relative difficulty of the Indian segmentation problem with respect to that of Europe (datasets such as Cityscape) and USA (datasets such as BDD). Often, objects that have less than 1000 pixels can be considered as small objects. The objects in IDD that often fall under the ambit of small objects are persons, animals, riders, motorcycles, auto-rickshaws, cars, trucks, buses, traffic signs, traffic lights, poles, and vegetation.

2.1 *Small Objects in Road Situations*

Solving perception problems (such as detection and segmentation) on Indian roads is key to the development of any accurate surveillance and road safety solutions, beyond that of autonomous driving. The following are some of the objects commonly seen in road scenes. Most of the objects mentioned below appear small at a distance. Reader may easily appreciate the need of segmenting such objects (Fig. 1).

- Vehicle registration numbers
- Vehicles at a distance
- Pedestrians at a distance
- Sharp turns at a distance
- Zebra crossing at a distance
- Damages to the road (e.g., potholes) at a distance
- Depth of water filled potholes and open manholes at a distance
- Breaking traffic rules at a distance
- Road blockade at a distance
- Small road markings

Fig. 1 Examples of images from Indian driving dataset [8] and the associated segmentation

Table 1 Number of images across train, val, and test in Indian Driving Dataset

Train	Val	Test	Total
6993	981	2029	10,003

- Traffic sign boards
- Helmets
- Seat belts
- Small and sharp objects on the road
- Traffic lights at a distance.

The distance mentioned in the above small objects in road situations depends on camera used and other external conditions such as clutter. Small objects may be understood in the context of object instances than object categories. Persons, animals, riders, motorcycles, auto-rickshaws, cars, trucks, and buses are important in autonomous driving. Accurate detection and segmentation of these objects is important, as they should not become obstacles or lead to accidents. Traffic signs and traffic lights are important in safety and also in communicating the regulations.

2.2 *Small Objects of Indian Driving Dataset*

IDD [8] has emerged as a challenging dataset for semantic segmentation and object detection in unconstrained road situations. The spread of images across train, val, and test can be seen in Table 1. There are many objects in a road scenes. Their size depends on the absolute size of the object and also the depth/location. IDD has pixels on the object annotated. There are many objects in such images.

The objects in each class of IDD can be classified into small, medium, and large categories based on the number of pixels present in the segmentation masks of each of the instances. We categorize the objects into small, medium, and large objects based on the no. of pixels in the segmentation mask.

- Small objects have less than 1000 pixels
- Medium size objects have pixels in the range 1000–8999
- Large objects have pixels ≥ 9000

2.3 *Deep Learning for Semantic Segmentation*

Early algorithms used clustering or graph-cut for segmentation. In the recent years, we have witnessed the success of the deep learning-based semantic segmentation algorithms. They get trained on large number of pixel-level labeled images. A loss function (or objective function) is used in the training process. In practice, back-propagation with an appropriate modification (such as regularization) is used in training.

The detection of small objects is considered challenging, and so is segmentation. Firstly, it is very hard to distinguish small objects from generic clutter in the background. This makes it hard for many of the standard detectors that rely on "objectness" due to the drastic increase in the number of possible locations. Secondly, the activations of small objects become smaller with each pooling layer as an image passes through a standard CNN architecture. Lastly, there is no large publicly available dataset for small objects [9] for studying this problem systematically.

Network Architecture In this work, we use CNN-based semantic segmentation architecture. Successive convolution layers act as encoder, and finally, the compact information is decoded into a segmented image. Additional skip connections are added. Such architectures are popularly known as U-Net. U-Net (Fig. 2) has been used as a backbone with fully convolutional network (FCN) as auxiliary head and decode head. We are motivated to use this architecture due to its popularity in semantic segmentation.

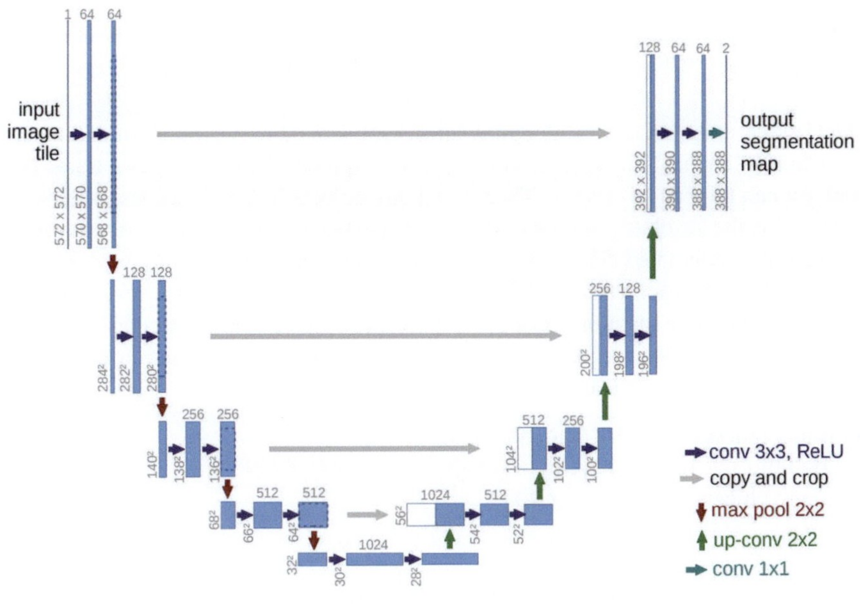

Fig. 2 U-net architecture (example for 32 × 32 pixels in the lowest resolution). Each blue box corresponds to a multichannel feature map. The number of channels is denoted on top of the box. The xy size is provided at the lower left edge of the box. White boxes represent copied feature maps. The arrows denote the different operations [3]

2.4 Loss Functions

The three loss functions considered for this study are: (i) cross-entropy loss, (ii) weighted cross-entropy loss, and (iii) dice loss. The following are the definitions of the three losses. The notation for the loss functions mentioned below is present in Table 2.

Cross-Entropy Loss: Cross-entropy (CE) measures the difference between two probability distributions for a given random variable.

$$L_{CE}(y, t) = -\sum_{n=1}^{N} \log(t_n \cdot y_n)$$

Weighted Cross-Entropy Loss: When dealing with imbalanced datasets, one approach to cross-entropy loss is to assign different weights to each class. This can help to balance the influence of each class on the overall loss and improve the performance of the model on the underrepresented classes.

$$L_{WCE}(y, t, w) = -\sum_{n=1}^{N} t_n \cdot w \log(t_n \cdot y_n)$$

Dice Loss: The dice loss originates from the dice coefficient which is a measure of similarity between two sets of data. It is commonly used in image segmentation to evaluate the overlap between a predicted segmentation mask and the target segmentation mask.

$$L_{dice} = 1 - \frac{1}{C}\sum_{c=0}^{C-1} \frac{2\sum_{n=1}^{N} t_n^c y_n^c}{\sum_{n=1}^{N}\left(t_n^c + y_n^c\right)}$$

Table 2 Notation for the loss functions [5]

Symbol	Description
N	Number of pixels
C	Number of target classes
t_N	One-hot encoding vector representing the target class of the nth pixel
y_N	Predicted class probabilities for n^{th} pixel
t_N^c	Binary indicator: 1 if the nth pixel belongs to class c, otherwise 0
y_N^c	Predicted class probabilities for n^{th} pixel

3 Experiments, Results, and Analysis

We first study the presence and distribution of small objects in the IDD. We start with objects and category distribution in the IDD dataset. Distributions of objects in IDD across labels can be seen in Fig. 3. We also study the distribution of pixels across categories. Figure 3e–h present the distribution of instances and pixels across labels in small, medium, and large object size categories of IDD.

3.1 Performance with Loss Functions

It is expected for a dataset to give different results when trained on different loss functions. One of the reasons for choosing a particular loss function is the nature of the dataset itself. Another important reason is the problem or task of interest. In the case of semantic segmentation of small objects, both these come together.

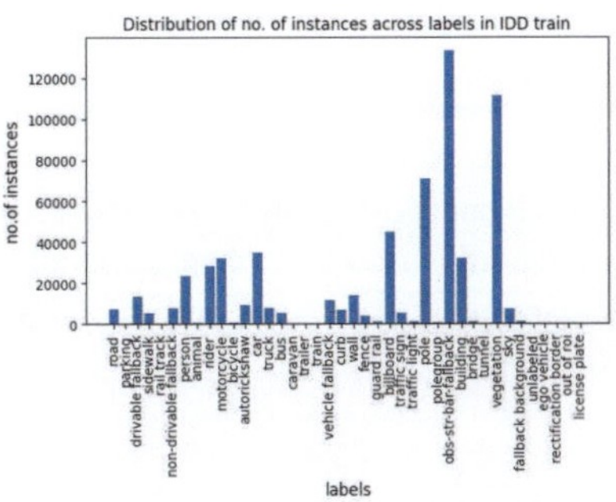

(a) Distribution of instances across labels in IDD train

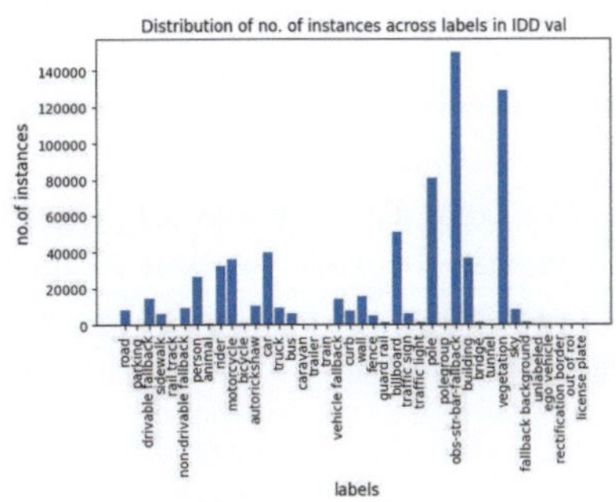

(b) Distribution of instances across labels in IDD val

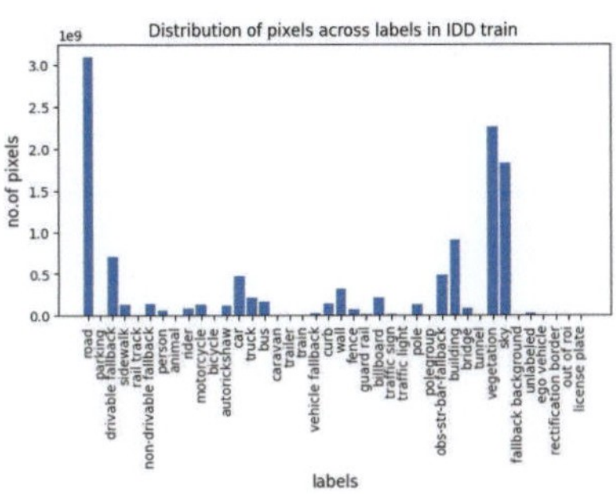

(c) Distribution of pixels across labels in IDD train

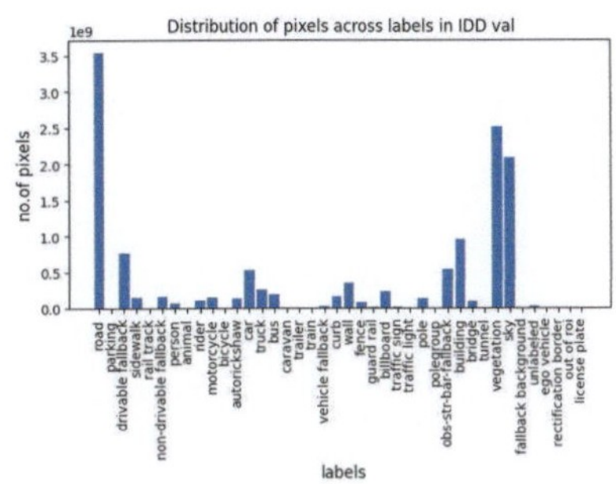

(d) Distribution of pixels across labels in IDD val

Fig. 3 IDD is split intro training, validation, and test splits. Here, we present the distribution of IDD in train and validation splits

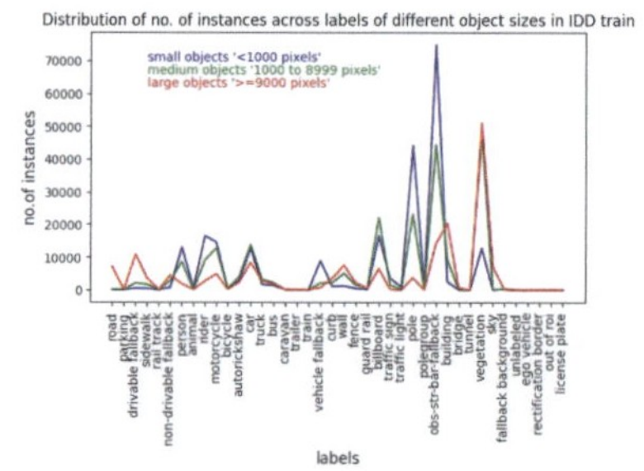

(e) Distribution of instances across labels in IDD train as per object size

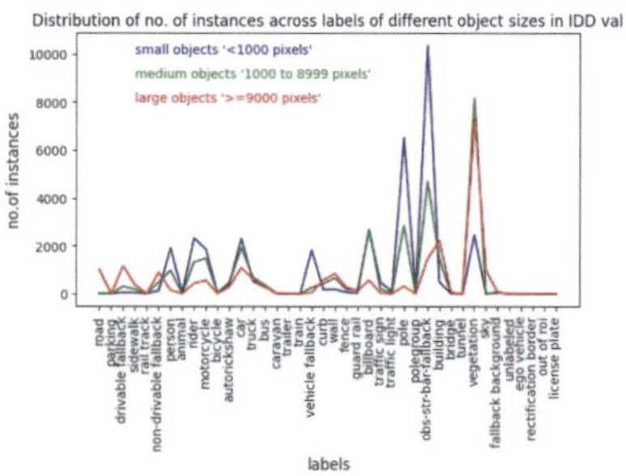

(f) Distribution of instances across labels in IDD val as per object size

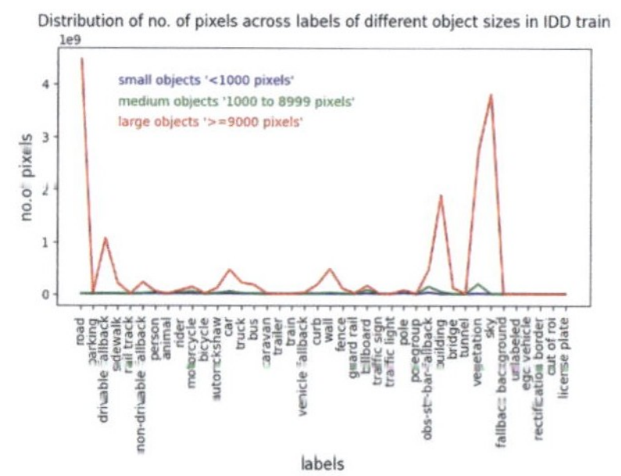

(g) Distribution of pixels across labels in IDD train as per object size

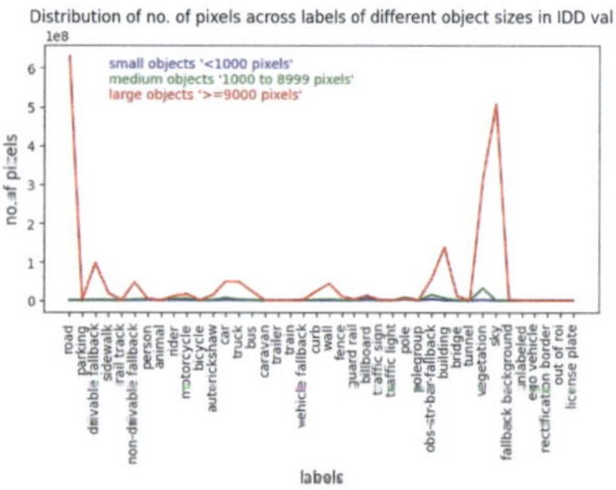

(h) Distribution of pixels across labels in IDD val as per object size

Fig. 3 (continued)

We systematically evaluate the performance with multiple loss functions. We start with cross-entropy loss and weighted cross-entropy loss. The former loss is chosen when the dataset is balanced with respect to classes, and the latter is chosen when the dataset has an imbalance with respect to classes. We also study another popular loss—dice loss.

We train using the Indian Driving Dataset with U-Net as backbone with FCN head as the decode head and auxiliary head. IDD provides pixel-level labeled images for training. The images in the Indian Driving Dataset consist of two different resolutions: (i) 1920 $\times$ 1080 and (ii) 1280 $\times$ 720. We train using three different losses cross-entropy loss, weighted cross-entropy loss, and dice loss and make some observations.

To maintain uniform image size for training, images are resized to a common size of 1024 $\times$ 512. The image crop size chosen is 512 $\times$ 256. We use SGD optimizer with an initial learning rate of 0.01, momentum coefficient of 0.9, and weight decay coefficient of 0.0005. Polynomial learning rate is used for the learning rate schedule with a minimum learning rate of 0.0001 and a power factor of 0.9. The performance graphs for the three loss functions can be seen in Fig. 4a–f. Objective performances are measured using mIoU (mean of the intersection over union). The mIoUs for the three losses across labels can be seen in Table 3. Table 4 presents the aggregate accuracy, mean intersection over union, and mean accuracy for the three losses. Table 3 gives the mIoUs obtained for each of the 40 classes in IDD.

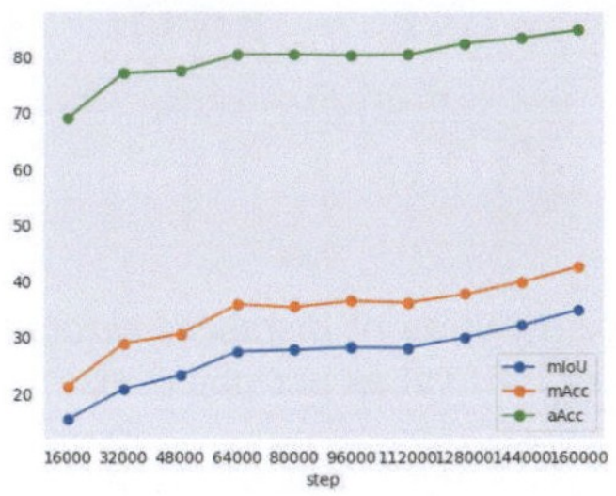

(a) Accuracies for Cross Entropy Loss for 160000 iterations

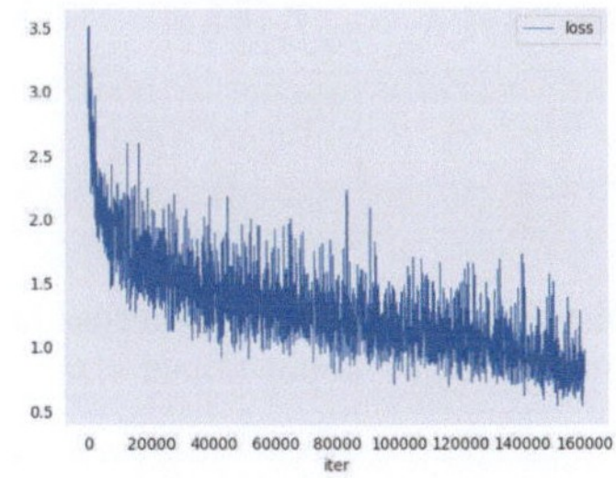

(b) Loss Plot for Cross Entropy Loss for 160000 iterations

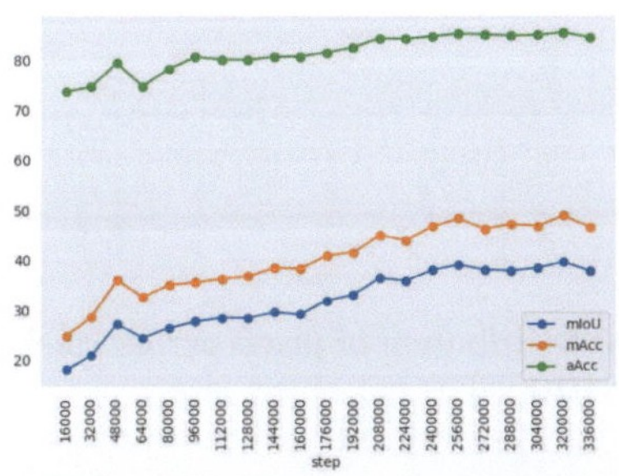

(c) Accuracies Plot for Weighted Cross Entropy Loss for 336000 iterations

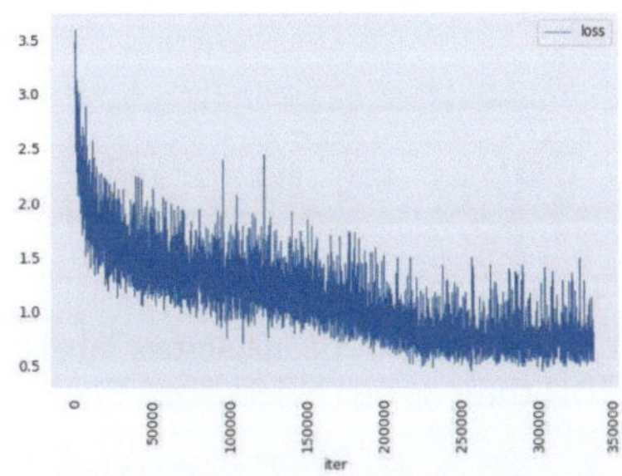

(d) Loss Plot for Weighted Cross Entropy Loss for 336000 iterations

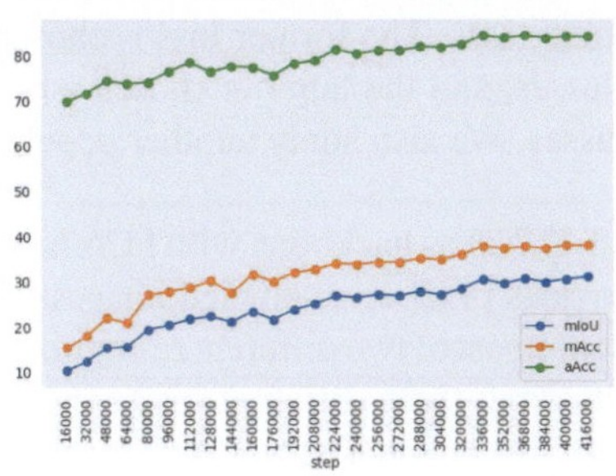

(e) Accuracies Plot for Dice Loss for 416000 iterations

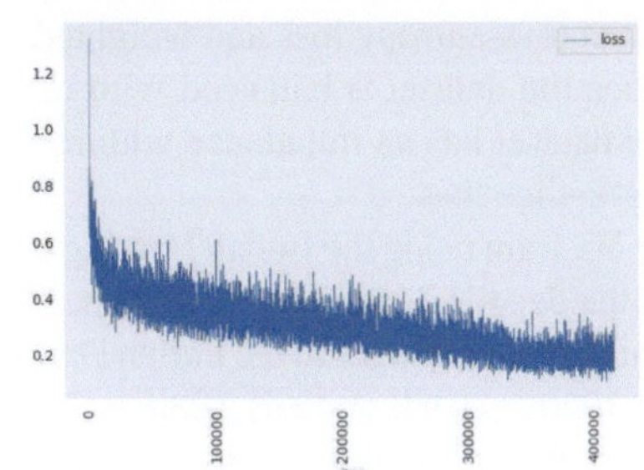

(f) Loss Plot for Dice Loss for 416000 iterations

Fig. 4 Performance and training evaluation across different loss functions. We present the performance and training process here. Plots of accuracies and losses

Table 3 mIoU for cross-entropy loss, weighted cross-entropy loss, and dice loss

Loss type	Road	Parking	Drivable fallback	Sidewalk	Rail track	Non-drivable fallback	Person	Animal	Rider	Motorcycle
WCE loss	91.66	0	54.72	51.14	Nan	45.65	53.15	0	64.38	68.03
CE loss	91.35	0	53.09	51.25	Nan	35.7	47.06	0	60.38	64.58
Dice loss	90.8	0	53.32	43.74	Nan	45.05	0	0	55.56	67.39
Loss type	Bicycle	Auto-rickshaw	Car	Truck	Bus	Caravan	Trailer	Train	Vehicle fallback	Curb
WCE loss	0	73.01	83.38	52.78	54.99	0	Na	Na	10.04	62.44
CE loss	0	67.48	81.2	54.38	58.75	0	Na	Na	0.03	58.33
Dice loss	0	65.56	81.15	52.9	46.53	0	Na	Na	0	61.9
Loss type	Wall	Fence	Guard rail	Billboard	Traffic sign	Traffic light	Pole	Polegroup	Obs-str-bar-fallback	Building
WCE loss	40.61	19.07	48.7	51.91	53.84	0	45.12	20.08	29.64	56.2
CE loss	40.65	12.5	44.63	49.72	10.6	0	43.08	0	27.05	50.55
Dice loss	41.32	0	0	52.42	0	0	40.61	0	28.79	55.45
Loss type	Bridge	Tunnel	Vegetation	Sky	Fallback background	Unlabeled	Ego vehicle	Rectification border	Out of ROI	License plate

(continued)

Table 3 (continued)

Loss type	Road	Parking	Drivable fallback	Sidewalk	Rail track	Non-drivable fallback	Person	Animal	Rider	Motorcycle
WCE loss	38.69	0	85.04	95.69	0	0	0	Na	Na	Na
CE loss	5.79	0	83.82	95.5	0	0	0	Na	Na	Na
Dice loss	0	0	84.67	94.56	0	0	0	Na	Na	Na

Table 4 Accuracies for losses on the validation set of IDD

Loss	Iterations	aAcc	IoU	mAcc
Cross-entropy loss	160,000	84.7400	34.9300	42.5600
Weighted cross-entropy loss	320,000	85.7000	39.7000	48.9500
Dice loss	416,000	84.3400	31.2300	38.1900

3.2 Discussions

The small objects of importance in IDD [8] are majorly persons, animals, riders, motorcycles, auto-rickshaws, cars, trucks, buses, traffic signs, and traffic lights. We briefly summarize the results and figures presented in the paper. For IDD, Fig. 3a, b depict the distribution of number of instances of labels in train and val across labels. Figure 3c, d depict the distribution of number of pixels of labels in train and val across labels. Figure 3e, f depict the distribution of number of instances of labels in train and val across labels for small, medium, and large object sizes. Figure 3g, h depict the distribution of number of pixels of labels in train and val across labels for small, medium, and large object sizes. Figure 4 depicts the accuracies and losses for the three losses across iterations. Figure 5 shows three images from IDD and predictions for dice loss for these three images.

The three losses (weighted cross-entropy loss, cross-entropy loss, and dice loss) perform comparably well on rider, motorcycle, auto-rickshaw, car, truck, and bus classes. The reason may be understood from Fig. 3. These classes have considerable instances and pixels (Fig. 3a–d) in the IDD [8]. They have considerable instances in smaller size objects, (Fig. 3e, f) though they have a lesser number of total pixels as smaller objects (Fig. 3g, h).

Weighted cross-entropy loss and cross-entropy loss perform well on the person class, but dice loss gives a 0 mIoU. Cross-entropy loss does not perform as well as weighted cross-entropy loss on traffic sign. The three losses perform equally poorly with a 0 mIoU on animal, bicycle, and traffic light classes. The reason why dice loss performs poorly with a 0 mIoU on person class may be observed from Fig. 3a–d, where the instances are considerable but the total pixels are less. Since the total pixels are less in person class, the dice loss performs poorly as it is a region-level loss, and in the initial stages of training, any wrong prediction of the region due to a lesser total number of pixels for the class may cause the future stages of training to be affected. On the contrary, cross-entropy loss and weighted cross-entropy loss are pixel-level losses, and hence, they have a chance for moving toward the correct class prediction at each pixel level giving a much better mIoU for the person class. A similar reason may be attributed to the poor mIoU for dice loss over traffic sign class. Relatively poorer performance of cross-entropy loss over weighted cross-entropy loss can be attributed to the imbalance in the dataset. Animal, bicycle, and traffic light classes have poor mIoUs for all the three losses in discussion as they have a lesser number of instances and total pixels (see Fig. 3a–d).

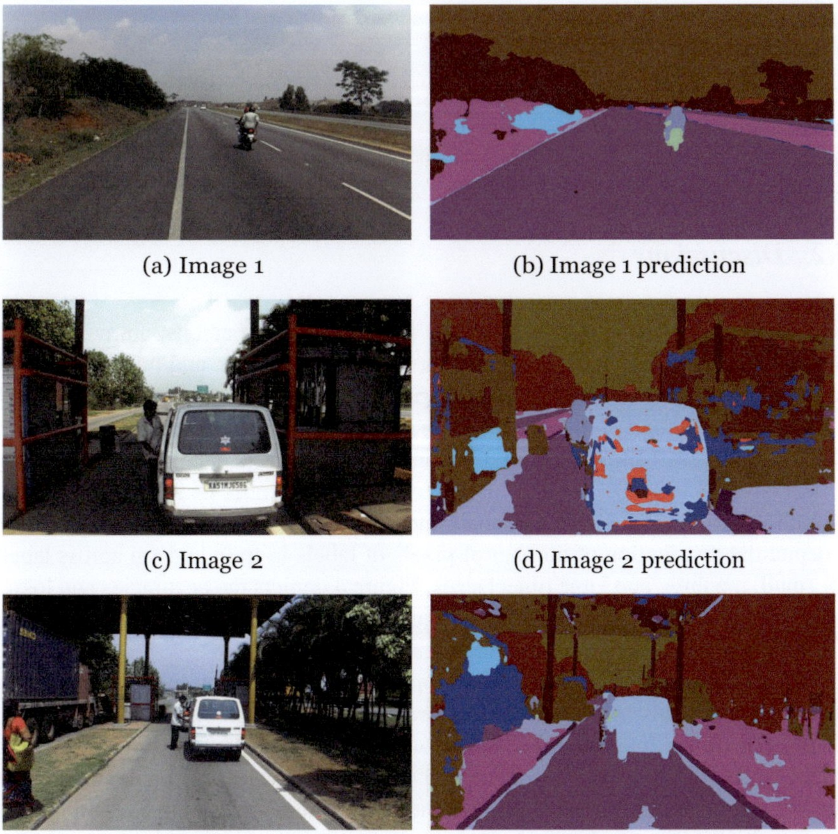

(a) Image 1 (b) Image 1 prediction

(c) Image 2 (d) Image 2 prediction

(e) Image 3 (f) Image 3 prediction

Fig. 5 Qualitative results with dice loss on three sample images

Figure 5 shows predictions for dice loss for three images. Clearly, one can see the person class is classified incorrectly, though the rider class, which belongs to the same class as humans, to which the person class belongs, has been segmented with a larger mIoU. The reason behind the rider class is detected with a larger mIoU and the person class with a lesser mIoU, by dice loss, though both belong to the same class of humans can be considered for future work.

Our analysis and experiments lead to the following observations: (i) Indian outdoor situations, especially the road settings, have large amount of small objects, and this makes the problem special. (ii) Semantic segmentation of small objects is challenging, and the average performance (mIoU) is poor compared to that of large objects; (iii) for driving and road safety problems, smaller objects have special importance, and this needs future dedicated effort. (iv) Performance of the semantic segmentation clearly depends on the loss function, and there is more work needed for design of specific loss functions or strategies for robust segmentation of road scenes.

4 Conclusions and Future Work

Segmenting small objects from road scenes is important for any AI system that is used in driving or surveillance. Often this is overlooked. We investigate this problem in the Indian setting based on a popular Indian Driving Dataset. We study the performance analysis of the segmentation with different loss functions and explore the potential directions for developing robust solutions. We see the need of more training data for obtaining higher accuracy in these object categories.

Potential future directions could include (i) design of novel loss functions and architectures that help in robust segmentation of smaller objects; (ii) exploitation of class hierarchies or cascading the segmentation (as two or three stages) to make the segmentation more accurate; (iii) addressing the class size imbalance (count of pixels and number of examples) could be another direction of research.

References

1. MirzaeiB, Nezamabadi-pour H, Raoof A, Derakhshani R (2023) Small object detection and tracking: a comprehensive review. Sensors 23(15):6887. https://doi.org/10.3390/s23156887
2. Jonathan L, Evan S, Trevor D (2015) Fully convolutional networks for semantic segmentation. In: 2015 IEEE conference on computer vision and pattern recognition (CVPR), pp 3431–3440. https://doi.org/10.1109/CVPR.2015.7298965
3. Olaf R, Philipp F, Thomas B (2015) U-Net: convolutional networks for biomedical image segmentation. In: Nassir Navab et al (eds)Medical image computing and computer-assisted intervention–MICCAI 2015. Cham, Springer International Publishing, pp 234–241. isbn: 978–3–319–24574–4.
4. Liang-Chieh C et al (2018) DeepLab: semantic image segmentation with deep convolutional nets, atrous convolution, and fully connected CRFs. In: IEEE transactions on pattern analysis and machine intelligence 40.4 (2018), pp 834–848
5. Reza Azad et al (2023) Loss functions in the era of semantic segmentation: a survey and outlook. arXiv: 2312.05391 [cs.CV]. url: https://arxiv.org/abs/2312.05391
6. Shruti J (2020) A survey of loss functions for semantic segmentation. In: 2020 IEEE Conference on Computational Intelligence in Bioinformatics and Computational Biology (CIBCB). IEEE
7. Yingjie L, Yueying H, Feng J (2022) Deep learning-based small object detection: a survey. In: Proceedings of the 8th international conference on computing and artificial intelligence. ICCAI '22. Tianjin, China: Association for Computing Machinery, pp 432–438
8. Varma G, Subramanian A, Namboodiri A, Chandraker M, Jawahar CV (2019) IDD: a dataset for exploring problems of autonomous navigation in unconstrained environments 1743–1751. https://doi.org/10.1109/WACV.2019.00190
9. Harish K, Jawahar CV (2017) Improving small object detection. In: 2017 4th IAPR Asian Conference on Pattern Recognition (ACPR), pp 340–345

Unveiling Dementia's Early Signals: Deep Learning Meets Image Processing

Maddela Likitha, Sunny Nalluri, and Inti Alekhya

Abstract Dementia is a progressive disease that impairs cognitive functions such as memory, thinking, and communication, affecting over 40 million individuals worldwide, according to the World Health Organization. Early diagnosis is essential for effective management and treatment. In this study, we present a Custom Convolutional Neural Network (CNN) model designed to analyze Magnetic Resonance Imaging (MRI) scans for dementia detection. The model classifies dementia into four stages: Non-demented, Very Mildly Demented, Mildly Demented, and Moderately Demented. Our system incorporates advanced preprocessing techniques, such as image resizing, Gaussian filtering for noise reduction, and normalization, to enhance image quality and ensure consistent inputs for the CNN. Performance evaluation of the Custom CNN was carried out using key metrics like accuracy, precision, recall, and the confusion matrix to measure classification performance. The Custom CNN achieved an accuracy of *98.18%*, demonstrating its effectiveness in accurately detecting dementia stages. The confusion matrix further confirmed the model's high precision in classification, showing significant improvements over existing models. This result underscores the potential of the proposed system to support early diagnosis, leading to timely interventions and improved patient outcomes. Our Custom CNN model offers a substantial advancement in dementia detection, contributing to both research and clinical practice in addressing the global challenge of dementia.

Keywords Dementia detection · Custom CNN · Magnetic Resonance Imaging (MRI) · Image preprocessing · Early diagnosis · Accuracy · Precision · Confusion matrix · Machine learning · Neural networks · Second keyword · Another keyword

M. Likitha (✉) · S. Nalluri · I. Alekhya
Department of Computer Science and Engineering, Velagapudi Ramakrishna Siddhartha Engineering College, Vijayawada, India
e-mail: likithamaddela145@gmail.com

S. Kumar et al. (eds.), *Artificial Intelligence for Next Generation Computing, Volume 1*,
Lecture Notes in Networks and Systems 1549,
https://doi.org/10.1007/978-981-95-0701-6_8

1 Introduction

Dementia is a significant global health issue, affecting millions of individuals and their families by causing cognitive decline and severely impacting quality of life. Although there is no cure for dementia, early and accurate diagnosis is crucial for effective management, allowing for timely interventions that can improve patient outcomes. This research explores the use of advanced deep learning methods, specifically a Custom Convolutional Neural Network (CNN), to classify dementia into four stages—Non-demented, Very Mildly Demented, Mildly Demented, and Moderately Demented—using Magnetic Resonance Imaging (MRI) scans. The key goal of this study is to enhance diagnostic accuracy in dementia detection by designing a CNN model tailored to the challenges of interpreting neuroimaging data. This model not only improves the detection process but also helps healthcare providers more accurately identify and classify the stages of dementia, ensuring patients receive appropriate care at the right time.

We assessed the performance of our proposed model by comparing it with several existing models, including MobileNet, AdaBoost, ResNet, AlexNet, and modified capsule networks (M-CapNets). While these models have shown varying levels of success in classifying dementia, our Custom CNN outperformed them, achieving an accuracy of 98.18.

This research makes key contributions to the field by:

1. Developing a specialized CNN architecture that significantly improves dementia stage classification accuracy.
2. Utilizing advanced preprocessing techniques such as Gaussian filtering and normalization to enhance MRI image quality and model performance.
3. Demonstrating the superior performance of the Custom CNN compared to both pre-trained models and hybrid approaches.

The structure of this paper is as follows: Sect. 2 reviews related work, outlining the performance of various models in dementia detection. Section 3 describes the methodology, including the architecture and training of the Custom CNN model. Section 4 presents the experimental setup and results. Finally, Sect. 5 summarizes the findings and offers suggestions for future research.

1.1 Basic Concepts

A Custom Convolutional Neural Network (CNN) is like a highly specialized brain optimized for processing images and grid-like data with remarkable efficiency. The network operates in distinct stages, each contributing to its overall functionality. At the core, Conv2D layers function as the network's "detectives," meticulously analyzing the input to identify local features such as edges, textures, and patterns. They achieve this by applying various filters that scan through the input data, extracting crucial details at different levels of abstraction. Following the Conv2D layers, Pooling layers plays a crucial role in simplifying the data while maintaining essential features. Pooling methods, such as maximum pooling and average pooling, reduce the dimensionality of the feature maps as we know. Max pooling selects the highest value from each local patch, ensuring that only the most significant features are retained, while average pooling calculates the average value, providing a more generalized representation of the data. This reduction helps in minimizing computational demands and reduces the risk of overfitting, making the model more robust. Finally in the network, fully connected layers take over, performing high-level reasoning by connecting every neuron from the previous layer. This dense connectivity helps in synthesizing the extracted features and converting them into a flat vector. This vector is then processed to generate the final predictions, such as class probabilities in classification tasks. By transforming the spatially reduced feature maps into a format suitable for final analysis, these layers enable the CNN to effectively grasp complex patterns and deliver accurate outcomes. This structured approach ensures the network's capacity to process complex data and deliver accurate predictions.

1.2 Motivation

The motivation to undertake this paper stems from the pressing global challenge of dementia. With millions affected worldwide, early detection is paramount for effective intervention. By harnessing advanced technology like Custom CNN models and MRI scans, we can enhance diagnostic accuracy, enabling timely treatment and support. Addressing dementia at its early stages not only enhances individual results but also alleviates the overall strain on healthcare systems and society, underscoring the importance for advancing public health.

1.3 Problem Statement

Addressing dementia detection challenges, we're crafting a Custom CNN model for MRI image analysis, categorizing dementia into four stages: Non-demented, Very Mild Demented, Mild Demented, and Moderate Demented. Utilizing MRI scan image data and advanced image processing, we strive to enhance early detection and intervention. Leveraging the adaptable CNN architecture, personalized layer configurations optimize disease classification precision. We are intended to use the flatten, dense layers in the custom CNN in order to enhance the results than others models.

1.4 Scope

1. Dementia classification is confined to four stages exclusively.
2. The proposed model is specifically centered on analyzing MRI scans/images for classification purposes.

1.5 Objective

1. To conduct a comparative analysis of various network models to assess their efficacy in detecting and classifying dementia.
2. To determine the severity of classified dementia within the designated stages.

1.6 Advantages

1. Accurate classification of dementia stages.
2. High accuracy in predicting Moderate Demented.
3. Proficiency in discerning subtle imaging differences.
4. Reliable tool for early detection.

2 Literature Review

In this methodology [1], the research incorporates pre-trained models—MobileNet for image data, YAMNet for audio, Mockingjay for speech, and BERT for text—by replacing their final layers with a custom fully connected layer tailored for dementia detection. It utilizes pre-trained models—MobileNet for images, YAMNet for audio, Mockingjay for speech, and BERT for text—customized with a new fully connected layer for dementia detection. Instead of using these models merely as feature extractors, all layers are fine-tuned on the ADReSS dataset to adapt to specific dementia features via back-propagation. Training employs the Adam optimizer with a reduced learning rate and customized batch sizes to address GPU limitations. This method leverages knowledge from larger datasets to enhance performance on limited dementia-specific data, improving diagnostic accuracy.

This methodology [2] employs the AdaBoost ensemble model, which achieves robust performance with approximately 83% accuracy, outperforming traditional models like ANN and SVM. It underscores the potential of machine learning methods to improve quick detection by leveraging cognitive, neuroimaging features. Previous research has varied in success, with cognitive features often proving more effective than neuroimaging alone. This system integrates both cognitive test scores and neuroimaging data, aiming to enhance early detection and intervention strategies for dementia. The study highlights the critical role of feature selection and model optimization in achieving high predictive accuracy, emphasizing their importance in advancing dementia detection capabilities.

The literature [3] highlights that dementia often results from factors like lack of exercise and social interaction, leading to symptoms such as memory loss. Early detection, especially during Mild Cognitive Impairment (MCI), is crucial to prevent progression. The paper presents DCGAN-based Augmentation and Classification model for dementia detection using MRI scans. It employs CNN with six convolutional layers, five pooling layers, batch normalization, dropout and four dense layers, achieving up to 94% training accuracy and 86% testing accuracy. The model predicts MCI with 74% accuracy and classifies dementia into four categories, using GradCAM for enhanced model interpretability.

The study in [4] explores various techniques for forecasting Alzheimer's Disease (AD) through neuroimaging techniques using Convolutional Neural Networks and EHR datasets for early dementia detection. The proposed methodology introduces a modified capsule network (M-CapNets) that simplifies the model while improving clinical applicability. It uses kernel principal component analysis (PCA) to filter features from the OASIS dataset, classifying subjects as demented or non-demented based on MRI images. The system achieves a prediction accuracy of 94.39%, addressing limitations of traditional CNNs, such as spatial information loss and pooling function issues. This approach aims to enhance early diagnosis and treatment in dementia care.

This Methodology [5] evaluates machine learning algorithms for diagnosing dementia using the OASIS dataset, emphasizing the effectiveness of deep learning models such as AlexNet and ResNet50. It also explores hybrid approaches that integrate these models with Support vector machine algorithms. It highlights the critical need for quick diagnosis in dementia and Alzheimers, given the lack of effective medications. Methodologically, the authors maintained the dataset, addressed missing values. The proposed hybrid models, particularly AlexNet+SVM, achieved high accuracy and sensitivity, demonstrating improved diagnostic performance. This study [6] presents systematic literature review (SLR)evaluating machine learning (ML) approaches for dementia prediction, including Alzheimer's Disease, Mild Cognitive Impairment, and other subtypes, using diverse data modalities such as images, scanning images, clinical pics, and voice data input. Adhering to PRISMA guidelines, the study analyzed research from 2011 to 2022, highlighting that image data generally offers superior sensitivity and specificity compared to clinical and voice data. The proposed model aims to enhance dementia prediction accuracy by integrating various data modalities.

The paper [8] presents various models for detecting Alzheimer's disease, including machine learning (ML) methods like SVM, logistic regression, and decision trees, as well as deep learning (DL) models such as BiLSTM and CNNs. The BiLSTM achieved the highest accuracy of 91.28%, outperforming other methods. However, the study identifies a gap in the computational cost and training time of DL models compared to ML methods, with Naïve Bayes performing the worst. Future research could focus on improving these aspects and exploring more efficient algorithms for better detection outcomes

The paper [11] explores dementia detection using machine learning (ML) and deep learning (DL) models. The Random Forest (RF) model achieved an accuracy of 87.6%, outperforming the PRCNN model at 85% . It emphasizes standalone speech features, addressing gaps in previous studies that combined linguistic and speech features. The study utilized 44 features for ML and 62 for DL, demonstrating efficiency and improved classification accuracy compared to earlier works. Overall, the research highlights the effectiveness of focused speech analysis in recognizing dementia .The surveyed literature reveals substantial advancements in machine learning applications for dementia detection, utilizing various models such as CNNs, AdaBoost, and hybrid systems that combine neural networks with Support Vector Machines. Many approaches demonstrate the effectiveness of multimodal data, with accuracy levels ranging from 82 to 94%. However, notable research gaps persist.

First, many models lack comprehensive preprocessing techniques, which can enhance input data quality. Additionally, while existing models achieve high accuracy, they often fail to distinguish between critical dementia stages, limiting their clinical utility. Furthermore, the complexity of some models hampers their practical applicability in clinical settings. Moreover, existing methods may not effectively adapt to small, dementia-specific datasets, restricting their generalizability.

In response, the proposed Custom Convolutional Neural Network (CNN) model addresses these gaps by incorporating advanced preprocessing techniques, achieving an accuracy of 98.18% across four distinct dementia stages. This model not only enhances classification precision but is also optimized for real-world clinical use, improving adaptability to limited datasets. Thus, it contributes significantly to advancing early detection and intervention strategies in dementia care.

2.1 Dataset Details

Dataset Source:-OASIS MRI brain images dataset (https://www.kaggle.com/datasets/ninadaithal/imagesoasis).

3 Proposed System

The proposed system shown in Fig. 1 ,for processing and analyzing dementia affected MRI scan images using a Custom Convolutional Neural Network model.

1. Input Image-Description:

The initial step involves acquiring MRI images of the brain. These images are the raw input for the system and will be subjected to various preprocessing steps.

2. Image Preprocessing- Resizing (220 × 224):

Fig. 1 Proposed architecture

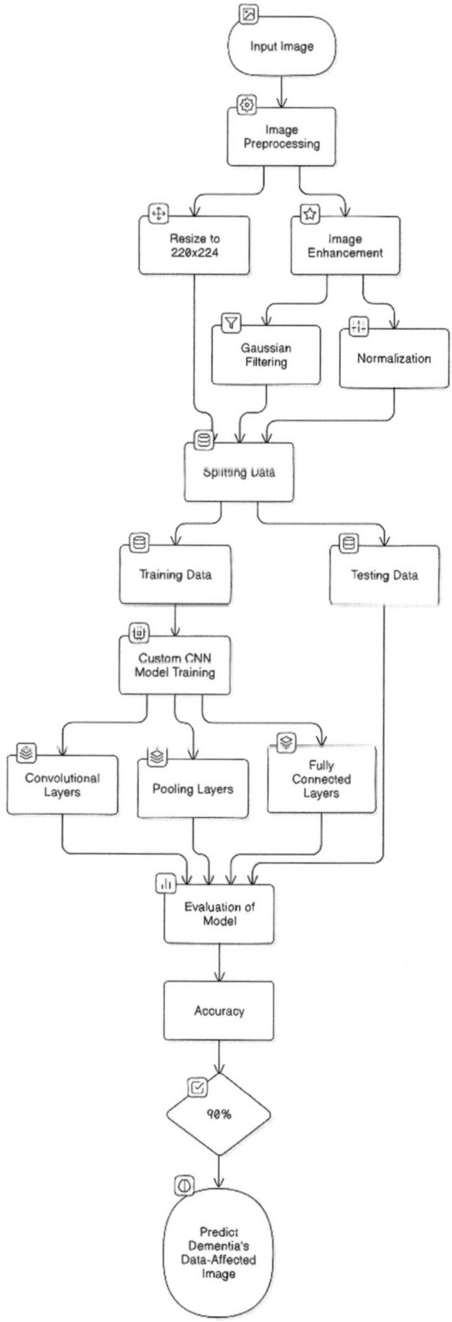

The acquired MRI images are resized to a uniform dimension of 220 × 224 pixels. This standardization facilitates consistent input to the CNN model, ensuring that all images are of the same size, which is critical for efficient and accurate processing by the neural network.

Filtering (Gaussian Filtering): This includes applying a Gaussian filter to the images to reduce extra noise and enhance image quality. Gaussian filtering smooths the image, which can help in emphasizing important features while diminishing irrelevant details.

- Normalization: Normalization adjusts the pixel values to a common scale, typically between 0,1 and which helps in fasting up the training process and improving model convergence. This dataset is not fit for training and used to assess the model's accuracy and generalizability to new, unseen data

3. Splitting Data

- Training Data: The preprocessed images are split into a training dataset that will be used to train the CNN model. This dataset typically comprises the majority of the images and is labeled to inform the model during the training

- Testing Data: A separate set of images, known as the testing dataset, is reserved to evaluate the performance of the trained model.

4. Custom CNN Model Training

- Convolutional Layers: The CNN model begins with convolutional layers that apply various filters to the input images. These filters detect features such as edges, textures, and patterns relevant to identifying dementia affected regions in the MRI images.

- Pooling Layers: Pooling layers comes after convolutional layers to reduce the spatial dimensions of the feature maps. This step mainly reduces the complexity and focuses on the most important features.

- Fully Connected Layers: After layers, the output is flattened and is send to fully connected layers. These layers perform the final classification task, determining whether an image shows signs of dementia.

5. Evaluation of Model

- Accuracy is 90%: The trained CNN model is evaluated using the testing dataset. The goal is to achieve an accuracy greater than 90%, indicating that the model is highly effective at distinguishing between dementia affected and non-affected images.

6. Predict Dementia's Data-Affected Image

- Prediction: After model is trained and calculated, it is used to predict new MRI images. The model will analyze these images and predict whether they are affected by dementia, providing correct insights for diagnosis and treatment. This structured approach ensures that the CNN model is well-prepared to analyze MRI images and detect signs of dementia with high accuracy.

4 Results and Discussion

To assess the effectiveness of the model, we employed several performance evaluation metrics, including accuracy, confusion matrix. Accuracy assesses the model's total correctness of predictions, whereas the confusion matrix describes the overall the performance of a model on a set of testing data.

This is essential for assessing how well the model detects the stage of dementia. It assists choosing the best design and training methods by enabling extensive comparisons between numerous model variations. At first, we obtain the input images from the dataset are collected from Kaggle dataset Repository as shown in Fig. 2.

Preprocessing for dementia detection images contains resizing of images to a specific dimension, normalizing the pixel values for uniformity, and applying augmentation techniques to enhance data diversity. Afterward these preprocessed data have sent into the proposed model which was created by using the Custom CNN architecture.

Later the performance is established by considering different parameters like accuracy and confusion matrix

The model's performance results are detailed above. Performance is assessed using accuracy as shown in Fig. 3. Which shows the total perfectness of the model's predictions, and precision, which shows how precise the model's predictions are. Recall calculates the ratio of actual positives correctly predicted by the model, and also the proportion of predicted positives that are truly positive.

Figure 4 represents the confusion matrix of Custom CNN Model of dementia detection which shows the classification of dementia

4.1 Comparison with Previous Implementation

Table 1 comparing different models and their performance metrics in the field of literature. The table lists various authors, the models they used (such as BERT, MobileNet, AdaBoost, MCapsNet, AlexNet, and ResNet50), and their respective accuracy percentages. Notably,"Custom CNN" model showed the highest accuracy of 98.18%, which is particularly interesting as it represents a significant achievement in model performance.

5 Conclusion and Future Work

This work details the development and evaluation of a Custom CNN model designed for the classification of demen tia using MRI scans. The primary objective is to accurately classify patients into four dementia stages: Non-demented, Very Mild

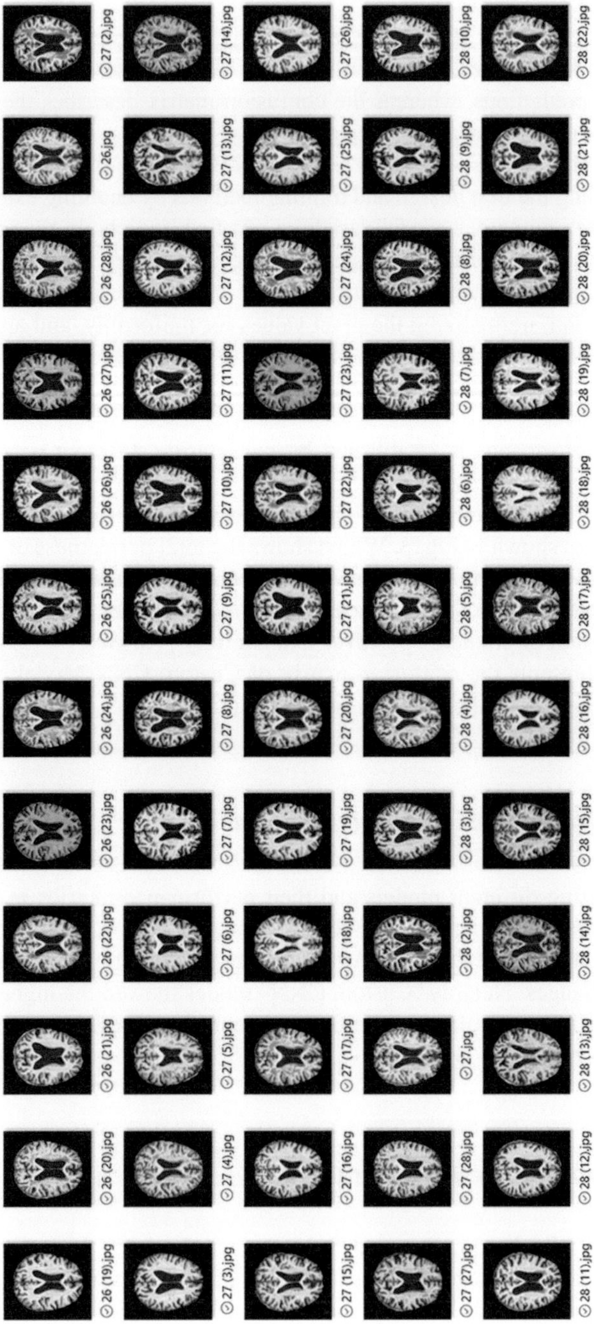

Fig. 2 Data collection

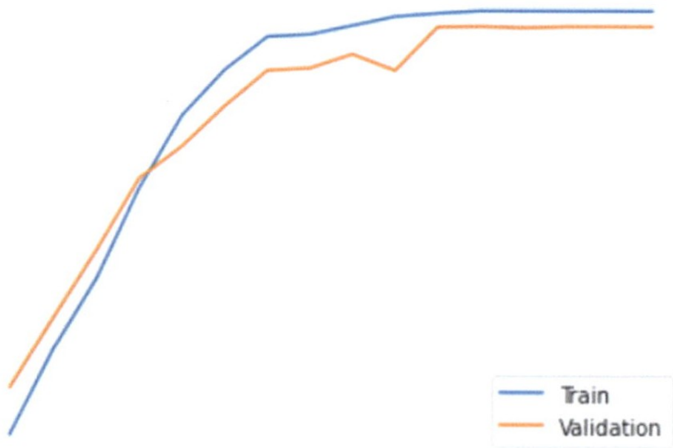

Fig. 3 Data collection

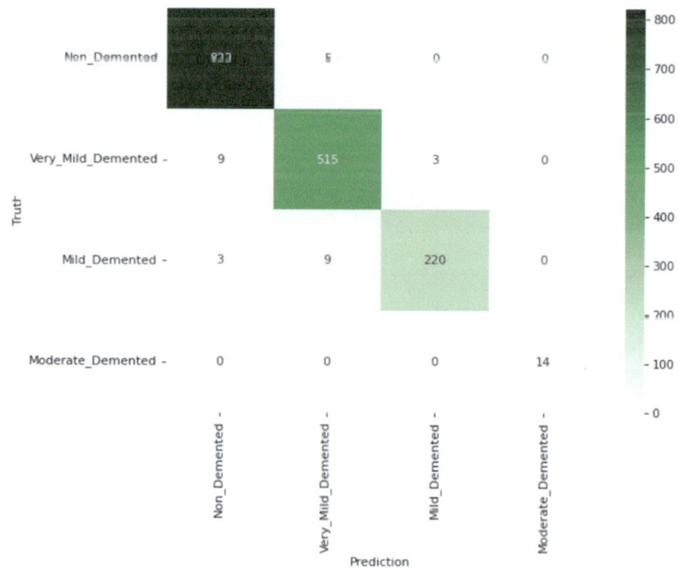

Fig. 4 Confusion matrix

Demented, Mild Demented, and Moderate Demented. The dataset utilized comprises MRI scans from individuals across these stages, which were used for both the training and the testing the model. A total of 1600 predictions were made, achieving a remarkable accuracy of 98.18 model demonstrated robust performance, effectively identify ing patients across all dementia stages, with particularly high accuracy in predicting Moderate Demented cases. Notably, the classifier showed proficiency in distinguishing between various dementia stages, highlighting its efficacy in capturing subtle differences in MRI patterns associated with disease progression. These results underscore the potential of the Custom CNN model as a reliable tool for

Table 1 Comparison of different models and their accuracy

Literature	Model	Performance metrics
Zhu Youxiang et al. [1]	BERT, MobileNet	Accuracy: 82.50%
Irfan Muhammad et al. [2]	AdaBoost	Accuracy: 83%
Jain et al. [3]	D-BAC	Accuracy: 85%
Basheer et al. [4]	MCapNet	Accuracy: 92.39%
Badiea Abdulkarem et al. [5]	AlexNet, ResNet50	Accuracy: 88%
Javeed et al. [6]	SVM & BiLSTM	Accuracy: 91.28%
Marwa EL-Geneedy et al. [7]	ResNet50, VGG16	Accuracy: 95.68%
Kia Dashtipour et al. [8]	BOF, SVM	Accuracy: 93%
Our work	Custom CNN	Accuracy: 98.18%

dementia diagnosis based on MRI scans. The model has high accuracy and ability to discern nuanced differences in imaging data suggest promising applications in clinical settings for early detection and intervention.

References

1. Zhu Y, Liang X, Batsis JA, Roth RM (2021) Exploring deep transfer learning techniques for Alzheimer's dementia detection. Front Comput Sci 3:624683
2. Irfan M, Shahrestani S, Elkhodr M (2023) Enhancing early dementia detection: a machine learning approach leveraging cognitive and neuroimaging features for optimal predictive performance. Appl Sci 13:10470. https://doi.org/10.3390/app131810470
3. Jain V, Nankar O, Jerrish DJ, Gite S, Patil S, Kotecha K (2021) A novel AI-based system for detection and severity prediction of dementia using MRI. IEEE Access 9:154324–154346. https://doi.org/10.1109/ACCESS.2021.3127394
4. Basheer S, Bhatia S, Sakri SB (2021) Computational modeling of dementia prediction using deep neural network: analysis on OASIS dataset. IEEE Access 9:42449–42462. https://doi.org/10.1109/ACCESS.2021.3066213
5. Mohammed BA, Senan EM, Rassem TH, Makbol NM, Alanazi AA, Al-Mekhlafi ZG, Almurayziq TS, Ghaleb FA (2021) Multi-method analysis of medical records and MRI images for early diagnosis of dementia and Alzheimer's disease based on deep learning and hybrid methods. Electronics 10(22):2860. https://doi.org/10.3390/electronics10222860
6. Javeed A, Dallora AL, Berglund JS, Ali A, Ali L, Anderberg P (2023) Machine learning for dementia prediction: a systematic review and future research directions. J Med Syst 47(1)
7. El-Geneedy M, Moustafa HED, Khalifa F, Khater H, AbdElhalim E (2023) An MRI-based deep learning approach for accurate detection of Alzheimer's disease. Alex Eng J 63:211–221. https://doi.org/10.1016/j.aej.2022.07.062
8. Dashtipour K, Taylor W, Ansari S, Zahid A, Gogate M et al (2021) Detecting Alzheimer's disease using machine learning methods. In: Procceding EAI, Glasgow, UK (2021). https://hal.science/hal-03381752
9. van der Velden BHM, Kuijf HJ, Gilhuijs KGA, Viergever MA (2022) Explainable artificial intelligence (XAI) in deep learning-based medical image analysis. Med Image Anal 79:102470. https://doi.org/10.1016/j.media.2022.102470

10. Patil BM, Raigonda MR, Anakal S, Bhadrashetty A (2023) Early detection of dementia using deep learning and image processing. Int J Eng Manuf 13(1)
11. Kumar MR, Vekkot S, Lalitha S, Gupta D, Govindraj VJ, Shaukat K, Alotaibi YA, Zakariah M (2022) Dementia detection from speech using machine learning and deep learning architectures. Sensors 22:9311. https://doi.org/10.3390/s22239311
12. Bansal D, Khanna K, Chhikara R, Dua RK, Malhotra R (2020) Classification of magnetic resonance images using bag of features for detecting dementia. Procedia Comput Sci 167:131–137. https://doi.org/10.1016/j.procs.2020.03.190
13. Valcour VG, Masaki KH, Curb JD, Blanchette PL (2021) The detection of dementia in the primary care setting. Arch Intern Med 160(19):2964–2968
14. Welsh KA, Breitner JCS, Magruder Habib KM (2023) Detection of dementia in the elderly using telephone screening of cognitive status. Cogn Behav Neurol 6(2):103–110
15. Davis DHJ, Creavin ST, Yip JLY, Storr AHN, Brayne C, Cullum S (2021) Montreal cognitive assessment for the detection of dementia. Cochran Database Syst Rev 7
16. Isaacs B, Kennie AT (2023) The Set test as an aid to the detection of dementia in old people. Br J Psychiatr 123(575):467–470
17. Giaquinto S, Parnetti L (2016) Early detection of dementia in clinical practice. Mech Ageing Dev 127(2):123–128
18. Rizzuto D, Feldman AL, Karlsson IK, Dahl Aslan AK, Gatz M, Pedersen NL (2018) Detection of dementia cases in two Swedish health registers: validation study. J Alzheimers Dis 61(4):1301–1310
19. Li R, Wang X, Lawler K, Garg S, Bai Q, Alty J (2022) Applications of artificial intelligence to aid early detection of dementia: a scoping review on current capabilities and future directions. J Biomed Inform 127:104030
20. Mitchell AJ (2019) A meta-analysis of the accuracy of the mini-mental state examination in the detection of dementia and mild cognitive impairment. J Psychiatr Res 43(4):411–431

An Approach for Cyclone Tracking and Monitoring

C. A. Rishikeshan, R. Jayanthi, Snehasis Ghosh, Navoneel Mondal, and Srijanbroto Deb

Abstract Accurate monitoring and forecasting of tropical cyclones is essential to reduce their devastating impacts on vulnerable coastal regions. This study presents an approach using convolutional neural network (CNN) that uses deep learning to detect and classify cyclones from remote sensing images. It uses a custom designed CNN architecture; the model effectively captures the distinct features between cyclone activity and normal weather conditions. The network architecture consists of convolutional layers that extract high-level semantic features which is followed by a fully connected layer for final classification. Performance evaluation is conducted through accuracy measurements and confusion matrix analysis. The proposed CNN model is novel compared to existing methods for its ability to retrieve meaningful contextual information (high-level features) from remotely sensing images, effectively distinguishing cyclone activity from normal weather patterns, offering improved accuracy (92.5%) in detection compared to traditional approaches. Despite challenges such as limited labelled storm data, the model shows strong potential to improve storm detection in different regions. This research helps in advancing deep learning applications in disaster risk management, providing a valuable tool to enhance storm monitoring and preparedness efforts.

Keywords Convolutional neural network (CNN) · Cyclone · Machine learning

1 Introduction

Tropical storms rank among the most destructive and costly natural disasters. It causes substantial damage to coastal regions worldwide. Accurate monitoring and surveillance are vital to mitigate impacts, save lives and protect infrastructure. According to the World Meteorological Organization (WMO), cyclone monitoring is a crucial component in reducing natural disaster risks [1]. Traditional cyclone

C. A. Rishikeshan (✉) · R. Jayanthi · S. Ghosh · N. Mondal · S. Deb
Vellore Institute of Technology (VIT), Chennai, India
e-mail: rishikeshan.ca@vit.ac.in

S. Kumar et al. (eds.), *Artificial Intelligence for Next Generation Computing, Volume 1*, Lecture Notes in Networks and Systems 1549,
https://doi.org/10.1007/978-981-95-0701-6_9

tracking methods depend heavily on human expertise and the interpretation of meteorological data. This can be labour-intensive, time-consuming and inconsistent [2]. Therefore, developing reliable and automated techniques for monitoring tropical cyclones are essential to enhance forecasting and preparedness.

Remote sensing technology, particularly satellite observations, has become indispensable for monitoring tropical cyclones. It is due to broad coverage, frequent revisit times and ability to capture storm characteristics from various locations [3, 4]. The primary methods for cyclone tracking using remote sensing data include feature extraction and pattern recognition techniques, such as edge detection, texture analysis and thresholding [5, 6]. However, these conventional methods often struggle with the complex and rapidly evolving nature of cyclones and the changing conditions in remote sensing images [7, 8].

Recently, deep learning techniques, especially convolutional neural networks (CNNs), have shown significant success in various computer vision tasks, including object detection, segmentation and tracking [9, 10]. CNNs can automatically learn hierarchical feature representations from raw data. It often outperforms traditional methods in many applications [11]. Consequently, there is a growing interest in using CNNs to track and analyse tropical cyclones from remote sensing data [12, 13]. Existing CNN-based cyclone tracking methods fall into two categories: object detection frameworks and segmentation-based methods. Object detection frameworks, like Faster R-CNN [14] and YOLO [15], aim to locate and classify cyclones in satellite images or sequences. These methods have shown promising results. But they can be computationally intensive and struggle with obscured or partially visible cyclones [16, 17]. Segmentation-based approaches, on the other hand, separate cyclone regions from background features, allowing for detailed analysis of storm structure and behaviour [18, 19].

In the context of cyclone tracking and monitoring, CNNs are favoured. It is due to their ability to efficiently process fixed-size inputs and outputs. Unlike recurrent neural networks (RNNs), which are designed to process variable input and output lengths, CNNs process fixed sizes, significantly improving computational speed and overall efficiency. This fixed structure allows CNNs to perform calculations faster. It makes them particularly well suited to tasks such as image recognition, pattern recognition and spatio-temporal data analysis, which are essential for cyclone tracking. When applying CNNs to satellite imagery and atmospheric data, the fixed input size ensures consistent processing. It reduces the computational time variability that can arise with RNNs' ability to process arbitrary-length sequences. This consistency allows for faster forecasting, which is critical for real-time cyclone tracking. Additionally, the CNN's layered architecture helps capture the spatial characteristics of weather patterns more efficiently, allowing for accurate tracking of cyclone movement, intensity and formation.

Among the difficulties faced by traditional cyclone monitoring include expensive processing costs, problems with accuracy and inadequate real-time performance. Techniques that rely on the manual interpretation of satellite data are costly and time-consuming. They frequently introduce discrepancies and miss important characteristics like storm severity. Emergency reactions are also delayed by these methods' inability to produce forecasts in a timely manner. CNNs, on the other hand, improve computing efficiency and automate feature extraction, enabling analysis in close to real-time. CNNs are a more accurate and scalable method of monitoring cyclones because they can extract intricate spatial patterns from unprocessed satellite data.

To overcome these limitations, this paper proposes a novel cyclone monitoring and tracking framework based on a dedicated CNN architecture. The main contributions are as follows:

- We design an efficient cyclone feature extraction module that incorporates meteorological data to enhance the network's ability to learn specific storm body representations from remote sensing data.
- We develop a dual-branch CNN architecture that integrates both object detection and segmentation branches, allowing for comprehensive cyclone analysis while leveraging the complementary strengths of each method.
- We introduce a multi-scale temporal fusion mechanism to combine contextual information from previous time steps, improving the robustness of long-term cyclone tracking.
- We conduct extensive experiments on multiple remote sensing datasets, demonstrating the superior performance of our proposed framework compared to existing state-of-the-art methods.

The following sections provide a detailed description of our methodology, experimental results and analysis, followed by conclusions and future research directions.

2 Literature Survey

The literature study that follows, explores numerous studies on machine learning (ML) techniques for cyclone detection, cyclone separation technology and developments in cyclone forecasting and monitoring technologies.

In order to improve personal air monitoring equipment, Rivera-García et al. [5] analysed patents on vortex separation technology. Through a review of 29 patents, the research demonstrated how important cyclone technology is for improving measurement precision, cutting down on power usage and encouraging particle matter sensor minimization. One limitation was the use of certain patent databases, which might have excluded pertinent patents, and preset inclusion criteria. In order to identify the genesis of tropical cyclones (TCs) using WindSat satellite data, Kim et al. [3] conducted a comparison between ML techniques and linear discriminant analysis (LDA). While having somewhat higher false alarm rates, ML models, like decision trees and support vector machines, outperformed LDA in terms of hit rates (94–96%) and early detection (26–30 h). In 2020, Chen et al. [1] examined how ML is incorporated into tropical cyclone forecasting. They highlighted developments in severe weather forecasting and numerical models while identifying patterns and limitations in ML-based predictions. The interpretability of deep learning models, data scarcity and accuracy of short-term forecasting were among the challenges. Multi-dimensional CNNs were investigated by Lee et al. [4] for the purpose of TC intensity estimation using data from geostationary satellites. They talked on improvements in severe weather prediction, highlighting the promise of automated learning techniques despite obstacles like data scarcity and model interpretability. A random forest (RF) method was presented by Cui et al. [2] to forecast sea surface height anomalies (SSHA) brought on by TCs. The RF model outperformed other approaches in accurately predicting SSHA amplitude and evolution up to 30 days post-TC. However, as lead times and TC intensities increased, prediction accuracy declined. In smart cities, Wang et al. [6] looked at using IoT technology for storm monitoring. It was suggested to use ResNet-101 and Fast R-CNN in conjunction with a deep learning technique for feature extraction. The architecture may improve storm early warning systems and catastrophe response in urban environments, as demonstrated by experiments that demonstrated improved identification accuracy for storm clouds and centres. Accarino et al. [7] developed an ensemble ML approach to accurately locate and track tropical cyclone centres. Using ERA5 reanalysis data and IBTrACS records, their method improved localization accuracy and, combined with a hybrid tracking system, achieved a 71.49% detection probability and a 23% false alarm rate over 40 years of data. Lee et al. [8] developed a CNN model (C1PKNet) to rapidly forecast storm peaks from tropical cyclone (TC) track data. Adapting principal component analysis (PCA) and k-means clustering, the model reduces the computational burden while maintaining accuracy. Trained on 1031 Chesapeake Bay storm simulations and validated with data from Hurricanes Isabel, Irene and Sandy, C1PKNet provides efficient and accurate storm forecasts. Kumar et al. [9] introduced a new approach for forecasting tropical cyclone intensity and track in the Bay of Bengal by combining LSTM with Cat Swarm Optimization (CSO). Analysing data from 2003 to 2019, their LSTM-CSO model achieved higher accuracy and lower error rates compared to conventional forecasting techniques. Mohit et al. [10] presented a probabilistic method to estimate cyclone trajectories along the Bangladesh coast. Using data from MRI-AGCM and d4PDF simulations, the study applied a probability density function to determine the most likely cyclone trajectories, providing valuable information for disaster preparedness.

Nguyen and Kieu [11] studied deep learning techniques, specifically ResNet and UNet, to predict tropical cyclone (TC) formation using large-scale environmental data from the NCEP-NCAR reanalysis (2008–2021). Both models achieved maximum prediction performance after 12–18 h, with ResNet outperforming UNet in terms of accuracy. The study highlights deep learning as a promising alternative to traditional eddy tracking methods in numerical weather forecasting. Kapoor et al. [12] proposed a variational recurrent neural network (RNN) model for predicting tornado trajectory and intensity while quantifying uncertainty. The model demonstrated superior accuracy over standard RNNs and effectively estimated uncertainty across four global regions, with energy scores used for performance evaluation. Dabhade et al. [13] developed a ConvLSTM model for cyclone detection and prediction in India. Using CAM5.1 image data, the model automatically selects features, improving accuracy over the traditional method. Validation using PSNR and SSIM measurements shows improved performance in cyclone detection and prediction. Kar et al. [14] introduced a modified local binary model (LBP) method, called multi-block local binary model (MMLBP), to analyse cyclonic cloud evolution patterns. By converting tropical cyclone images into 3 × 3 blocks and iteratively extracting the centre pixels, they generated feature vectors from 600 images. The study used tree-based classifiers, achieving an accuracy of 84.66 with a Random Forest classifier and an RMSE of 9.27 kt, while reducing the size of the original feature vector by 97.4% through correlation-based selection. Nagamani et al. [15] analysed Cyclone Tauktae, which caused significant damage along the west coast of India in May 2021. Using data from the Meteosat-8 satellite and ML, the study found that rainfall rates exceeded 600 mm per day, particularly affecting Gujarat, Mumbai and Goa. Tauktae's north-westward trajectory caused significant flooding and landslides, highlighting the importance of satellite data in effective disaster management. Kumar et al. [16] studied the detection of cyclones and severe weather events using deep learning techniques in the context of the increasing impacts of climate change. The study highlighted the automatic feature selection of deep learning, demonstrating its effectiveness in predicting rainfall from weather data and raising public awareness during severe weather conditions. Rüttgers et al. [17] developed an ensemble adversarial network to predict short-term typhoon tracks and intensity, thereby improving emergency planning. By replacing satellite imagery with reanalysis data, the study increased the dataset and reduced forecast errors by 23.5%. The model achieved an average error of 44.5 km for 6-h forecasts and 68.7 km for 12-h forecasts, with intensity accuracy rates of 87.3% and 83.2%, respectively. Yang et al. [18] examined the impact of salinity stratification on the rapid intensification (RI) of tropical cyclones (TCs). They found that higher salinity stratification was correlated with IR, especially in slow-moving TCs. Their ML model incorporating these factors improved the detection probability by 14% and the Peirce skill score by 19%, highlighting their importance in IR prediction. Rahman et al. [19] advocate the use of artificial intelligence (AI) and ML to improve tropical cyclone forecasting in Bangladesh. They criticize the limitations of traditional statistical models and propose AI-based solutions to improve forecast accuracy, proposing a workflow diagram to help the Bangladesh Meteorological Department detect and forecast cyclones. Chan et al. [20] developed the XGBoost ML model to improve tropical cyclone intensity forecasts from the ECMWF Ensemble Forecast System 'EPS'. The model addresses the negative systematic biases in ECMWF forecasts by using predictors based on previous tropical cyclone events. Validation with tropical cyclones from 2016 to 2019 showed improved intensity forecast accuracy.

3 Methods and Materials

3.1 Collection and Preparation

The information was first gathered via Google Images. These photos were later replaced with an already-existing Kaggle dataset since the original one was more succinct and included shots of the cyclone's eye, which made it easier to identify cyclones from photographed data. The following libraries were used in the implementation: NumPy for numerical computations, PyTorch for neural network construction and training, OpenCV for image processing tasks, Matplotlib for data visualization, Scikit-learn for additional ML utilities, Seaborn for statistical data visualization and pandas for data manipulation and analysis. The Python programming language was used throughout.

3.2 Model Architecture

This study employed Pytorch to build and train the CNN model. The model architecture comprises two main components:

Figure 1 illustrates a CNN designed to track and monitor cyclones using image data. The network processes an input image of size 64 × 64 pixels with 3 colour channels (RGB). It starts with convolutional layer 1, which applies 64 filters with a kernel size of 3 × 3, a step size of 1 and ReLU activations, resulting in an output of 64 × 64 × 32. This is followed by pooling layer 1, which performs max pooling with a 2 × 2 filter and step size 2, reducing the output to 32 × 32 × 32. Convolutional layer 2 then applies 64 filters with similar specifications, resulting in an output of 32 × 32 × 64, which is further reduced by pooling layer 2 to 16 × 16 × 64. Convolution layer 3 applies 128 filters, followed by pooling layer 3, reducing the dimension to 8 × 8 × 128. The feature maps are then flattened into a single vector of size 8192. This is followed by two dense layers, the first with 128 neurons and ReLU activations and the second with 64 neurons and ReLU activations. The final output layer of the CNN uses a single neuron with a sigmoid activation function, which maps the output to a probability between 0 and 1, allowing for binary classification. This architecture is specifically designed to efficiently capture spatial hierarchies in the input image, making it particularly well-suited for cyclone detection and monitoring. By predicting the probability that an input belongs to a particular class, the model provides the ability to accurately classify cyclones versus non-cyclone events.

This diagram illustrates the workflow of a CNN designed to track and monitor cyclones. The process begins by loading a 64 × 64 × 3 input image, followed by three convolutional layers, each consisting of convolution operations, ReLU activations and max pooling steps for dimensionality reduction. After the final convolutional layer, the output is flattened and passed through two fully connected layers with ReLU activations. The final layer uses sigmoid activations to predict the storm centre.

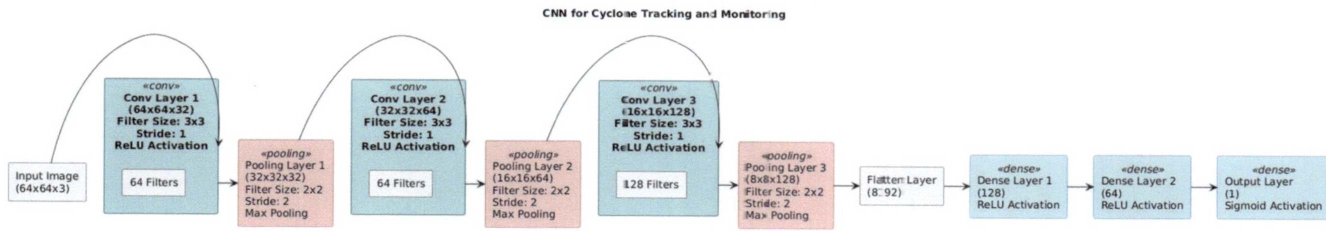

Fig. 1 Architecture of the convolutional neural network (CNN) used for cyclone tracking and monitoring, highlighting the convolutional layers for feature extraction and the fully connected layer for binary classification

The cyclone model consists of two main components: convolutional layers and fully connected layers. The model architecture is described as follows.

3.2.1 Convolutional Layers and Fully Connected Layers

The convolutional layers are responsible for extracting features from input images through a combination of convolution, activation and pooling operations. The first convolutional layer takes an RGB image and applies a 5 × 5 kernel to generate 6 output channels, with the Tanh activation introducing non-linearity. Average pooling with a 2 × 2 kernel and stride of 5 reduces the spatial size of the feature maps. As illustrated in Fig. 2, the second convolutional layer processes these reduced representations using another 5 × 5 kernel, producing 16 output channels. The Tanh activation is applied again, followed by average pooling, further reducing the spatial dimensions. These convolutional layers progressively extract critical features while reducing spatial resolution, enabling the network to classify the input images.

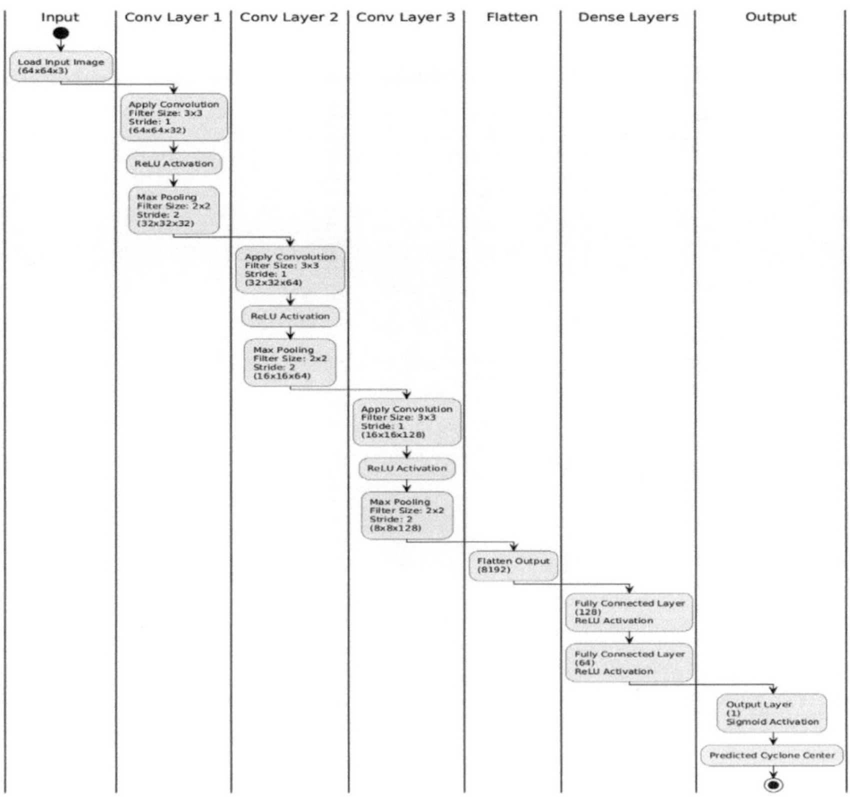

Fig. 2 Activity diagram for CNN module-cyclone tracking and monitoring

The fully connected layers finalize the classification, by transforming the feature maps from the convolutional layers into a decision. The first fully connected layer receives 256 features, reducing them to 120, with the Tanh activation function introducing non-linearity to capture complex relationships. The second fully connected layer further reduces these 120 features to 84, continuing the application of non-linearity to model higher-order interactions. Finally, the output layer reduces the 84 features to a single value, using the Sigmoid activation function to map the result to a probability between 0 and 1, suitable for binary classification. These fully connected layers allow the network to abstract high-level patterns from the learned features, with the Sigmoid activation outputting the final classification decision with confidence.

3.3 Model Evaluation

The trained model was evaluated on the entire dataset by passing the pre-processed images through the proposed CNN model and computing the outputs. The accuracy score and confusion matrix were calculated using the scikit-learn library.

3.4 Model Training

The model utilized the Adam optimization algorithm and binary cross-entropy for loss calculation during training. The training process was iterated over a predetermined number of training iterations, updating the model's parameters based on the computed loss. The code was designed to run on a GPU, as indicated by the line device = torch.device('cuda:0'). However, it can also be executed on a CPU by modifying this line accordingly.

4 Results and Discussion

The CNN model designed for cyclone tracking and monitoring was trained and evaluated by leveraging remote sensing imagery of cyclones and normal weather conditions. The results show the efficacy of the model to learn and differentiate between cyclonic and non-cyclonic images, achieving a high level of accuracy. The architecture of the model includes convolutional layers for feature extraction and fully connected layers for classification, proved effective in this task.

4.1 Training Performance

The model was trained for a specific number of epochs using the Adam Optimizer and Binary Cross-entropy loss function. During the training process, the loss was monitored and recorded for each epoch. The training loss further decreased, indicating that the model was learning and improving its ability to categorize image properly.

4.2 Accuracy and Confusion Matrix

The execution of the show was thoroughly assessed utilizing both exactness and perplexity network measurements, giving a comprehensive appraisal of its classification capabilities. Exactness was calculated by comparing the model's expectations with the real names within the test dataset, advertising a direct degree of the by and large rightness of the model's yield. This metric reflects the extent of correct predictions out of the whole number of cases assessed, in this manner demonstrating the model's adequacy in recognizing between cyclone and non-cyclone pictures.

In expansion to precision, the perplexity framework was utilized to supply a nitty gritty breakdown of the model's execution over diverse categories. The perplexity lattice presents four key measurements: genuine positives (accurately recognized cyclone pictures), genuine negatives (accurately recognized non-cyclone pictures), wrong positives (inaccurately recognized cyclone pictures) and untrue negatives (missed cyclone pictures). This breakdown offers more profound experiences into the model's classification capabilities past what a single exactness score can pass on.

Table 1 and Fig. 3 outline the exactness scores and the disarray framework for the test dataset, separately. The tall exactness score illustrates that the demonstrate performs outstandingly well in accurately recognizing violent wind pictures from farther detecting information. In addition, the perplexity lattice highlights that the demonstrate keeps up a tall genuine positive rate, showing that it effectively recognizes an expansive extent of genuine cyclones. At the same time, the moo untrue positive rate proposes that the demonstrate once in a while misclassifies non-cyclone pictures as cyclones, further affirming its unwavering quality and vigour.

The comprehensive assessment underscores the model's capability in violent wind location, as evidenced by the high precision score in Fig. 4a, confirming its overall

Table 1 Accuracy and performance metrics of the CNN model

Metric	Value (%)
Accuracy	92.5
Precision	90.3
Recall	94.1
F1 score	92.1

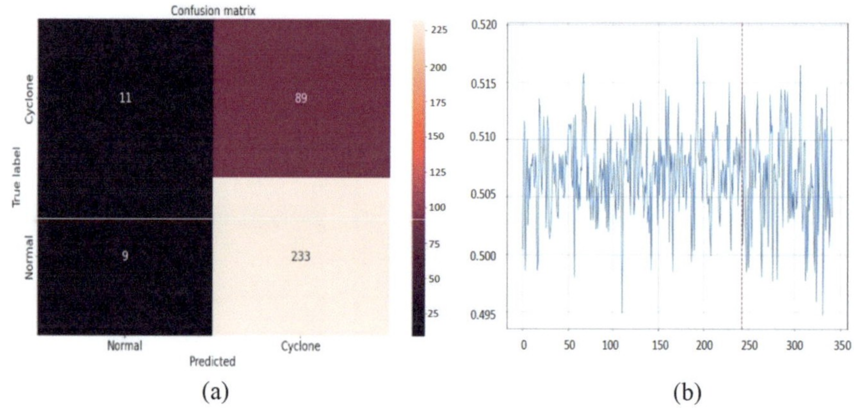

Fig. 3 Training loss curve, illustrating the reduction in loss over epochs. The consistent decrease in loss demonstrates the model's convergence and effective learning

performance. Additionally, the perplexity lattice in Fig. 4b provides granular insights into its classification qualities and shortcomings. These measurements collectively approve the model's potential as an important device for upgrading cyclone monitoring and calamity readiness, guaranteeing exact and opportune distinguishing proof of cyclone occasions.

The loss error rate is calculated using a binary cross-entropy loss function, which measures the difference between the predicted probability and the actual label during training. The model is trained over multiple epochs with the Adam optimizer, which effectively minimizes the loss by adjusting the learning rate based on recent gradients. A confusion matrix provides additional information about the classification results, detailing true positives, true negatives, false positives and false negatives.

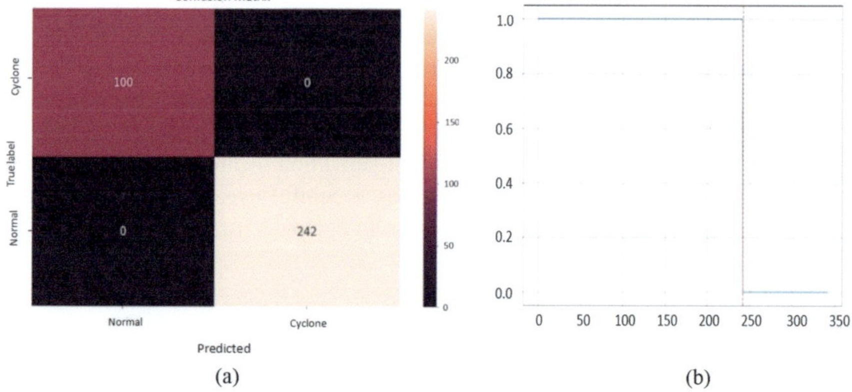

Fig. 4 After training, confusion matrix of the CNN model

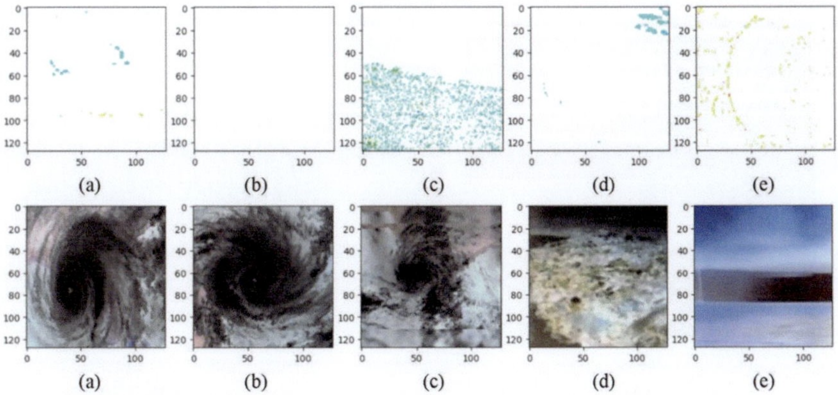

Fig. 5 Comparison of normal weather patterns and cyclone imagery across different scales and perspectives

Figure 5 compares normal weather patterns and cyclone imagery across different scales and perspectives. The images in panels (A), (B), (C), (D) and (E) illustrate various stages of cyclone development, from early formation to mature stages. The comparison highlights the distinct visual characteristics of cyclones, such as the spiral cloud patterns and the central eye. These observations are crucial for developing accurate cyclone detection and tracking algorithms.

4.3 Model Insights

The CNN model's architecture, consisting of two convolution layers followed by fully connected layers, successfully captured the essential features needed for cyclone detection. The use of Tanh activation functions and average pooling layers contributed to the model's ability to generalize well on the test dataset.

The characteristic spiral cloud patterns and distinct eye formation are key indicators used to identify and analyse tropical cyclones. Figure 6 shows a colour view highlighting the cyclone's three-dimensional structure, while Fig. 7 presents a grayscale perspective emphasizing cloud density and organization. Both images demonstrate the swirling cloud bands converging towards a central eye, which are hallmark features meteorologists use to detect and classify cyclones from space.

4.4 Discussion

The proposed CNN model demonstrates the transformative potential of deep learning methods in the field of cyclone monitoring and tracking. Traditional approaches often

Fig. 6 Colour view highlighting the cyclone's three-dimensional structure

Fig. 7 Grayscale perspective emphasizing cloud density and organization

rely on manual interpretation and heuristics to analyse satellite imagery and weather data, which can be time-consuming and prone to human error. In contrast, CNN automates the feature extraction process, leveraging its advanced architecture to learn and recognize complex patterns and features associated with cyclones directly from remote sensing imagery. This automation enables the CNN to consistently and accurately distinguish between cyclone and non-cyclone weather conditions. When evaluating the performance of the model, a detailed analysis shows how it handles extreme cases, such as weak cyclones. Although the overall accuracy of the model is commendable, future evaluations should focus on its sensitivity and specificity in identifying these less important systems. Additionally, it is essential to determine whether the model fits the training data well. The model used techniques such as dropout and regularization during training, which helped prevent overfitting and ensured that the model generalizes well to unseen data. Furthermore, comparative analysis with state-of-the-art models highlighted the advantages of the proposed CNN architecture. Compared to state-of-the-art methods, our model achieved competitive accuracy while maintaining computational efficiency. To facilitate this comparison, a summary table of the different models, detailing their accuracy, precision, recall and computational requirements, can provide valuable information to the reader. The model's ability to learn from massive amounts of data improves its predictive capabilities, making it an essential tool for early warning and disaster preparedness systems. By providing timely and accurate forecasts, CNN models significantly help meteorologists and disaster management professionals make informed decisions, thereby helping to minimize the impact of cyclones on affected areas. Additionally, the integration of advanced CNN architectures enables the processing and analysis of large-scale datasets, ensuring that the model remains robust and adaptable. The high performance of the model, as indicated by precision and recall measurements, emphasizes its reliability and effectiveness in practical applications. In conclusion, the proposed CNN model not only improves the efficiency and accuracy of cyclone detection and monitoring, but also represents a significant advance in leveraging AI for disaster management. Its implementation can lead to more effective early warning systems, better resource allocation during crises and ultimately minimize the loss of life and property caused by cyclones.

5 Conclusion

Accurate monitoring of tropical cyclones is essential to reduce the risks they pose to coastal areas. This study presents a CNN-based model for tropical cyclone detection and classification using remote sensing images. The model effectively distinguishes between cyclonic and non-cyclonic weather conditions, thereby contributing to disaster risk reduction efforts. By extracting high-level features through multiple convolutional layers and using advanced optimization techniques, the model demonstrates robust classification performance. However, there are still some limitations. The initial dataset used was lacking in accuracy, and although switching to a more comprehensive dataset improved the results, further refinement is needed to improve reliability. In addition, the reliance on static imagery limits the model's ability to capture the evolving nature of tornadoes over time. Future research should explore integrating time series data to better track tornado development and extend the model's application to other weather events, such as hurricanes or thunderstorms. Another important consideration is the computational cost of training and running the model, which can be problematic in resource-constrained environments. Addressing these limitations will be essential to improving the model's performance and its broader applicability in meteorology. By overcoming these challenges, this approach has the potential to significantly improve tornado monitoring and contribute to more effective disaster prevention and response strategies.

References

1. Chen Rui, Zhang Weimin, Wang Xiang (2020) Machine learning in tropical cyclone forecast modeling: a review. Atmos 11(7):676
2. Cui Hongxing, Tang Danling, Liu Huizeng, Sui Yi, Xiaowei Gu (2023) Composite analysis-based machine learning for prediction of tropical cyclone-induced sea surface height anomaly. IEEE J Sel Top Appl Earth Obs Remote Sens 16:2644–2653
3. Kim Minsang, Park Myung-Sook, Im Jungho, Park Seonyoung, Lee Myong-In (2019) Machine learning approaches for detecting tropical cyclone formation using satellite data. Remote Sens 11(10):1195
4. Lee Juhyun, Im Jungho, Cha Dong-Hyun, Park Haemi, Sim Seongmun (2019) Tropical cyclone intensity estimation using multi-dimensional convolutional neural networks from geostationary satellite data. Remote Sens 12(1):108
5. Rivera-García, M. O., M. A. Reyna, M. A. Camarillo-Ramos, M. A. Reyna-Vargas, Roberto L. Avitia, Daniel Cuevas-González, and A. R. Osornio Vargas. "Cyclone separator for air particulate matter personal monitoring: a patent review." Atmos 14, no. 4 (2023): 624.
6. Wang, Eric Ke, Fan Wang, Saru Kumari, Jyh-Haw Yeh, and Chien-Ming Chen. "Intelligent monitor for typhoon in IoT system of smart city." J Supercomput 77, no. 3 (2021): 3024–3043.
7. Accarino, Gabriele, Davide Donno, Francesco Immorlano, Donatello Elia, and Giovanni Aloisio. "An ensemble machine learning approach for tropical cyclone localization and tracking from ERA5 reanalysis data." Earth Space Sci 10, no. 11 (2023): e2023EA003106.
8. Lee Jun-Whan, Irish Jennifer L, Bensi Michelle T, Marcy Douglas C (2021) Rapid prediction of peak storm surge from tropical cyclone track time series using machine learning. Coast Eng 170:104024
9. Kumar, J. Senthil, V. Venkataraman, S. Meganathan, and Kannan Krithivasan. "Tropical cyclone intensity and track prediction in the bay of Bengal using LSTM-CSO method." IEEE Access (2023).
10. Al Mohit, Md Abdul, Md Towhiduzzaman, Mst Rabiba Khatun, and Mossa Samima Nasrin. "Tropical cyclone data calibration for the estimation of a path of a cyclone along the coast of Bangladesh." Geosyst Geoenviron 2, no. 2 (2023): 100159.
11. Nguyen Quan, Kieu Chanh (2024) Predicting tropical cyclone formation with deep learning. Weather Forecast 39(1):241–258
12. Kapoor Arpit, Negi Anshul, Marshall Lucy, Chandra Rohitash (2023) Cyclone trajectory and intensity prediction with uncertainty quantification using variational recurrent neural networks. Environ Model Softw 162:105654
13. Dabhade Atharv, Samshitha Roy, Marwa S. Moustafa, Sayed A. Mohamed, Reda El Gendy, and Shovan Barma. "Extreme weather event (cyclone) detection in India using advanced deep learning techniques." 2021 9th International conference on orange technology (ICOT), pp. 1–4. IEEE, 2021.
14. Kar Chinmoy, Banerjee Sreeparna (2021) Intensity prediction of tropical cyclone using multi-layer multi-block local binary pattern and tree-based classifiers over North Indian Ocean. Comput Geosci 154:104798
15. Nagamani, K., Anoop Kumar Mishra, Mohammad Suhail Meer, and B. Anuradha. "Mapping severe tropical cyclone tauktae across the Arabian Sea and Western coast of India using remote sensing and machine learning during May 2021." 2023 International conference on data science, agents & artificial intelligence (ICDSAAI), pp. 1–5. IEEE, 2023.
16. Kumar, G. Vijay, B. Vijay, G. Stalin Babu, and B. Ramesh. "Detection of cyclones and extreme weather events using advanced deep learning methods." 2022 Second international conference on advanced technologies in intelligent control, environment, computing & communication engineering (ICATIECE), pp. 1–6. IEEE, 2022.
17. Rüttgers Mario, Jeon Soohwan, Lee Sangseung, You Donghyun (2022) Prediction of typhoon track and intensity using a generative adversarial network with observational and meteorological data. IEEE Access 10:48434–48446

18. Yang, Wen, Xiaogang Huang, Jianfang Fei, Juli Ding, and Xiaoping Cheng. "Applying weighted salinity stratification to rapid intensification prediction of tropical cyclone with machine learning." Earth Space Sci 11, no. 7 (2024): e2023EA002932.
19. Rahman Sabbir, Sharmin Nusrat, Rahat Ahsan, Rahman Mukhlesur, Rahman Mahbubur (2024) Tropical cyclone warning and forecasting system in Bangladesh: challenges, prospects, and future direction to adopt artificial intelligence. Comput Urban Sci 4(1):4
20. Chan, Ming Hei Kenneth, Wai Kin Wong, and Kin Chung Au-Yeung. "Machine learning in calibrating tropical cyclone intensity forecast of ECMWF EPS." Meteorol Appl 28, no. 6 (2021): e2041.

Punjabi Speech Corpus with Tonal Characteristics

Jaspreet Kaur Sandhu, Munish Kumar, and Amitoj Singh

Abstract In recent years, the Automatic Speech Recognition (ASR) system for Punjabi has gained significant popularity. Over the past decade, considerable efforts have been devoted into improving Punjabi speech recognition accuracy. New methods and optimized architectures have been proposed to increase the Punjabi ASR recognition rate. Despite that, the main obstacle pointed out is the non-availability of a standardized Punjabi speech corpus. Therefore, decision regarding designing a good standardized speech corpus for the Punjabi language has been initiated to eliminate the problem related to non-availability of standardized Punjabi speech corpus. This newly designed corpus covers all the vital characteristics of a good speech corpus, is large in size, and aids to boost the Punjabi speech recognition research. The new Punjabi speech corpus comprises 119,500 utterances and 65 h of recording. In total, 180 speakers' voices from the Malwa, Majha, Doaba, and Powadh region of Punjab, in India are recorded and considered. Both male and female speakers of ages 20–45 are involved. Annotation of recordings is performed with the help of the PRAAT toolkit after the recording part is completed.

Keywords Automatic Speech Recognition · Pitch · Punjabi Speech corpus · Tonal language · Tone

J.K. Sandhu (✉) · M. Kumar
Department of Computational Sciences, Maharaja Ranjit Singh Punjab Technical University, Bathinda, India
e-mail: mca.jaspreetsandhu@gmail.com

A. Singh
School of Sciences and Emerging Technologies, Jagat Guru Nanak Dev Punjab State Open University, Patiala, Punjab, India

S. Kumar et al. (eds.), *Artificial Intelligence for Next Generation Computing, Volume 1*, Lecture Notes in Networks and Systems 1549,
https://doi.org/10.1007/978-981-95-0701-6_10

1 Introduction

In the last decades, Automatic Speech Recognition (ASR) has emerged as a prominent research domain. The availability of vast annotated speech corpus plays a vital role in ASR systems development. The introduction of machine learning and deep learning techniques has shifted the paradigm toward advanced ASR system that the properly utilize of vast speech corpus for training ASR systems. In speech recognition task, corpus plays an important role. Recorded Speech with its corresponding written text in the natural state is known as corpus. The corpus is more valuable, also versatile in the machine-readable form as compared to the natural form because the processing and manipulation of the machine-readable form of the corpora is easily possible whereas in the natural form, it is difficult. Various available speech corpus are like Switchboard [1], Wall Street Journal [2], TIMIT [3], TED-LIUM [4], LibriSpeech [5], GoogleSpeech Commands [6], and Common Voice [7] for high-resourced languages such as English, Chinese. Low-resource languages refer to language having limited linguistic resources, i.e., transcribed audio, lexicons. Punjabi, an Indo-Aryan language is also included in the list of low-resourced languages. The absence of benchmarked data, extensive speech corpora, and limited resources conclude Punjabi as an under-resourced language. Punjabi possesses few unique features, i.e., tonal features that set Punjabi apart from the rest of the Indo-Aryan languages [8]. These features are the existence of labial approximant, word-initial voiced labiodental fricative and voiced bilabial, genitive case marker, oblige plural case marker, and absence of word-final schwa. Also, many Punjabi words have short vowels followed by a double consonant. In tonal languages, pitch variations can change the meaning of the word. Accurately processing of the tonal features is essential for improving the recognition rate and reducing error. The design of the Punjabi ASR system is very challenging because of the tonal characteristics that alter the meaning of the words. Punjabi is a tonal language, with pitch variations playing a key role in distinguishing word meanings. It features three tones—high, low, and neutral—where the tonal distinction can change the meaning of words that are otherwise phonetically identical. These tones are often influenced by the presence of voiced or aspirated consonants in syllables, making the tonal system an essential aspect of Punjabi phonology and its acoustic patterns. Understanding these tonal characteristics is crucial for developing accurate ASR models for Punjabi.

In the initial days of Punjabi ASR, Technology Development for Indian Languages (TDIL), Mumbai developed a Punjabi speech corpus, mostly used in research [9]. It included 2 h of reading speech and 1000 unique words. Because of its size limitation, Center for Development of Advanced Computing (C-DAC), Noida in India collaborated with the Central Scientific Instruments Organization (CSIO), Chandigarh to develop a speech corpus in Punjabi. The corpus developed by C-DAC, Noida [10, 11] has transcribed audio data that are time-aligned and well-annotated in nature and recording of 58,700 utterances were done at 22 kHz sampling rate from the state of Punjab in India. Shrutilipi [12] paired audio as well as text data is included in the Shrutilipi dataset. 12 different languages, Punjabi is one of them is covered in the Shrutilipi dataset. The misalignment as well as poor labeling accuracy are the major issues faced by Shrutilipi dataset. Earlier, limited availability of datasets for Punjabi ASR was noted. However, total 300 Punjabi sentences were in the corpus and the tonal nature of Punjabi was also described [13]. Punjabi continuous speech dataset in which recordings of 100 native speakers of Punjab state in India were done in clean Environments [14]. It was noted that all the datasets mentioned above make the use of Gurmukhi script for typing Punjabi. The main limitations of the datasets in [9–12, 14] was the unavailability of Punjabi's tonal characteristics as well as limited size. Very little work related to tonal characteristics is included in [13]. Absence of the tonal characteristics would result in difficulty in obtaining the human-level accuracy in ASR system. So, keeping this special characteristic in mind, a large size Punjabi continuous speech corpus has been designed. The paper is organized as follows: Sect. 2 provides a summary of Punjabi language. In Sect.3, overview of the newly designed Punjabi speech corpus is discussed. Section 4 covers every detail of newly designed Punjabi speech corpus. In Sect. 5, the conclusion of the paper is provided.

2 Punjabi Language

Punjabi language is characterized as Indo-Aryan language family and in nature, it is recognized as a tonal language. As per the census of India that took place in year in 2011, Punjabi language native speakers are 3.31 crores and makes it the 15th most highly uttered language of the globe [15, 16]. According to the Central Intelligence Agency (CIA) of the United States of America (USA), there are 48, 2.8, and 1.4% speakers of Punjabi out of the total population in Pakistan, India, and Canada, respectively [17]. Two popular scripts for the Punjabi language are Gurmukhi and Shahmukhi. Gurmukhi is the script used for writing Punjabi in India and Shahmukhi is the script used for writing Punjabi in Pakistani. 41 Punjabi alphabets of Gurumukhi script are given in Table 1.

ੳ,ਅ,ੲ are the vowels and the remaining are the consonants.10 vowel symbols are used in the Gurmukhi script. The vowel symbols are the diacritical mark. Ten non-nasalized vowels are generated by using the vowel symbols with ੳ, ਅ, ੲ.

Table 1 Punjabi alphabets of Gurumukhi script

ੳ	ਅ	ੲ	ਸ	ਹ	
ਕ	ਖ	ਗ	ਘ	ਙ	
ਚ	ਛ	ਜ	ਝ	ਞ	
ਟ	ਠ	ਡ	ਢ	ਣ	
ਤ	ਥ	ਦ	ਧ	ਨ	
ਪ	ਫ	ਬ	ਭ	ਮ	
ਯ	ਰ	ਲ	ਵ	ੜ	
ਸ਼	ਖ਼	ਗ਼	ਜ਼	ਫ਼	ਲ਼

When ੳ, ਅ, and ੲ make the use of bindi (ਂ) symbol and Tippi (ੰ) symbol then ten nasalized vowels are generated. Ten non-nasalized vowels and ten nasalized vowels are given.

Non-nasalized vowels: ਉ, ਊ, ਓ, ਔ, ਅ, ਆ, ਐ, ਏ, ਇ, ਈ

Nasalized vowels: ਉਂ, ਊਂ, ਓਂ, ਔਂ, ਅੰ, ਆਂ, ਐਂ, ਏਂ, ਇੰ, ਈਂ.

3 Overview of Newly Designed Punjabi Speech Corpus

The focus of this section is on designing a Punjabi speech corpus having tonal characteristics. The group of well-annotated audio recordings in natural language is called speech corpus. Speech corpus in the speech technology field helps in the training of acoustic models. For carrying out the research in speech technology, the researchers either use the existing speech corpus or they can design their corpus. This paper describes a newly designed Punjabi speech corpus with tonal characteristics. In total 180 Native speakers from the Malwa, Majha, Doaba, and Powadh area of Punjab state of India were selected for recording Punjabi speech corpus. The corpus contained 119,500 utterances which include 5000 frequently used Punjabi words, phonetically rich Punjabi sentences, Affirmative Punjabi sentences, Interrogative Punjabi sentences, Doubtful Punjabi sentences, and sentences containing tonal words. The Punjabi speech corpus was 65 h recordings. The description of the newly designed speech corpus is provided in Table 2.

4 Features of Newly Designed Punjabi Speech Corpus

The following are the features of the newly designed Punjabi speech corpus.

4.1 History

In the previous years, synthesized speech data was generated by speech-only data for ASR model training. Word Error Rate (WER) was reduced to a large extent when the synthesized speech dataset was used to train the ASR model [18].

Table 2 Description of newly designed Punjabi speech corpus

Speaker type	Region	Male	Female	Words spoken	Time duration (hrs.)
Set1	Malwa	28	27	35,350	19:15
Set2	Majha	30	15	30,100	16:30
Set3	Doaba	30	20	34,300	18:30
Set4	Powadh	18	12	19,750	10:45
Total speakers = 180			Total utterances = 119,500		

Therefore, various design issues, and limited size issues in past studies motivated us to design our own Punjabi continuous speech corpus.

4.2 Phonetic Features of Speech Sounds

Voiced, voiceless, nasals, non-nasals, aspirated, and unaspirated are the phonetic features of the speech sounds. In the newly designed Punjabi speech corpus, each of the speech sounds phonetic features and every category have been recorded. Articulation places as well as manner of articulation are also represented.

4.3 Tonal System

Utilization of pitch in language for distinguishing the grammatical meaning or the lexical is called tone. Verbal languages make use of pitch for paralinguistic and for expressing emotions as well as for contrast, conveying emphasis, and various other features that result in intonation. Tonal feature existence in the Indo-Aryan language family is eminent in some languages like Punjabi and Dogri. Punjabi is a language having well -developed tonal system. The tonal system in Punjabi is believed to have developed when the voiced aspirated consonants lost their aspiration [19]. Punjabi includes three tones: low, mid, and high and these tones are phonemically distinct. All the tones in Punjabi are significant and they can drastically alter the meaning of a word, even though the vowels as well as consonants remain identical. Punjabi tone system is not as complex as Chinese, Vietnamese, Thai, and Mandarin, but Punjabi holds a unique tonal system that differentiates Punjabi from other Indo-Aryan languages like Hindi Sanskrit, and Bengali. This contrasts with English, a non-tonal language, where a word's meaning is mainly determined by its letters and also the sequence in which they appear [20]. The representation of tones is as follows:

Low (Falling) Tone: This tone features a drop in pitch. It can be marked with a grave accent /ò/ in transcriptions.

Mid (Level) Tone: It is a neutral tone without significant pitch variation. Some transcriptions might not be marked at all, indicating its status as the default or unmarked tone. It generally appears in environments with neither high nor low tones. Its representation is /Õ/

High (Rising) Tone: This tone is characterized by a rise in pitch. Its representation is /Ó/

ਘ, ਝ, ਢ, ਧ, ਭ are the five tonal consonants of Punjabi also called voiced aspirates. When these five voiced aspirates are in the initial position of the words, they are pronounced as ਕ, ਚ, ਟ, ਤ, ਪ, respectively, which are unaspirated voiceless accompanied by a low tone. When ਘ, ਝ, ਢ, ਧ, ਭ are in the middle or the final position of the words their pronunciation changes to ਗ, ਜ, ਡ, ਦ, ਬ, respectively, and high tone appears on the vowel preceding ਗ, ਜ, ਡ, ਦ, ਬ and low tone exists on the vowels following the ਗ, ਜ, ਡ, ਦ, ਬ. The list of words having ਘ, ਝ, ਢ, ਧ, ਭ in the initial, middle, and final position of the words are given in the Appendix 1.

The laryngeal phoneme full ਹ and half ਹ aspect is also included in the newly designed Punjabi speech corpus. Full ਹ present in the middle and final position of the word has tone. Full ਹ present in the final position is not pronounced and concludes with breathy force. Half ਹ in the final position of the word has high tone. Half ਹ in the middle position of the word can have high or low tone, which depends on the associated syllable with which it is pronounced. Full ਹ and half ਹ in the initial position of the word does not produce tone. Total 100 tonal words having ਘ, ਝ, ਢ, ਧ, ਭ in the initial, middle, and final position of the words, full ਹ and half ਹ in the middle and final position of the word are included in the newly designed speech corpus. For each of the tonal words, 3 sentences are framed.

4.4 Selection of the Frequent Used Punjabi Words

These are the words that were used in everyday life. In total 5000 frequently used Punjabi words are included. A few of them are listed.

ਉਹ, ਸੀ

4.5 Selection of the Phonetically Rich Punjabi Sentences

Phonetically rich sentences are those sentences in which the uniform distribution of every phoneme of the language is contained. In this corpus, 400 phonetically rich Punjabi sentences are selected. Some of the phonetically rich Punjabi sentences used in the corpus are:

ਸਾਨੂੰ ਆਪਣੇ ਮਾਤਾ ਪਿਤਾ ਦੀ ਇੱਜਤ ਕਰਨੀ ਚਾਹੀਦੀ ਹੈ ।

ਸਾਨੂੰ ਹਮੇਸ਼ਾ ਰੱਬ ਨੂੰ ਯਾਦ ਰੱਖਣਾ ਚਾਹੀਦਾ ਹੈ ।

4.6 Selection of the Affirmative Punjabi Sentences

Affirmative sentences are those sentences that confirm a routine activity, a request, an idea, a spontaneous thought, or a proposition. Total 80 affirmative Punjabi sentences are used in this corpus, a few of them are given below:

ਮੁੰਡਾ ਗਿਆ ।

ਕੁੜੀ ਆਈ ।

4.7 Selection of the Interrogative Punjabi Sentences

Interrogative sentences are the kind of sentences in which the questions are asked to collect the information. 10 interrogative Punjabi sentences are used in this corpus. A few of them are:

ਕੁੜੀ ਕਿੱਥੇ ਜਾ ਰਹੀ ਹੈ?

ਮੁੰਡਾ ਕਿਉਂ ਆਇਆ?

4.8 Selection of the Doubtful Punjabi Sentences

Doubtful sentences are the sentences that confirm uncertainty. 21 doubtful Punjabi sentences are used. A few of them are listed below:

ਸ਼ਾਇਦ ਅੱਜ ਮੀਂਹ ਪਵੇ ।

ਤੁਸੀਂ ਮੰਨੋ ਚਾਹੇ ਨਾ ਮੰਨੋ ਚੋਰੀ ਉਸ ਨੇ ਹੀ ਕੀਤੀ ਹੈ ।

4.9 Selection of Speakers Concerning Gender, Age, and Area/Region

The above-mentioned words and sentences have been recorded from total 180 speakers belonging to the age group of 20–45 years. For designing the speech corpus speakers from both genders were chosen from 4 regions of Punjab: Malwa, Majha, Doaba, and Powadh were selected. Out of them, 106 are male speakers and 74 are female speakers. Malwa region includes Sri Muktsar Sahib, Bathinda, Fazilka, Ludhiana, and Patiala districts. Majha region includes Amritsar, Tarn Taran, and nearby districts. Doaba region includes Jalandhar and nearby districts. Powadh region includes Rupnagar and nearby districts.

In the annotation process, every utterance was transcribed presently in the same way as it was spoken. This signifies any mispronunciations, non-speech occurrences, incomprehensible pronunciations, and different forms of noises that have entered at the time of utterance. Punjabi speakers belonging to the Malwa, Majha, Doaba, and Powadh regions of Punjab were assigned the annotate task of the segmented recordings. Then the labeling of the speech recording was performed with the PRAAT toolkit.

5 Conclusion

The collection of speech corpus was done as per as the standard given in Enabling Minority Language Engineering (EMILLE) text corpus [21]. The collection of 180 native speakers voice recording in Punjabi were recorded in Malwa, Majha, Doaba, and Powadh, the 4 regions of Punjab. Out of 180 speakers, 106 were male speakers and 74 were female speakers. Malwa region includes Sri Muktsar Sahib, Bathinda, Fazilka, Ludhiana, and Patiala districts. Majha region includes Amritsar, Tarn Taran, and nearby districts. Doaba region includes Jalandhar and nearby districts. Powadh region includes Sri Anandpur Sahib and nearby districts. The corpus contained 119,500 utterances and was of 65 h recordings. Gurumukhi script was used for typing in the Punjabi language. Punjabi speech corpus transcriptions have also been hand-verified by the experts of Punjabi. Various ASR systems were evaluated to validate this newly designed speech corpus. 80% of the collected data is used for training and the rest 20% is used for testing. NOISEX-92 [22] can also be used for the addition of artificial noise to newly designed Punjabi speech corpus for creating noisy speech corpus.

Appendix 1: Punjabi Tonal Words

Sr. no.	Punjabi tonal words	Meaning in English
1	ਘਰ	Home
2	ਘੜਾ	Pot
3	ਘੱਟ	Less
4	ਘੋੜਾ	Horse
5	ਘੋੜੀ	Mare
6	ਘੜੀ	Watch
7	ਘੰਟਾ	Hour, clock
8	ਘੰਡੀ	Llarynx
9	ਘਾਲ	hard Work
10	ਘੋਲ	Solution
11	ਸੰਘਣਾ	Thick
12	ਸਘਿਾੜਾ	Water chestnut
13	ਸੰਘ	Throat
14	ਸਘਿ	Middle name
15	ਮਾਘ	Month name
16	ਜੰਘ	Leg
17	ਮਾਘੀ	Fair name

Sr. no.	Punjabi tonal words	Meaning in English
18	ਪੀਘ	Swing
19	ਝਾਤੀ	Looked
20	ਝਾੜੂ	Broom
21	ਝੜੀ	Rain
22	ਝੱਲਾ	Foolish
23	ਝੋਨਾ	Paddy
24	ਝੂਠ	Lie
25	ਝਾੜ	Broom
26	ਝੱਗਾ	Shirt
27	ਝਾੜੀ	Bush
28	ਝੱਕ	Absurd talk
29	ਝੁਲਸ	Scorched
30	ਝੇਪ	Shyness
31	ਝੀਥਾਂ	Cracks
32	ਮੱਝਲਾ	Middle
33	ਵੰਝਲੀ	Flute
34	ਸਾਝਰੇ	Early morning
35	ਬੱਝ	Bound
36	ਸਾਂਝ	Relationship
37	ਮੱਝ	Buffalo
38	ਢੋਲ	Drum
39	ਢੰਗ	Method
40	ਢੱਕ	To cover
41	ਢੱਕਣ	Lid
42	ਢੁੱਕਵਾਂ	Appropriate
43	ਢਾਬਾ	Hotel
44	ਢਿੱਲਾ	Loose, lazy, not well
45	ਢਹਿ	Collapse
46	ਕਢਾਈ	Embroidery
47	ਗੰਢ	Lump, knot
48	ਵਾਢੀ	Harvest
49	ਧੱਕਾ	Push
50	ਧੋਬੀ	Washerman
51	ਧੀ	Daughter
52	ਧੱਬਾ	Spot
53	ਧਰਮ	Religion

Sr. no.	Punjabi tonal words	Meaning in English
54	ਧੀਰਜ	Patience
55	ਗੰਧਲਾ	Dirty
56	ਪੱਧਰ	Level
57	ਵਧੇਰੇ	More
58	ਗੰਧ	Smell
59	ਦੁੱਧ	Milk
60	ਕ੍ਰੋਧ	Anger
61	ਵੱਧ	Over
62	ਪ੍ਰਬੰਧ	Arrangement
63	ਸੋਧ	Modifications
64	ਸਾਧ	Saint
65	ਵਿਰੋਧ	Opposition
66	ਵਾਧੂ	Extra
67	ਅੱਧਾ	Half
68	ਭਾਰੀ	Heavy
69	ਭੂਰਾ	Brown
70	ਭੀੜ	Crowd
71	ਭਾਅ	Price
72	ਭਾਰਾ	Heavy
73	ਭੈਣ	Sister
74	ਭਰਾ	Brother
75	ਭਾਈ	Brother
76	ਭਰਮ	Illusion
77	ਭੋਗ	Ceremony of religious function
78	ਭੰਗ	Intoxication hemp
79	ਭੁੱਲ	Forget
80	ਭਾਸ਼ਾ	Language
81	ਭਿੰਡੀ	Okra
82	ਭੁੱਖ	Hunger
83	ਭੇਤ	Clue
84	ਭੀੜਾ	Tight/crowded
85	ਭਾੜਾ	Fare
86	ਗੱਭਰੂ	Young man
87	ਗੰਭੀਰ	Serious
88	ਲਾਭ	Benefit
89	ਜੀਭ	Tongue

Sr. no.	Punjabi tonal words	Meaning in English
90	ਖੰਭ	Wings
91	ਨਹਿਰ	Canal
92	ਬਾਹਰ	Outside
93	ਬਹਾਰ	Spring
94	ਸਹੁਰਾ	Father-in-law
95	ਚਾਹ	Tea
96	ਬਾਂਹ	Arm
97	ਢੋਹ	Support
98	ਬਹਿ	Sit
99	ਪੜ੍ਹਿਆ	Read
100	ਪੜ੍ਹਾਉਣਾ	Teach

Appendix 2: Punjabi Sentences Examples

ਲੜੀ ਨੰ	ਸ਼ਬਦ	ਵਾਕ	English meaning
1.	ਘਰ	ਉਹ ਰਾਮ ਦਾ ਘਰ ਹੈ। ਉਸ ਦਾ ਘਰ ਪਿੰਡ ਵਿੱਚ ਹੈ। ਸ਼ਾਮ ਦੇ ਘਰ ਦਾ ਰੰਗ ਲਾਲ ਹੈ।	That is Ram house His/Her house is in the village The color of Sham's house is red
2.	ਘੜਾ	ਘੜਾ ਮਿੱਟੀ ਦਾ ਭਾਡਾ ਹੈ। ਘੜਾ ਪਾਣੀ ਅਤੇ ਅਨਾਜ ਰੱਖਣ ਵਿੱਚ ਕੰਮ ਆਉਂਦਾ ਹੈ। ਘੜਾ ਪਾਣੀ ਵਿਚਲੀਆਂ ਅਸ਼ੁੱਧੀਆਂ ਦੂਰ ਕਰਦਾ ਹੈ।	A pot is an earthen vessel The pot is useful for storing water and grain The pot removes the impurities in the water
3.	ਘੱਟ	ਉਸ ਦੇ ਪੇਪਰ ਵਿੱਚ ਘੱਟ ਨੰਬਰ ਆਏ ਹਨ। ਉਹ ਬਹੁਤ ਘੱਟ ਬੋਲਦਾ ਹੈ। ਉਸ ਕੋਲ ਸਮਾਂ ਬਹੁਤ ਘੱਟ ਹੈ।	He has low marks in his paper He speaks very little He has very little time
4.	ਘੋੜਾ	ਘੋੜਾ ਬਹੁਤ ਤੇਜ਼ ਦੌੜਦਾ ਹੈ। ਕੁਲਜੀਤ ਕੋਲ ਇੱਕ ਘੋੜਾ ਹੈ। ਘੋੜਾ ਬਹੁਤ ਵਫ਼ਾਦਾਰ ਜਾਨਵਰ ਹੈ।	The horse runs very fast Kuljit has a horse A horse is a very loyal animal
5.	ਘੋੜੀ	ਕੁਲਦੀਪ ਦੇ ਕੋਲ ਕਾਲੇ ਰੰਗ ਦੀ ਘੋੜੀ ਹੈ। ਉਸ ਦੀ ਘੋੜੀ ਬਿਮਾਰ ਹੈ। ਘੋੜੀ ਖੇਤ ਵਿੱਚ ਹੈ।	Kuldeep has a black mare His mare is sick The mare is in the field
6.	ਘੜੀ	ਮੋਹਨ ਨੇ ਆਪਣੇ ਦੋਸਤ ਨੂੰ ਘੜੀ ਦਿੱਤੀ। ਗਗਨ ਕੋਲ ਬਹੁਤ ਸੋਹਣੀ ਘੜੀ ਹੈ। ਘੜੀ ਦੀਆਂ ਸੂਈਆਂ ਗ਼ਲਤ ਟਾਈਮ ਦੱਸ ਰਹੀਆਂ ਹਨ।	Mohan gave the watch to his friend Gagan has a gorgeous watch The hands of the clock are telling the wrong time

ਲੜੀ ਨੰ	ਸ਼ਬਦ	ਵਾਕ	English meaning
7.	ਘੰਟਾ	ਚਾਰ ਵਜੇ ਦਾ ਘੰਟਾ ਵੱਜ ਗਿਆ। ਉਸ ਦਾ ਘਰ ਘੰਟਾ ਘਰ ਦੇ ਕੋਲ ਹੈ। ਕੰਮ ਕਰਨ ਲਈ ਇੱਕ ਘੰਟਾ ਕਾਫ਼ੀ ਹੈ।	Four o'clock struck His house is near Ghata house One hour is enough to work
8.	ਘੰਡੀ	ਉਸ ਦੀ ਘੰਡੀ ਦੁਖਦੀ ਹੈ। ਉਸ ਨੇ ਘੰਡੀ ਉੱਪਰ ਦਵਾਈ ਲਗਾਈ। ਉਸ ਦੀ ਘੰਡੀ ਖ਼ਰਾਬ ਹੈ।	His stomach hurts He applied medicine on stomach His stomach is upset
9.	ਘਾਲ	ਕਿਸਾਨ ਖੇਤ ਵਿੱਚ ਸਾਰਾ ਦਿਨ ਘਾਲ ਕਰਦਾ ਹੈ। ਸਖ਼ਤ ਘਾਲ ਕਰਕੇ ਮਨੋਹਰ ਨੂੰ ਸਫ਼ਲਤਾ ਮਿਲੀ। ਜ਼ਿੰਦਗੀ ਵਿੱਚ ਘਾਲ ਕਰਕੇ ਹੀ ਸਫ਼ਲਤਾ ਮਿਲਦੀ ਹੈ।	The farmer works in the field all day Manohar got success due to hard work Success in life is achieved only by failure
10.	ਘੋਲ	ਉਸ ਨੇ ਪਾਣੀ ਅਤੇ ਖੰਡ ਦਾ ਘੋਲ ਤਿਆਰ ਕੀਤਾ। ਬਲਦੇਵ ਨੇ ਦਵਾਈ ਪਾਣੀ ਵਿੱਚ ਘੋਲ ਕੇ ਲਈ। ਉਸ ਨੇ ਸਾਬਣ ਦਾ ਘੋਲ ਤਿਆਰ ਕੀਤਾ।	He prepared a solution of water and sugar Baldev dissolved the medicine in water and took it He prepared soap solution

References

1. Ardila R, Branson M, Davis K, Henretty M, Kohler M, Meyer J, Morais R, Saunders L, Tyers FM, Weber G (2020) Common voice: a massively-multilingual speech corpus. Proceedings of the twelfth language resources and evaluation conference. European Language Resources Association, Marseille, France, pp 4218–4222. https://aclanthology.org/2020.lrec-1.520/
2. Warden P (2018) Speech commands: a dataset for limited-vocabulary speech recognition. arXiv: 1804.03209
3. Panayotov V, Chen G, Povey D, Khudanpur S (2015) Librispeech: an ASR corpus based on public domain audio books. In: Proceeding IEEE International Conference on Acoustics, Speech and Signal Processing (ICASSP), South Brisbane, QLD, Australia, pp 5206–5210
4. Rousseau A, Deléglise P, Estéve Y (2012) TED-LIUM: an automatic speech recognition dedicated corpus. In: Proceeding 8th International Conference on Language Resources and Evaluation (LREC), Istanbul, Turkey, pp 125–129
5. Godfrey J, Holliman EC, McDaniel J (1992) SWITCHBOARD: telephone speech corpus for research and development. In: Proceeding IEEE international conference on acoustics, speech, and signal processing, pp 517–520
6. Paul DB, Baker J (1992) The design for the wall street journal-based CSR corpus. In: Proceeding speech and natural language: a workshop, Harriman, New York, pp 23–26
7. Garofolo JS, Lamel LF, Fisher WM, Fiscus JG, Pallett ADS (1993) DARPA TIMIT acoustic-phonetic continuous speech corpus CD-ROM, NIST Speech Disc 1–1.1. NASA STI/Recon Technical Report
8. Grierson GA (1961) Linguistic survey of India. vol 9, Indo-aryan Family, Central Group Part-3

9. Jha GN (2010) The TDIL program and the Indian Language Corpora Initiative (ILCI). In: Proceeding 7th International Conference on Language Resources and Evaluation (LREC), Valletta, Malta
10. Choudhary N, Ramamoorthy L (2019) LDC-IL raw speech corpora: an overview. In: Linguistic resources for AI/NLP in Indian languages central institute of Indian languages, Mysuru, pp 1–10
11. Ramamoorthy L, Choudhary N, Dhillon P, Kaur S (2019) Punjabi raw speech corpus. Central Institute of Indian Languages, Mysore
12. Bhogale K, Raman A, Javed T, Doddapaneni S, Kunchukuttan A, Kumar P, Khapra M M (2023) Effectiveness of mining audio and text pairs from public data for improving ASR systems for low-resource languages. In: Proceeding IEEE International Conference on Acoustics, Speech and Signal Processing (ICASSP), Rhodes Island, Greece, pp 1–5
13. Dhanjal S, Bhatia SS (2014) PUNJARPAbet: a new phonetic alphabet for speech processing in the Punjabi language. J Circ Syst Comput 23(05):1450070
14. Kadyan V, Mantri A, Aggarwal RK (2018) Refinement of HMM model parameters for Punjabi automatic speech recognition (PASR) system. IETE J Res 64(5):673–688
15. Zaidi A (2014) Exiled in its own land: diasporification of Punjabi in Punjab. South Asian Diaspora 6(2):209–224
16. Passricha V, Singhal S (2019) Hindi and Punjabi Continuous Speech Recognition Using CNSVM. Int J Adv Pervasive Ubiquit Comput (IJAPUC) 11(4):1–15
17. The World Factbook—Central Intelligence Agency (2018) Retrieved from https://www.cia.gov/library/publications/the-world-factbook/fields/402.html
18. Singh S, Wang R, Hou F, Ma Z (2022) Enhancing end-to-end automatic speech recognition for low-resource punjabi language using synthesized datasets. Available at SSRN 4181844
19. Lata S, Arora S (2012) Exploratory analysis of Punjabi tones in relation to orthographic characters: a case study. In: Proceeding workshop on Indian language and data: resources and evaluation workshop programme, p 76
20. Lata S, Prashant V, Simerjeet K (2018) Acoustic characteristics of Schwa Vowel in Punjabi. In: Proceeding 6th workshop on spoken language technologies for under-resourced languages (SLTU 2018), pp 85–89
21. http://www.emille.lancs.ac.uk
22. Varga A, Steeneken HJM (1993) Assessment for automatic speech recognition: II. NOISEX-92: a database and an experiment to study the effect of additive noise on speech recognition systems. Speech Commun 12(3):247–325

Leaf Disease Detection using Deep Learning: A Survey

Pradeep Nazareth, S. U. Deepika, Rakshitha, M. N. Varshini, and Vidyalaxmi D. Navalur

Abstract Demand for global food productivity increased from past many years. At the same time, agriculture is facing number of challenges from pests and diseases. Thus, it is necessary to predict and detect various diseases in agriculture. The efficiency and accuracy of predicting and detecting agricultural diseases will play a vital role in improving productivity. Technology play a major role in predicting and detecting agricultural diseases. Deep learning (DL) technology is one of the potential technology, that can be used in the field of agriculture to predict and detect various diseases. This paper focuses on reviewing the use of various DL techniques to predict and detect the crop leaf diseases. Further, discusses the on current trends, challenges, and future research directions in using deep learning for crop leaf disease identification and broader plant disease and pest management strategies.

Keywords Deep learning · Leaf disease · CNN · ANN · Decision trees

1 Introduction

Identifying plant diseases is crucial for maintaining crop yield and quality. It involves studying visually observable patterns in plants. Monitoring plant health and detecting diseases are essential for sustainable agriculture [1]. However, it is exceedingly difficult to monitor plant diseases on a regular basis. It demands a significant amount of work, requiring significant expertise and processing time [2]. Plant disease outbreaks significantly harm agricultural production, causing increased food anxiety when diseases are not promptly detected. Farmers rely on subjective and inefficient methods based on experience to identify crop diseases and pests [3].

P. Nazareth (✉) · S. U. Deepika · Rakshitha · M. N. Varshini · V. D. Navalur
Department of Artificial Intelligence and Machine Learning, Alva's Institute of Engineering and Technology, Moodbidri, India
e-mail: pradeepn@aiet.org.in

S. Kumar et al. (eds.), *Artificial Intelligence for Next Generation Computing, Volume 1*,
Lecture Notes in Networks and Systems 1549,
https://doi.org/10.1007/978-981-95-0701-6_11

1.1 General Steps for Detecting Crop Disease

Detecting crop disease using deep learning and machine learning entails multiple phases. Initially, crop input photographs are obtained with drones or cameras. These photos are preprocessed to improve quality, reduce noise, and segment regions of interest, specifically leaves or stems. The important features, such as color, texture, and shape, are then extracted. The processed photos are then saved in a database, creating a full dataset. The dataset is divided into training and testing subsets before being used to construct models. The training images are used to train a deep learning model, often a Convolutional Neural Network (CNN), to determine if an image is normal or abnormal. The model detects and classifies defective regions. The resulting model, once evaluated and tested, accurately distinguishes between healthy and diseased crops, allowing for precise and timely interventions. Figure shows the preprocessing step for the input image [4].

1.2 Role of Machine Learning and Deep Learning

ML is designed to parse data and learn from it. Based on the requirement, they applied for the decision. Several methods were created to handle diverse tasks such as classification, clustering, association rule mining, outlier identification, and so on. Deep learning is an extension of machine learning that addresses many types of datasets in a suitable manner. CNN is widely used in deep learning to perform image identification tasks. The structure's likeness aligns with the human recognition process from pupils to information that brain cells can understand. The amount of layers in the architecture impacts how different CNN models function [5].

Plant diseases can be categorized via machine learning based on a variety of factors. Preprocessing, such as picture enhancement, color modification, and segmentation, is required prior to effective feature extraction. Following feature extraction, a variety of classifiers may be applied. Some popular classifiers include K-nearest neighbor (KNN), Support Vector Machine (SVM), Decision Tree, Random Forest (RF), Naïve Bayes (NB), Logistic Regression (LR), Artificial Neural Networks (ANNs), and Convolution Neural Networks (CNNs) [6].

2 Challenges of Leaf Disease Detection

1. Noise in image processing can significantly reduce efficiency and effectiveness. This form of interruption is caused by technological devices and illumination. Noise in leaf images, including Gaussian and pulse noise, can hinder prediction accuracy.

2. Defocused photographs captured by electronic gadgets are a key hurdle in this investigation. Defocused images get hazy, making them difficult to view properly [7].
3. Typically, researchers use datasets available in many repositories. Few researchers collect photos of infected leaves from agricultural fields for experiments, but the method is a hurdle.
4. Researchers face challenges in dealing with various weather conditions, picture capture directions, distance between devices and infected leaves, and occlusion-based images, among other factors.
5. Choosing lots of image for a research study is another challenging task. While huge images can provide significant information, blurring can be a challenge in detecting leaf disease [8].

2.1 Performance Matrix

In crop disease detection, metrics like as accuracy, recall, precision, and F1-score are critical for assessing the usefulness of machine learning models.

1. Accuracy: Accuracy is the percentage of correctly categorized occurrences (both healthy and ill) out of all instances and the accuracy is given by Eq. 1.

$$\text{Accuracy} = \frac{\text{True Positive} + \text{True Negative}}{\text{Total Predictions}} = \frac{\text{TP} + \text{TN}}{\text{TP} + \text{TN} + \text{FP} + \text{FN}} \tag{1}$$

   Why It is Used: Accuracy is a generic measure of how well the model distinguishes between healthy and unhealthy crops. However, in imbalanced datasets (with fewer damaged crops), high accuracy can be misleading because the model is biased toward predicting the majority class [9].
2. Recall (sensitivity): What It Measures: Recall is the fraction of actual positive cases (diseased crops) that were accurately detected by the model is given by Eq. 2.

$$\text{Recall} = \frac{\text{True Positive}}{\text{True Positive} + \text{False Negative}} \tag{2}$$

   Why It is Used: High recall is crucial in disease detection since it shows how well the model recognizes damaged crops. A high recall model reduces the danger of missing damaged crops, which is critical for early detection and disease prevention.
3. Precision: What It Measures: Precision measures the proportion of expected positive instances (diseased crops) that turn out to be positive given by Eq. 3 [10]

$$\text{Precision} = \frac{\text{True Positive}}{\text{True Positive} + \text{False Positive}} \tag{3}$$

Why It is Used: Precision is essential for ensuring that the model's predictions are reliable. High precision indicates that when the model forecasts a crop is unhealthy, it is more likely to be correct. This contributes to fewer false positives, which can result in unneeded treatments.

4. F1-Score: What It Measures: This statistic balances false positives and false negatives by calculating the harmonic mean of precision and recall given by Eq. 4.

$$\text{F1 Score} = 2 \times \frac{\text{Precision} \times \text{Recall}}{\text{Precision} + \text{Recall}} \tag{4}$$

Why It is Used: The F1-score is very important when working with imbalanced datasets, as neither precision nor recall can provide a whole picture. It integrates all measurements into a single measure, making it easy to grasp the trade-offs between detecting all infected crops and reducing false alarms [11].

3 Leaves Disease Detection Using Deep Learning

To address these challenges, research on using image processing techniques for plant disease recognition has gained significant attention. This technology aims to improve disease identification, reducing inaccuracies and enhancing agricultural productivity.

Dubey et al. introduced a K-means clustering algorithm for leaf infection segmentation, followed by multi-class SVM classification. A probabilistic neural network was employed by Yun et al. to extract statistical and meteorological information from cucumber plants that were afflicted with blights, anthracnose, and downy mildew. Traditional disease recognition methods, such as those by Liu et al. using K-means clustering and SVM, remain prevalent despite their time-consuming nature. Recent advancements in machine learning and deep learning have significantly enhanced plant disease recognition, enabling feature extraction and more accurate classification. These innovations leverage datasets, GPU technology, and sophisticated software to transition from traditional to deep learning approaches [12].

Mekonnen et al. conducted a study on the application of various machine learning techniques in agricultural sensor data analytics. They found that clustering is widely used, but selecting algorithms often involves trial and error rather than a standard approach [13]. As shown in Fig. 1 the study reviewed several ML methods in agriculture including regression, Decision Trees (DT), ensemble learning, Bayesian models, Support Vector Machines (SVM), and Artificial Neural Networks (ANNs), highlighting their diverse applications and challenges in optimizing algorithm selection for agricultural data analysis [14].

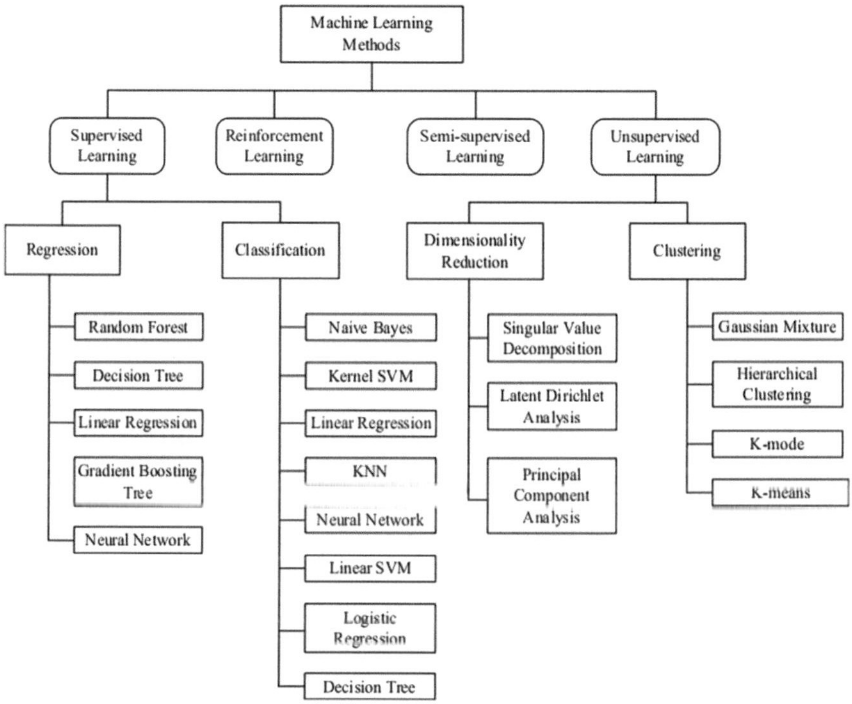

Fig. 1 Machine learning methods

3.1 Decision Trees (DT)

The Decision Tree is a supervised learning tool for classification and regression applications. It expresses decisions and their potential repercussions using a tree-like structure, with each internal node representing a test on an attribute, each branch representing the result of the test, and each leaf node representing a class label or conclusion. Decision Trees, unlike neural networks or SVMs, are straightforward, simple to read, and need little data preprocessing. In leaf disease identification, Decision Trees are preferred due to their simplicity and efficacy in dealing with complicated datasets with non-linear correlations, making them suited for effectively finding patterns and classifying diverse leaf states [15].

Advantages: Decision Trees have several benefits in leaf disease prediction, including their simplicity, interpretability, and capability to deal with data that is numerical as well as categorical. They require little data preprocessing and can handle complicated data sets with non-linear correlations successfully.

Disadvantages: Decision Trees do have drawbacks, such as the potential to overfit, particularly with noisy data. They can be sensitive to tiny changes in data, resulting in diverse tree architectures. Furthermore, they may struggle to capture interactions between characteristics until substantially developed, complicating the model and reducing interpretability. Despite these problems, their simplicity makes them useful in agricultural applications [16].

3.2 Support Vector Machine (SVM)

Support Vector Machine (SVM) is a supervised machine learning approach used for classification and regression. It works by identifying the hyperplane that best separates the data into multiple classes, maximizing the margin between the nearest points of each class, also known as support vectors. Unlike other techniques such as Decision Trees or neural networks, SVM is successful in high-dimensional areas and can handle non-linear data using kernel functions. SVM is used in leaf disease detection because of its robustness in dealing with complicated patterns and capacity to deliver excellent classification even with limited training data, which is critical for detecting subtle disease symptoms [17].

Advantages: Support Vector Machines (SVMs) have various advantages in leaf disease diagnosis. They excel at managing high-dimensional data and can accurately classify complicated patterns, making them ideal for detecting subtle illness indications based on leaf attributes. SVMs are highly resistant to overfitting, which is very useful when dealing with small datasets typical in agricultural situations.

Disadvantages: SVMs can be computationally costly, particularly for huge datasets, and the kernel function used has a considerable impact on performance. Furthermore, SVMs can have difficulty with datasets where classes coincide considerably or when the amount of features exceeds the number of samples, necessitating careful parameter adjustment and preprocessing to produce the best results [18].

3.3 Artificial Neural Networks (ANNs)

Artificial Neural Networks (ANNs) are a type of machine learning model that draws inspiration from the neuronal structure of the human brain. They are made up of interrelated nodes arranged into layers, each of which processes and transforms data to generate an output. ANNs excel at recognizing patterns and non-linear relationships, setting them apart from traditional techniques such as statistical models or Decision Trees, which can struggle with complicated, non-linear data. In leaf disease detection, ANNs provide benefits by autonomously learning detailed patterns from picture data, allowing for precise disease categorization based on leaf attributes. This capability drives their progress, providing improved effectiveness and accuracy in crop disease management via automated and early detection approaches [19].

Advantages: Artificial Neural Networks (ANNs) have multiple benefits in leaf disease detection, most notably their ability to learn complicated patterns from vast datasets, resulting in reliable disease categorization. They can handle relationships that are not linear and adapt to a variety of input kinds, including picture data, making them adaptable.

Disadvantages: ANNs take significant computational resources to train and may experience overfitting if the dataset is not balanced or representative. Because ANNs are sometimes regarded as "black-box" models, the interpretability of their results might be difficult. Despite these shortcomings, their efficacy in automated detection and classification tasks demonstrates their potential for increasing precision agriculture methods [20].

3.4 Convolutional Neural Network (CNNs)

Convolutional Neural Networks (CNNs) are a type of deep learning model that processes structured grid-like data, such as photographs. CNNs, unlike typical neural networks, use convolutional layers to apply filters to input data, allowing them to detect spatial hierarchy and learn hierarchical representations of information. This technique is especially useful for image processing applications since it can automatically extract essential features from raw pixel data, eliminating the need for the extraction of features manually. CNNs are motivated in leaf disease identification by their capacity to recognize subtle textures and patterns within leaf images, allowing for precise and quick disease categorization based on visual symptoms [21].

Advantages: CNNs have significant advantages in leaf disease detection since they can automatically learn hierarchical characteristics from photos, allowing for reliable categorization based on visual patterns. They excel at handling enormous datasets and generalizing to new, previously viewed images, making them suitable for real-world applications.

Disadvantages: CNNs take significant computer resources to train, and their success is strongly dependent on the quality and volume of training data. Furthermore, because CNNs are complicated black-box models, it might be difficult to explain their decisions. Despite these issues, CNNs continue to be an effective technique for detecting leaf disease and considerably improving automated plant health monitoring [22].

3.5 Comparative Analysis of ANN, CNN, DT, and SVM

The Artificial Neural Network, or ANN, is a simple neural network model that performs well with smaller datasets and specified feature extraction. However, it is not as effective with high-dimensional data sets such as photographs. Although it is computationally expensive and requires strong hardware, like as GPUs, Convolutional Neural Networks (CNNs) are perfect for leaf disease detection because they excel at picture classification tasks and automatically capture complicated patterns. Decision Trees work quickly on smaller datasets and are easy to understand, but they struggle to perform well on high-dimensional picture data due to overfitting. Support Vector Machines, or SVMs, work well in high-dimensional spaces and with smaller datasets, but they are computationally demanding on larger datasets and may necessitate extensive feature engineering.

4 Analysis of existing technique

Detecting leaf diseases is critical for crop health and production optimization, and several strategies have been developed to meet this need. Traditional visual inspection relies on human skill to identify disease symptoms, but it is restricted by subjective variability and unsuitable for large-scale use. Microscopy provides high-resolution insights into cellular changes, but it is labor-intensive and limited to lab settings. Chemical analysis and molecular techniques, such as PCR, offer precise and early identification of pathogens, but need costly equipment and expertise, which makes them less practical for field application. Combining these strategies can improve illness identification by harnessing their strengths while addressing individual limitations. Advancements in technology, particularly AI and machine learning, can lead to easier and more effective disease detection methods.

Table 1 gives the information about the analysis of leaf diseases which includes Common Name, Scientific Name, Common Disease, Machine Learning Techniques, and accuracy of some plant leaves.

Table 1 Analysis of leaf disease

Sl. No.	Common name	Scientific name	Common diseases	Machine learning techniques	Accuracy
1.	Grapevine	*Vitis vinifera*	Powdery mildew	CNN	85–90
2.	Tea plant	*Camellia sinensis*	Blister blight	Random forest	85–90
3.	Apple tree	*Malus domestica*	Apple scab	SVM	75–85
4.	Rice plant	*Oryza sativa*	Rice blast	CNN	85–95
5.	Neem	*Azadirachta indica*	Leaf spot	SVM	85–90
6.	Mango	*Mangifera indica*	Anthracnose	Random forest	85–90
7.	Tulasi	*Ocimum tenuiflorun*	Downy mildew	SVM	78–82
8.	Guava	*Psidium guajava*	Bacterial spot	CNN	85–92

4.1 Grape

The principal pathogens causing harm to grape harvests are listed below.

(i) Black Measles: Symptoms include small, round, dark spots separated by a brown-purple ring.
(ii) Black rot: The principal sign of black rot is a local infection produced by bacteria entering the leaves via hydathodes. The diseased tissue begins wilted and pale green, then turns yellow, finally brown, and dies.
(iii) Blight: Symptoms appear as dark brown-purplish patches on diseased berries, rachis, and bunch stalks just below the shoot attachment point [23].

The work used a publicly available grape leaf disease dataset, which may be downloaded from Kaggle. The grapes dataset includes 4062 256 $\times$ 256 pixel photos grouped into four categories: black rot (1180 images), Esca-black-measles (1383 images), healthy (423 images), and leaf-blight-isariopsis-leafspot (1076 images). During preprocessing, we reduced noise using three methods: background removal, intensity averaging, and bilateral filtering. CNNs are a key component of deep learning and are commonly employed in computer vision. CNN detects visual patterns across picture pixels with little preprocessing. As a result, they've been used widely in image-based detection research. This section explains the CNN models used in the study. The RF is a classifier that applies numerous Decision Trees to distinct subsets of the input dataset and averages them to improve the dataset's predicted accuracy. It is based on ensemble learning, which involves combining several classifiers to solve a complex problem and improve the model's performance. RF prevents Decision Tree overfitting and is more robust to noise and outliers. Furthermore, it has a short processing time and remains effective even when dealing with large amounts of data. The proposed ensemble model consists of CNN-based two-way models as feature extractors and RFs as classifiers [24].

4.2 Tea

The existence of tea leaf diseases is one of the most significant elements that might have a detrimental impact on the end product's quality and production. Machine vision and artificial intelligence approaches are two methods that can help achieve this goal. Tea leaf photos can be automatically processed to detect illnesses. Preprocessing is an important step in building a precise approach for diagnosing diseases in tea leaves. The suggested method begins with preprocessing images of tea leaves. The preprocessing stage aims to remove unnecessary information from images to prevent errors during detection [25].

CNN is a deep learning technique that applies biases and weights to items in an image or dataset to differentiate them from each other. The input image is sent to the feature retrieval network, which then delivers signals from the obtained features to the

NN for recognition. The network generates output based on the image's properties. In a neural network, the feature extraction process comprises convolution and pooling layers. The term "convolutionlayer" refers to the technique of finding features in an image using digital filters. The pooling layer connects nearby pixels to form a single pixel, resulting in reduced image size [26].

4.3 Apple

Apple trees are particularly sensitive to a variety of diseases, which can have a substantial impact on fruit quality and productivity.Detecting and accurately identifying infections in apple orchards is vital for preserving their health. Deep learning offers great potential for automating and improving the detection of apple leaf disease [27].

Common illnesses affecting apple yields sometimes cause size variations and geometric abnormalities. Traditional CNNs have poor performance in modeling apple disease geometric alterations due to their constant receptive fields in each convolution layer. The deformable convolution network (DCN) can extract dense spatial transformations by augmenting sampling locations with infinite offsets, making it more suitable for this task. In the apple leaf disease identification task, Brown Spot and Rust are invariably larger than Alternaria and Mosaic, whereas Mosaic and Aphis typically cover a wider area. The DCN offers a more flexible convolution model than standard models, however its infinite offset may not detect diseases that are too vast or small. To address this issue, we offer a new module termed dual-constrained deformable convolution.

DCN adds various offset limit ranges to the two restricted deformable convolution branches. The convolution kernel's sampling position might vary according on disease size and geometry, resulting in distinct receptive fields between the two branches. Different disease traits can be retrieved dependent on the offset range of convolution sampling positions, even if the convolution kernels in both branches have the same weight. This improves the network's capacity to distinguish diseases of different sizes and deformable shapes without needing additional weight factors [28].

4.4 Rice

Previous research on rice leaf disease classification relied heavily on image processing and machine learning. Images of sick leaves are preprocessed to extract relevant information. These characteristics include the color, texture, and shape of lesions, as well as any other anomalies. After extracting features, machine learning algorithms like SVM, Decision Trees, K-nearest neighbors, and Random Forests are used to train a classification model. Once trained, the model can categorize new diseased leaf images into various disease groups. Deep learning, a subtype of machine learning, uses Artificial Neural Networks to learn from data. These networks consist of interconnected nodes that perform simple mathematical operations.

Rice leaf disease classification techniques include CNN-based approaches, transfer learning with pre-trained CNN models, ensemble learning, and hybrid methods [29].

The Dynamic Mode Decomposition (DMD)-based preprocessing technique can extract unhealthy regions from rice leaf images, which are subsequently fed into a transfer learning model for classification in the first experiment. Figure 1 depicts the architecture of the transfer learning approach with DMD preprocessing. The effectiveness of DMD preprocessing was evaluated based on the classification model's performance. The extracted deep features from Deep CNNs (DCNN) trained on prepossessed leaf photos were supplied into Machine Learning models, such as Random Forests, Naive Bayes, and SVM, for classification [30].

5 Conclusion

This study aims to review research using machine learning and deep learning techniques. The complexity and variability of real-world scenarios pose significant challenges in creating representative datasets with sufficient labeled samples, a crucial obstacle for advancing practical machine learning methods. Deep learning models offer advantages like speed, precision, and universal applicability but face obstacles such as accuracy in complex environmental conditions. Modern approaches with faster Convolutional Neural Networks show superior robustness and parameter efficiency compared to traditional methods. Integrating hyperspectral/multispectral imaging enhances disease detection capabilities, promoting sustainable agriculture by reducing pesticide use. In order to improve agricultural productivity and sustainability, future research should standardize deep learning models across a variety of diseases, improve dataset robustness, and investigate real-time disease detection applications. We also plan to investigate additional algorithms across a range of datasets in order to evaluate consistency in the outcomes. To improve algorithmic accuracy, we will also look into techniques for data augmentation and image preprocessing.

References

1. Muruganantham P, Wibowo S, Grandhi S, Samrat NH, Islam N (2022) A systematic literature review on crop yield prediction with deep learning and remote sensing. Remote Sens 14(9):1990
2. Domingues T, Brandão T, Ferreira JC (2022) Machine learning for detection and prediction of crop diseases and pests: a comprehensive survey. Agriculture 12(9):1350
3. Balafas V, Karantoumanis E, Louta M, Ploskas N (2023) Machine learning and deep learning for plant disease classification and detection. IEEE Access 14:1134
4. Jackulin C, Murugavalli S (2022) A comprehensive review on detection of plant disease using machine learning and deep learning approaches. Measur Sens 24:100441
5. Sujatha R, Chatterjee JM, Jhanjhi N, Brohi SN (2021) Performance of deep learning versus machine learning in plant leaf disease detection. Microprocess Microsyst 80:103615
6. Ahmed I, Yadav PK (2023) A systematic analysis of machine learning and deep learning based approaches for identifying and diagnosing plant diseases. Sustain Oper Comput 4:96–104
7. Sarkar C, Gupta D, Gupta U, Hazarika BB (2023) Leaf disease detection using machine learning and deep learning: review and challenges. Appl Soft Comput 145:110534
8. Sunil C, Jaidhar C, Patil N (2023) Systematic study on deep learning-based plant disease detection or classification. Artif Intell Rev 56(12):14955–15052
9. Abd-Algani YM, Caro OJM, Bravo LMR, Kaur C, Al-Ansari MS, Bala BK (2023) Leaf disease identification and classification using optimized deep learning. Measur Sens 25:100643
10. Sajitha P, Andrushia AD, Anand N, Naser MZ (2024) A review on machine learning and deep learning image-based plant disease classification for industrial farming systems. J Ind Inform Integr 34:100572
11. Praba V, Krishnaveni K (2024) An enhanced random forest classifier to detect crop disease with texture and shape features of corn images (erfcts). Educ Admin Theory Pract 30(5):3720–3727
12. Ali AH, Youssef A, Abdelal M, Raja MA (2024) An ensemble of deep learning architectures for accurate plant disease classification. Ecol Inform 81:102618
13. Hassan RJ, Abdulazeez AM (2021) Plant leaf disease detection by using different classification techniques: comparative. Asian J Res Comput Sci 8(4):1–11
14. Omaye JD, Ogbuju E, Ataguba G, Jaiyeoba O, Aneke J, Oladipo F (2024) Cross-comparative review of machine learning for plant disease detection: apple, cassava, cotton and potato plants. Artif Intell Agricult
15. Shafik W, Tufail A, Namoun A, De Silva LC, Apong RAAHM (2023) A systematic literature review on plant disease detection: motivations, classification techniques, datasets, challenges, and future trends. IEEE Access 11:59174–59203
16. Hessane A, El Youssefi A, Farhaoui Y, Aghoutane B, Amounas F (2023) A machine learning based framework for a stage-wise classification of date palm white scale disease. Big Data Min Anal 6(3):263–272
17. Sahu SK, Pandey M (2023) An optimal hybrid multiclass svm for plant leaf disease detection using spatial fuzzy c-means model. Exp Syst Appl 214:118989
18. Prabu M, Chelliah BJ (2023) An intelligent approach using boosted support vector machine based arithmetic optimization algorithm for accurate detection of plant leaf disease. Pattern Anal Appl 26(1):367–379
19. Bangari S, Rachana P, Gupta N, Sudi PS, Baniya KK (2022) A survey on disease detection of a potato leaf using CNN. In: Proceedings of the 2022 second international conference on artificial intelligence and smart energy (ICAIS). IEEE, pp 144–149
20. Alshammari HH, Taloba AI, Shahin OR (2023) Identification of olive leaf disease through optimized deep learning approach. Alexand Eng J 72:213–224
21. Fenu G, Malloci FM (2023) Classification of pear leaf diseases based on ensemble convolutional neural networks. AgriEngineering 5(1):141–152
22. Pavithra A, Kalpana G, Vigneswaran T (2023) Deep learning-based automated disease detection and classification model for precision agriculture. Soft Comput 14:1–12

23. Rao US, Swathi R, Sanjana V, Arpitha L, Chandrasekhar K, Naik PK et al (2021) Deep learning precision farming: grapes and mango leaf disease detection by transfer learning. Global Trans Proceed 2(2):535–544
24. Ishengoma FS, Lyimo NN (2024) Ensemble model for grape leaf disease detection using CNN feature extractors and random forest classifier. Heliyon 14:e33377
25. Heng Q, Yu S, Zhang Y (2024) A new AI-based approach for automatic identification of tea leaf disease using deep neural network based on hybrid pooling. Heliyon 10(5):1224
26. Datta S, Gupta N (2023) A novel approach for the detection of tea leaf disease using deep neural network. Proced Comput Sci 218:2273–2286
27. Kannan E, Begum A, Hemalatha D et al (2024) Deep learning techniques advancements in apple leaf disease detection. Proced Comput Sci 235:713–722
28. Liu B, Huang X, Sun L, Wei X, Ji Z, Zhang H (2024) Mcdcnet: multi-scale constrained deformable convolution network for apple leaf disease detection. Comput Electr Agricult 222:109028
29. Simhadri CG, Kondaveeti HK, Vatsavayi VK, Mitra A, Ananthachari P (2024) Deep learning for rice leaf disease detection: a systematic literature review on emerging trends, methodologies and techniques. Inform Process Agricult
30. Sudhesh K, Sowmya V, Kurian S, Sikha O (2023) Ai based rice leaf disease identification enhanced by dynamic mode decomposition. Eng Appl Artif Intell 120:105836

Machine Learning Integration in Bamboo Treatment and Seasoning for RCC Compound Wall Panels

Shilpa Kewate and Pankaj Kewate

Abstract This research explores the use of machine learning (ML) to optimize bamboo treatment and seasoning for reinforced cement concrete (RCC) compound wall panels. The study focuses on improving key processes such as drying, splitting, and bending of bamboo, along with its treatment for waterproofing, anti-termite protection, and enhanced bonding with concrete. By analyzing environmental data and monitoring in real-time, ML models help to fine-tune these processes, ensuring consistent quality and durability of bamboo reinforcement. Predictive modeling enables the adjustment of treatment conditions and parameters, reducing waste and improving energy efficiency. Machine learning can enhance the effectiveness of bamboo processing, leading to stronger and more reliable RCC panels. This approach not only supports sustainable construction practices but also improves the performance and longevity of bamboo-reinforced concrete structures.

Keywords Machine learning · Bamboo treatment · Compound wall panels

1 Introduction

The construction industry continually seeks innovative methods to enhance the performance and sustainability of building materials. Bamboo, known for its rapid growth and strength, is emerging as a viable reinforcement material in reinforced cement concrete (RCC) compound wall panels. However, the effective utilization of bamboo in construction requires precise treatment and seasoning to ensure its durability and compatibility with concrete.

S. Kewate (✉)
Pillai HOC College of Engineering and Technology, Rasayani, India
e-mail: shilpakewate@mes.ac.in

P. Kewate
Gyproc & Insulation Business, Saint-Gobain Industries India Pvt. Ltd., Mumbai, India
e-mail: pankaj.kewate@saint-gobain.com

S. Kumar et al. (eds.), *Artificial Intelligence for Next Generation Computing, Volume 1*,
Lecture Notes in Networks and Systems 1549,
https://doi.org/10.1007/978-981-95-0701-6_12

Traditional methods for treating and seasoning bamboo involve manual processes that may lack consistency and efficiency. Advances in machine learning (ML) offer a transformative approach to optimize these processes. By using ML, we can analyze extensive data on environmental conditions, treatment methods, and bamboo characteristics to improve key stages such as drying, splitting, bending, and chemical treatments.

This research focuses on integrating ML techniques to enhance the treatment and seasoning of bamboo for RCC compound wall panels. We explore how ML can refine drying and treatment conditions, predict optimal processing parameters, and ensure uniform quality in bamboo reinforcement. The application of ML aims to reduce waste, improve energy efficiency, and enhance the bonding between bamboo and concrete, leading to stronger and more reliable construction materials. The paper outlines the methodologies employed, the benefits of ML integration, and the potential impact on sustainable construction practices. By optimizing bamboo processing, this research contributes to advancing eco-friendly building solutions and improving the performance of bamboo-reinforced concrete structures.

2 Research Problem Overview

The construction industry is increasingly seeking sustainable and eco-friendly alternatives to conventional building materials. Bamboo, known for its high strength-to-weight ratio and rapid growth, presents a promising option as a reinforcement material in RCC compound wall panels. However, the traditional processes of bamboo treatment and seasoning, which include drying, splitting, and bending, are labor-intensive, time-consuming, and often inconsistent in quality. This variability in bamboo properties can lead to performance issues in the final RCC panels, affecting their strength, durability, and overall reliability.

The integration of ML into the bamboo treatment and seasoning processes offers a potential solution by optimizing these processes for better consistency, efficiency, and performance. However, the challenge lies in developing an effective ML-based approach that can account for the natural variability in bamboo, predict optimal treatment conditions, and automate the seasoning process while ensuring the treated bamboo meets the required engineering standards for use in RCC panels.

This research aims to address these challenges by developing and validating a ML model that can enhance the bamboo treatment and seasoning process, thereby improving the quality and performance of bamboo-reinforced RCC compound wall panels.

3 Technology Trends from Literature Survey

3.1 Machine Learning in Material Processing

Recent literature highlights the growing use of ML to optimize material processing across various industries. In construction and material science, ML algorithms have been applied to predict and enhance the properties of materials such as concrete and steel [1]. Advances in ML techniques, including deep learning and neural networks, are now being explored for optimizing the treatment and processing of natural materials like bamboo. These techniques offer the potential to improve consistency, reduce waste, and enhance material performance [2].

3.2 Bamboo as a Sustainable Construction Material

The use of bamboo in construction is gaining momentum due to its sustainability, rapid growth, and strength [3, 4]. Studies have shown that bamboo can effectively reinforce concrete, contributing to more eco-friendly building practices [5]. Recent research emphasizes the importance of proper treatment and seasoning of bamboo to maximize its effectiveness as a reinforcement material. Innovations in bamboo processing technology, including advanced drying and treatment methods, are being developed to improve the performance and durability of bamboo in construction applications [6]. Table 1 shows mechanical properties.

Bamboo is natural grass; it is having around 1200 species across the globe that can have the ability to act as reinforcement in concrete instead of steel, due to its various qualities [7]. Remarkable tensile strength, rapid growth rate, economy, less weight, and energy efficiency are the upsides of bamboo as compared to mild steel. For decades people are using bamboo as a building material in rural areas. The north-eastern states of India prefer to use bamboo reinforcement houses which highlight its great potential. In the case of bamboo, there is a very less amount of CO_2 emission

Table 1 Mechanical properties of steel, bamboo and concrete

Properties	Steel	Bamboo	Concrete
Density (kg/m^3)	7850	693	2400
Young's modulus (Pa)	2×10^{11}	1.73×10^{10}	3×10^{10}
Poisson's ratio	0.3	0.384	0.18
Bulk modulus (Pa)	1.6667×10^{11}	2.4856×10^{10}	1.5625×10^{10}
Shear modulus (Pa)	7.6923×10^{10}	6.25×10^{9}	1.2712×10^{10}
Tensile ultimate strength (Pa)	5.45×10^{8}	2.26×10^{8}	5×10^{6}
Compressive yield strength	2.5×10^{8} Pa	–	3×10^{7} Pa

Fig. 1 Brown color bamboo. *Source* https://images.app.goo.gl/eCv6TcoSUBJ2RY3X6

in the atmosphere, like that of cement and steel. Throughout the phase of growth of bamboo, around 1000 kg of carbon dioxide is consumed by bamboo plants [8].

3.3 Bamboo Treatment and Seasoning

The literature indicates significant advancements in the treatment and seasoning of bamboo to enhance its properties for construction use [9]. Techniques such as pressure treatment, chemical impregnation, and controlled drying are being optimized to address issues like moisture absorption and susceptibility to pests. Recent trends involve the use of automated systems and sensors to monitor and control these processes, ensuring uniformity and efficiency. ML models are increasingly being employed to analyze data from these processes and make real-time adjustments to improve outcomes. Figure 1 shows brown color bamboo and Fig. 2 shows application epoxy on bamboo.

3.4 Integration of Machine Learning with Construction Materials

Research papers reveal a trend toward integrating ML with various aspects of construction material processing [10]. Innovations include the use of ML for predicting material properties, optimizing manufacturing processes, and improving quality control. In the context of bamboo-reinforced concrete, ML is being explored to enhance the precision of treatment and seasoning processes, as well as to predict and optimize the bonding between bamboo and concrete.

Fig. 2 Applying epoxy coating on bamboo surface

3.5 Methodological Details

Neural networks (NNs) as the ML model would be used to optimize and predict the bamboo treatment and seasoning process for RCC compound wall panels. Neural networks are well-suited for this application due to their ability to model complex, nonlinear relationships between multiple input variables such as temperature, humidity, airflow, and bamboo properties like moisture content and density.

The model includes:

1. Input Layer: Features such as initial bamboo moisture content, drying temperature, airflow rate, and time.
2. Hidden Layers: These layers capture complex interactions between the input features, with nonlinear activation functions to help the model learn intricate patterns in the seasoning process.
3. Output Layer: Provides predictions such as the moisture level of bamboo at specific points during the seasoning or recommendations for optimal seasoning time.

The data used for training the neural network consists of historical bamboo seasoning data would be collected from both lab-controlled experiments and real-world production processes.

This dataset includes:

1. Environmental conditions: Temperature, humidity, and airflow rates recorded at regular intervals during the bamboo treatment process.
2. Bamboo properties: Moisture content, thickness, and density of different bamboo samples at various stages of seasoning.

3. Time series data: Moisture reduction over time, captured for each batch of bamboo, providing dynamic information to the model.

The data would be split into three subsets: 70% for training, 15% for validation, and 15% for testing. Success metrics include mean squared error (MSE) to quantify prediction accuracy and *R*-squared (R^2) to assess the model's explanatory power regarding moisture reduction. These metrics will be rigorously calculated and reported to provide a clear evaluation of the model's effectiveness, enhancing the research's credibility and demonstrating the advantages of the ML approach in bamboo treatment.

3.6 Comparative Analysis

The ML-driven approach using neural networks offers significant advantages over traditional bamboo treatment methods, such as air-drying and kiln-drying. Traditional methods are time-consuming, inconsistent, and energy-intensive, relying on fixed durations and manual monitoring, leading to inefficiencies and variability in moisture levels. In contrast, the ML approach provides real-time predictions and optimizes the seasoning process by dynamically adjusting temperature, humidity, and airflow based on bamboo characteristics and environmental conditions. This results in faster, more consistent, and energy-efficient treatment.

3.7 Scalability and Practical Application

The ML-driven approach using neural networks can be scaled up for industrial bamboo treatment by automating real-time monitoring and control across multiple drying units. As more data is collected, the model's accuracy improves, making it ideal for large-scale operations. However, challenges include initial infrastructure costs, integration with existing systems, and training personnel to manage the new technology. Additionally, handling environmental variations in real-world conditions may require further model refinement. Figure 3 shows ML integration process in bamboo treatment and seasoning.

4 Output Resulting from Invention

Authors have filed patent for this research work and patent application number is 202121049746. Final output of this research work would be 'Precast Prefab Bamboo Reinforced Cement Concrete Compound Wall Panels.' Figure 4 shows newly developed product. The compound wall panel will be designed based on the required span,

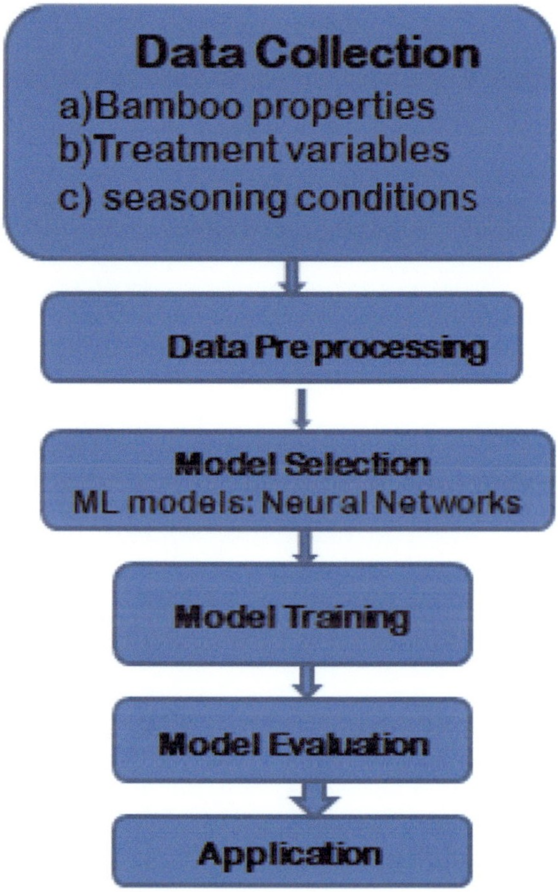

Fig. 3 ML integration process in bamboo treatment and seasoning

with fixed sizes for optimal performance. The concrete mix will be tailored using locally available materials, ingredients, and admixtures to ensure cost-effectiveness. Local bamboo will undergo a seasoning process, including drying, splitting, and bending, to prepare it for reinforcement. To enhance its durability, the bamboo will be coated with coal tar or epoxy and treated with an anti-termite solution for long-lasting protection.

The materials used, like cement, aggregates, and bamboo reinforcement, should meet the required standards and be tested accordingly. The mold will be crafted based on the design, and to meet aesthetic needs, the inside of the mold can feature an architectural pattern for a stylish finish. Table 2 shows mix design of M30 grade of concrete. Figure 5 shows SWOT analysis of Newly developed product Precast prefab Bamboo RCC compound wall.

Fig. 4 Newly developed product precast prefab bamboo-reinforced cement concrete compound wall

Table 2 Mix design of M30 grade

M30 grade	Materials	kg/m^3
1	Cement	360
2	Water	144
3	Fine aggregate	812
	Sand	751.10
	E-waste powder (7.5%)	60.90
4	Coarse aggregate	1284
	20 mm coarse aggregate	645
	10 mm coarse aggregate	639
5	Admixture	1.8
6	Water cement ratio	0.40

5 Bond Stress Analysis

The reinforcing pull-out test was used to determine the bonding between concrete and bamboo for three different types of products. There are three different product categories: Coir Rolled Bamboo Rebar, G.I. Rolled Bamboo Rebar, and Sand Rolled Bamboo Rebar. The test was performed in accordance with IS: 11309–1985. Three distinct cubes measuring 15 cm × 15 cm × 15 cm have been reinforced with materials that have various bonding characteristics, as illustrated in Fig. 6 and Table 3 shows bond shear stress of bamboo subjected to pull-out test.

Figure 7 shows bond shear stress of bamboo subjected to pull-out test. The results of the test indicate that G.I. rolled bamboo rebar and sand rolled bamboo rebar offer higher bonding shear stresses than coir rolled bamboo rebar; Specifically, the bonding shear stress of sand rolled rebar is 1.001 times higher than that of G.I. rolled rebar and 1.519 times higher than that of coir rolled rebar.

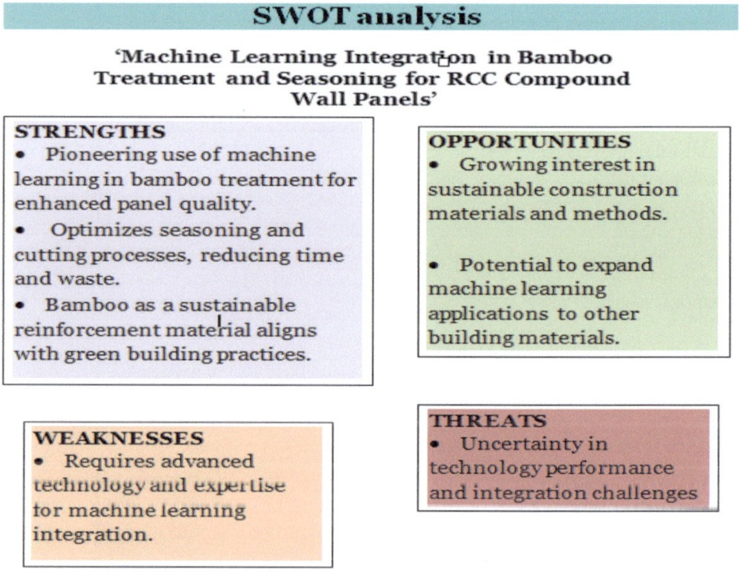

Fig. 5 SWOT analysis of newly developed product precast prefab bamboo-reinforced cement concrete compound wall

Fig. 6 Specimens for pull-out test

Table 3 Bond shear stress of bamboo subjected to pull-out test

Sr. No	Specimen ID mark	Diameter of bar (mm)	Rebar anchorage penetration depth (mm)	Applied pull-out load (kN)	Bonding Shear Stress, τb (N/mm^2)
1	Sand rolled bamboo rebar	16	75	11.37	9.475
2	Coir rolled bamboo rebar	16	75	7.47	6.225
3	G.I. wire rolled bamboo rebar	16	75	11.35	9.458

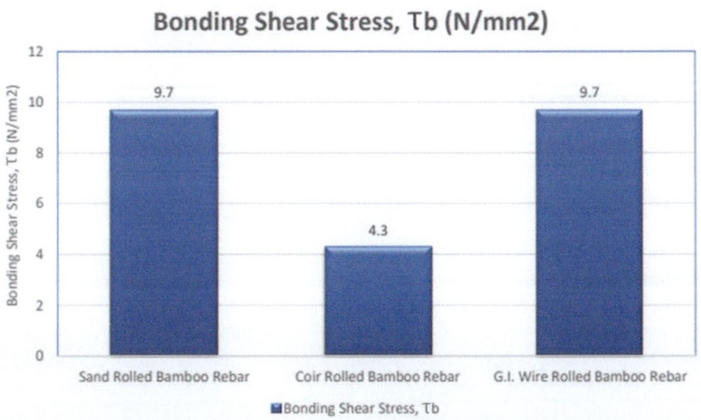

Fig. 7 Bond shear stress of bamboo subjected to pull-out test

6 Expected Impact

- Reduced Processing Time: ML could reduce bamboo seasoning and cutting time by up to 30–40%. For instance, if the traditional process takes 30 days, ML could cut it down to 18–21 days.
- ML optimization might lower waste by 20–30%. If the current waste is 10% of material, this could reduce waste to 7–8%.
- By optimizing the process, operational costs might decrease by 15–25%. For example, if the current cost is ₹1,00,000 per batch, the cost could drop to ₹75,000–85,000.
- Improved efficiency can reduce the need for excess materials, potentially saving 10–20% on raw material costs.
- Panel Strength: Bamboo-reinforced RCC panels may exhibit a 10–15% increase in strength and durability compared to traditional panels. For instance, if traditional panels have a compressive strength of 20 MPa, the improved panels could reach 22–23 MPa.

- ML can lead to a 20–30% improvement in consistency of panel quality, reducing defects and variability.
- Enhanced efficiency and reduced waste may lower the carbon footprint of the panel production by 10–20%. For instance, if the current carbon footprint is 1000 kg CO_2 per batch, the new process could reduce it to 800–900 kg CO_2.
- Using ML to optimize bamboo treatment could lead to a 15–20% increase in resource conservation by maximizing the use of bamboo and reducing the need for additional materials.

7 Conclusion

The integration of ML into the bamboo treatment and seasoning processes for RCC compound wall panels represents a significant advancement in sustainable construction. By optimizing bamboo processing, this research not only enhances the efficiency and consistency of panel production but also contributes to reducing waste and operational costs. The resulting bamboo-reinforced panels offer improved strength and durability, aligning with the growing demand for eco-friendly building materials. This innovative approach demonstrates the potential for ML to revolutionize traditional construction methods, paving the way for broader adoption of sustainable practices in the industry. The outcomes of this research could have far-reaching implications, driving both technological progress and environmental stewardship in construction.

References

1. Kaffayatullah F, Waqas A, Muhammad A (2022) A systematic review of the research development on the application of machine learning for concrete. Materials 15(13):4512. https://doi.org/10.3390/ma15134512
2. Yaser G (2023) Machine learning in concrete technology: a review of current researches, trends, and applications. Front Built Environ 9. https://doi.org/10.3389/fbuil.2023.1145591
3. Kewate S, Jamgade M, Sinha M (2022) Design of cost-effective bamboo reinforced manhole cover; A step toward sustainable development. Lecture Notes in Mechanical Engineering | Conference paper. https://doi.org/10.1007/978-981-19-0296-3_28
4. Kitti C, Supanit W, Pannipa C (2021) Bamboo as a sustainable building material—culm characteristics and properties. Sustainability 13(13):7376. https://doi.org/10.3390/su13137376
5. Shilpa K, Pankaj K Precast prefab Bamboo reinforced cement concrete compound wall panels. Indian Patent application number: 202121049746
6. Anu B, Supratic G (2023) Engineered bamboo and bamboo-reinforced concrete elements as sustainable building materials: a review. Constr Build Mater 394(29):132116. https://doi.org/10.1016/j.conbuildmat.2023.132116
7. Zhao W, Zhou J, Bu G (2012) Application technology of bamboo reinforced concrete in building. Appl Mech Mater 195–196:297–302
8. Abhishek D, Vishal P (2019) Bamboo bonding in concrete: a critical research. Int J Innov Technol Explor Eng (IJITEE) ISSN:2278–3075, 8(11S)

9. Bhopatrao S, Kewate S (2023) Manufacturing of sustainable manhole covers designed for pedestrian networks. AIP conference proceedings | Conference paper. https://doi.org/10.1063/5.0189456
10. Muhammad S, Muhammad J, MachineFahid A, Hisham A (2022) Learning modeling integrating experimental analysis for predicting the properties of sugarcane bagasse ash concrete. Constr Build Mater Part A, 314: 125634. https://doi.org/10.1016/j.conbuildmat.2021.125634

Biorthogonal Wavelet-Based Image Enhancement for Accurate Pneumonia Diagnosis from Lung CT Scans Using Deep Learning

S. Rohith Kumar, S. Vineth Ligi, R. Kumar, and Samiappan Dhanalakshmi

Abstract Identifying pneumonia in lung CT scans is complex as the hidden symptoms of the disease are not evident and barely distinguishable by the naked eye, leading to probable misdiagnosis or loss of cases. Further, it is rather challenging to distinguish between the diseases because of the complex structure of the lungs and the variety of their pathology. This study's originality is found in an image preprocessing approach that enhances the diagnosis of pneumonia from lung radiographs. Specifically, a wavelet-based image enhancement technique is used to improve the visibility of infection details in the images, the preprocessed images are fed into several deep learning networks, which classify the outputs into three categories: Pneumonia due to COVID-19, pneumonia due to other etiologies, and no findings. Furthermore, to measure the effectiveness of the suggested technique and the outcomes of this proposed preprocessing method, it experiments on various deep learning models such as DenseNet, ResNet50, and GoogLeNet. As a result, DenseNet achieved a better accuracy of 99.20%, which is higher than ResNet50 97.93%, and GoogLeNet 97.59%. The observations indicate that the proposed preprocessing leads to greater sensitivity and better results in diagnosing pneumonia, further stressing the applicability of this method to aid in early and accurate diagnosis potentially.

Keywords Pneumonia · Biorthogonal wavelet · DenseNet · ResNet50 · GoogLeNet · Medical imaging

1 Introduction

In the vast majority of instances, infection is caused by bacteria, viruses, and fungi, which are the root cause of pneumonia [1], which is an inflammatory disorder that affects the lungs. Lobar pneumonia mainly affects the alveoli [2] or commonly

S. Rohith Kumar · S. Vineth Ligi · R. Kumar · S. Dhanalakshmi (✉)
Department of Electronics and Communication Engineering, SRM Institute of Science and Technology, Kattankulathur, Chennai, Tamil Nadu, India
e-mail: dhanalas@srmist.edu.in

S. Kumar et al. (eds.), *Artificial Intelligence for Next Generation Computing, Volume 1*, Lecture Notes in Networks and Systems 1549,
https://doi.org/10.1007/978-981-95-0701-6_13

referred to as the little air sacs. The manifestations may range from very mild to potentially lethal [3] depending on the cause, the type of pathogen, the patient's age, and general state of health. Pneumonia is said to be a fatal disease given that the WHO [4] has indicated that about 800,000 children under the age of five die from it accounting for fifteen percent of all children under the age of five years. The possible causes of pneumonia include some categories one of which is coronavirus otherwise known as SAR-COV-2 virus. Consequently, differentiation in disease type characteristics, singularity of each type's pathogens, and the difficulties that both diagnosis and treatment entail make the outcomes of the patients important.

The diagnostic methods currently used for pneumonia [5] include physical examinations, blood tests, sputum tests, chest X-rays, and computed tomography (CT) scans. While physical and laboratory testing provides initial clues, the use of imaging tools is crucial for a definitive diagnosis. Chest X-rays are frequently used because they are widely accessible and cost-effective. Conversely, they have limitations in detecting pneumonia in its first phases and differentiating between similar conditions. CT scans offer higher picture resolution, [6] providing more detailed information on lung architecture and the severity of illnesses. As a result, they are the favored option for accurate diagnosis. Deep learning (DL) and convolutional neural networks (CNNs) have greatly revolutionized the field of medical image analysis [7–11]. These technologies outperform traditional methods by independently accumulating information and identifying complex patterns in medical images, leading to improved accuracy and efficiency in detecting pneumonia [12–14].

CNN [15] which is a type of deep learning model that has been shown to have a lot of potential in enhancing pneumonia detection from CT scans' accuracy. CNNs are particularly designed to work and recognize [16] patterns and characteristics such as edge detection [17] using connectionist neurons arranged in layers. Successively the layers increase data abstraction until the obtainable meaningful information is achieved. Various regularization techniques can be used to improve the performance of CNN [18]. By feeding CNNs with large amounts of data containing lung CT scans, they can easily differentiate between healthy lungs and other different types of pneumonia which includes bacterial, viral, and fungal pneumonia. This approach to diagnosis is free from human errors, reduces the time taken in the diagnosis process, and improves accuracy. This paper describes an improved approach to image preprocessing determining the applicability of wavelet-based image processing in [19] enhancing the visibility of the infection features. This method seeks to aid CNN in its undertaking of correctly distinguishing cases of pneumonia from lung radiographs This paper focuses on a classification of the type of pneumonia by surveying the existing literature in Sect. 2, followed by methodology and comparative study of different networks in Sect. 3 to enhance the classification accuracy of pneumonia type using the proposed preprocessing strategy with result and discussion in Sect. 4 and conclusion in Sect. 5.

2 Literature Review

The purpose of this particular literature review is to offer a comprehensive view of all the prior work done in the field relating to the detecting and categorizing of pneumonia using medical images and deep learning. Its primary aim is to explain how imaging processes such as chest X-rays, as well as computed tomography (CT) scans, have progressed in disease diagnosis. It will also seek to examine how the latest subfield of deep learning primarily CNN can improve the diagnosis of pneumonia and the duration it would take. It also examines how image enhancement techniques based on wavelets can be used in enhancing the visibility of infection features in lung images which can assist the doctors in making better decisions.

Ghaderzadeh et al. [20] have developed the most optimal classifier using a deep neural network (NN) based on a Neural Search Architecture Network (NASNet) feature extractor to classify the type of lesion to Air Space Consolidation, Ground Glass Opacity or Negative, achieving an accuracy of 99.6%, specificity of 98.6%, and sensitivity of 99.9%. Subramanian et al. [30] have explained the difference in performance between the custom models and pretrained models on both CXR and CT. The works [20, 30] show that well-developed custom models perform better when developed optimally. Ibrahim et al. [31] carried out a comprehensive evaluation of different deep learning architectures such as VGG19-CNN, ResNet152V2, ResNet152V2 + Gated Recurrent Unit (GRU), and ResNet152V2 + Bidirectional GRU (Bi-GRU) using public digital CXR and CT data. On experimentation, it was found that the VGG19 + CNN model outperforms the three other proposed models with 98.05% accuracy, 98.43% precision, 98.05% sensitivity, and 99.5% specificity. Furthermore, other similar models [7] that were tested against VGG19 are DenseNet169, VGG-16, ResNet-50, and InceptionV3. The VGG-19 again proved to be superior with an accuracy of 94.52% as compared to all other deep learning models. The ResNet-5021 model pre-trained using the ImageNet dataset achieved a perfect recall score of 100% [23]. The model was built on top of UNet++ and was trained using the highest dataset image count of 46,096 CT scan images. The pneumonia classification approaches followed in the preceding works are given in Fig. 1a, b and the summary of the related review is tabulated in Table 1.

3 Methodology

The proposed work describes an approach to enhance the quality of images obtained through a computed tomography scan for improved pneumonia classification. A wavelet-based image enhancement technique is proposed in this work (as shown in Fig. 2) to enable discrimination between different pneumonia. The processed images are then subjected to the classification stage to categorize pneumonia into three groups: COVID-19 pneumonia, other pneumonia, and normal.

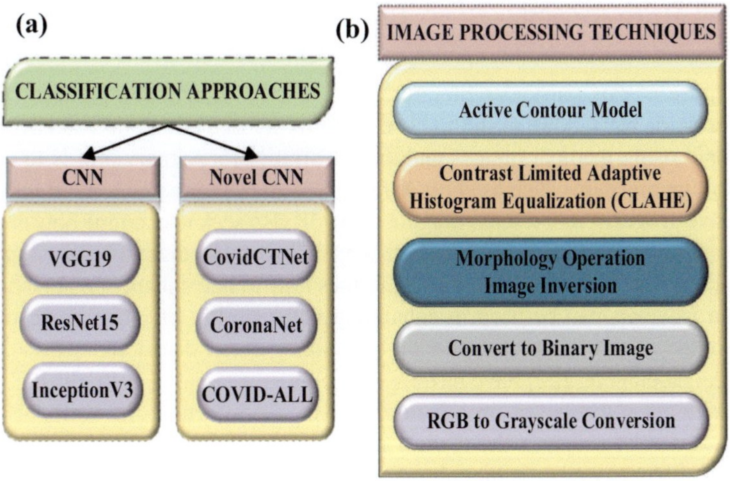

Fig. 1 **a** Pneumonia classification methods, **b** general image enhancement methods

Table 1 Summary of research on COVID-19 infection type detection

Preceding works	Dataset	Models used	Accuracy	Sensitivity	Specificity	F1-score
[20]	10,153 CTs	NASNet	99.60	99.90	98.60	NA
[21]	33,676 CXRs and CTs	VGG19 + CNN	98.05	98.05	99.5	98.24
[22]	738 CTs	VGG-19	94.52	NA	NA	NA
[23]	46,096 CTs	ResNet-50	92.59	100	81.82	NA
[24]	748 CTs	CoronaNet	92	NA	NA	87.35
[25]	962 CTs	COVID-AL	86.6	NA	NA	NA
[26]	16,750 CTs	CovidCTNet	95	74.00	NA	80.00
[27]	CTs from 522 patients	XGBoost	94.6	75.7	96.4	NA
[28]	2446 CTs	Custom CNN	NA	90	83	NA
[29]	6475 CTs and 8581 CXRs	Custom CNN	96.68	96.24	95.65	NA

The aim of the proposed work is to enhance the CT data from the dataset in order to improve the classification process. The image enhancement stage begins by applying a nonlocal means (NLM) filter [32] for denoising, which forms the mean of the intensity of similar neighborhoods to eliminate noises while retaining important features of structures. Preserving detail is critical in medical images, especially in diagnosis, meaning that the outline algorithm must preserve image detail. By using the NLM filter in CT scans, a good signal is obtained while removing random noise

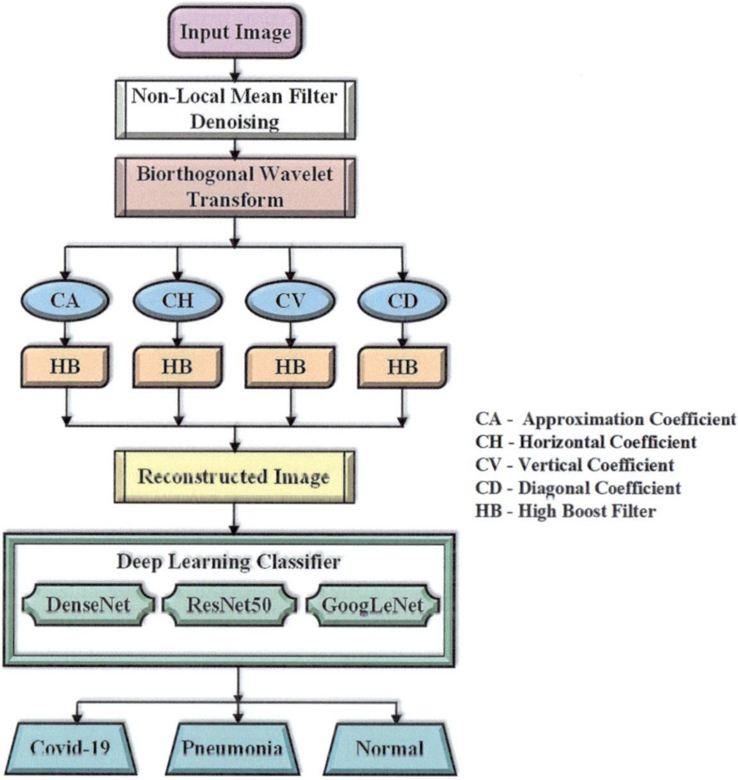

Fig. 2 Proposed methodology for pneumonia classification

that is not important in detecting the change in diseases in the lungs and retaining edges and textures that are important in displaying those changes. Higher image quality of obtained images leads to the higher performance of the subsequent image analysis techniques and deep learning methods used for pneumonia classification; therefore, the accuracy and reliability of this process is increased. The NLM filter is characterized by,

$$\mathrm{NL}[u](x) = \sum_{i \in I} \frac{e^{-(d(u(x),u(i))^2/h^2)}}{Z(x)} u(i) \tag{1}$$

where $u(x)$ is the intensity value at pixel (x), $d(u(x), u(i))$ is the Euclidean distance between patches centered at x and i, h is the filtering parameter, and $Z(x)$ is the normalizing constant. The denoised images are then subjected to the image enhancement phase. A biorthogonal wavelet transform-based image enhancement method is proposed in this work. The biorthogonal wavelet transform is mostly used in image processing as not only it decomposes a picture into various frequency subbands but also is capable of preserving some important structural features; therefore, it is used in medical imaging. This transformation is used in the proposed work for enhancing the quality of the images acquired through the CT scan for the diagnosis of pneumonia as it allows to work with many characteristics of the images at different levels of detail. The biorthogonal wavelet transforms [33] gives a flexible way of analysis that conveys the picture both at the high level, in the hunt of the edges and texture as well as at the low level in search of smooth areas. This disintegration is necessary to highlight the differences in the pulmonary tissue that could indicate the presence of pneumonia regardless of its roots in COVID-19 or other viruses. The biorthogonal wavelet transform is precisely defined as follows:

$$W\varphi(f)(a,b) = \frac{1}{\sqrt{|a|}} \int_{-\infty}^{\infty} f(t)\varphi\left(\frac{t-b}{a}\right)\mathrm{d}t \tag{2}$$

where $f(t)$ is the original image, (a) is the scaling parameter, and (b) is the translation parameter. There is also a necessity to emphasize the infection features, and hence, the high-boost filter is applied to the coefficients of the wavelet transform: approximation, horizontal, diagonal, and vertical. In the field of CT scan of pneumonia, a high-boost filter serves the purpose of increasing the contrast of infection-related features which are possibly weak and concealed in the primary scan. With the help of the high-boost filter [34], the small differences in the lung tissues are amplified, which helps in the distinction between various types of pneumonia and regular lung disorders.

$$I_{\text{hb}} = I_{\text{orig}} + k \cdot \left(I_{\text{orig}} - I_{\text{blur}}\right) \tag{3}$$

where I_{hb} is the high-boosted image, I_{orig} is the original image, I_{blur} is the blurred version of the image, and k is the boost factor. The next step involves the reconstruction of the high-boost filtered wavelet coefficients into spatial domains using the inverse wavelet transform which enhances the images' features. The enhanced images are then fed to the deep learning classifiers such as DenseNet [35], GoogleNet [36], and ResNet50 [37] models as shown in Figs. 3, 4, and 5.

The application of residual and deep design models to the pneumonia classification system makes it possible to identify the subtle features of the diseases, which are crucial to diagnosis and treatment. The COVID-19&Normal&Pneumonia_CT_Images data set [38] used for the study contains 7544 CT scan slices among which 2035 belong to COVID-19 pneumonia infections, 3090 images of other pneumonia, and 2119 images of normal cases. To experiment with the proposed method, 6000 images are used from the dataset with 2000 images per class. The image set is divided into 70% for training, 10% for validating, and 20% for testing. The training process was performed in all the aforementioned networks with the specified training options: Stochastic gradient descent with moment (SGDM) of 0.9, mini-batch size of 32, and initial learning rate of 1e-4. The results obtained from the trained, validated, and tested DL models and their corresponding performance measures are discussed in the following section.

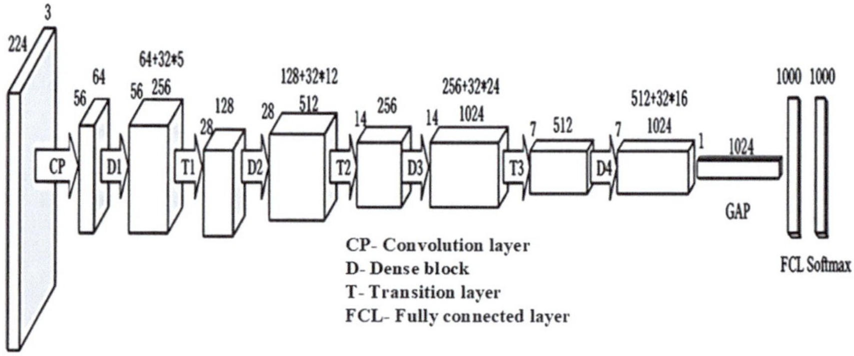

Fig. 3 The architecture of DenseNet 121 [35]

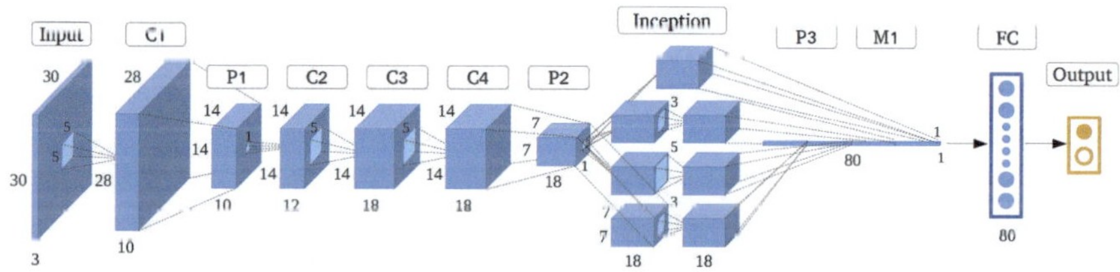

Fig. 4 Architecture of GoogLeNet (Inception-v1) [36]

4 Results and Discussion

The results obtained for the proposed pneumonia classification method are analyzed and discussed in this section. The various performance measures evaluated for the proposed pneumonia classification method are tabulated in Table 2. From the table, it can be inferred that DenseNet achieved the highest accuracy of 99.20%, outperforming ResNet50 and GoogLeNet by 1.27% and 1.61%, respectively. Sensitivity, which measures the model's ability to correctly identify positive cases, is 1.15% higher than ResNet50 and 2.08% higher than GoogLeNet when compared to DenseNets performance. DenseNet again leads in specificity with a score of 99.62%, indicating its superior ability to correctly identify negative cases with a percentage difference of 0.66% for ResNet50 and 1.04% for GoogLeNet. The F1-Score which balances precision and recall, was also the highest for DenseNet which is 1.21% higher than ResNet50 and 1.54% higher than GoogLeNet. DensNet achieved PPV of 1.24% higher than ResNet and 0.92% higher than GoogLeNet, Similarly for NPV, DenseNet outperforms ResNet50 and GoogLeNet, with 0.69% each. The region of the curve obtained for pneumonia classification on all three models is depicted in Fig. 6 (Note: In Fig. 6a The ROC curves for the other two models (normal CT and COVID CT) have almost identical performance to the pneumonia CT model, their ROC curves are overlapped. This would result in them being plotted on top of each

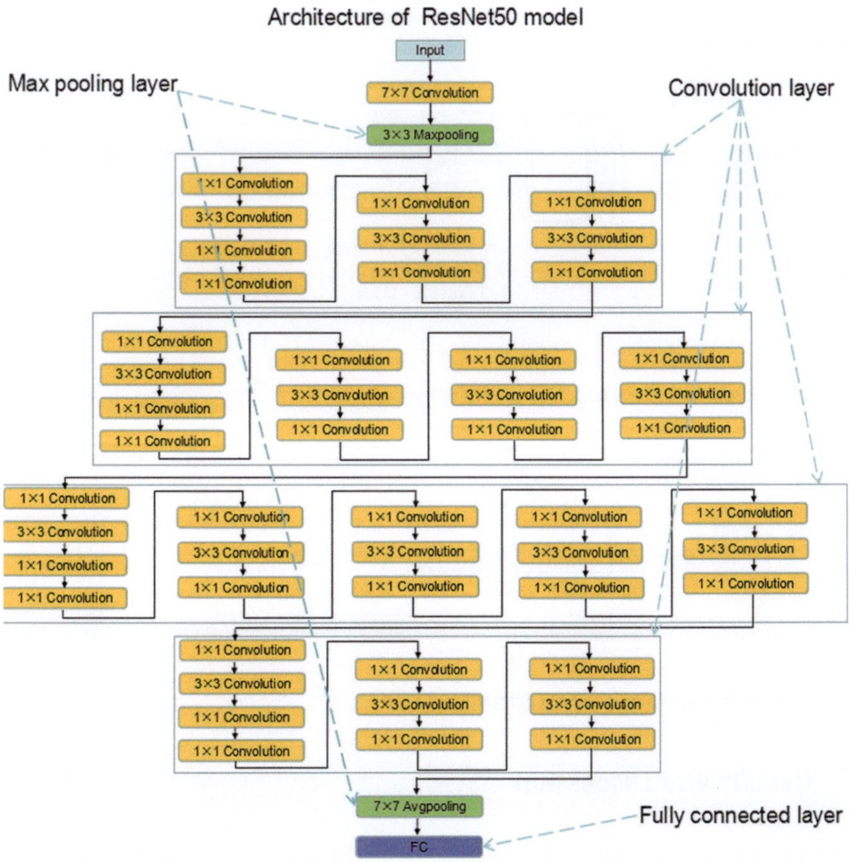

Fig. 5 Architecture of ResNet50 [37]

other, making them indistinguishable from the visible curve). Based on the above inference, DenseNet's superior performance is proven with respect to all performance metrics discussed.

Table 2 Performance metrics of the proposed pneumonia classification method using different DL models

Performance measures	DenseNet	ResNet50	GoogLeNet
Accuracy (%)	99.20	97.93	97.59
Sensitivity (%)	99.27	98.12	97.19
Specificity (%)	99.62	98.96	98.58
F1-score (%)	99.18	97.97	97.64
Positive predictive value (PPV) (%)	99.09	97.85	98.17
Negative predictive value (PPV) (%)	99.56	98.87	98.87

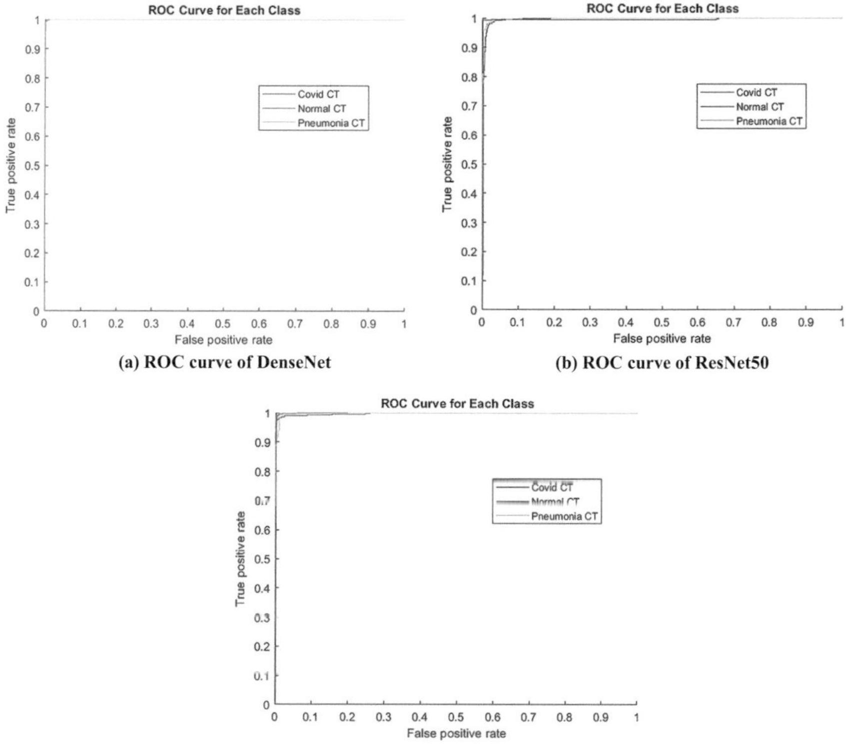

(a) ROC curve of DenseNet

(b) ROC curve of ResNet50

(c) ROC curve of GoogLeNet

Fig. 6 ROC comparison of pneumonia classification using DenseNet, ResNet50, and GoogLeNet

The preprocessing method proposed in this work using biorthogonal wavelets tends to produce beneficial results as the wavelet transform can effectively characterize the texture of various regions in different scales. The proposed pneumonia classification works appreciably well when compared to the aforementioned preceding work's performance measures tabulated in Table 3. Hence, the biorthogonal wavelet-based image enhancement method proposed in this work can be further studied and applied to various other similar medical applications to improve diagnostic accuracy. Moreover, the proposed enhancement method can be hybridized with various other image-processing strategies to improve its efficiency.

Table 3 Comparison of the proposed method with preceding works

Preceding works	Methodology	Accuracy (%)	Sensitivity (%)	Specificity (%)	F1-score (%)
[31]	VGG19 + CNN	98.05	98.05	99.30	98.24
[23]	ResNet-50	92.59	100	81.82	–
[39]	F-RRN-LSTM	95.05	91.63	94.21	95.19
[40]	ImageNet	96.4	99.62	–	–
[41]	ALbAE	97	–	–	98
Proposed	DenseNet	99.20	99.27	99.62	99.18

5 Conclusion

A deep learning-based pneumonia classification method is proposed in this work with a wavelet-based image enhancement technique to enhance the visibility of infection details in lung CT scans. Additionally, the usage of a high-boost filter improves the details of the decomposed wavelet components. The enhanced images were tested for classification on various networks DenseNet, ResNet50, and GoogLeNet. The performance metrics obtained indicate that Densenet's performance was higher with a classification accuracy of 99.20% than that of ResNet50 and GoogLeNet. The proposed image enhancement method provided better discrimination between the different pneumonia infections and the normal cases. As the results are appreciable, the wavelet-based enhancement method can assist the medical team in early disease detection and further treatments.

References

1. White RJ, Harrison J, Clarke S (1981) Causes of pneumonia presenting to a district general hospital [Online]. Available: http://thorax.bmj.com/
2. Chitra S, Mani (2018) 34 - Acute pneumonia and its complications, principles and practice of pediatric infectious diseases (5th, edn.). Elsevier, pp 238–249.e4, ISBN 9780323401814. https://doi.org/10.1016/B978-0-323-40181-4.00034-7. (https://www.sciencedirect.com/science/article/pii/B9780323401814000347)
3. Ewig S, Schäfer H, Torres A (2000) Severity assessment in community-acquired pneumonia. Eur Respir J 16(6):1193–201. https://doi.org/10.1034/j.1399-3003.2000.16f27.x. PMID: 11292126
4. Revised WHO classification and treatment of childhood pneumonia at health facilities. Evidence Summaries 2014. https://www.who.int/publications/i/item/9789241507813
5. Bartlett JG (2011) Diagnostic tests for agents of community-acquired pneumonia. Clin Infect Dis 52(SUPPL):4. https://doi.org/10.1093/cid/cir045
6. Mary Shyni H, Chitra E (2022) A comparative study of X-ray and CT images in COVID-19 detection using image processing and deep learning techniques. Elsevier B.V. https://doi.org/10.1016/j.cmpbup.2022.100054

7. Shah V, Keniya R, Shridharani A, Punjabi M, Shah J, Mehendale N (2021) Diagnosis of COVID-19 using CT scan images and deep learning techniques. Emerg Radiol 28(3):497–505. https://doi.org/10.1007/s10140-020-01886-y
8. Puttagunta M, Ravi S (2021) Medical image analysis based on deep learning approach. Multimed Tools Appl 80(16):24365–24398. https://doi.org/10.1007/s11042-021-10707-4
9. Vineth Ligi S, Kundu SS, Kumar R, Narayanamoorthi R, Lai KW, Dhanalakshmi S (2022) Radiological analysis of COVID-19 using computational intelligence: a broad gauge study. Hindawi Limited. https://doi.org/10.1155/2022/5998042
10. Stokes K et al (2022) The use of artificial intelligence systems in diagnosis of pneumonia via signs and symptoms: a systematic review. Elsevier Ltd. https://doi.org/10.1016/j.bspc.2021.103325
11. Exarchos KP, Gkrepi G, Kostikas K, Gogali A (2023) Recent advances of artificial intelligence applications in interstitial lung diseases. Multidisciplinary Digital Publishing Institute (MDPI). https://doi.org/10.3390/diagnostics13132303
12. Samiappan D, Latha S, Rao TR, Verma D, Sriharsha C (2020) Enhancing machine learning aptitude using significant cluster identification for augmented image refining. Intern J Pattern Recognit Artif Intell 34(09):2051009. https://doi.org/10.1142/S021800142051009X
13. Sathya A, Priya L, Dhanalakshmi SA (2022) Hybrid algorithm for face recognition system based smart attendance system. ECS Trans 107(1):11487–11495. https://doi.org/10.1149/10701.11487ecst
14. Woan Ching SL et al (2022) Multiclass convolution neural network for classification of COVID-19 CT images. Comput Intell Neurosci 2022:1–15. https://doi.org/10.1155/2022/9167707
15. Sarvamangala DR, Kulkarni RV (2022) Convolutional neural networks in medical image understanding: a survey. Springer Science and Business Media Deutschland GmbH. https://doi.org/10.1007/s12065-020-00540-3
16. Lo S-CB, Lin J-SJ, Freedman MT, Mun SK (1998) Application of artificial neural networks to medical image pattern recognition: detection of clustered microcalcifications on mammograms and lung cancer on chest radiographs. J VLSI Signal Process Syst Signal Image Video Technol 18(3):263–274
17. Li X, Jiao H, Wang Y (2020) Edge detection algorithm of cancer image based on deep learning. Bioengineered 11(1):693–707. https://doi.org/10.1080/21655979.2020.1778913
18. Ligi SV, Kumar R, Dhanalakshmi S (2023) COVID-19 detection from pulmonary CT images using neural networks based on dropout-driven hidden layers. In: 2023 international conference on intelligent systems for communication, IoT and security (ICISCoIS). IEEE, pp 310–316. https://doi.org/10.1109/ICISCoIS56541.2023.10100361
19. Beladgham M, Bessaid A, Lakhdar AM, Taleb-Ahmed A (2011) Improving quality of medical image compression using biorthogonal CDF wavelet based on lifting scheme and SPIHT coding
20. Ghaderzadeh M, Asadi F, Jafari R, Bashash D, Abolghasemi H, Aria M (2021) Deep convolutional neural network-based computer-aided detection system for COVID-19 using multiple lung scans: design and implementation study. J Med Internet Res 23(4):e27468. https://doi.org/10.2196/27468
21. Ibrahim DM, Elshennawy NM, Sarhan AM (2021) Deep-chest: multi-classification deep learning model for diagnosing COVID-19, pneumonia, and lung cancer chest diseases. Comput Biol Med 132:104348. https://doi.org/10.1016/j.compbiomed.2021.104348
22. Shah V, Keniya R, Shridharani A, Punjabi M, Shah J, Mehendale N (2021) Diagnosis of COVID-19 using CT scan images and deep learning techniques. Emerg Radiol, 28(3). https://doi.org/10.1007/s10140-020-01886-y
23. Chen J et al (2020) Deep learning-based model for detecting 2019 novel coronavirus pneumonia on high-resolution computed tomography. Sci Rep 10(1):19196. https://doi.org/10.1038/s41598-020-76282-0
24. Bhansali R, Kumar R, Writer D (2020) CoronaNet: a novel deep learning model for COVID-19 detection in CT scans. J Stud Res 9(2):1–9. https://doi.org/10.47611/jsrhs.v9i2.1246
25. Wu X, Chen C, Zhong M, Wang J, Shi J (2021) COVID-AL: the diagnosis of COVID-19 with deep active learning. Med Image Anal 68:101913. https://doi.org/10.1016/j.media.2020.101913

26. Javaheri T et al (2021) CovidCTNet: an open-source deep learning approach to diagnose Covid-19 using small cohort of CT images. NPJ Digit Med 4(1):29. https://doi.org/10.1038/s41746-021-00399-3
27. Feng Z et al (2021) Machine learning based on clinical characteristics and chest CT quantitative measurements for prediction of adverse clinical outcomes in hospitalized patients with COVID-19. Eur Radiol 31(10):7925–7935. https://doi.org/10.1007/s00330-021-07957-z
28. Mortani Barbosa EJ et al (2021) Machine learning automatically detects COVID-19 using chest CTs in a large multicenter cohort. Eur Radiol 31(11):8775–8785. https://doi.org/10.1007/s00330-021-07937-3
29. Gilanie G et al (2021) Coronavirus (COVID-19) detection from chest radiology images using convolutional neural networks. Biomed Signal Process Control 66:102490. https://doi.org/10.1016/j.bspc.2021.102490
30. Subramanian N, Elharrouss O, Al-Maadeed S, Chowdhury M (2022) A review of deep learning-based detection methods for COVID-19. Elsevier Ltd. https://doi.org/10.1016/j.compbiomed.2022.105233
31. Ibrahim DM, Elshennawy NM, Sarhan AM (2021) Deep-chest: multi-classification deep learning model for diagnosing COVID-19, pneumonia, and lung cancer chest diseases. Comput Biol Med 132:104348. https://doi.org/10.1016/j.compbiomed.2021.104348
32. Yang J et al (2016) Local statistics and non-local mean filter for speckle noise reduction in medical ultrasound image. Neurocomputing 195:88–95. https://doi.org/10.1016/j.neucom.2015.05.140
33. Prakash O, Park CM, Khare A, Jeon M, Gwak J (2019) Multiscale fusion of multimodal medical images using lifting scheme based biorthogonal wavelet transform. Optik (Stuttg) 182:995–1014. https://doi.org/10.1016/j.ijleo.2018.12.028
34. Srivastava R, Gupta JRP, Parthasarthy H, Srivastava S (2009) PDE based unsharp masking, crispening and high boost filtering of digital images
35. Wang SH, Zhang YD (2020) DenseNet-201-based deep neural network with composite learning factor and precomputation for multiple sclerosis classification. ACM Trans Multimed Comput Commun Appl 16(2s):1–9. https://doi.org/10.1145/3341095
36. Guo Z, Chen Q, Wu G, Xu Y, Shibasaki R, Shao X (2017) Village building identification based on ensemble convolutional neural networks. Sensors (Switzerland) 17(11):2487. https://doi.org/10.3390/s17112487
37. Peng J et al (2020) Residual convolutional neural network for predicting response of transarterial chemoembolization in hepatocellular carcinoma from CT imaging. Eur Radiol 30(1):413–424. https://doi.org/10.1007/s00330-019-06318-1
38. COVID-19&Normal&Pneumonia_CT_Images. Accessed: 17 July 2024 [Online]. Available: https://www.kaggle.com/datasets/anaselmasry/covid19normalpneumonia-ct-images/data
39. Goyal S, Singh R (2023) Detection and classification of lung diseases for pneumonia and Covid-19 using machine and deep learning techniques. J Ambient Intell Humaniz Comput 14(4):3239–3259. https://doi.org/10.1007/s12652-021-03464-7
40. Chouhan V et al (2020) A novel transfer learning based approach for pneumonia detection in chest X-ray images. Appl Sci 10(2):559. https://doi.org/10.3390/app10020559
41. Braveen M et al (2023) ALBAE feature extraction based lung pneumonia and cancer classification. Soft Comput. https://doi.org/10.1007/s00500-023-08453-w

YOLOV8 for Urban EV Detection: Sustainable Mobility Management

Tarushi Singh and Samiappan Dhanalakshmi

Abstract This paper explores the use of YOLOv8 algorithm for object detection to identify electric vehicles (EVs) from other conventional diesel/petrol vehicles in real time in urban parking infrastructure. By leveraging a dataset with 1200 high-resolution images, augmented to 3353 images, the model was trained and evaluated under various conditions like diverse lighting, background clutter, and different vehicle orientations. The results showcased the algorithm's remarkable performance, with a mean average precision (mAP) of 91% and precision and recall rates of 91.8%. These metrics underscore YOLOv8's validity and accuracy in intricate, real-world scenarios, making it a compelling and powerful tool for optimizing parking facilities and managing EV charging stations. The framework also plays a critical role in developing sustainable urban mobility by effectively assigning dedicated charging stations and facilitating the harmonious integration of EVs into current infrastructure. This paper highlights the rising significance of EV detection in environment friendly urban development, providing insights into practical applications and future research directions to further increase intelligent parking solutions. YOLOv8 provides an encouraging solution to the challenges of real-time EV detection, contributing substantially to sustainable urban planning and transportation systems.

Keywords Electric vehicle detection · YOLOv8 · Real-time inference · Sustain urban mobility

1 Introduction

The universal shift toward electric vehicles (EVs) and sustainable transportation is gaining significant momentum, necessitating the enhancement of parking environments and the development of charging infrastructure to accommodate the rising

T. Singh · S. Dhanalakshmi (✉)
Department of Electronics and Communication Engineering, SRM Institute of Science and Technology, Kattankulathur, Chennai, Tamil Nadu, India
e-mail: dhanalas@srmist.edu.in

S. Kumar et al. (eds.), *Artificial Intelligence for Next Generation Computing, Volume 1*, Lecture Notes in Networks and Systems 1549,
https://doi.org/10.1007/978-981-95-0701-6_14

demand for EVs [1]. However, this transition introduces specific challenges, particularly in accurately distinguishing EVs from conventional petrol/diesel engine vehicles within parking spaces [2]. Figure 1 shows one of the key indicators for identifying EVs in many regions: the presence of green license plates, which sets them apart from traditional vehicles. Effectively addressing these challenges is critical for the efficient administration and utilization of charging and parking resources.

The main purpose of this project is to tackle these challenges by leveraging the capabilities of the cutting-edge YOLOv8 object detection algorithm [3]. This advanced algorithm is recognized for its state-of-the-art performance in real-time object detection. By utilizing YOLOv8, the project aims to establish a sophisticated framework for the accurate and real-time detection of electric vehicles, specifically by focusing on identifying green license plates within parking facilities [4]. Such a framework will significantly streamline the allotment of dedicated charging stations, thereby enhancing the management quality and durability of parking infrastructure [5].

Furthermore, deploying this state-of-the-art technology in this context makes a significant contribution to the ongoing transition toward eco-friendly and environmentally sound mobility solutions [6]. By facilitating the efficient and reliable operation of EV infrastructure through the detection of green license plates, this project not only supports the increasing adoption of electric vehicles but also aligns with

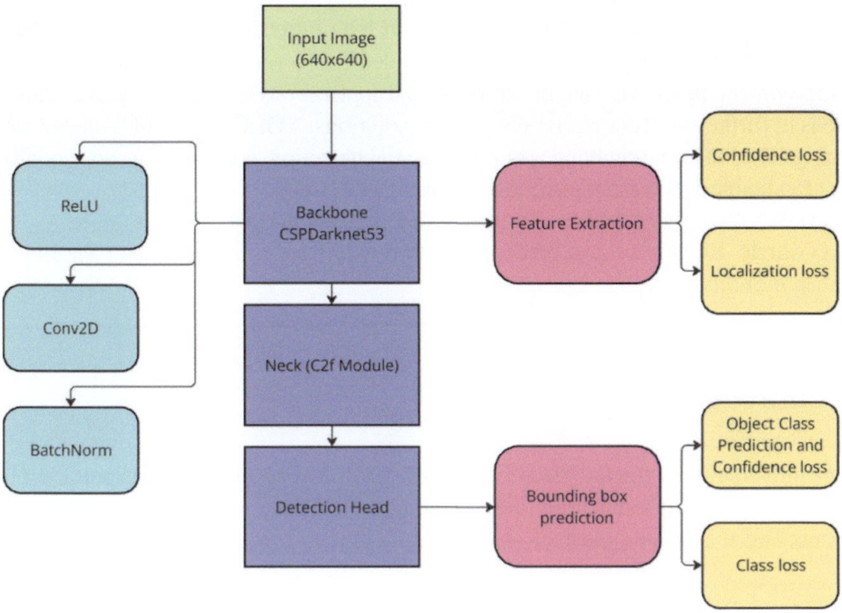

Fig. 1 Architecture of the YOLOV8 model

global efforts to reduce carbon emissions and promote sustainable urban development. This initiative, therefore, represents a critical step forward in the pursuit of a more sustainable and environmentally responsible transportation ecosystem.

2 Literature Review

As more and more people choose ecologically friendly and renewable forms of transportation, the shift to electric vehicles (EVs) is gathering steam. This quick mobility emphasizes how urgently effective infrastructure and identification and detecting systems are needed to handle the increasing number of electric vehicles on the road [7]. The demand for more parking spaces and charging stations rises along with the acceptance and adoption of electric vehicles. Ensuring the seamless integration of electric vehicles (EVs) into current transportation networks requires addressing the related infrastructural concerns. Accurately differentiating electric cars (EVs) from traditional diesel and gasoline vehicles is a major difficulty in this transition, especially in parking facilities where charging station assignment is crucial. Despite the urgent need, there is a significant research vacuum about the ability of advanced detection algorithms to guide the design of urban infrastructure, particularly concerning the best location for EV charging stations. Comprehending the correlation between accurate identification of electric vehicles and the well-planned placement of charging stations may enable a smoother assimilation of electric vehicles into the existing transportation infrastructure. Furthermore, more research is necessary to determine how real-time EV detection data affect urban management and planning. Advanced detection algorithms, such as those from the YOLO (You Only Look Once) family, have emerged as pivotal tools in addressing these challenges, significantly advancing the field of real-time object detection [8].

YOLOv8, developed by Ultralytics, that is upon the success of its predecessor, YOLOv5, with various improvements, provides further enhancement of these capabilities, offering refined performance, better efficiency, and greater accuracy in detecting a wide range of objects, including vehicles [9]. The model has several enhancements, including the modification of the backbone network and anchor-free detection head, making it more efficient and adaptable for diverse computer vision tasks. YOLOV8 is particularly good at identifying moving objects because of its quick picture processing speed, which makes it a good fit for real-time applications [10]. Additionally, it presents a fresh loss function that improves object localization, making it especially skilled at discerning slight distinctions between EVs and conventional vehicles in busy parking areas [11]. Recent studies have successfully applied YOLO for vehicle detection, exhibiting its capability to accurately differentiate between EVs and traditional vehicles in various scenarios, thereby advancing the effective management of EV-specific infrastructure [12].

YOLOv8 performs better than competing models in a variety of object detection settings, including SSD and Faster R-CNN. Accuracy is one of its strong points; it routinely outperforms SSD and Faster R-CNN in situations with similar object and background colors and variable lighting conditions. Furthermore, YOLOv8 is especially well suited for real-time applications because of its higher inference speed, which facilitates prompt decision-making in dynamic environments. Moreover, YOLOv8's reduced hardware needs add to its effectiveness and affordability, making it a more feasible option for implementation than the resource-intensive Faster R-CNN, which might not be able to handle the demands of time-sensitive applications [13].

For this detection algorithm to operate at its immense potential, the quality and diversity of the datasets used for training matter significantly. Vigorous datasets, with a wide variety of vehicle types and conditions, are of utmost importance to ensure the model's ability to generalize well in real-world situations. The selection of dataset is a critical factor in determining the efficacy and dependability of detection algorithms when it comes to electric vehicle (EV) identification. Although well-known datasets offer a solid foundation for training and assessment, the algorithms' ability to generalize can be strongly impacted by the intrinsic limits of these datasets. For example, the performance of these models can decline in tough categories, emphasizing the challenges associated with tiny, occluded, and truncated object detection. Strictly depending on these popular datasets may limit the algorithms' applicability in a variety of real-world situations [14]. Thus, creating reliable EV detection systems that can operate consistently in a range of environmental circumstances requires a large and varied dataset [15]. Additionally, advanced training techniques, such as data augmentation, are often employed to improve model accuracy by artificially expanding the dataset through modifications like rotations, scaling, and brightness adjustments [16].

The performance of these deep learning models is typically evaluated using parameters such as precision, recall, and mean average precision (mAP). High precision and recall rates are desirable in vehicle detection tasks, as they indicate the model's ability to correctly identify vehicles while minimizing false positives and negatives. Numerous studies in the field have reported impressive precision and recall rates, reinforcing the efficacy of deep learning models like YOLOv8 in the context of vehicle detection and their potential application in distinguishing EVs from conventional vehicles [17].

In conclusion, as the world continues to shift toward electric vehicles, the development and deployment of advanced detection algorithms like YOLOv8 are crucial to ensuring that the necessary infrastructure keeps pace with this transition. By leveraging the power of deep learning and sophisticated data processing techniques, these algorithms offer a promising solution to the challenges of managing EV infrastructure, ultimately contributing to a more efficient, sustainable, and environmentally friendly transportation system.

3 Methodology

3.1 Yolov8 Architecture

Figure 1 presents the head, neck, and backbone of the YOLOv8 architecture, which marks a departure from the CSPDarknet53 framework utilized in previous iterations like YOLOv4 and YOLOv5 [18]. This updated architecture is strategically designed to enhance the flow of information across the network's stages, ensuring an optimal balance between speed and accuracy when processing complex image inputs [18].

The head and neck components of YOLOv8 leverage sophisticated methods for predicting bounding boxes, objectness scores, and class probabilities [18]. These components serve as the model's decision-making modules, converting the features extracted by the backbone into precise object detection outputs.

A notable advancement in YOLOv8 is the integration of a self-attention mechanism, which enables the model to dynamically focus on pertinent areas of an image, thus improving its capacity to detect objects with varying sizes, orientations, and complexities [19]. Furthermore, YOLOv8 incorporates adaptive training, a dynamic approach that refines the loss function and optimizes the learning rate throughout the training process, leading to improved model performance across diverse datasets and tasks [18, 19]. The availability of pre-trained models in YOLOv8 further streamlines deployment and facilitates transfer learning, providing a robust foundation for users to build upon in their projects [20, 21].

3.2 Dataset and Model Training

This section presents the process employed in our research for developing a real-time electric vehicle (EV) detection system within parking facilities. Leveraging the YOLOv8 object detection algorithm, our objective is to accurately discern EVs from conventional diesel or petrol vehicles, facilitating the streamlined allocation of dedicated charging stations [22].

Dataset Collection: A dataset comprising 1200 high-resolution images was collected, featuring both EV and diesel/petrol vehicles within parking lots. Each image was meticulously annotated with bounding boxes around the vehicle number plates and labeled accordingly for supervised learning [9]. The dataset exhibits diversity in lighting conditions, vehicle orientations, and background clutter, ensuring robust model training [23]. The dataset was split into 70% training, 15% validation, and 15% testing, ensuring that the diversity of lighting, vehicle orientation, and background clutter was evenly represented across all sets.

Model Architecture and Training: The YOLOv8 object detection architecture was selected for its efficiency and effectiveness in real-time applications [24]. Transfer learning was employed by initializing the model with pre-trained weights, followed

Table 1 Completed dataset after all the augmentations

Augmentations	
Flip	Horizontal
Rotation	Between − 15° and + 15°
Shear	± 10° Horizontal, ± 10° Vertical
Saturation	Between − 25% and + 25%
Brightness	Between − 15% and + 15%
Exposure	Between − 10% and + 10%
Noise	Up to 0.5% of pixels
Dataset	
Total	3353 images
Training set	3204 images
Validation set	97 images
Test set	52 images

by fine-tuning our annotated dataset [25]. To enhance model generalization, advanced data augmentation techniques, including rotation, scaling, saturation, brightness, exposure, and noise augmentation, were applied as shown in Table 1 [26].

The neural network model consists of 168 layers and 11,126,358 trainable parameters, enabling it to learn complex patterns. Its computational performance is 28.4 GFLOPs, indicating efficiency for real-time applications. The model was trained for 20 epochs, with validation loss and precision monitored to prevent overfitting. Images were resized to 800 × 800 pixels for optimal detail, using a batch size of 16 based on available GPU memory. The initial learning rate was set to 0.001, with a reduction of 10% every 5 epochs to improve convergence.

Evaluation: The trained model underwent rigorous evaluation using a separate validation dataset. To evaluate the model's potency in EV detection, parameters like precision, mean average precision (mAP), and recall were calculated [27]. Cross-validation was conducted to ensure consistent performance across different dataset subsets, affirming the model's reliability for deployment in real-world scenarios [28].

4 Results and Discussion

The YOLOv8 object detection algorithm showcased excellent performance in detecting electric vehicles (EVs) within parking facilities. During the training phase, the model achieved a notable accuracy of 91.0%, underscoring its capability to learn and generalize the distinguishing features of EVs in complex environments. This high accuracy indicates the model's proficiency in recognizing and differentiating EVs from conventional vehicles, even under varied lighting conditions and angles typically encountered in parking lots. To ensure the model's effectiveness, an extensive and diverse dataset was used during the training process. This dataset contained photographs taken from different angles, in different lighting settings, and in different weather scenarios that are typical of parking lots in real life. As a result, the YOLOv8 model was able to generalize well, adapting to the nuances of different environmental factors that might affect EV detection.

Furthermore, the algorithm demonstrated strong performance metrics on the validation dataset, with precision and recall rates for EV detection reaching 91.8% and 91.8%, respectively. The precision rate showcases the model's ability to accurately identify true positives, ensuring that the detected EVs are indeed EVs. The recall rate, on the other hand, reflects the model's effectiveness in detecting most of the EVs present, minimizing the occurrence of missed detections.

Additionally, the high precision and recall rates indicate that the model maintains a good balance between correctly identifying EVs and detecting as many as possible. This balance is crucial in real-world scenarios where the misidentification of vehicles could lead to inefficiencies in parking management, such as improper allocation of charging stations or mismanagement of parking spaces.

The model's mean average precision (mAP) of 91.0% further underscores its effectiveness and reliability in real-world applications. The mAP metric, which accounts for both precision and recall across various threshold levels, provides an overall evaluation of the model's performance. This high mAP score suggests that YOLOv8 is not only precise but also consistent in its detection capabilities across different scenarios.

These performance attributes collectively demonstrate the effectiveness of the YOLOv8 model in accurately distinguishing EVs from conventional vehicles, validating its potential for real-world deployment in parking lot management systems [29]. By effectively identifying and tracking EVs, this model could play a role in optimizing parking management, ensuring efficient use of space, and supporting the growing adoption of electric vehicles in urban environments. Moreover, the YOLOv8 algorithm's performance in various challenging scenarios, such as low-light conditions and extreme angles, highlights its robustness and adaptability. The model's ability to maintain high accuracy despite these challenges suggests that it can handle the dynamic nature of real-world environments effectively. This capability is particularly important for modern parking management systems, which must accommodate a wide range of conditions and vehicle types. The comprehensive nature of the YOLOv8 model's feature extraction and its advanced architecture position it as a highly effective tool for EV detection, contributing to the advancement of intelligent parking solutions and facilitating the integration of electric vehicles into urban infrastructure.

Figure 2 demonstrates the YOLOv8 model's ability to accurately detect green license plates for electric vehicles and conventional plates from various angles and under different lighting conditions. The model excels in challenging scenarios, including low-light environments, extreme angles, and partial occlusions, showcasing its advanced architecture and vigorous feature extraction. YOLOv8's high accuracy and efficiency ensure reliable detection across diverse real-world applications, making it a standout choice for object detection tasks. Due to its grid-based architecture, the YOLOv8, like previous iterations, has trouble identifying small objects, which could cause some issues. This restriction is especially noticeable in crowded metropolitan environments where it is more common for small cars or objects to be missed. Furthermore, complex scenes with high densities and occlusion may provide difficulties for the algorithm, leading to a rise in false positives as well as missing detections [30]. Rain, fog, and dim lighting can also have an impact on the model's functionality. Another problem could be that biases could be highlighted if the training dataset is not sufficiently varied or does not fairly represent the real world. It might not function well in some other regions, for instance, if it was trained only using pictures of cars from one region. Data augmentation can help address these issues by increasing the diversity of the training dataset, particularly for tiny cars. Through multi-scale training, the model can adapt to varying resolutions and focus loss prioritizes situations that are challenging to identify. When combined, these techniques get around the algorithm's limitations and yield more accurate detection even in challenging situations.

Fig. 2 YOLOV8 results distinguishing EVs from non-EVs

4.1 Performance Metrics and Loss Functions

Performance metrics and loss functions are useful tools for assessing a deep learning model's performance. A loss function is employed to optimize the model's parameters during training by calculating the discrepancy between the model's expected and projected outputs. Reducing this disparity is the aim of training. In contrast, the model is assessed using a performance indicator after training. It aids in figuring out how well the model can forecast accurately and generalize to new data. Comparing several models or configurations to determine which performs best is made easier with the help of performance measurements [31].

Figure 3 shows the metrics used by YOLOV8 object detection algorithm to assess its performance. The consistent downward trend observed in both the training and validation loss curves signifies that the model is effectively learning and improving throughout the training process. Concurrently, the upward progression in evaluation metrics, including precision, recall, and mean average precision (mAP), further demonstrates enhanced performance on the validation set.

Precision and recall were used to measure the accuracy of positive predictions and the proficiency to detect all relevant instances of objects:

Finally, an evaluation criterion for gauging the model's overall performance was calculated: the mean average precision (mAP).

$$\mathrm{mAP} = \sum_{n} (R_n - R_{n-1}) P_n \quad (1)$$

Here, P_n and R_n are the precision and recall at the nth threshold.

4.2 Confusion Matrix

Figure 4 displays the confusion matrix of our model. It gives a thorough analysis of the execution of an object detection model, specifically in predicting the classes "green license plate," and "license plate." Here, "green license" indicates electric vehicles (EVs), and "license" indicates other conventional diesel/petrol vehicles. The evaluation results, presented in Table 2, indicate that the proposed model performs better on EV samples in terms of precision and accuracy, whereas the non-EV samples show higher F1 score values.

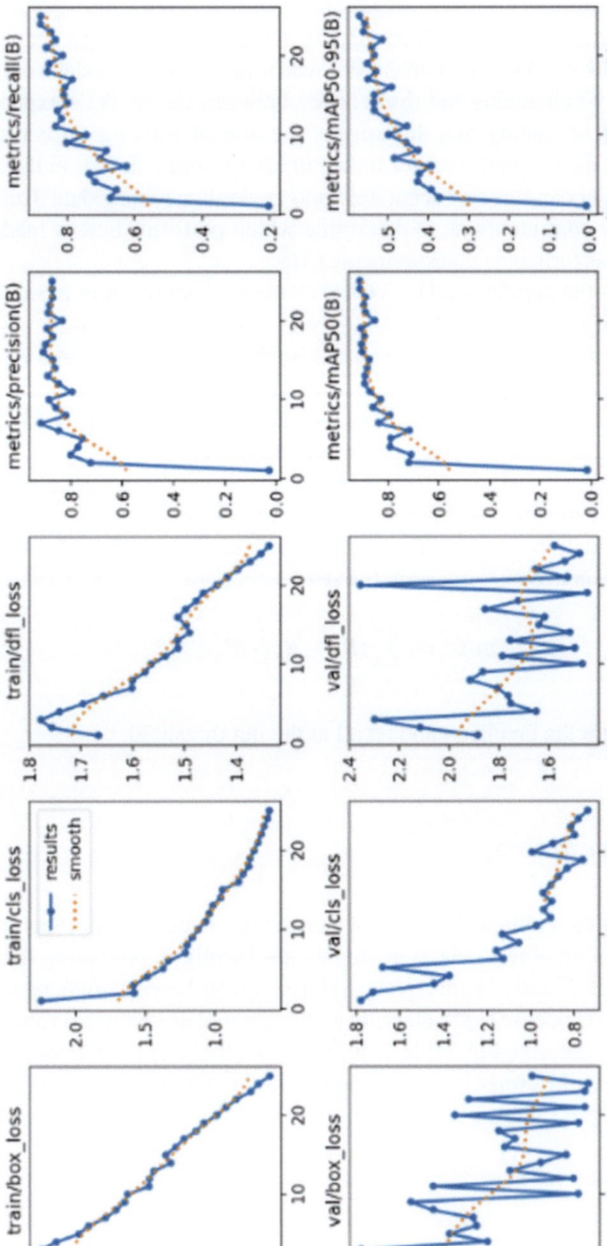

Fig. 3 Performance metrics and loss functions

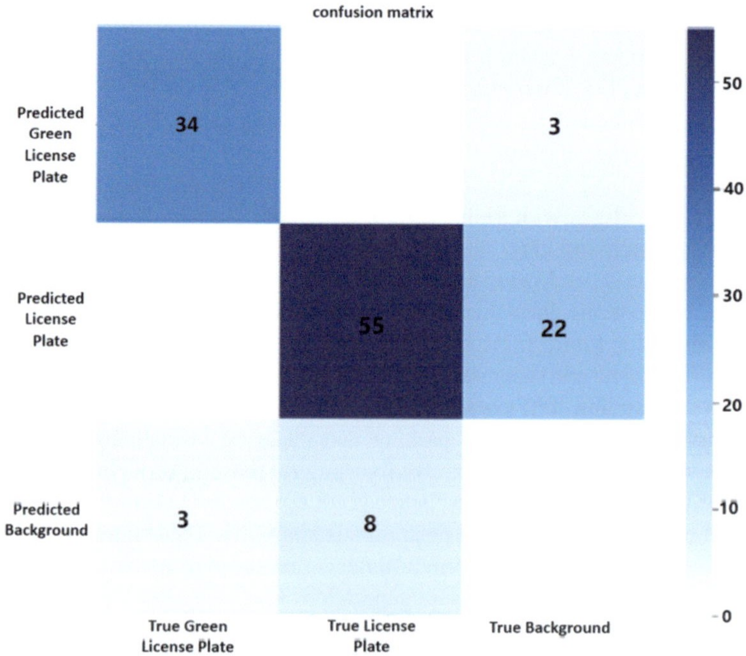

Fig. 4 Confusion matrix of the model

Table 2 Performance metrics for EV and non-EV classification

Metric	Formula	EV	Non-EV
Precision	$\frac{TP}{TP+FP}$	0.9189	0.7142
Recall	$\frac{TP}{TP+FN}$	0.9189	0.873
F1 score	$\frac{2}{\frac{1}{Recall}+\frac{1}{Precision}}$	0.787	0.919
Accuracy	$\frac{TP+TN}{TP+TN+FP+FN}$	0.952	0.763

5 Conclusion

The widespread adoption of electric vehicles (EVs) is set to transform urban mobility, offering significant environmental benefits while posing challenges that must be addressed to ensure a successful transition. Central to this transition is the development of effective parking and charging infrastructure, which is crucial for integrating EVs into urban environments. Without efficient infrastructure, the full potential of EVs in promoting a sustainable and efficient urban landscape may not be realized [32].

As cities increasingly embrace EVs, the enhancement of parking and charging facilities becomes a priority. The strategic placement and accessibility of charging stations are essential to support the growing number of EVs. Urban planners and policymakers must carefully integrate these facilities with existing infrastructure to facilitate a smooth transition for users, benefiting both the environment and the urban ecosystem [33]. Improving EV parking and charging options is a key strategy in reducing greenhouse gas emissions and decreasing reliance on fossil fuels. By making charging stations more accessible and convenient, cities can encourage more drivers to switch to EVs, promoting greener transportation solutions on a larger scale. This shift aligns with global efforts to combat climate change and supports sustainable urban living [34].

As urban areas expand, prioritizing the advancement of EV infrastructure becomes increasingly important. Investment in these areas addresses current challenges and prepares cities for future transportation demands. This foresight is important for developing a resilient and sustainable urban landscape capable of adapting to evolving needs and the ongoing shift toward environmentally conscious mobility [35].

In conclusion, enhancing EV parking and charging infrastructure is vital for creating a sustainable urban future. By focusing on these improvements, cities can reduce their environmental impact, better serve the needs of current and future EV users, and contribute to the global push for sustainability. These initiatives will lead to cleaner, more functional urban environments that support modern mobility while preserving the environment for future generations.

Author Contributions Conceptualization, Samiappan Dhanalakshmi; methodology, Tarushi Singh, Samiappan Dhanalakshmi; validation, Tarushi Singh, Samiappan Dhanalakshmi; formal analysis, Tarushi Singh; investigation, Tarushi Singh, Samiappan Dhanalakshmi; data curation: Tarushi Singh; writing-original draft preparation, Tarushi Singh; writing—review and editing, Samiappan Dhanalakshmi; supervision, Samiappan Dhanalakshmi; All authors have read and agreed to the published version of the manuscript.

References

1. Jeong J-Y, Cho M, Lee M-H (2022) An analysis of the effect of non-parking facilities in parking-only buildings on the traffic inducement rate. Sustainability 14(7):4250
2. Jacoutot P, Scaccabarozzi A, Nodari D, Panidi J, Qiao Z, Schiza A, Gasparini N (2023) Enhanced sub-1 eV detection in organic photodetectors through tuning polymer energetics and microstructure. Sci Adv 9
3. Terven JR, Córdova Esparza DM, Romero-González J (2023) A comprehensive review of YOLO architectures in computer vision: from YOLOv1 to YOLOv8 and YOLO-NAS. Mach Learn Knowl Extr 5(4):1680–1716
4. Lou H, Duan X, Guo J, Liu H, Gu J, Bi L, Chen H (2023) DC-YOLOv8: small-size object detection algorithm based on camera sensor. Electronics 12(10):2323
5. Graber G, Calderaro V, Galdi V, Piccolo A (2020) Battery second-life for dedicated and shared energy storage systems supporting EV charging stations. Electronics 9(6):939
6. Al-Thani H, Koç M, Isaifan R, Biçer Y (2022) A review of the integrated renewable energy systems for sustainable urban mobility. Sustainability 14(17):10517
7. Zhang X, Li Z, Luo L, Fan Y, Du Z (2022) A review on thermal management of lithium-ion batteries for electric vehicles. Energy 238:121652
8. Carion N, Massa F, Synnaeve G, Usunier N, Kirillov A, Zagoruyko S (2020) End-to-end object detection with transformers. ArXiv abs/2005.12872

9. Terven JR, Córdova Esparza DM, Romero-González J (2023) A comprehensive review of YOLO architectures in computer vision: from YOLOv1 to YOLOv8 and YOLO-NAS. Mach Learn Knowl Extr 5:1680–1716
10. Terven J, Córdova-Esparza DM, Ramirez-Pedraza A, Chavez-Urbiola EA, Romero-Gonzalez JA (2024) Loss functions and metrics in deep learning. arXiv:2307.02694v3
11. Sohan M, Ram T, Ch V (2024) A review on YOLOv8 and its advancements
12. Yang Y, Chang X, Wang Z (2019) YOLO-based vehicle detection in traffic surveillance systems. J Comput Vis 13(3):45–58
13. Wang CY et al (2020) YOLOv4: optimal speed and accuracy of object detection. arXiv:2004.10934
14. Kaliappan VK, Manjusree SV, Shanmugasundaram K, Ravikumar L, Hiremath GB (2023) Performance analysis of YOLOv8, RCNN, and SSD object detection models for precision poultry farming management. In: 2023 IEEE 3rd international conference on applied electromagnetics, signal processing, & communication (AESPC). Bhubaneswar, India, pp 1–6
15. Hasan I, Liao S, Li J, Akram SU, Shao L (2021) Generalizable pedestrian detection: the elephant in the room. In: Proceedings of the IEEE/CVF conference on computer vision and pattern recognition, pp 11328–11337
16. Cubuk ED, Zoph B, Shlens J, Le QV (2019) Randaugment: practical automated data augmentation with a reduced search space. In: 2020 IEEE/CVF conference on computer vision and pattern recognition workshops (CVPRW), pp 3008–3017
17. Rasley J, Rajbhandari S, Ruwase O, He Y (2020) DeepSpeed: system optimizations enable training deep learning models with over 100 billion parameters. In: Proceedings of the 26th ACM SIGKDD international conference on knowledge discovery & data mining
18. Jocher G, Chaurasia A et al (2024) YOLOv8: next-generation object detection. Available at: https://github.com/ultralytics/ultralytics. Accessed: Aug 2024
19. Afonso MHF, Teixeira EH, Cruz MR, Aquino GP, Boass ECV (2023) Vehicle and plate detection for intelligent transport systems: performance evaluation of models YOLOv5 and YOLOv8. https://doi.org/10.13140/RG.2.2.11022.95042
20. Roboflow. YOLOv8 object detection and image classification (n.d.)
21. Lin T-Y et al (2014) Microsoft COCO: common objects in context. In: European conference on computer vision (ECCV), pp 740–755
22. Mahaadevan V, Rajamanickam N, Panda S, Dutta S, Dooly G (2024) AViTRoN: advanced vision track routing and navigation for autonomous charging of electric vehicles. IEEE Access 12:12402–12423. https://doi.org/10.1109/ACCESS.2024.3355018
23. Luo S, Tan Y, Huang L, Li J, Zhao H (2023) Latent consistency models: synthesizing high-resolution images with few-step inference. arXiv.org
24. Alexanderson S, Henter G (2020) Robust model training and generalisation with studentising flows. ArXiv abs/2006.06599
25. Wang G, Chen Y, An P, Hong H, Hu J, Huang T (2023) UAV-YOLOv8: a small-object-detection model based on improved YOLOv8 for UAV aerial photography scenarios. In: Italian national conference on sensors
26. Zhuang F, Qi Z, Duan K, Xi D, Zhu Y, Zhu H, Xiong H, He Q (2019) A comprehensive survey on transfer learning. In: Proceedings of the IEEE
27. Howard J, Ruder S (2018) Universal language model fine-tuning for text classification. In: Annual meeting of the association for computational linguistics
28. Law C, Chen Y, Shi W, Smyth G (2014) Voom: precision weights unlock linear model analysis tools for RNA-seq read counts. Genome Biol 15:R29
29. Irizarry R (2019) Cross validation. In: Introduction to data science
30. MMYOLO: YOLOv8 description (n.d.)
31. Mohya H, Rasheed K (2022) The Ecological dimension in the formation of resilient urban landscape
32. Karimi Z (2021) Confusion matrix
33. Zhang X, Li Z, Luo L, Fan Y, Du Z (2022) A review on thermal management of lithium-ion batteries for electric vehicles. Energy 238:121652

34. Dimitriadou K, Rigogiannis N, Fountoukidis S, Kotarela F, Kyritsis A, Papanikolaou N (2023) Current trends in electric vehicle charging infrastructure; opportunities and challenges in wireless charging integration. Energies 16(4):2057
35. Psaraftis H (2016) Green transportation logistics: the quest for win-win solutions

Enhanced Alzheimer's Disease Classification Using CNN and SMOTE with Optimizer Evaluation

Chitresh Singhal, Nilesh Kumar Pandey, and Rekh Ram Janghel

Abstract Alzheimer's disease is a significant neurological condition that result in memory loss and a steady deterioration in cognitive function. It primarily affects those 65 and older. Early diagnosis is crucial for efficient illness management and better patient outcomes. This paper presents a through deep learning approach that begins with the over sampling synthetic minority method (SMOTE) to balance the Alzheimer's Disease Neuroimaging Initiative's (ADNI) dataset which improves the model's ability to learn from minority classes. Following that, a 2D convolutional neural network (CNN) is used to categorize MRI images into four groups: non-demented, very mild demented, mild demented, and moderate demented. The model underwent rigorous experimentation, assessing various optimizers, dropout rates, and architectural configurations. Finally, the 4 Convo + 4 Pool layer configuration combined with a dropout rate of 0.3 and Adam optimizer achieved an impressive classification accuracy of 99.4%. These findings illustrate the effectiveness of the proposed methodology in accurately detecting Alzheimer's disease and demonstrate its potential for clinical implementation.

Keywords Deep learning · Parameter optimization · Convolutional neural networks and · Alzheimer's disease

1 Introduction

Degenerative neurological illnesses, which primarily affect the elderly, include Alzheimer's disease (AD) and are linked to cognitive decline, memory loss, and functional deficits [1]. An accurate and timely diagnosis is essential since Alzheimer's disease can significantly affect treatment plans and patients' quality of life. The

C. Singhal (✉) · N. K. Pandey · R. R. Janghel
National Institute of Technology, Raipur, India
e-mail: chitreshkumarsinghal1@gmail.com

R. R. Janghel
e-mail: rrjanghel.it@nitrr.ac.in

S. Kumar et al. (eds.), *Artificial Intelligence for Next Generation Computing, Volume 1*,
Lecture Notes in Networks and Systems 1549,
https://doi.org/10.1007/978-981-95-0701-6_15

development of machine learning and deep learning models for the identification and categorization of Alzheimer's disease has advanced dramatically in recent years, with a focus on neuroimaging data such as MRI and PET scans [2]. The application of deep learning for Alzheimer's disease (AD) detection has significantly evolved, with various methodologies employed to enhance accuracy and efficiency. One of the promising approach involves the use of DenseNet models with transfer learning on the ADNI dataset resulted in a classification accuracy of 93.18% [3]. Another approach integrated attention mechanisms and generative adversarial networks (GANs) with CNNs, attaining an accuracy of 97.43% for multistage AD classification [4]. Incorporating transfer learning has proven to be highly beneficial for Alzheimer's disease detection, with deep convolutional neural networks (DCNN) and models like VGG-19, ResNet50, and DenseNet121 achieving a classification accuracy of 91% [5]. Transfer learning, coupled with anomaly detection models, has also been explored, utilizing reconstruction-based anomaly detection with contextual, latent, and adversarial losses to further refine accuracy [6].

Additional methods include using artificial neural networks (ANN) in conjunction with principal component analysis (PCA) for dimensionality reduction and classification, which produces an accuracy of 89.22% on the OASIS dataset [7]. Additionally, CNNs with transfer learning have been applied for early AD detection, achieving 95.17% accuracy [8]. Deep learning-based hybrid models have also shown significant promise. Another hybrid model combined Inception V3 and CNN for AD detection, with an accuracy of 97.31% [9]. Ensemble learning methods have also proven effective, with techniques like bagging, boosting, and stacking achieving a classification accuracy of 92.7% [10]. Adaptive voting with multiple classifiers has also shown promise for early AD detection, achieving an accuracy of 90.23% [11]. A model combining VGG-16, Inception V3, and random forest classifiers achieved 91.10% accuracy [12]. Another study evaluated CNN, VGG-16, and VGG-19 on the Kaggle Alzheimer's dataset, finding that VGG-19 outperformed the others with a 77% accuracy rate [13]. A multilevel classification model using MobileNetV2 and LSTM achieved an accuracy of 90.03% [14]. A study that used SVM for classification with AlexNet, DenseNet, and VGG-16 for feature extraction produced an accuracy of 99.23% [15]. Another study applied transfer learning on VGG-16 and CNN models trained on segmented data, resulting in an accuracy of 99.05% [16].

A framework combining CNN and KNN achieved 95.60% accuracy on MRI images [17]. Patch-based classifiers have also shown promise, with an approach using CNNs and manually localized hippocampus regions achieving 90.05% accuracy [18]. In another study, 2D CNN and VGG-16 models applied to MRI images yielded 70.30% accuracy, with VGG-16 performing better [19]. Feature extraction techniques, such as HAAR cascades combined with SVM, KNN, decision trees, and naive Bayes, have also demonstrated effectiveness, with the highest accuracy of 94.10% achieved using SVM [20]. A study that used an ensemble of VGG-16, EfficientNet, and ResNet152V2 through transfer learning reached 96% accuracy [21].

Additional studies highlight the effectiveness of segmentation, transfer learning, and anomaly detection techniques. For instance, ResNet50 with additional CNN layers achieved an accuracy of 97.49% [2]. The ADASYN technique was used for balancing datasets, leading to an accuracy of 95.45% [22]. Other approaches, such as SVM and region of interest (ROI) segmentation, resulted in accuracies of 90.66% and 92.40%, respectively [23, 24]. Finally, transfer learning for AD detection using customized CNN models on segmented images reported an accuracy of 97.84%. These advancements demonstrate how deep learning and transfer learning can revolutionize Alzheimer's disease diagnosis.

With these developments in mind, this study presents an innovative deep learning approach that maximizes classification accuracy in Alzheimer's disease detection by combining the over-sampling synthetic minority method (SMOTE) with a 2D convolutional neural network (CNN). Section 2 will detail the proposed methodology, outlining the steps taken for data preparation and model implementation. Section 3 will present the results and discussion, focusing on the experiments conducted with various parameters and optimizers. Finally, Sect. 4 will summarize the findings and their implications.

2 Proposed Methodology

The meticulous process of developing a deep learning model that can divide MRI images of Alzheimer's disease into four groups is described in this section: very mildly demented, mildly demented, moderately demented, and non-demented as shown in Fig. 1. The methodology comprises five main stages: dataset gathering, data preprocessing, CNN model training, Alzheimer's disease classification, and performance evaluation.

2.1 Alzheimer MRI Dataset

The Alzheimer MRI Dataset serves as the beginning point for the process, consisting of 6400 MRI images that are initially classified into four stages of dementia. Each category represents a distinct stage of the disease: non-demented, very mild

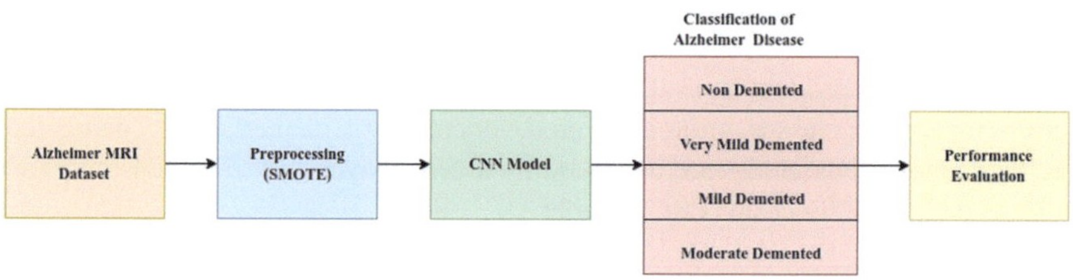

Fig. 1 Flowchart of the proposed system

demented, mild demented, and moderate demented. However, the class distribution in the original dataset is imbalanced, which could result in biased model predictions. To address this issue, we use a data balancing strategy during the preprocessing step.

2.2 *Data Preprocessing*

We apply the over-sampling synthetic minority method (SMOTE) to rectify the class imbalance in the MRI dataset. To guarantee that every minority class has an equal number of images, SMOTE [25] generates synthetic samples. With 3200 samples per class, the total number of images increased to 12,800 after applying SMOTE, enabling more efficient model training. The dataset in divided into 70% (training) and 30% (15% for validation and 15% for testing) ratio and then implemented CNN model.

2.3 *CNN Model*

We use the expanded dataset to train a convolutional neural network (CNN). The MRI pictures are scaled to a standard dimension of 176 × 176 pixels before being entered into the model. The CNN is composed of various layers, including convolutional, pooling, and fully connected layers, which extract and learn specific information associated with different phases of Alzheimer's disease.

2.3.1 Architecture of Proposed CNN Model

The four convolutional layers in the suggested CNN architecture—which uses ReLU activation and 3 × 3 kernels—have 32, 64, 128, and 256 filters, respectively, in order to recognize MRI pictures of Alzheimer's disease. A MaxPooling layer (2 × 2 pool size) that downsamples the feature maps while preserving crucial spatial information comes after each convolutional layer. Padding is set to same (preserving the spatial dimensions of the input). The 2D feature maps are converted into a 1D vector via a flatten layer after the convolutional blocks. This 1D vector is then passed through a fully connected (dense) layer with 1024 neurons, and in order to minimize overfitting, it is followed by a dropout layer (0.3 dropout rate). Ultimately, the probability distributions for the four classes—non-demented, very mildly demented, mildly demented, and moderately demented—are produced by a dense layer with four neurons and a softmax activation function (Fig. 2; Table 1).

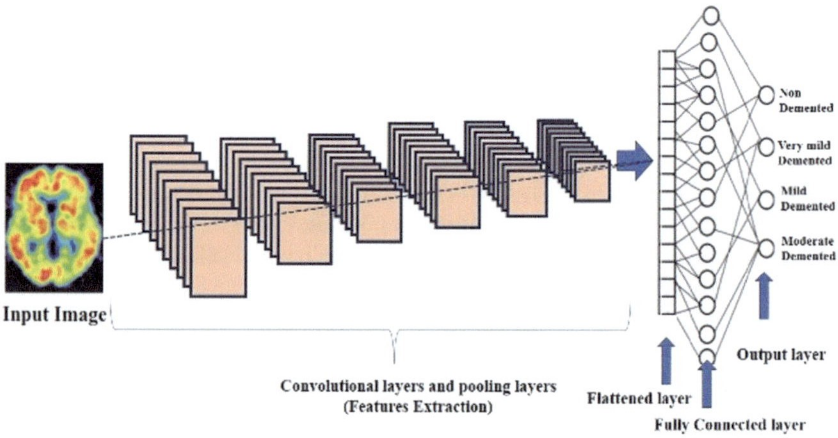

Fig. 2 Proposed architecture of CNN model

Table 1 Model summary of proposed CNN architecture

Layers	Filters	Filter size	Stride	Padding	Size of featured map	Activation function
Conv2D-1	32	3 × 3	2	Same	88 × 88 × 32	ReLU
MaxPooling2D-1	–	2 × 2	2	Same	44 × 44 × 32	–
Conv2D-2	64	3 × 3	2	Same	22 × 22 × 64	ReLU
MaxPooling2D-2	–	2 × 2	2	Same	11 × 11 × 64	–
Conv2D-3	128	3 × 3	2	Same	6 × 6 × 128	ReLU
MaxPooling2D-3	–	2 × 2	2	Same	3 × 3 × 128	–
Conv2D-4	256	3 × 3	2	Same	2 × 2 × 256	ReLU
MaxPooling2D-4	–	2 × 2	2	Same	1 × 1 × 256	-
Flatten	–	–	–	–	256	-
Dense-1	1024	–	–	–	1024	ReLU
Dropout	–	–	–	–	1024	–
Dense-2 (Output)	4	–	–	–	4	Softmax

2.4 *Classification of Alzheimer Disease*

Four categories comprise the outputs generated by the trained CNN model: mildly demented, moderately demented, very mildly demented, and non-demented. Each MRI picture is categorized using these categories based on the qualities that were learned during training.

2.5 Performance Evaluation

Finally, a number of metrics, including accuracy, recall, F1-score, and precision, are used to assess the model's performance. These metrics provide a thorough evaluation of the model's classification performance in each of the four classes, guaranteeing that the classifier operates efficiently and is not prejudiced in favor of any one category. These measures, which are obtained from the confusion matrix, shed light on several elements of the model's functionality, including precision, recall, and F1-score [26]. These metrics assess the model's ability to discover all relevant examples, categorize positive instances appropriately, and balance the two. By taking this step, you may be sure that the classifier is impartial and works well in all classifications. Below are the mathematical formulas for each of these metrics.

Accuracy evaluates the model's overall accuracy.

$$\text{Accuracy} = \frac{\text{TP} + \text{TN}}{\text{TP} + \text{FP} + \text{TN} + \text{FN}} \tag{1}$$

Precision is the percentage of genuine positive cases irrespective of all predicted positives.

$$\text{Precision} = \frac{\text{TP}}{\text{TP} + \text{FP}} \tag{2}$$

Recall calculates the percentage of correctly detected true positives.

$$\text{Recall} = \frac{\text{TP}}{\text{Tp} + \text{FN}} \tag{3}$$

F1-Score provides a harmonic mean of the precision and recall.

$$\text{F1 - Score} = 2 * \frac{\text{Precision} * \text{recall}}{\text{Precision} + \text{recall}} \tag{4}$$

In the above formulations, we have indicated true positives (TP), false positives (FP), false negatives (TN), and true positives (TP). The comprehensive evaluation utilizing these measures guarantees that the model's strengths and limitations are fully captured, allowing for a balanced assessment of its categorization performance.

3 Results and Discussion

In the Results and Discussion section, we evaluate our proposed model's performance through a set of experiments. Section 3.1 describes the class distribution of the Alzheimer's Disease Neuroimaging Initiative's (ADNI) dataset both before and after the SMOTE method was applied, along with the number of samples in training, validation, and testing dataset. Sections 3.2 and 3.3 present the impact of different CNN architectures without and with SMOTE, respectively, highlighting that the 4 Convo + 4 Pool (with SMOTE) configuration achieved the best accuracy of 99.4%. Section 3.4 details the performance metrics for the best model, while Sects. 3.5 and 3.6 explore the effect of different optimizers and dropout rates. Finally, Sect. 3.7 compares our proposed model with the existing techniques, demonstrating its superior performance.

3.1 Datasets

This study employed the Alzheimer's dataset, which came from https://www.kagglc.com/datasets/chitreshsinghal/alzheimer/data. It is a 2D picture and slice taken from the Alzheimer Disease Neuroimaging Initiative (ADNI) database [27]. Most investigations have made use of it, either on its own or in combination with other data sets. The primary goal of the ADNI project is to gather an increasing amount of relevant data and assess how well MRI and other indicators work for identifying Alzheimer's disease development. Sixty-four hundred of the MRI pictures in the dataset are utilized in the study. The dataset contains four types of data: mildly demented, moderately demented, very mildly demented, and non-demented (Table 2).

The total number of samples are divided into training, validation, and testing sets in the ratio of 70% (training) and 30% (15% of validation and 15% of testing) before and after SMOTE as shown in Table 3.

Table 2 Number of samples in ADNI dataset before SMOTE and after SMOTE

Class	Number of samples (Before SMOTE)	Number of samples (After SMOTE)
Non-demented	3200	3200
Mild demented	896	3200
Moderate demented	64	3200
Very mild demented	2240	3200
Total	6400	12,800

Table 3 Number of training, validation, and testing dataset

Class	No. of samples (Before SMOTE)	No. of samples (After SMOTE)
Training dataset	4480	8960
Validation dataset	960	1920
Testing dataset	960	1920

Table 4 Different CNN architectures without SMOTE

Configuration	Number of conv layers	Number of pool layers	Stride	Dropout	Optimizer	Accuracy (%)
1 Convo + 1 Pool	1	1	2	0.3	Adam	96.6
2 Convo + 2 Pool	2	2	2	0.3	Adam	95.3
3 Convo + 3 Pool	3	3	2	0.3	Adam	95.5
4 Convo + 4 Pool	4	4	2	0.3	Adam	94.7
5 Convo + 5 Pool	5	5	2	0.3	Adam	93.5
6 Convo + 6 Pool	6	6	2	0.3	Adam	92.6
7 Convo + 7 Pool	7	7	2	0.3	Adam	91.7
8 Convo + 8 Pool	8	8	2	0.3	Adam	91.6

3.2 Experimentation with Different CNN Architectures Without SMOTE Technique

Various CNN architectures were tested to evaluate their impact on classification accuracy (without SMOTE), as shown in Table 4. The configurations included:

3.3 Experimentation with Different CNN Architectures with SMOTE Technique

Various CNN architectures were tested to evaluate their impact on classification accuracy (with SMOTE), as shown in Table 5. The configurations included:

The results demonstrate that increasing the number of layers improves accuracy up to a point, with the 4 Convo + 4 Pool architecture yielding the best performance, i.e., 99.4%.

Table 5 Different CNN architectures with SMOTE

Configuration	Number of conv layers	Number of pool layers	Stride	Dropout	Optimizer	Accuracy (%)
1 Convo + 1 Pool	1	1	2	0.3	Adam	99.0
2 Convo + 2 Pool	2	2	2	0.3	Adam	99.0
3 Convo + 3 Pool	3	3	2	0.3	Adam	99.1
4 Convo + 4 Pool	4	4	2	0.3	Adam	99.4
5 Convo + 5 Pool	5	5	2	0.3	Adam	98.7
6 Convo + 6 Pool	6	6	2	0.3	Adam	96.7
7 Convo + 7 Pool	7	7	2	0.3	Adam	96.9
8 Convo + 8 Pool	8	8	2	0.3	Adam	95.8

Table 6 Testing result of the proposed model

Classification	Precision	Recall	F1-score
Non-demented	1.0	1.0	1.0
Very mild demented	1.0	1.0	1.0
Mild demented	0.99	0.99	0.99
Moderate demented	0.99	0.99	0.99
Overall accuracy	0.99	0.99	0.99

3.4 An Assessment of the Proposed CNN Model's Performance

With the optimal setup (4 Conv + 4 Pool), the proposed CNN model produced a 99.4% classification accuracy. Recall, precision, and F1-score were among the evaluation metrics that were also calculated and are displayed in Table 6.

3.5 Comparison with Different Optimizers

Table 7 shows the effect of different optimizers on the model's performance.

The Adam optimizer consistently outperformed others due to its adaptive learning rate, which allows for faster convergence and better handling of the sparse gradients often encountered in deep learning models.

Table 7 Experiment with different optimizers of proposed model

Optimizers	SGD	RmsProp	Adamax	Adam
Accuracy (%)	98.02	98.64	99.1	99.4

Table 8 Test around with various dropout rates

Dropout rates	0.1	0.2	0.3	0.4
Accuracy (%)	97.7	97.77	**99.4**	98.69

3.6 Comparison with Different Dropout Rates

To investigate the effects of dropout rates on accuracy, the following results were obtained as shown in Table 8.

A dropout rate of 0.3 provided the best trade-off between regularization and model performance, effectively reducing overfitting while maintaining high accuracy.

Thus, we have experimented with different CNN architecture, different optimizers, and different dropout rates. We are getting maximum accuracy with CNN architecture with SMOTE (4 Convo + 4 Pool layers) with a dropout rate of 0.3 and Adam optimizer.

3.7 Comparative Analysis with Existing Models

In order to contextualize the performance of the new model, it was compared with models that already existed in the literature, as Table 9 illustrates.

Table 9 Comparison of accuracy with existing models on ADNI dataset

Technique	Accuracy (%)
CNN [12]	92.50
MobilenetV2 + LSTM [15]	90.03
InceptionV3 + CNN [9]	97.31
Renet50 [22]	97.49
SVM [25]	92.40
VGG-16-CNN [17]	99.05
CNN (SMOTE + 4 Convo) proposed model	99.4

4 Conclusion

This paper offers a novel method for diagnosing Alzheimer's illness by combining the over-sampling synthetic minority method (SMOTE) with a 2D convolutional neural network (CNN). By effectively balancing the dataset and optimizing architectural settings, the proposed model's astounding 99.4% classification accuracy was attained. The results demonstrate that a 0.3 dropout rate, the Adam optimizer, and the selected CNN architecture significantly enhance model performance. These results contribute to the expanding corpus of knowledge in Alzheimer's research, highlighting the utility of deep learning techniques in clinical settings for early detection and intervention. Future research could look into different datasets and hybrid models to improve diagnostic accuracy.

References

1. Meyer-Baese A, Schmid V (2014) Foundations of neural networks. Pattern Recognit Signal Anal Med Imaging 197–243. https://doi.org/10.1016/B978-0-12-409545-8.00007-8
2. Hanoon FS, Hassin Alasadi AH (2022) A modified residual network for detection and classification of Alzheimer's disease. Int J Electr Comput Eng 12(4):4400–4407. https://doi.org/10.11591/ijece.v12i4.pp4400-4407
3. Islam J, Zhang Y (2017) An ensemble of deep convolutional neural networks for Alzheimer's disease detection and classification [Online]. Available: http://arxiv.org/abs/1712.01675
4. Wong PC, Abdullah SS, Shapiai MI (2024) Double-enhanced convolutional neural network for multi-stage classification of Alzheimer's disease. Adv Sci Technol Eng Syst J 9(2):9–16. https://doi.org/10.25046/aj090202
5. Dawood AS (2024) A comparative study using deep learning models and transfer learning for detection and classification of Alzheimer's disease. Iraqi J Comput Commun Control Syst Eng 24(1):57–70. https://doi.org/10.33103/uot.ijccce.24.1.5
6. Baydargil HB, Park J, Ince IF (2024) Anomaly-based Alzheimer's disease detection using entropy-based probability positron emission tomography images. ETRI J 46(3):513–525. https://doi.org/10.4218/etrij.2023-0123
7. Mahmood R, Ghimire B (2013) Automatic detection and classification of Alzheimer's disease from MRI scans using principal component analysis and artificial neural networks
8. Helaly HA, Badawy M, Haikal AY (2022) Deep learning approach for early detection of Alzheimer's disease. Cognit Comput 14(5):1711–1727. https://doi.org/10.1007/s12559-021-09946-2
9. Rana MM et al (2023) A robust and clinically applicable deep learning model for early detection of Alzheimer's. IET Image Process 17(14):3959–3975. https://doi.org/10.1049/ipr2.12910
10. Buyrukoglu S (2021) Improvement of machine learning models performances based on ensemble learning for the detection of Alzheimer disease. In: Proceedings—6th International Conference on Computer Science and Engineering UBMK 2021, pp 102–106. https://doi.org/10.1109/UBMK52708.2021.9558994
11. Irfan M, Shahrestani S, Elkhodr M (2023) Early detection of Alzheimer's disease using cognitive features: a voting-based ensemble machine learning approach. IEEE Eng Manage Rev 51(1):16–25. https://doi.org/10.1109/EMR.2022.3230820
12. Saim M, Amel F (2022) Classification and diagnosis of Alzheimer's disease based on a combination of deep features and machine learning. In: 2022 7th international conference on image and signal processing and their applications ISPA 2022—proceedings, pp 1–6. https://doi.org/10.1109/ISPA54004.2022.9786318

13. Ajagbe SA, Amuda KA, Oladipupo MA, Afe OF, Okesola KI (2021) Multi-classification of Alzheimer disease on magnetic resonance images (MRI) using deep convolutional neural network (DCNN) approaches. Int J Adv Comput Res 11(53):51–60. https://doi.org/10.19101/ijacr.2021.1152001
14. Rajesh Khanna M (2023) Multi-level classification of Alzheimer disease using DCNN and ensemble deep learning techniques. Signal Image Video Process 17(7):3603–3611. https://doi.org/10.1007/s11760-023-02586-z
15. Naz S, Ashraf A, Zaib A (2022) Transfer learning using freeze features for Alzheimer neurological disorder detection using ADNI dataset. Multimed Syst 28(1):85–94. https://doi.org/10.1007/s00530-021-00797-3
16. Tanveer M, Rashid AH, Ganaie MA, Reza M, Razzak I, Hua KL (2022) Classification of Alzheimer's disease using ensemble of deep neural networks trained through transfer learning. IEEE J Biomed Health Inform 26(4):1453–1463. https://doi.org/10.1109/JBHI.2021.3083274
17. Lanjewar MG, Parab JS, Shaikh AY (2023) Development of framework by combining CNN with KNN to detect Alzheimer's disease using MRI images. Multimed Tools Appl 82(8):12699–12717. https://doi.org/10.1007/s11042-022-13935-4
18. Ahmed S et al (2019) Ensembles of patch-based classifiers for diagnosis of Alzheimer diseases. IEEE Access 7:73373–73383. https://doi.org/10.1109/ACCESS.2019.2920011
19. Mggdadi E, Al-Aiad A, Al-Ayyad MS, Darabseh A (2021) Prediction Alzheimer's disease from MRI images using deep learning. In: 2021 12th International Conference on Information and Communication Systems ICICS 2021, pp 120–125.https://doi.org/10.1109/ICICS52457.2021.9464543
20. Shankar VG, Sisodia DS, Chandrakar P (2022) A novel discriminant feature selection–based mutual information extraction from MR brain images for Alzheimer's stages detection and prediction. Int J Imaging Syst Technol 32(4):1172–1191. https://doi.org/10.1002/ima.22685
21. Sadat SU, Shomee HH, Awwal A, Amin SN, Reza MT, Parvez MZ (2021) Alzheimer's disease detection and classification using transfer learning technique and ensemble on convolutional neural networks. In: Conference proceeding—IEEE international conference on systems, man, and cybernetics, pp 1478–1481. https://doi.org/10.1109/SMC52423.2021.9659179
22. El-Assy AM, Amer HM, Ibrahim HM, Mohamed MA (2024) A novel CNN architecture for accurate early detection and classification of Alzheimer's disease using MRI data. Sci Rep 14(1):1–19. https://doi.org/10.1038/s41598-024-53733-6
23. Ben Rabeh A, Benzarti F, Amiri H (2016) Diagnosis of Alzheimer diseases in early step using SVM (Support Vector Machine). In: Proceeding—computer graphics, imaging and visualization new tech. trends, CGiV 2016, pp 364–367. https://doi.org/10.1109/CGiV.2016.76
24. Morra JH, Tu Z, Apostolova LG, Green AE, Toga AW, Thompson PM (2010) Comparison of adaboost and support vector machines for detecting Alzheimer's disease through automated hippocampal segmentation. IEEE Trans Med Imaging 29(1):30–43. https://doi.org/10.1109/TMI.2009.2021941
25. Enhancing Classification Accuracy for Imbalanced Image Data Using SMOTE | by Fatimazahra Belharar | Medium. Accessed: 3 Oct 2024 [Online]. Available:https://medium.com/@fatimazahra.belharar/enhancing-classification-accuracy-for-imbalanced-image-data-using-smote-41737783a720
26. LabelF.AI (2024) What is accuracy, precision, recall, and F1 score? [Online]. Available:https://www.labelf.ai/blog/what-is-accuracy-precision-recall-and-f1-score. Accessed: 3 Oct 2024
27. Alzheimer's Disease Neuroimaging Initiative (ADNI): ADNI Data. https://adni.loni.usc.edu/data-samples/access-data/

Deep Learning for Brain Tumor Detection Xception CNN-based MRI Image Classification

Narsingh Nath Bauddha and Rakesh Kumar Khare

Abstract Premature recognition of brain tumors is crucial. Only biopsies can classify brain cancers, requiring invasive brain surgery. Brain malignancies can be identified and classified by medical professionals with the use of computationally focused procedures. Here, we present a deep learning approach based on a comparison of several recent studies that used a variety of machine learning techniques to diagnose three different tumor types using magnetic resonance brain images: glioma, meningioma, and pituitary gland in addition, no tumors were also included in the analysis. This makes it possible for doctors to accurately diagnose malignancies in their early stages. In this study, 7023 MRI brain pictures were utilized in this investigation. The Xception network creates a novel convolution neural network (CNN) once the images have been preprocessed and enhanced. Multiple convolution layers with 3 * 3 kernel functions are present in the network. To avoid overfitting, batch normalization layers were utilized, and the ReLU function served as each layer's activation function. The Adamax optimization function was used with a 0.001 rate of learning to maximize efficiency. The suggested model's validation accuracy was determined to be 99.70%, while its training accuracy was found to be 99.98%. The current investigation demonstrates that the suggested CNN Xception Net model has achieved the best classification accuracy for brain tumors. This model obtained excellent performance and optimal execution time compared with other CNN and machine learning approaches used in earlier research. Radiologists and doctors can utilize this recommended network in healthcare facilities to diagnose brain tumors.

Keywords Convolutional neural network (CNN) · Brain tumor · Machine learning · Medical imaging

N. N. Bauddha (✉) · R. K. Khare
Department of Computer Science Engineering, Shri Shankaracharya Institute of Professional Management and Technology, Raipur, Chhattisgarh, India
e-mail: n.bauddha@ssipmt.com

R. K. Khare
e-mail: rk.khare@ssipmt.com

S. Kumar et al. (eds.), *Artificial Intelligence for Next Generation Computing, Volume 1*,
Lecture Notes in Networks and Systems 1549,
https://doi.org/10.1007/978-981-95-0701-6_16

1 Introduction

Brain tumors are a serious illness caused by the unregulated and abnormal expansion of brain cells, which affects the brain's function. These growths might be noncancerous (benign) or cancerous (malignant). Symptoms include headaches, seizures, vision problems, and balance issues, but can vary. Brain tumors are widely recognized as the tenth highest cause of mortality globally [1] with an estimated 18,990 deaths expected in the United States in 2023 [2]. The origins of brain tumors remain elusive, necessitating early and accurate detection and treatment [3].

Unlike other types of cancer, early signs of brain tumors may not be apparent making timely identification difficult [4]. Approximately 24,810 new cases of brain tumors were projected to be diagnosed in the United States in 2023, with a gender split of 14,280 men and 10,530 women [2]. India has a major prevalence of brain tumors with 28,142 new cases per year (UNI, citing Globocan 2018) [5]. Deep learning has transformed clinical studies by providing algorithms that can learn from data with no need of an individual, leading to widespread adoption across various sectors, and provides a classification of images [3]. The work's objective is to evaluate brain MRI's dependence on various predictive CNN models to handle varying levels of complexity in brain MRI data. This will be achieved by contrasting the classification efficiency of related studies on various models, i.e., two-dimensional CN [6] N, ResNet50 [7], GoogleNet [8], InceptionV3 [7], M-SVM, and MobileNetV2 [9], and all these models were compared with proposed XceptionNet (Tested) model. The paper is structured as follows: Section 2 Literature review. Section 3 Methodology Sect. 4 Experimental results and analysis Sect. 5 Discussions and conclusions draw inferences from the results, as well as future remarks.

2 Literature Review

This section introduces a variety of advanced brain tumor classification approaches, including methods based on deep learning and transfer learning (TL). The combination of deep learning and transfer learning yields several tumor classification algorithms. These cutting-edge approaches may be classified into three broad groups: those based on deep learning, those built-in machine learning, and those that combine components of both. In this study [10], author aimed to assess the effectiveness of pre-trained transfer learning models classifiers-"InceptionResNetV2, InceptionV3, Xception, ResNet18, ResNet50, ResNet101, ShuffleNet, DenseNet201, and MobileNetV2" by employing a detailed classification approach for automatic identification and detection of brain tumors. The detailed classification experiment revealed that the Inception, ResNetV2, and TL algorithms outperformed others, achieving optimal tumor detection and classification. Consequently, it can be designated as the most effective classification algorithm, attaining an accuracy of 98.91%.

The authors [11] proposes a BT (brain tumor) classification method based on a CNN for computer-assisted diagnostics. Using an 18-layered CNN model, the experiments on three datasets achieved 94.74% accuracy in classifying brain tumors, and 90.35% accuracy in tumor grade classification. This study [4] used a transfer learning approach with deep learning to distinguish between malignant and benign brain tumor images, using a 696 T1-weighted dataset. The classification research featured a thorough comparison of the performance of multiple CNN models, including ResNet-101, ResNet-50, GoogleNet, AlexNet, and SqueezeNet. The AlexNet CNN model, which was pretrained via transfer learning, achieved a maximum accuracy of 99.04%, demonstrating its effectiveness in the classification test.

The authors [3] focuses on assessing how brain MRI classification depends on various CNN models while accounting for data complexity. The researchers experimented extensively with four CNN models: S-CNN (trained from scratch), ResNet50, InceptionV3, and Xception. These models were then compared to a newly suggested PCA-CNN model, which had the greatest accuracy (99.27% for the BT dataset). According to [9], linear contrast stretching was used as a processing step to improve image edge detail. A 17-layer CNN uses an entropy-based technique to choose features, employing a modified MobileNetV2 for feature extraction via transfer learning. The suggested technique achieves great accuracy using the M-SVM architecture, scoring 97.47–98.92% in brain tumor identification. The authors [12] focused on the detection of abnormalities, specifically classifying tumorous cells from magnetic resonance images (MRI). "The proposed deep learning model's (LeU-Net) performance was compared to that of basic CNN models, including Le-Net, U-Net, and VGG-16." The LeU-Net model demonstrated a remarkable accuracy of 98%.

The authors [7] initiated a study aiming to enhance the precision of brain cancer classification. The study utilized "VGG16, VGG19, ResNet50, ResNet101, and InceptionV3" for brain tumor classification, achieving individual accuracies of 96.72%, 96.17%, and 95.55%, respectively. Combining the top-performing three classifiers into an ensemble model, named EBTDM, resulted accuracy of 98.60%. The study [6] compared an auto-encoder network (95.63% accuracy) to a 2D CNN (96.47% accuracy) for brain tumor classification, and both achieved high average recall values (94–95%). Notably, K-nearest neighbors (KNN) demonstrated the highest accuracy (86%), while multilayer perceptron (MLP) had the lowest accuracy (28%), with a significant difference in means ($p < 0.05$).

In this study [10], the EfficientNet-B0 model was fine-tuned with custom layers to enhance the classification and detection of brain tumor images. The proposed approach outperformed other CNN models, achieving superior accuracy, precision, recall, and AUC values. With an overall classification accuracy of 98.87%, the fine-tuned EfficientNet-B0 surpassed state-of-the-art models, demonstrating its effectiveness in brain tumor detection (Table 1).

Table 1 Research on identifying brain tumors

Refs.	Title	Model	Research gap
[3]	"Data complexity based evaluation of the model dependence of brain MRI images for classification of brain tumor"	ResNet50, inceptionV3	Performance scores for the less complicated data are greater than those for the more complex data.
[4]	"Multi-classification of brain tumor MRI images using deep convolutional neural network with fully optimized framework"	AlexNet, Inceptionv3, ResNet-50, VGG-16 and GoogleNet.	Try to achieve similar accuracy using different network model
[6]	"MRI-based brain tumor detection using convolutional deep learning methods and chosen machine learning technique"	2D CNN, auto encoder	There is a need to improve classification accuracy and provide timely, accurate brain tumor diagnosis.
[8]	"Brain tumor classification using deep CNN features via transfer learning"	GoogLeNet	Insufficient training data led to overfitting
[13]	"Classification of brain tumor from magnetic resonance imaging using vision transformers ensembling"	ImageNet-based ViT	Dimensions and position of the brain tumor

3 Material and Methods

3.1 Dataset

Figure 1 presents the methodology used in this study. The primary processes in this study are selecting brain tumor datasets, preparing MRI images for analysis, extracting relevant features, and classifying tumors using multiple classification algorithms. This work applies deep learning to diagnose brain tumors using XceptionNet correctly. Noise and MRI artifacts are removed before processing the input pictures. Next, training and testing sets are created using the MRI image dataset. Here, 1311 MR images were used for testing or validation and 5712 pictures for training to classify the data into four groups. The applied image-based dataset, which is publicly available on Kaggle, has 7023 pictures [14]. In this dataset, there were four distinct picture types: meningioma (1339), glioma (1321), pituitary gland tumor (1457), and no tumor (1595). The images were taken in the coronal, axial, and sagittal planes. Figure 2 presents illustrations of different tumor types and anatomical planes.

Fig. 1 Flow diagram of proposed methodology

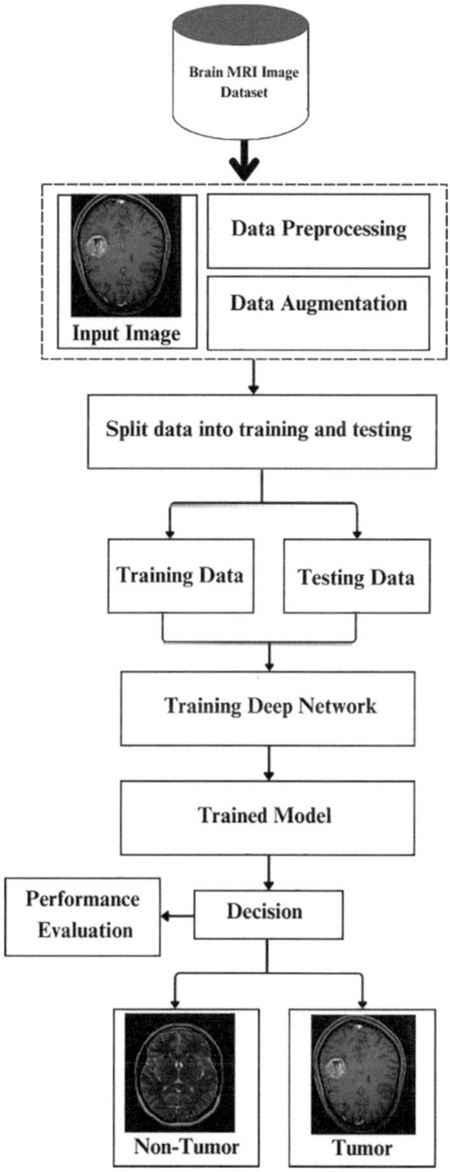

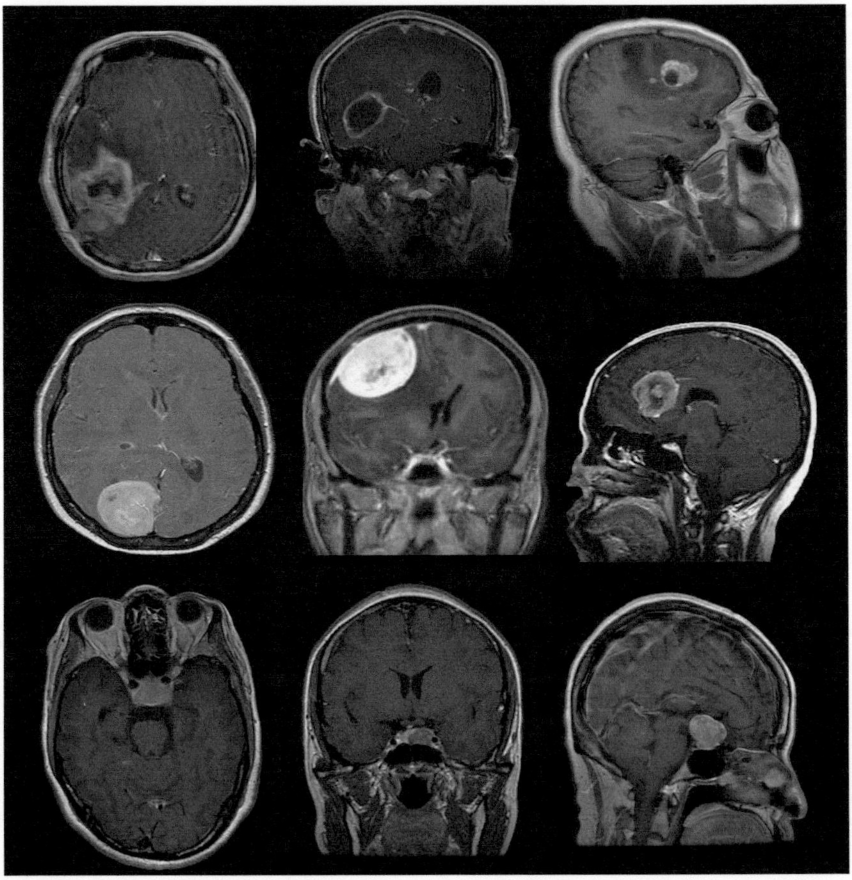

Fig. 2 Various MRI brain tumor

3.2 Data preparation and enhancement

MRI images were sourced from the dataset has 512 * 512 pixels and resized to 299 * 299 pixels. Each image was rescaled pixel value in the image by dividing it by 255. This normalization ensures all pixel intensities fall within a common range, potentially improving the training process. Brightness adjustment to the images during training. It randomly picked a value between 0.8 and 1.2 and multiplied each pixel value in the image by that factor, the model learned to be more robust and helped prevent overfitting to the specific lighting conditions in the original dataset. The model processed a batch size of 32 images together before updating its weights.

3.3 Proposed Approach

The CNN Xception Net's proposed design is seen in Fig. 3. This study used a collection of 7023 data, of which 80% (5712) were used as training data and 20% (1311) as testing data. The network's underlying layer is called Xception Functional. The term "Xception model" refers to an efficient convolutional neural network (CNN) architecture. Features are extracted from the input picture using this pretrained model. It accepts an input picture of size (299, 299, 3), where 3 is the number of RGB color channels and 299 is the image height and width (assuming square images). The proposed layer output shape to the batch size of 32, which can change to 16 based on the number of features (or dimensions) in the output vector 2048 depending on the way the model is employed. The Xception model contains 21069740 trainable parameters in total.

The structure of a convolutional network, often called a neural network, is hierarchical. Flatten: this layer transforms the 2D output into a 1D vector, most likely carrying spatial information. While the overall number of components (2048 features from each image) remains constant, the output form changes to (None, 2048). Dropout: To avoid overfitting, this layer randomly removes a specific proportion of neurons (units) during training. There are no more trainable parameters on it (0). Dense (Dense): With 128 neurons, this layer is completely linked. It creates a new vector with 128 features by using the previous layer's flattened vector and transforming it. Here, there are 262,272 trainable parameters. The initial dropout layer uses a 128-dimensional vector for operation but likewise randomly removes neurons during training. It contains zero extra parameters. dense_1 (Dense): The last layer consists of four neurons and is also completely linked at Table 2. It converts the preceding layer's 128-dimensional vector into a four-dimensional output vector. Here, the weights connecting each neuron in the preceding layer to each neuron in the output layer total 516 trainable parameters. For feature extraction, this neural network architecture makes use of a pre-trained Xception model. The output is then flattened, regularization is applied using dropout, and four fully connected layers

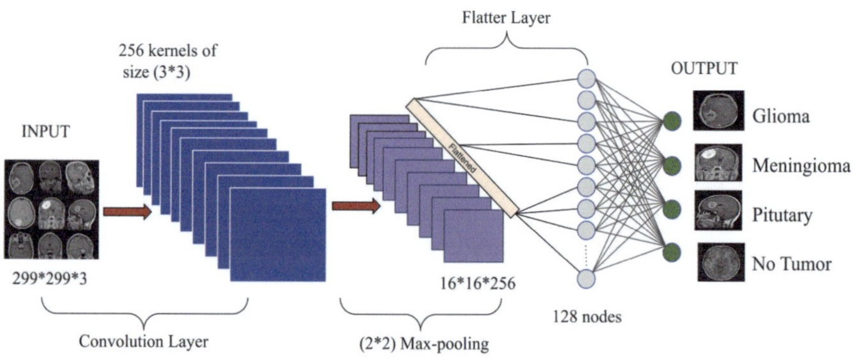

Fig. 3 Xception CNN architecture

Table 2 CNN Xception network architecture for the classification of four class

Layer (type)	Output Shape	Parameters
Xception	(0, 2048)	20861480
Flatten	(0, 2048)	Nil
Dropout	(0, 2048)	Nil
Dense	(0, 128)	262272
Dropout_1	(0, 128)	Nil
Dense_1	(0, 4)	516
Total parameters: 21124268 (80.47 MB) Trainable parameters: 21069740 (80.37 MB) Nontrainable parameters: 54528 (213.00 KB)		

with dropout in between are used to possibly categorize the input image into four categories. A softmax activation function was employed to categorize the data. The architecture consisted of 128 completely connected layers with four fully connected layers. Batch normalization layers were included to mitigate overfitting. Dropout layers with a rate of 0.25 were inserted after the completely connected and max-pooling layers. The ReLU activation function was applied to all layers except the final fully connected layer. The learning rate parameter was tested using the Adamax as an optimizing function with a value of 0.001 to maximize efficiency. Additionally, 0.0006 was shown to be the greatest figure with minimal error. The training method was verified after 15 epochs. A batch comprising 32 samples was employed. Each training cycle consumed roughly 1904 s. Extracted features from the convolutional layer were fed into the initial, fully connected layer containing 128 neurons. The quantity of weights (Wconv) within the convolutional layer was determined by the dimensions of the preceding convolutional layer output ($x_1.x_2$), the filter count (z), and the neuron count in the fully connected layer. Consequently, the convolutional layer's weight calculation was [6, 15]: $W_{\text{conv}} = x_1 * x_2 * z * U_{\text{fc}} = 32 * 32 * 2 * 128 = 262144$. The total parameters for the initial fully connected layer amounted to 262144 plus 128 bias terms, equaling 262272. The details of system hardware specification are shown in Table 3.

3.4 Performance evaluation metrics

In this paper, the performance of deep neural networks was evaluated using a number of critical metrics, including accuracy, precision, recall, and F1-score [16]. Every statistic was calculated using pre-established algorithms to give a thorough evaluation of the models' efficacy.

$$\text{Accuracy} = \frac{\text{TP} + \text{TN}}{\text{TP} + \text{TN} + \text{FP} + \text{FN}} \tag{1}$$

Table 3 System hardware specification

Hardware	Description
GPU	GPU with CUDA support (NVIDIA GPUs are commonly used) NVIDIA GeForce GTX 1060
CPU	Quad-core processor or higher i7-8700 CPU @ 3.20GHz
RAM	16 GB
DISK	256 GB SSD
OS	Window 10
Development tool	Python 3.7, Jupyter Notebook
Library	TensorFlow, Numpy, Pandas, Matplotlib, Seaborn, Scikit-learn, Pytorch, Keras

$$\text{Precision} = \frac{\text{TP}}{\text{TP} + \text{FP}} \tag{2}$$

$$\text{Recall or Sensitivity} = \frac{\text{TP}}{\text{TP} + \text{FN}} \tag{3}$$

$$\text{F - measure} = 2 \times \frac{\text{Precision} \times \text{Recall}}{\text{Precision} + \text{Recall}} \tag{4}$$

True-negatives (TN), true-positives (TP), false-positives (FP), and false-negatives (FN).

4 Experimental results and analysis

Table 4 shows the outcomes of our suggested CNN Xception network. The suggested model's validation accuracy was determined to be 99.70%, while its training accuracy was found to be 99.98%. Table 5 summarizes the F-measure, recall, and accuracy for each of the four classes that were acquired via the Xception network. The suggested model's training, validation, and loss assessments concerning epoch counts are displayed in Fig. 4, value of instances is represented by each column, while the actual (true) data points is represented by each row. This matrix explains its name, which helps us identify whether the findings are vague or the classes overlap. Improving accuracy and precision in the modeling process is vital for clinical research. Figure 5 illustrates the effect of FP and FN rates. Table 6 presents an analysis of performance between the proposed CNN architecture and comparative models. Xception achieved the highest mean precision of 99.75% compared to the other classifiers. EfficientNet-B0 demonstrated precision values more adjacent to Xception, ResNet50 exhibited the lowest precision among all. Based solely on precision, Xception and EfficientNet-B0 are superior choices. Figure 6 illustrates the graph of the above comparisons.

Table 4 Four-class classification results of Xception

Model	Training-accuracy	Test accuracy	Train-loss	Test-loss
2D CNN	0.9647	0.9344	0.09329	0.28095
Convolution auto-encoder	0.95637	0.9092	0.1147	0.32122
Proposed model	0.9998	0.9970	0.0006	0.0235

Table 5 Xception model performance metrics

Tumor (type)	Precision	Recall	F-measure	Support
Glioma	1.00	0.99	1.00	150
Meningioma	0.99	0.99	0.99	153
No tumor	1.00	1.00	1.00	203
Pituitary gland tumor	1.00	1.00	1.00	150
Average	0.9975	0.995	0.9975	

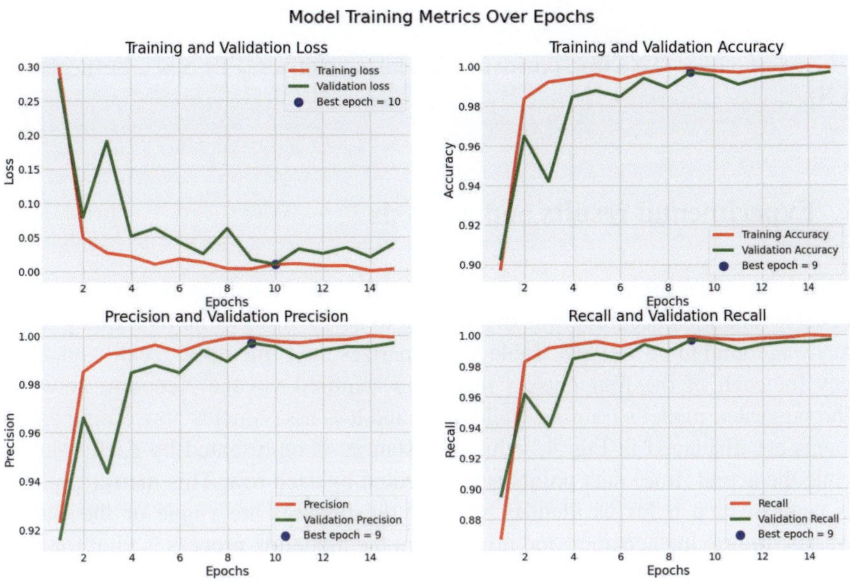

Fig. 4 Model performance Over 15 epochs

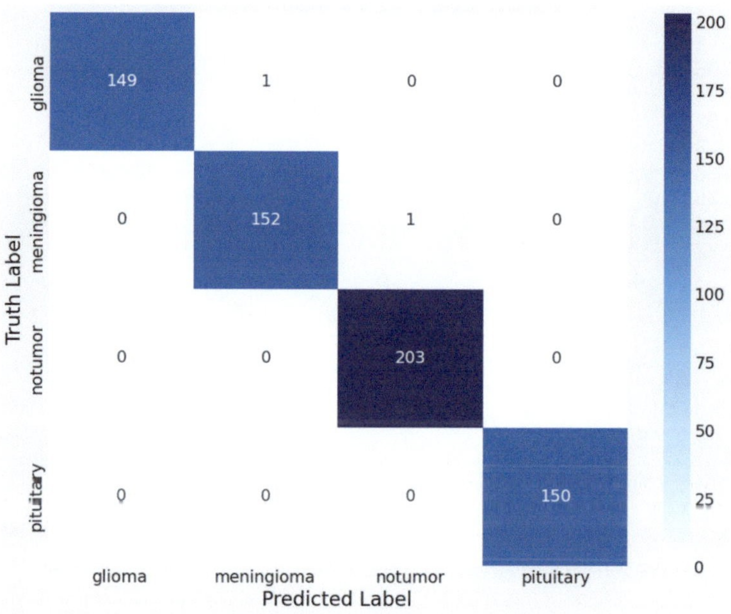

Fig. 5 Confusion matrix Xception model performance

Table 6 Model performance comparison

Model	Precision	Recall	F-measure
2D CNN [6]	0.9475	0.9575	0.95
Auto-encoder [6]	0.9425	0.9425	0.94
S-CNN [3]	0.9712	0.976	0.9736
ResNeT50[7]	0.8838	0.9229	0.9003
InceptionV3[7]	0.9832	0.9806	0.982
GoogleNet[8]	0.973	0.976	0.9896
EfficientNet-B0[10]	0.994	0.995	0.989
Xception (Proposed model)	0.9975	0.995	0.9975

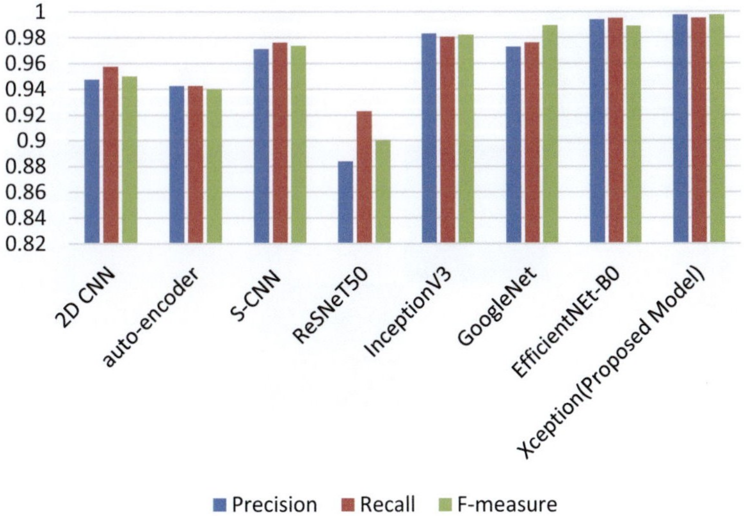

Fig. 6 Model comparison graph

5 Discussions and Conclusions

The core purpose of this research was to design deep learning model to get the best result and compare with the outcomes of different deep learning models as well as ML techniques applied to classify MRI images into four classes of brain tumors (glioma, meningioma, no tumor, and pituitary gland tumor) with the previous studies. The image dataset used was accessible to the public at Kaggle with 7023 MRI images. Labeling medical images requires a great deal of work and expertise as well as the system requires a considerable amount of time to process the datasets during training and testing. Upgrading the hardware is required to ensure seamless operation; alternatively, switching to Google Collab online should be taken into consideration. To identify brain tumors more quickly and accurately, future research will focus on developing deeper neural networks and working on huge datasets with upgraded tools that render more effective results. This may need using DL and ML model for improvements. Our proposed model may also be used to the identification of various malignancies using MRI images. The suggested model's validation accuracy was determined to be 99.70%, while its training accuracy was found to be 99.98%. The current investigation demonstrates that the suggested CNN Xception Net model has achieved the best classification accuracy for brain tumors. This model obtained excellent performance and optimal execution time when compared to previous CNN and ML techniques from previous studies.

References

1. Alsaif H et al (2022) A novel data augmentation-based brain tumor detection using convolutional neural network. Appl Sci 12(8):3773. https://doi.org/10.3390/app12083773
2. Siegel RL, Miller KD, Wagle NS, Jemal A (2023) Cancer statistics, 2023. CA Cancer J Clin 73(1):17–48. https://doi.org/10.3322/caac.21763
3. Kujur A, Raza Z, Khan AA, Wechtaisong C (2022) Data complexity based evaluation of the model dependence of brain MRI images for classification of brain tumor and Alzheimer's disease. IEEE Access 10:112117–112133. https://doi.org/10.1109/ACCESS.2022.3216393
4. Irmak E (2021) Multi-classification of brain tumor MRI images using deep convolutional neural network with fully optimized framework. Iran J Sci Technol-Trans Electr Eng 45(3):1015–1036. https://doi.org/10.1007/s40998-021-00426-9
5. Brain Tumor: Statistics | Cancer.Net. (2024). Accessed: 19 Feb 2024 [Online]. Available: https://www.cancer.net/cancer-types/brain-tumor/statistics
6. Saeedi S, Rezayi S, Keshavarz H, Niakan Kalhori SR (2023) MRI-based brain tumor detection using convolutional deep learning methods and chosen machine learning techniques. BMC Med Inform Decis Mak 23(1):16. https://doi.org/10.1186/s12911-023-02114-6
7. Islam AT et al (2022) An efficient deep learning approach to detect brain tumor using MRI images. In: Proceedings of 2022 25th international conference on computer and information technology, ICCIT 2022, pp 143–147. https://doi.org/10.1109/ICCIT57492.2022.10054999
8. Deepak S, Ameer PM (2019) Brain tumor classification using deep CNN features via transfer learning. Comput Biol Med 111:103345. https://doi.org/10.1016/j.compbiomed.2019.103345
9. Maqsood S, Damaševičius R, Maskeliūnas R (2022) Multi-modal brain tumor detection using deep neural network and multiclass SVM. Medicina (Lithuania) 58(8):1090. https://doi.org/10.3390/medicina58081090
10. Shah HA, Saeed F, Yun S, Park JH, Paul A, Kang JM (2022) A robust approach for brain tumor detection in magnetic resonance images using finetuned EfficientNet. IEEE Access 10:65426–65438. https://doi.org/10.1109/ACCESS.2022.3184113
11. Işin A, Direkoğlu C, Şah M (2016) Review of MRI-based brain tumor image segmentation using deep learning methods. In: Procedia computer science. Elsevier B.V., pp 317–324. https://doi.org/10.1016/j.procs.2016.09.407
12. Rai HM, Chatterjee K (2021) 2D MRI image analysis and brain tumor detection using deep learning CNN model LeU-Net. Multimed Tools Appl 80(28):36111–36141. https://doi.org/10.1007/s11042-021-11504-9
13. Tummala S, Kadry S, Bukhari SAC, Rauf HT (2022) Classification of brain tumor from magnetic resonance imaging using vision transformers ensembling. Curr Oncol 29(10):7498–7511. https://doi.org/10.3390/curroncol29100590
14. Cheng J (2024) "brain tumor dataset," figshare. Accessed: 23 Feb 2024 [Online]. Available: https://figshare.com/articles/dataset/brain_tumor_dataset/1512427
15. Riazi R, Asrardel M, Shafaee M, Vakilipour S, Zare H, Veisi H (2017) A data mining study on combustion dynamics and NOxemission of a swirl stabilised combustor with secondary fuel injection. Int J Heavy Veh Syst 24(3):215–238. https://doi.org/10.1504/IJHVS.2017.084865
16. Sokolova M, Japkowicz N, Szpakowicz S (2006) Beyond accuracy, F-score and ROC: a family of discriminant measures for performance evaluation. In: AAAI workshop—technical report, vol WS-06-06, pp 24–29. https://doi.org/10.1007/11941439_114

Machine Learning Model to Study Single-Cell Osteoarthritic Chondrocytes via Raman Spectroscopy

Gavish Uppal, Nisha, Tarun Goyal, Anup Kumar, and Rajesh Kumar

Abstract Raman spectroscopy (RS) has become an imperative instrument in biomedical research. It is a nondestructive and label-free laser spectroscopic technique, which can identify chemical composition of a biological sample by detecting multiple molecular components simultaneously without the need or minimal sample preparation under physiological condition. A major limitation while analyzing Raman spectra of a biological sample is the overlapping nature of vibrational modes of different molecules at a single spectral peak. Machine learning (ML) techniques encompass various multivariate and statistical analysis methods, which can be broadly classified into supervised and unsupervised techniques. Multivariate curve resolution-alternating least square (MCR-ALS) technique is a multivariate ML model that can potentially deconvolute overlapping Raman vibrational modes and extract the meaningful spectra that can be associated with biochemical components along with their concentrations. In this study, Raman spectral data were acquired from osteoarthritic chondrocytes. The spectra of three cellular components associated with DNA, proteins, and lipids were extracted using MCR-ALS technique, and it was observed that concentration of DNA and proteins decreased while concentration of lipids increased with progression of osteoarthritic grades. The study indicated that combined approach of Raman-MCR analysis can open new possibilities for improved Raman spectral analysis, which might assist further in providing the meaningful interpretation of complex Raman spectra originated from the biological samples.

G. Uppal · Nisha · R. Kumar (✉)
Department of Biomedical Engineering, Indian Institute of Technology Ropar, Rupnagar, Punjab, India
e-mail: rajeshkumar@iitrpr.ac.in

R. Kumar
Department of Physics, Indian Institute of Technology Ropar, Rupnagar, Punjab, India

T. Goyal
Department of Orthopedics, All India Institute of Medical Sciences Bathinda, Bathinda, Punjab, India

A. Kumar
Department of Orthopedics, All India Institute of Medical Sciences Patna, Patna, Bihar, India

S. Kumar et al. (eds.), *Artificial Intelligence for Next Generation Computing, Volume 1*, Lecture Notes in Networks and Systems 1549,
https://doi.org/10.1007/978-981-95-0701-6_17

Keywords Raman spectroscopy · Osteoarthritis · Chondrocytes · MCR-ALS · Machine learning

1 Introduction

Raman spectroscopy (RS) a nondestructive and label-free laser spectroscopy technique, which can identify chemical composition of a biological sample under physiological condition. RS can detect multiple molecular components simultaneously with minimal or no sample preparation. RS has been used extensively for research in the field of chemical biology for biochemical analysis of tissues and cells to monitor the variations in their molecular distribution with the progression of a disease [1, 2]. The use of near infrared (NIR) lasers as light source allowed capturing intracellular biochemical information with little auto-fluorescence generated from biological sample. Other majorly used analytical spectroscopic techniques such as infrared spectroscopy (IR) require sample preparation as the vibrations associated with water molecules can hinder the interpretation of data while mass spectroscopy (MS) is a destructive technique and nuclear magnetic resonance (NMR) lacks spatial resolution. Optical techniques such as fluorescence spectroscopy have also been used extensively to understand the disease at molecular level. However, this technique has a few limitations: Labeling with a fluorescent dye leads to alteration in physiological state of biological sample, and there is a restriction to number of colors that can be detected simultaneously [3, 4].

One of the cutting-edge applications of RS is the analysis of single cell to detect their biomolecular composition. RS could be an appropriate approach for single-cell analysis in the context of scale value because the measurements can be carried out at micrometer (μm)-scale level. The cell comprising the biomolecules such as lipids, proteins, nucleic acids can provide Raman spectral fingerprint revealing their changes and concentrations during disease progression [5, 6]. These spectral features can be extracted through different machine learning models and can further analyzed to detect early changes with enhanced accuracy. A major limitation while analyzing Raman spectra of a biological sample is that a spectral peak might be overlapped with the signal originated from a vibrational mode of different biomolecules. Therefore, such overlapping vibrational information from bio-sample may lead to misinterpretation of content of multiple constituents present in the sample.

Machine learning (ML) encompasses multivariate and various statistical techniques that can be categorized into supervised and unsupervised models. In unsupervised ML models, there are no training and test data while in supervised ML models, a particular model is skilled using data containing training information, and to determine the accuracy of that supervised ML model, test data are fed to the model [7]. In general, MCR grouped under supervised ML approach but can be executed in an unsupervised manner without the use of constraints in spectral/concentration direction [8, 9]. Multivariate curve resolution-alternating least square (MCR-ALS) is an algorithm which can unmix the Raman data containing mixed spectral information into the spectral signatures of interpretable biochemical components and their corresponding concentrations [10]. MCR-ALS is based on bilinear multivariate ML model in which Raman spectra are assumed to be linear combination of spectral signature of different chemical components along with their concentrations (Fig. 1) [7]. Various algorithms have been used to unmix the overlapping spectral peaks into the meaningful spectral profile of pure chemical components. Algorithms such as vertex component analysis (VCA) and N-FINDR were applied to unmix complex Raman spectra [11]. MCR-ALS might be a more appropriate algorithm compared to other algorithms as it is more flexible. The first reason is that algorithms other than MCR-ALS assume the existence of "pure" components in the spectral data matrix that is difficult to achieve in the case of biomedical samples due to the complex Raman spectra caused by the overlapping or mixing of molecular components. The second reason is that other models need to use non-negativity and closure constraints while MCR model is flexible in context of inclusion of multiple constraints such as selectivity, non negativity, and local rank [11, 12]. Another advantage of MCR approach is that it does not require any prior information about the sample composition. Other multivariate techniques such as principal component analysis (PCA) which falls under unsupervised ML model have also been compared to MCR-ALS on the same spectral data. Compared to PCA, MCR analysis provides more interpretable spectra of molecular constituents along with their concentrations, which can be attributed to the flexibility of the MCR-ALS algorithm [11]. Thus, the MCR-ALS approach might be more suitable for in-depth spectral analysis of biological samples.

As spectral overlapping of a vibrational mode of different molecules is one of the major challenges in analyzing the spectra originating from biological samples, in this work, MCR-ALS has been utilized to perform spectral unmixing of Raman data acquired from osteoarthritic chondrocytes. Raman-MCR has been coupled and utilized to detect biomolecular changes in chondrocytes isolated from three different grades of osteoarthritic cartilage (ICRS grades I, II, and III), and the relative quantitative changes in biochemical constituents at the cellular level were analyzed.

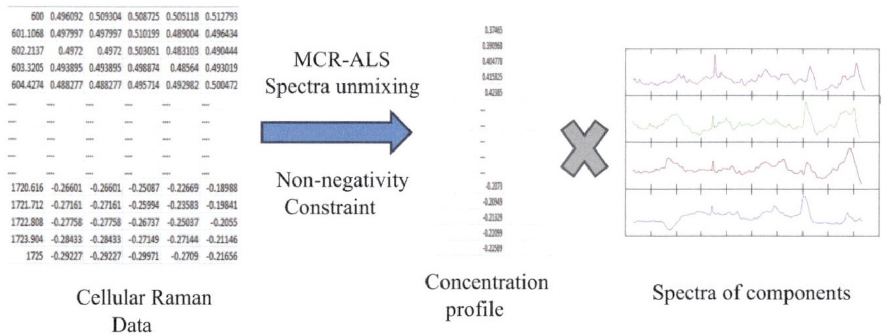

Fig. 1 Schematic of MCR-ALS algorithm

2 Materials and Methods

2.1 *Sample Details*

Osteoarthritic articular cartilage samples were taken from femoral condyle of patients undergoing knee arthroplasty. A total of 15 osteoarthritic cartilage tissue sections were collected (five sections of each ICRS grade I, II, and III) for the pilot study. Chondrocytes were isolated from these tissue sections according to the protocol described in our previous study [13]. The data used in this study were acquired in previous work. The Regional Committee for Medical Research Ethics (2013/265 REK, Norway) approved the acquisition of human samples.

2.2 *Confocal Raman Measurement*

A confocal Raman microscope (LabRAM HR800, HORIBA) equipped with a laser of wavelength 632.10 nm was used to acquire spectra from the nucleus of each osteoarthritic chondrocyte. Thirty cells were randomly selected from each sample, and a single spectrum was recorded for each cell. Thus, for each ICRS grade, 150 spectra (30 spectra/sample for five samples) were recorded, resulting in a total of 450 spectra (150 spectra $\times$ 3 grades $=$ 450 spectra). The spectra were recorded in the region of Raman shift covering 600–1725 cm^{-1}. The cells were exposed to a laser power of 8 mW, and each spectrum was recorded for 90 s.

2.3 *Data Analysis*

The Raman spectra were preprocessed before MCR analysis. The preprocessing steps were as follows: First cosmic rays were removed from the spectra. Then, the background noise due to fluorescence and the phosphate buffer solution was reduced. To smoothen the spectral data, a Savitzky–Golay filter was used by applying third-order polynomial and nine-point smoothing. Finally, the spectra were normalized with the peak 1452 cm^{-1} used as the reference for normalization.

MCR-ALS and SVD

MCR-ALS is a bilinear multivariate ML model that assumes the matrix containing Raman spectral data is a linear combination of a matrix containing spectral signature of specific biochemical components and another matrix containing the concentrations of those biochemical components (Fig. 1).

$$A_{(z,n)} = X_{(z,h)} Y_{(h,n)} + E_{(z,n)} \tag{1}$$

In this algorithm, the original data matrix $A_{(z,n)}$ consisting of n spectra taken from different locations of a sample with z wavenumber points can be decomposed into two matrices: $X_{(z,h)}$ and $Y_{(h,n)}$. The rows of matrix $X_{(z,h)}$ represent spectra of h components, while columns of matrix $Y_{(h,n)}$ represent the concentrations of h components. $E_{(z,n)}$ represents the error or residue (Eq. 1).

MCR-ALS algorithm is applied to the above equation, and it is solved iteratively till the Frobenius norm $||A - XY||^2$ is minimized and the value of error E becomes negligible. The compulsory non-negativity restraints $X \geq 0$ and $Y \geq 0$ are applied during the minimization procedure to ensure meaningful interpretation. These constraints arise from the fact that neither the Raman spectra of the component nor its concentration can be negative.

To apply MCR-ALS, the total number of components to be used for analysis must first be determined. The number of components is determined from the singular value decomposition (SVD). The SVD method decomposes the original matrix $z \times n$ into $L\Sigma M^T$. Matrix L has dimensions $z \times z$, while matrix Σ has dimensions $z \times n$, and M^T matrix has dimensions $n \times n$. Matrix Σ is a diagonal matrix which represents singular values. The number of components selected for MCR analysis are based on the significant drop observed in eigenvalues or until the eigenvalues are approximately zero. Therefore, SVD can also be used for reducing the dimensionality of large datasets [14].

The lack of fit (LOF) and percentage of variation (R^2) are the quality parameters obtained from MCR model (Eqs. 2 and 3).

$$\mathrm{LOF} = 100\sqrt{\frac{\sum_{i=1}^{n}\sum_{j=1}^{m}B_{i,j}^2}{\sum_{i=1}^{n}\sum_{j=1}^{m}A_{i,j}^2}} \tag{2}$$

$$R^2 = 100\frac{\sum_{i=1}^{n}\sum_{j=1}^{m}A_{i,j}^2 - \sum_{i=1}^{n}\sum_{j=1}^{m}B_{i,j}^2}{\sum_{i=1}^{n}\sum_{j=1}^{m}A_{i,j}^2} \tag{3}$$

Here, $B_{i,j} = A_{i,j} - \hat{A}_{i,j}$, $A_{i,j}$ denotes original data matrix, and $\hat{A}_{i,j}$ is calculated from MCR-ALS algorithm.

Although the MCR-ALS algorithm is effective for analyzing Raman spectra, it has some limitations. It is sensitive to noise and baseline variations, relies on linearity assumptions, and can be influenced by the choice of initial guesses for pure spectra. The potential ambiguity in interpreting results from complex spectra can present further challenges. Additionally, the number of components selected needs to be optimized, as this can affect the shape of the MCR loadings (the spectra of components) and the ratio of the MCR scores (concentration profiles).

3 Results and Discussion

Figure 2 shows the Raman spectra acquired from the nucleus of chondrocytes isolated from human osteoarthritic cartilage of ICRS grades I, II, and III.

A total of 450 spectra were shown in Fig. 2 (150 spectra of each grade—I, II, and III). The spectra acquired contained multiple spectral peaks that represent various biomolecular components present within the sample. The tentative assignments of Raman bands associated with particular vibrational mode originating from the biomolecules were shown in Table 1. The acquired spectra were then preprocessed (removal of cosmic rays, background correction, smoothing, and normalization) and were subjected to MCR-ALS analysis.

Figure 3 shows the SVD analysis performed on the acquired Raman data to determine the initial number of components to input into MCR algorithm. As shown in Fig. 3 that with approximately five components, the eigenvalue approaches near to zero. The number of components selected for MCR model significantly impacts the shape of MCR loadings (spectral profiles) and ratio of MCR scores (concentration profile). Additionally, the number of components was varied from four to seven to compare their effect on loading and scores. The analysis revealed that a five-component MCR model was optimal for generating the most appropriate shape of spectral loadings. The spectral loadings generated by MCR were used for ALS optimization. The number of iterations for ALS optimization was set to 50, with a convergence value of 0.1.

The MCR-ALS analysis generated spectral loadings for five components, which are shown in Fig. 4. These spectral loadings were assigned to specific biochemical constituents based on a comparison of their spectral shapes to those of pure chemical spectra. Among those five components, spectra of components 2, 3, and 4 were selected because their spectral loading contained interpretable spectral features specific to a biochemical component. The spectra of components 1 and 5 could not be

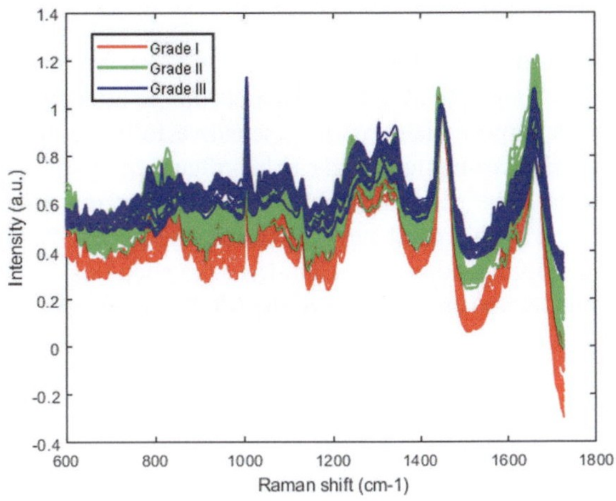

Fig. 2 Set of Raman spectra acquired from chondrocytes isolated from three ICRS grades—I, II, and III of osteoarthritic cartilage

Table 1 Raman peak assignment [15–17]

Wavenumber (cm^{-1})	Peak assignment	Macromolecules
784	Thymine, cytosine ring breath	DNA
830	Tyrosine ring breathing vibration	Proteins
850	Tyrosine ring breathing vibration	Proteins
1004	Symmetric ring breathing mode	Phenylalanine
1032	Symmetric ring breathing mode	Phenylalanine
1085	P–O stretch	Phospholipids
1094	PO_2 backbone	DNA
1128	C–N stretch	Proteins
1178	C–H bend, C–C stretch	Proteins
1210	C–C_6–H_5 stretch, phenylalanine	Proteins
1240	Thymine ring breath	DNA
1305	CH_2 twist	Lipids
1320	Adenine, Guanine	Nucleic acids
1340	Adenine ring breath, CH deformation in proteins	DNA, proteins
1378	Thymine, adenine, guanine ring breath	Nucleic acids
1452	CH_2 bend	Proteins, lipids
1580	–N_7–H, Guanine, Adenine	Nucleic acids
1660	C=C stretch, amide I	Proteins

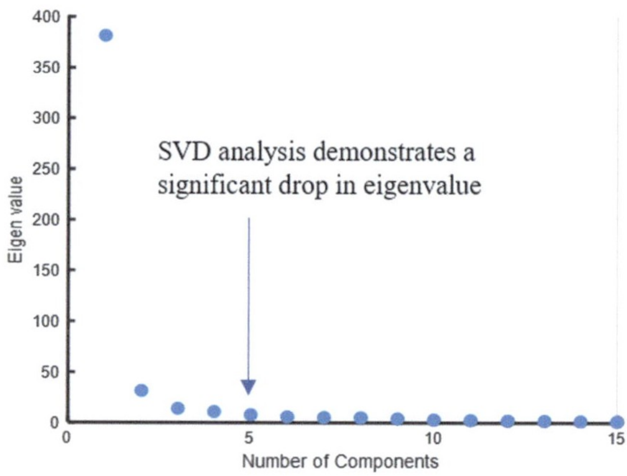

Fig. 3 Plot of eigenvalues obtained from SVD analysis to determine number of components to be selected for MCR analysis

interpreted, as they did not contain useful information for assigning them to specific cellular components and were therefore excluded from the study.

Figure 5 shows that the assignment of biochemical components (i.e., lipid, DNA, and protein) to the spectral profiles generated by MCR was made based on the

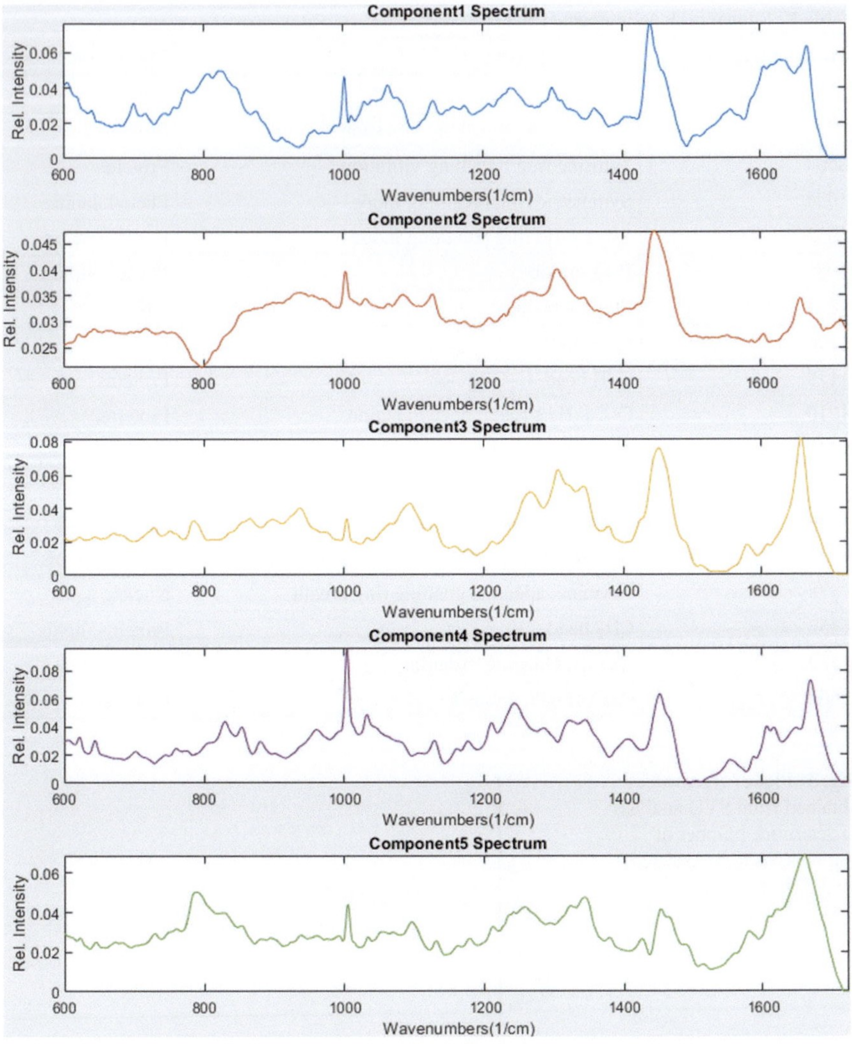

Fig. 4 Spectral loadings of five components generated by MCR-ALS algorithm

different spectral peaks observed in the spectra of these components. MCR component 2 was identified as lipids, with major peaks at 1080 cm^{-1}, 1305 cm^{-1}, 1452 cm^{-1} (Fig. 5a). MCR component 3 was identified as DNA, with major peaks at 784 cm^{-1}, 1085 cm^{-1}, 1320 cm^{-1}, 1340 cm^{-1}, 1378 cm^{-1}, 1585 cm^{-1} (Fig. 5b). These peaks from cell nuclei predominantly represent contributions from nucleic acids. MCR component 4 depicts proteins present in nucleus, with major peaks present at 830 cm^{-1}, 850 cm^{-1}, 1004 cm^{-1}, 1032 cm^{-1}, 1128 cm^{-1}, 1178 cm^{-1}, 1210 cm^{-1}, 1340 cm^{-1}, 1452 cm^{-1}, 1660 cm^{-1} (Fig. 5c). The spectral interpretation of these peaks that were associated with components 2, 3, and 4 which were interpreted as lipids, DNA, and protein, respectively, is mentioned in Table 1. The total variance explained by MCR model was 99.896%, and the LOF was 0.6902%.

It was observed that the concentration of components 3 and 4 (i.e., DNA and proteins) decreased with the progression of osteoarthritic grade, while the concentration of component 2 (i.e., lipids) increased with the progression of osteoarthritic grade (Fig. 6). The results also indicated that chondrocyte death is enhanced as the disease progresses, because the variations in the contributions of these biochemical components follow a similar trend during apoptosis in cells. These results are consistent with previous findings where it was observed that contributions of DNA and protein were lower while accumulation of lipids was higher in apoptotic cells [18–20].

Figure 7 demonstrates the peak area analysis for specific peaks corresponding to DNA, lipid, and protein. The area of peak at 1305 cm^{-1}, calculated over the region 1300–1310 cm^{-1}, represents the lipid peak. The area of peak at 784 cm^{-1}, calculated over the region 780–800 cm^{-1}, represents the DNA peak. The area of peak at 1004 cm^{-1}, calculated over the region 997–1011 cm^{-1}, represents the protein (phenylalanine) peak. The statistical analysis of peak area demonstrated that content of DNA and protein decreased, while the content of lipid increased with progression of osteoarthritic grades.

4 Conclusion

The application of combined Raman-MCR approach has been found to be useful for detecting intracellular biomolecular components. The Raman data acquired from osteoarthritic chondrocytes which contained mixed spectral information were analyzed by MCR-ALS machine learning algorithm to extract spectra of meaningful biochemical components along with their relative concentrations. It was observed that concentration profile of DNA and protein decreased while lipid increased with progression of osteoarthritic grades. The study indicated that combining Raman spectroscopy with MCR-ALS analysis could enhance the Raman spectral analysis, potentially leading to a better interpretation of complex spectra acquired from biological samples. This approach opens new possibilities for studying molecular differences, which may be helpful in improving prognosis, early prevention, and the clinical diagnostic applications of Raman spectroscopy.

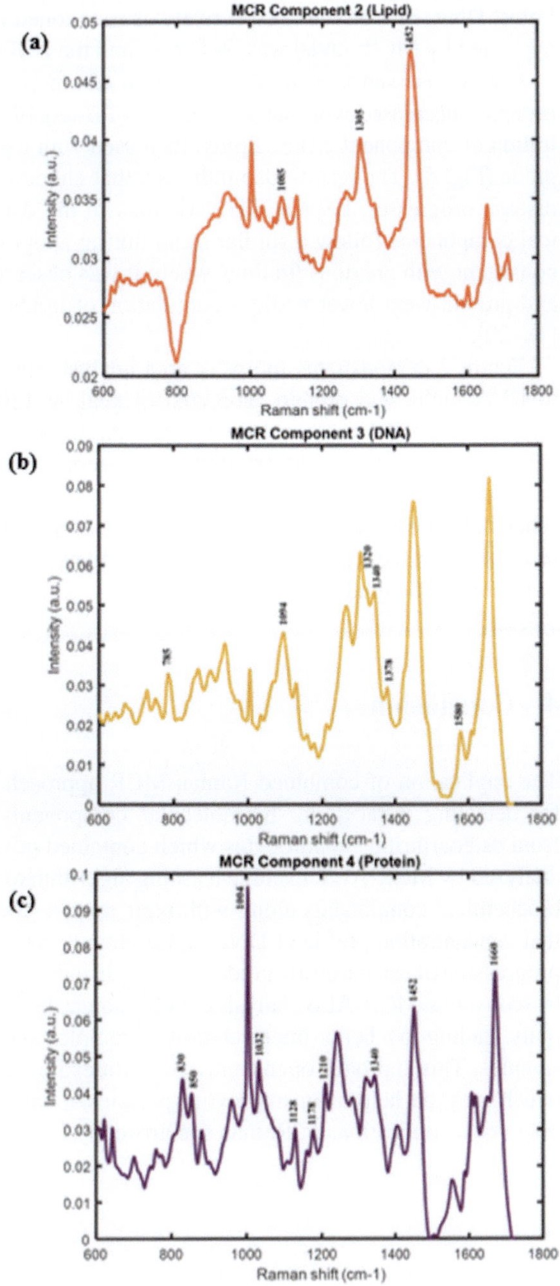

Fig. 5 Using MCR-ALS analysis, three components **a** lipid, **b** DNA, **c** protein were selected, and their respective loadings were plotted

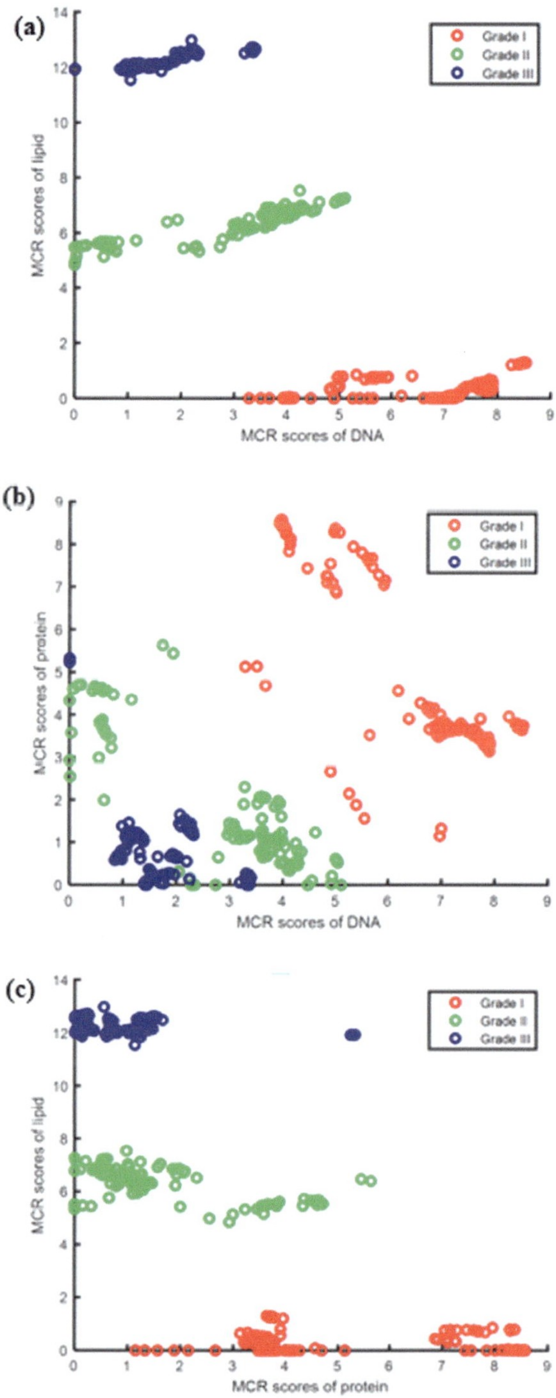

Fig. 6 Using MCR-ALS analysis, three components were selected and their respective scores **a** lipid and DNA, **b** protein and DNA, **c** lipid and protein were plotted

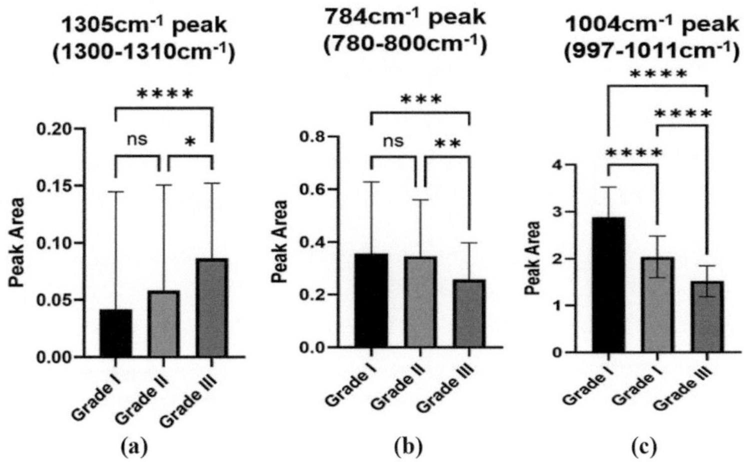

Fig. 7 Analysis of spectral peak area of **a** 1305 cm^{-1} (lipid), **b** 784 cm^{-1} (DNA), and **c** 1004 cm^{-1} (protein)

References

1. Brazhe NA, Vlasova AD, Osipov SD, Sudarev VV, Ryzhykau YL, Bogorodskiy AO, Zinovev EV, Rogachev AV, Manukhov IV, Borshchevskiy VI, Kuklin AI, Pokorn J, Sosnovtseva O, Maksimov GV, Gordeliy VI (2020) Raman scattering: from structural biology to medical applications. Crystals 10
2. Solodskikh SA, Gryaznova MV, Dvoretskay YD, Gureev AP, Panevina AV, Maslov AY, Serzhantova OV, Mikhailov AA, Chinopoulos C, Popov VN (2019) Postgenomic technologies for genomic and proteomic analysis in biological and medical research. Ukr J Ecol 9(4):765–776
3. Geraldes CFGC (2025) Introduction to infrared and raman-based biomedical molecular imaging and comparison with other modalities. Molecules 25(23):5547. https://doi.org/10.3390/molecules25235547
4. Downes A, Elfick A (2010) Raman spectroscopy and related techniques in biomedicine. Sensors 10:1871–1889. https://doi.org/10.3390/s100301871
5. Ghita A, Pascut FC, Sottile V, Denning C, Notingher I (2015) Applications of Raman microspectroscopy to stem cell technology: label-free molecular discrimination and monitoring cell differentiation. EPJ Tech Instrum 2:1–14. https://doi.org/10.1140/epjti/s40485-015-0016-8
6. Khristoforova Y, Bratchenko L, Bratchenko I (2023) Raman-based techniques in medical applications for diagnostic tasks: a review. Int J Mol Sci 24(21):15605. https://doi.org/10.3390/ijms242115605
7. Liu YJ, Kyne M, Wang C, Yu XY (2020) Data mining in Raman imaging in a cellular biological system. Comput Struct Biotechnol J 18:2920–2930. https://doi.org/10.1016/j.csbj.2020.10.006
8. Jaumot J, Gargallo R, De Juan A, Tauler R (2005) A graphical user-friendly interface for MCR-ALS: a new tool for multivariate curve resolution in MATLAB. Chemom Intell Lab Syst 76:101–110. https://doi.org/10.1016/j.chemolab.2004.12.007
9. Jaumot J, de Juan A, Tauler R (2015) MCR-ALS GUI 2.0: New features and applications. Chemom Intell Lab Syst 140:1–12. https://doi.org/10.1016/j.chemolab.2014.10.003
10. de Juan A, Jaumot J, Tauler R (2014) Multivariate curve resolution (MCR). Solving the mixture analysis problem. Anal Methods 6(14):4964–4976. https://doi.org/10.1039/c4ay00571f

11. Marro M, Nieva C, De Juan A, Sierra A (2018) Unravelling the metabolic progression of breast cancer cells to bone metastasis by coupling Raman spectroscopy and a novel use of MCR-ALS algorithm. Anal Chem 90:5594–5602. https://doi.org/10.1021/acs.analchem.7b04527
12. Felten J, Hall H, Jaumot J, Tauler R, De Juan A, Gorzsás A (2015) Vibrational spectroscopic image analysis of biological material using multivariate curve resolution-alternating least squares (MCR-ALS). Nat Protoc 10:217–240. https://doi.org/10.1038/nprot.2015.008
13. Kumar R, Singh GP, Grønhaug KM, Afseth NK, de Davies CL, Drogset JO, Lilledahl MB (2015) Single cell confocal Raman spectroscopy of human osteoarthritic chondrocytes: a preliminary study. Int J Mol Sci 16(5):9341–9353. https://doi.org/10.3390/ijms16059341
14. Santana IM, Breitkreitz MC, Pinto L (2021) Multivariate curve resolution alternating least squares applied to chromatographic data: from the basics to the recent advances. Braz J Anal Chem 8:22–44. https://doi.org/10.30744/BRJAC.2179-3425.RV-30-2021
15. Notingher I, Verrier S, Romanska H, Bishop AE, Polak JM, Hench LL (2002) In situ characterisation of living cells by Raman spectroscopy. Spectroscopy 16:43–51. https://doi.org/10.1155/2002/408381
16. Bik E, Dorosz A, Mateuszuk L, Baranska M, Majzner K (2020) Fixed versus live endothelial cells: the effect of glutaraldehyde fixation manifested by characteristic bands on the Raman spectra of cells. Spectrochim Acta—Part A: Mol Biomol Spectrosc 240:118460. https://doi.org/10.1016/j.saa.2020.118460
17. Zhao Y, Tang P, He X, Xie Y, Cheng W, Xing X, Xing M, Lu X, Liu S, Zhong L (2020) Study on the precise mechanism of mitoxantrone-induced Jurkat cell apoptosis using surface enhanced Raman scattering. Spectrochim Acta—Part A: Mol Biomol Spectrosc 228:117718. https://doi.org/10.1016/j.saa.2019.117718
18. Uzunbajakava N, Lenferink A, Kraan Y, Volokhina E, Vrensen G, Greve J, Otto C (2003) Nonresonant confocal Raman imaging of DNA and protein distribution in apoptotic cells. Biophys J 84:3968–3981. https://doi.org/10.1016/S0006-3495(03)75124-8
19. Ong YH, Lim M, Liu Q (2012) Comparison of principal component analysis and biochemical component analysis in Raman spectroscopy for the discrimination of apoptosis and necrosis in K562 leukemia cells. Opt Express 20:22158. https://doi.org/10.1364/oe.20.022158
20. Brauchle E, Thude S, Brucker SY, Schenke-Layland K (2014) Cell death stages in single apoptotic and necrotic cells monitored by Raman microspectroscopy. Sci Rep 4:1–9. https://doi.org/10.1038/srep04698

Optimizing Heart Disease Prediction with Ensemble Learning and Behavioral Patterns

Channabasappa Muttal, Padmshree Desai, and Prathit Kulkarni

Abstract Heart disease remains a leading cause of morbidity and mortality worldwide, driven by various Behavioral Patterns such as age, smoking habits, and exercise frequency. This study explores the predictive power of these Behavioral Patterns in forecasting heart disease through ensemble learning techniques. Using a comprehensive dataset, we applied multiple machine learning models, including Random Forest, Extra Trees, LightGBM, CatBoost, and XGBoost, to build a robust predictive model. Our analysis revealed significant correlations between lifestyle factors and heart disease, with age, smoking, and exercise being strong predictors. The ensemble approach, particularly the LightGBM model, demonstrated superior performance with an accuracy of 91.86% and the fastest training time. Visualization of the relationships between different Behavioral Patterns and heart disease provided clear insights into the risk factors. This study underscores the potential of leveraging ensemble learning models to enhance predictive accuracy and offers valuable insights for targeted health interventions.

Keywords Ensemble learning · Behavioral patterns · Heart disease prediction · Machine learning models · Predictive accuracy

1 Introduction

Different types of cardiovascular diseases continue to be major contributors to death and illness rates among people irrespective of their ages. Even with advances in medical research and technology, it can still be difficult to diagnose heart disease

C. Muttal (✉) · P. Desai · P. Kulkarni
School of Computer Science and Engineering, KLE Technological University, Hubli, India
e-mail: channabasappa.muttal@kletech.ac.in

P. Desai
e-mail: padmashri@kletech.ac.in

P. Kulkarni
e-mail: prathitk2003@gmail.com

S. Kumar et al. (eds.), *Artificial Intelligence for Next Generation Computing, Volume 1*,
Lecture Notes in Networks and Systems 1549,
https://doi.org/10.1007/978-981-95-0701-6_18

early and accurately. Heart disease prevention is another important public health issue. Behavioral factors—including nutrition, exercise routines, smoking status, and stress levels—are critical in determining a person's risk of heart disease. In the current scenario of machine learning and massive data sets, using these Behavioral factors to predict cardiac disease presents a promising way to enhance patient care and lessen the strain on healthcare systems. A popular machine learning technique called ensemble learning combines several models' predictive strengths to improve prediction accuracy. This method is helpful when working with intricately interconnected Behavioral characteristics in complicated and diverse datasets. By employing ensemble learning methods, we may leverage the collective wisdom of various models, each tailored to capture distinct facets of the heart disease prediction problem.

This study aimed to investigate the potential of ensemble learning techniques for early heart disease detection using the person's day-to-day Behavioral. The goal is to create a prediction model that will improve the accuracy of heart disease diagnosis and provide insightful information about the Behavioral factors that majorly impact the risk of developing heart disease. An ensemble model is used for the prediction of heart disease by using the Behavioral patterns.

We intend to advance the fields of customized healthcare and preventive medicine, bringing about a paradigm change in the fight against heart disease beyond therapeutic to preventive strategies.

Section 2 discusses the different investigations by different authors on predictive models for heart disease prediction. Section 3, presents the proposed methodology, data set description, features, and models used. Section 4 provides the results of the analysis of the datasets. Section 5 provides the summary of the analysis and prediction obtained.

2 Background Study

Cardiovascular diseases, which encompass heart diseases, remain a pressing global health issue. Recently, there has been a surge in the utilization of machine learning and ensemble learning tactics in the forecasting of heart disease, particularly by incorporating Behavioral patterns. This literature review offers an outline of the essential research in this domain, underscoring the various methodologies and breakthroughs that have influenced the discipline. Machine and ensemble learning have proven to be effective tools for detecting heart disease at an early stage. In a study conducted by [1], the utility of ensemble machine learning in predicting cardiovascular disease was examined, highlighting its potential to improve prediction accuracy. The researchers acknowledged the intricacy of cardiovascular diseases and emphasized the importance of developing comprehensive predictive models that consider Behavioral factors.

The study by [2] underscored the crucial role of machine learning in predicting all-cause mortality among individuals with suspected coronary artery disease.

The authors highlighted its potential for risk stratification, aiming to facilitate early intervention and personalized patient care. Additionally, [3], emphasized the significant impact of artificial intelligence in precision cardiovascular medicine, noting that machine learning models offer valuable insights for personalized healthcare, particularly in risk assessment and management. Machine learning has been utilized not only for risk prediction but also for the early detection of heart diseases. Al'Aref et al. [4] and Kavitha et al. [5] emphasized the potential of machine learning-assisted electrocardiograms for identifying cardiac contractile dysfunction. This approach demonstrates the adaptability of machine learning in cardiac diagnostics. Ensemble learning, which is a highly effective method for improving prediction models, has had a significant effect. Ahmad et al. [6] and Shorewala [7] used an ensemble of weak learners to predict heart disease, highlighting the complementary nature of ensemble techniques in enhancing prediction accuracy [8]. Similarly, [9, 10]emphasized the potential of ensemble models when combined with feature selection.Ensemble learning can be applied to under sampling techniques and model combinations. Yewale et al. [11] created an ensemble model for predicting heart disease by combining machine learning algorithms, illustrating the potential for more accurate risk assessment. Chaurasia and Chaurasia [12] also examined ensemble methods, highlighting the significance of feature selection in heart disease prediction models. The study conducted by [13] evaluated predictive models for heart disease, with a focus on their potential for clinical use. Their findings highlight the importance of medical expert systems in cardiac risk stratification and have significant implications for the development of clinical decision support systems as machine learning applications in cardiac risk prediction extend beyond specific data types. For instance, [14] utilized 2D/3D visual markers to predict cardiovascular events, thereby expanding the scope of machine learning to include imaging data. The potential for harnessing real-world data for improved predictive models was demonstrated by [15, 16], who utilized electronic health records and large datasets to conduct personalized risk prediction and survival analysis with Medicare claims data. The integration of these resources has created new opportunities to advance medical research and patient care.

In addition to conventional machine learning techniques [17], deep learning methods [18] have gained considerable momentum. Dernoncourt et al. [19] emphasized the significance of privacy and data security in healthcare applications by focusing on de-identifying patient notes through the use of recurrent neural networks. In doing so, these authors drew attention to the profound impact of deep learning approaches in this particular domain.

The main drawback of the existing works is researchers focused on using machine learning models individually to predict heart disease based on parameters. The ensemble machine learning model uses few models to predict heart disease. Our proposed approach uses seven machine learning models to predict the results and analyze the performance of the system.

3 Methodology

The suggested methodology provides a thorough and sturdy technique for forecasting heart disease by utilizing ensemble learning strategies that take into account Behavioral patterns.

3.1 Dataset Description

The Health and Behavioral Dataset used in this study was obtained from a public health repository. The data were collected through medical records and patient surveys conducted over a five-year period. The dataset consists of 308,855 records, each representing an individual patient. There are 19 features in the dataset. These features include a range of demographic, medical, and Behavioral patterns such as age, gender, frequency of general health checkups, exercise frequency, presence of heart disease, presence of skin cancer, height, weight, vaccination status, alcohol consumption, smoking habits, presence of chronic conditions, current medication usage, presence of other types of cancer, blood pressure readings, cholesterol levels, diabetes status, family history of heart disease [20].

3.2 Data Preprocessing

During the data preprocessing step, we employed cleaning, normalization, and encoding of categorical variables to prepare the dataset for analysis. Missing values were substituted with zeros and used the average values of the columns to compensate for these values, ensuring the general statistical features of the data were maintained. Numerical features, including age, height, weight, blood pressure, and cholesterol levels, were normalized to standardize the range of the data. Categorical variables, such as sex, frequency of general health checkups, exercise frequency, vaccination status, smoking habits, medication usage, and diet, were encoded using one-hot encoding to convert them into a numerical format suitable for machine learning models.

3.3 Feature Selection

During the advanced phase of feature engineering, our goal was to enhance the predictive power of our model beyond the capabilities provided by basic feature engineering. Building on this foundation, we identified and introduced a set of features that were deemed most relevant for predicting heart disease. These features included the BMI category, checkup frequency, Behavioral score, healthy diet score, smoking and alcohol habits, checkup and exercise frequency, height-to-weight ratio, fruits and vegetables intake, depression and exercise correlation, healthy diet and Behavioral score, financial stress impact, and combined alcohol and fried potato consumption. Each feature was selected based on its importance and contribution to the overall model performance, providing unique insights into various dimensions of the patients' health and Behavioral that are pertinent to heart disease risk. This meticulous selection process ensured a more comprehensive understanding of the factors influencing heart disease, thereby improving the model's effectiveness.

3.4 Models

We now have a well-organized and uncluttered dataset that is ideal for predictive modeling. Our plan is to use a variety of machine learning models to predict the desired outcome. As we gathered each new piece of data, we smoothly added it to the dataset, enhancing its value with useful information.

Ensemble learning is a powerful machine learning paradigm that combines multiple models, or "learners," to improve overall performance and robustness compared to individual models. By aggregating the predictions of various models, ensemble methods can significantly reduce the likelihood of errors and enhance generalization. This is achieved through techniques like bagging, boosting, and stacking as illustrated in the Fig. 1.

In our research, we deliberately selected a diverse array of machine learning models, such as Random Forest, Extra Trees, XGBoost, CatBoost, and LightGBM, Voting and Stacking. We chose these models based on their suitability for achieving

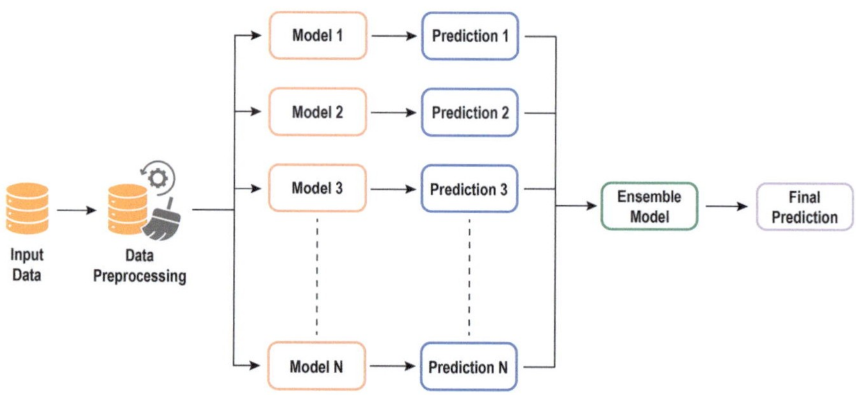

Fig. 1 Ensemble learning: the input data is processed by multiple base models. Each base model produces its own predictions, which are then combined by an ensemble model to produce the final output

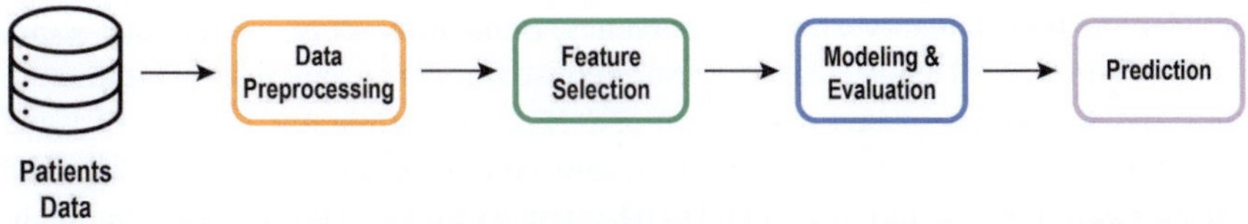

Fig. 2 Architecture diagram of the proposed methodology

our research objectives. Figure 2 demonstrates the proposed methodology. Below, we describe each method, how it works, and the relevant mathematical formulas.

Random Forest Random Forest is an ensemble learning method that creates multiple decision trees and aggregates their results for improved accuracy and robustness. Each tree is built from a random subset of the training data and features.

$$\hat{y} = \frac{1}{N} \sum_{i=1}^{N} T_i(x) \tag{1}$$

where $T_i(x)$ is the prediction of the i-th decision tree and N is the number of trees.

Extra Trees Extremely Randomized Trees (Extra Trees) is similar to Random Forest but differs in how the trees are constructed. Extra Trees use the entire sample for each tree and randomize the splitting points of features.

$$\hat{y} = \frac{1}{N} \sum_{i=1}^{N} ET_i(x) \tag{2}$$

where $ET_i(x)$ is the prediction of the i-th extra tree and N is the number of trees.

XGBoost Extreme Gradient Boosting (XGBoost) is a gradient boosting algorithm that builds additive models in a forward stage-wise manner. It optimizes a differentiable loss function by adding weak learners.

$$\hat{y} = \sum_{k=1}^{K} f_k(x), \quad f_k \in \mathcal{F} \tag{3}$$

where $\mathcal{F}$ is the space of regression trees, and K is the number of trees.

CatBoost CatBoost is a gradient boosting algorithm that handles categorical features effectively. It uses ordered boosting to reduce overfitting and target leakage.

$$\hat{y} = \sum_{k=1}^{K} f_k(x), \quad f_k \in \mathcal{F} \tag{4}$$

where $\mathcal{F}$ is the space of regression trees, and K is the number of trees.

LightGBM Light Gradient Boosting Machine (LightGBM) is a gradient boosting framework that uses tree-based learning algorithms. It is designed for efficiency and scalability.

$$\hat{y} = \sum_{k=1}^{K} f_k(x), \quad f_k \in \mathcal{F} \tag{5}$$

where $\mathcal{F}$ is the space of regression trees, and K is the number of trees.

Voting Voting is an ensemble technique that combines the predictions of multiple models. It can be implemented as hard voting (majority voting) or soft voting (average of probabilities).

$$\hat{y} = \text{mode}(\hat{y}_1, \hat{y}_2, \ldots, \hat{y}_M) \tag{6}$$

For soft voting, the prediction is:

$$\hat{y} = \arg\max \left(\frac{1}{M} \sum_{i=1}^{M} P_i \right) \tag{7}$$

where $\hat{y}_i$ is the prediction of the i-th model, M is the number of models, and P_i is the probability prediction of the i-th model.

Stacking Stacking is an ensemble technique that combines multiple models by training a meta-model on the predictions of the base models.

$$\hat{y} = g(\hat{y}_1, \hat{y}_2, \ldots, \hat{y}_M) \tag{8}$$

where g is the meta-model, and $\hat{y}_i$ is the prediction of the i-th base model.

4 Experimental Results

4.1 *Data Visualization and Analytics*

In this section, we present detailed statistical analysis and visualizations that demonstrate the relationships between various Behavioral patterns and heart disease.

Figure 3 shows the distribution of patients based on age and sex. The data suggests a possible gender disparity, with females being more prominently represented in most age groups than males. The most significant populations fall within the age ranges of 55–59 and 65–69 years, indicating a considerable number of individuals above the age of 55. This may have implications for healthcare, social security, and pension schemes.

Figure 4 illustrates the self-rated general health status of respondents. The majority rated their health as "very good" (35.7%), followed by "good" (29.0%), "excellent"

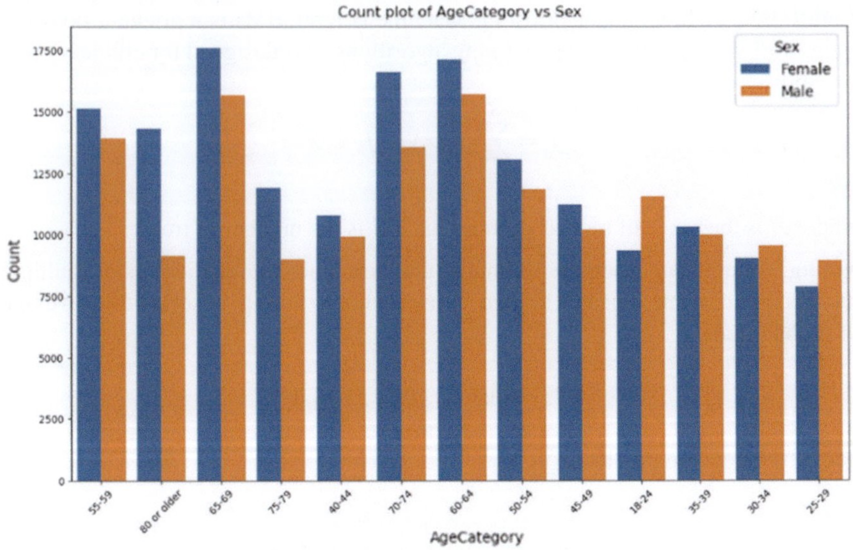

Fig. 3 Total number of patient w.r.t age and sex

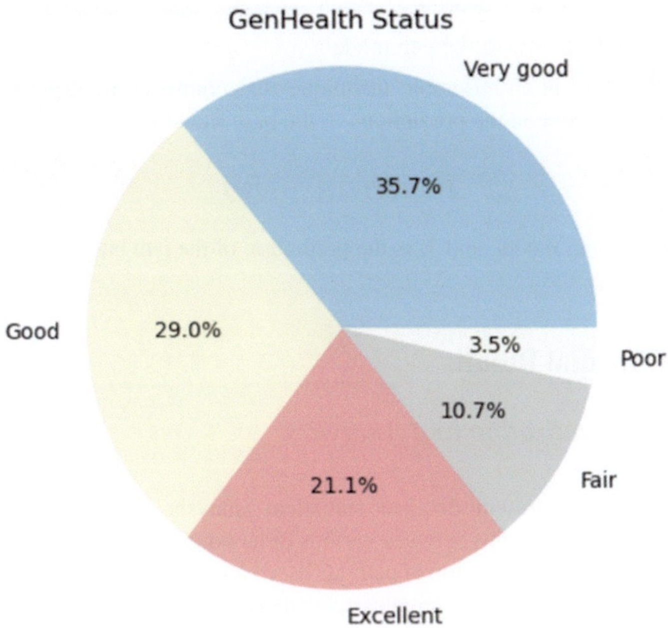

Fig. 4 Self-rating of general health status

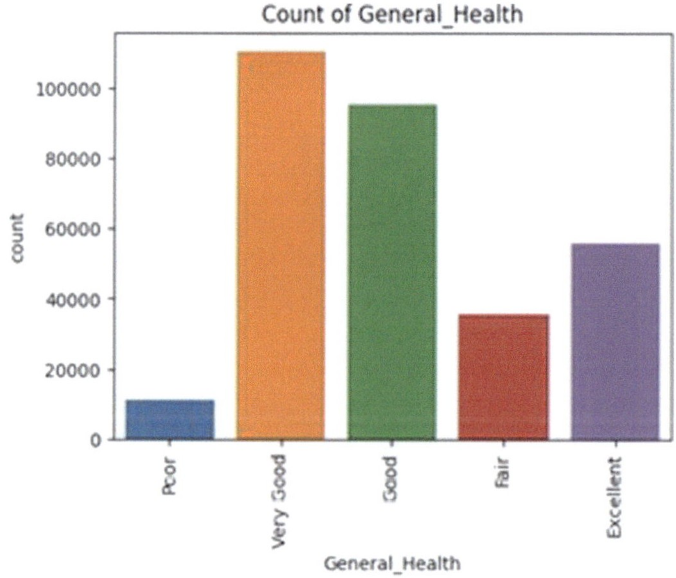

Fig. 5 Count of people within different categories of general health

(21.0%), "fair" (10.0%), and "poor" (3.5%). These ratings provide insights into the overall life satisfaction and health perceptions of the population.

Figure 5 shows the distribution of individuals within different categories of general health. The majority fall into the "very good" and "good" categories, indicating a generally positive perception of health among the population.

Figure 6 explores the relationship between general health ratings and the presence of heart disease. It helps identify trends and correlations that may exist between perceived health status and actual health outcomes. Figure 7 examines the correlation between exercise frequency and the incidence of heart disease. It provides insights into how physical activity levels impact heart disease risk. Figure 8 analyzes the relationship between gender and heart disease prevalence. It helps identify any significant differences in heart disease rates between males and females. Figure 9 examines the impact of smoking habits on heart disease prevalence. It underscores the significance of smoking as a major risk factor for heart disease.

4.2 *Performance comparison of learning models*

Table 1 provides an overview of the performance of various ensemble learning models in predicting heart disease based on multiple evaluation metrics, including the F1 score, precision, recall and accuracy. This study provides insightful information about the performance of each model. XGBoost emerged as the most efficient

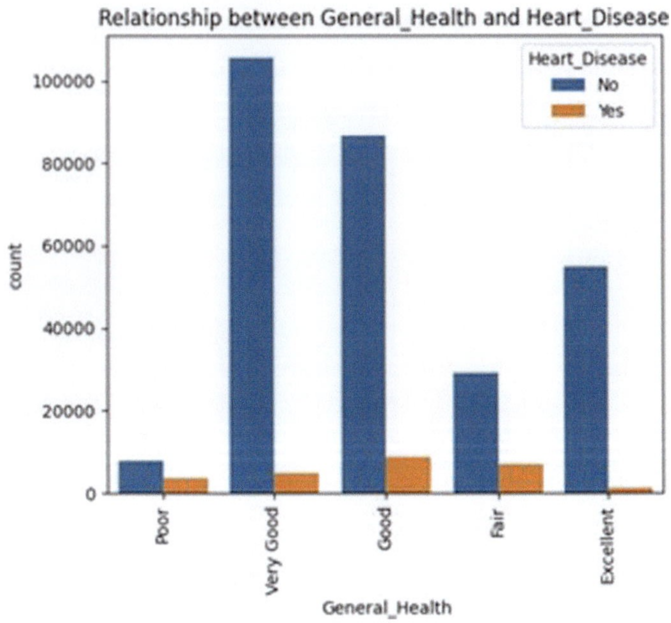

Fig. 6 General health versus heart disease

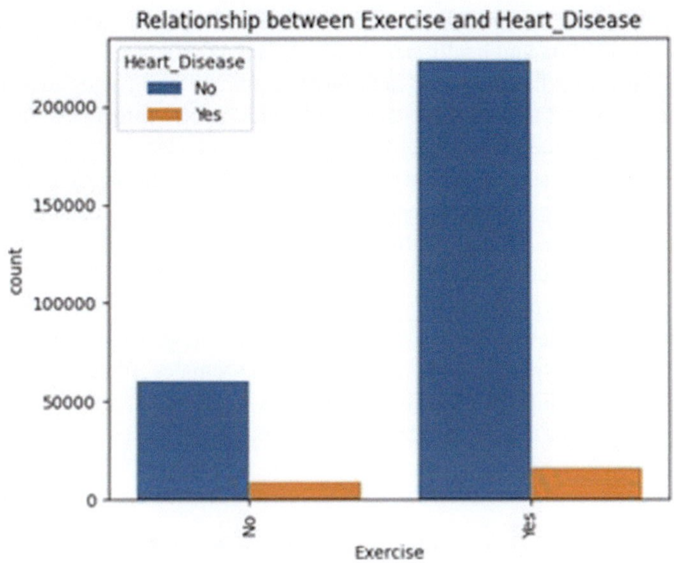

Fig. 7 Physical exercises versus hear disease

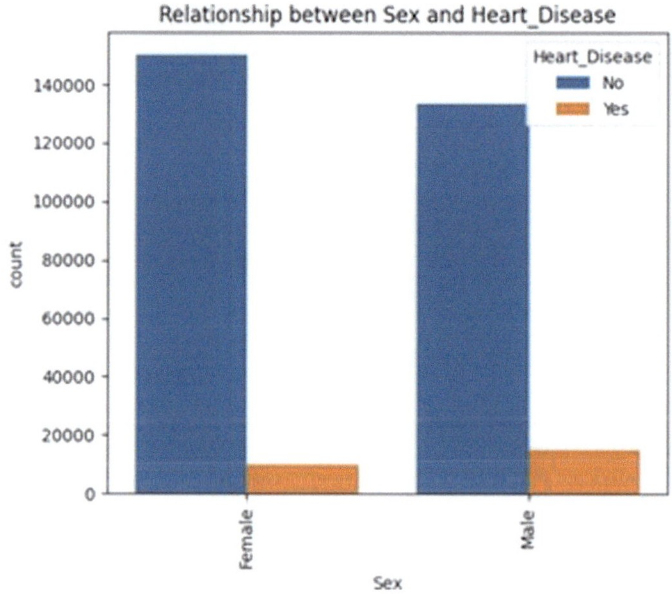

Fig. 8 Sex versus heart disease

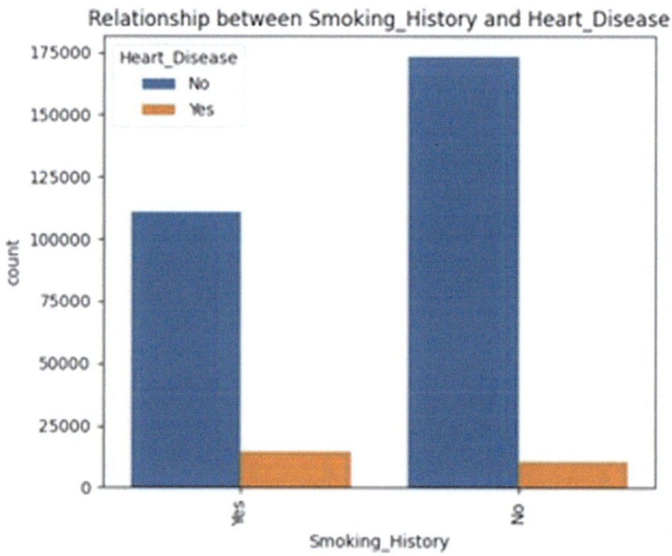

Fig. 9 Smoking habit versus heart disease

Table 1 Performance Comparison of learning models

Model name	F1 score	Precision	Recall	Accuracy (%)
Random Forest	0.49	0.75	0.43	92.63
Extra Trees	0.48	0.73	0.44	92.63
XGBoost	0.53	0.82	0.50	92.88
CatBoost	0.54	0.81	0.52	92.68
LightGBM	0.43	0.85	0.52	92.68
Voting	0.45	0.80	0.47	91.67
Stacking	0.54	0.74	0.45	92.69

model, achieving the highest accuracy (92.88). The LightGBM Classifier showed the highest precision (0.85), indicating its ability to identify true positives and low F1 scores accurately. LightGBM and CatBoost provided balanced performance with good accuracy, precision, and recall. Training time also varies for different models.

5 Conclusion

According to the assessment of the learning models performance, the CatBoost and Stacking models exhibit the highest F1 scores of 0.54, which indicates strong overall performance in balancing precision and recall. The CatBoost model also displays remarkable recall (0.52) and precision (0.81), making it one of the top performers. The XGBoost model follows closely behind with an F1 score of 0.53 and highest accuracy of 92.88% showcasing its robustness.

Although the LightGBM model boasts the highest precision (0.85), its F1 score is lower (0.43) due to a compromise with recall. The Random Forest and Extra Trees models perform similarly in terms of accuracy and precision but have lower F1 scores and recall. In conclusion, the CatBoost and XGBoost models are the most balanced and effective options, with the CatBoost model being slightly superior due to its combination of F1 score, precision, recall, and accuracy. This research will be extended with an image data set that consists of MRI scans of the heart to classify the severity as poor, fair, good, very good, etc. There are plenty of heart diseases that affect the anatomy and physiology of the heart, functioning in different ways. Hence, we would like to consider how machine learning and deep learning models will work for various abnormalities seen in heart functioning. This allows medical practitioners to make better decisions for patients.

Acknowledgements We acknowledge the team of Srusti, Shrilakshmi, Danish, and Atharva for the analysis of data set required for model building.

References

1. Ganie SM, Pramanik PKD, Malik MB, Nayyar A, Kwak KS (2023) An improved ensemble learning approach for heart disease prediction using boosting algorithms. Comput Syst Sci Eng 46(3):3993–4006
2. Motwani M, Dey D, Berman DS, Germano G, Achenbach S, Al-Mallah MH, Andreini D, Budoff MJ, Cademartiri F, Callister TQ et al (2017) Machine learning for prediction of all-cause mortality in patients with suspected coronary artery disease: a 5-year multicentre prospective registry analysis. Eur Heart J 38(7):500–507
3. Krittanawong C, Zhang HJ, Wang Z, Aydar M, Kitai T (2017) Artificial intelligence in precision cardiovascular medicine. J Am Coll Cardiol 69(21):2657–2664
4. Al'Aref SJ, Anchouche K, Singh G, Slomka PJ, Kolli KK, Kumar A, Pandey M, Maliakal G, Van Rosendael AR, Beecy AN et al (2019) Clinical applications of machine learning in cardiovascular disease and its relevance to cardiac imaging. Eur Heart J 40(24):1975–1986
5. Kavitha M, Gnaneswar G, Dinesh R, Rohith Sai Y, Sai Suraj R (2021) Heart disease prediction using hybrid machine learning model. In: 2021 6th international conference on inventive computation technologies (ICICT), pp 1329–1333
6. Ahmad GN, Fatima H, Abbas M, Rahman O, Alqahtani MS (2022) Mixed machine learning approach for efficient prediction of human heart disease by identifying the numerical and categorical features. Appl Sci 12(15):7449
7. Shorewala V (2021) Early detection of coronary heart disease using ensemble techniques. Inform Med Unlocked 26:100655
8. Ali MM, Paul BK, Ahmed K, Bui FM, Quinn JM, Moni MA (2021) Heart disease prediction using supervised machine learning algorithms: performance analysis and comparison. Comput Biol Med 136:104672
9. Dhiyanesh B, Ammal SG, Saranya K, Narayana KE (2024) Advanced cloud-based prediction models for cardiovascular disease: integrating machine learning and feature selection techniques. SN Comput Sci 5(5):572
10. Islam MR, Hoda MD, Rashid MA, Suha SA, Miya MT (2022) Data-driven heart disease prediction by ensemble feature selection and machine learning techniques. In: 2022 25th international conference on computer and information technology (ICCIT), pp 575–580
11. Yewale D, Vijayaragavan SP, Bairagi VK (2023) An effective heart disease prediction framework based on ensemble techniques in machine learning. Int J Adv Comput Sci Appl 14(2)
12. Chaurasia V, Chaurasia A (2023) Novel method of characterization of heart disease prediction using sequential feature selection-based ensemble technique. Biomed Mater Dev 1(2):932–941
13. Wynants L, Kent DM, Timmerman D, Lundquist CM, Van Calster B (2019) Untapped potential of multicenter studies: a review of cardiovascular risk prediction models revealed inappropriate analyses and wide variation in reporting. Diagn Progn Res 3:1–17
14. Liu F, Yang Y, Xu XS, Yuan M (2024) Mesbc: a novel mutually exclusive spectral biclustering method for cancer subtyping. Comput Biol Chem 109:108009
15. Samad MD, Ulloa A, Wehner GJ, Jing L, Hartzel D, Good CW, Williams BA, Haggerty CM, Fornwalt BK (2019) Predicting survival from large echocardiography and electronic health record datasets: optimization with machine learning. JACC: Cardiovasc Imaging 12(4):681–689
16. Lip GYH, Genaidy A, Estes C (2023) Cardiovascular disease (CVD) outcomes and associated risk factors in a medicare population without prior CVD history: an analysis using statistical and machine learning algorithms. Intern Emerg Med 18(5):1373–1383
17. Darapaneni N, Rao SR, Sagare DR, Paduri AR, DS BN, Desai S, BG S, Harsha R (2022) Machine learning based classification algorithms performance analysis for heart disease prediction. In: 2022 IEEE 9th Uttar Pradesh section international conference on electrical, electronics and computer engineering (UPCON), pp 1–8

18. Ali F, El-Sappagh S, Islam SMR, Kwak D, Ali A, Imran M, Kwak K-S (2020) A smart healthcare monitoring system for heart disease prediction based on ensemble deep learning and feature fusion. Inf Fusion 63:208–222
19. Dernoncourt F, Lee JY, Uzuner O, Szolovits P (2017) De-identification of patient notes with recurrent neural networks. J Am Med Inform Assoc 24(3):596–606
20. The Health and Behavioral Dataset (2020)

A Comprehensive Analysis on Machine Learning and Deep Learning Techniques for Anomaly Detection Using IoT Dataset

Sarvesh Sawant and Bhawana Rudra

Abstract Detection of anomalous behaviour in a network is crucial to ensure the safety of data. However, developing an effective anomaly detection mechanism is difficult. This is because these mechanisms have a comparatively higher tendency to raise false alarms. Thus, development of effective training mechanisms for predictive models such as machine learning and deep learning is of paramount importance. At the same time, recent datasets must be considered in order to train the predictive models. In this paper, we have considered recent datasets, tested them in balanced and unbalanced formats over various machine learning and deep learning models. In our analysis, we found that it is crucial to consider imbalanced data in order to generate an efficient anomaly detection model, even if it trained on balanced datasets.

Keywords Intrusion detection system · Anomaly detection · Machine learning · Deep learning

1 Introduction

Anomaly Detection in a network is currently one of the most vibrant research areas. The ability to identify an unforeseen event in a network has many technological and economic benefits. A direct beneficiary of anomaly detection is the intrusion detection system (IDS) since it helps the IDS predict attacks that have not been previously observed. IDS are mechanisms that detect any unwarranted network activity that may be caused by an adversary. They generate a pattern based on the data that they were exposed to during training and detect the malicious traffic. However, an anomaly detection mechanism is deployed in an IDS to detect zero-day attacks (unforeseen attacks). This is challenging because if an anomaly has occurred in the network

S. Sawant (✉) · B. Rudra
National Institute of Technology Karnataka, Surathkal, India
e-mail: svsawant.217it004@nitk.edu.in

B. Rudra
e-mail: bhawanarudra@nitk.edu.in

S. Kumar et al. (eds.), *Artificial Intelligence for Next Generation Computing, Volume 1*, Lecture Notes in Networks and Systems 1549,
https://doi.org/10.1007/978-981-95-0701-6_19

traffic, the reason need not be an adversary. It could be a genuine case of an anomaly in traffic. Thus, there is a high chance of false alarms.

In order to work in anomaly detection, it is crucial to have a clear overview of it. As discussed earlier, an anomaly is a deviation from known behaviour and, the techniques used to detect them fall under the category of anomaly detection. The various techniques for detecting an anomaly are statistical methods, machine learning, rule-based, time-series, graph-based, and domain-specific techniques.

However, these are the same methods that are used in the Intrusion Detection System to generate patterns and recognize a malicious packet. As the same methods are used, the training strategy must first be addressed. This will help enhance the capability of using these methods to detect zero-day attacks. Along with that, it is also very important to understand how to use these methods to detect anomalies. For that, we have to curate the dataset such that we can perform such analysis. In this paper, we focus on analysing various models with balanced and imbalanced datasets to perform this analysis. The use of an unbalanced dataset provides us with the necessary real-world touch to make the predictive mechanism more effective.

The rest of the paper is organized as follows. Section 2 discusses the related work with respect to anomaly detection in IDS. In Sect. 3, we present our proposed methodology. In Sect. 4, we discuss our implementation and results. Finally, in Sect. 6, we conclude by presenting our conclusion and future work.

2 Literature Survey

Sharafaldin et al. [1] addressed the need for robust datasets in Intrusion Detection Systems (IDSs) and Intrusion Prevention Systems (IPSs). They critiqued existing datasets like DARPA98 and KDD99 for being outdated and lacking diversity. Introducing IDS2017, they offered a meticulously crafted dataset to meet real-world criteria. Sato et al. [2] explored the limitations of signature-based Intrusion Detection Systems (IDSs) in detecting unknown attacks and proposed an automated method for identifying such threats from anomaly-based IDS alerts. Their approach, utilized 10 extracted features and one-class SVM on the Kyoto2006+ dataset. Song et al. [3] addressed the challenge of detecting 0-day attacks in Intrusion Detection Systems (IDSs) and proposed a generalized feature extraction method from IDS alerts, enabling detection across various IDS products. By defining seven new features based on basic IDS alert attributes and applying One-class SVM, their method demonstrated efficacy in detecting multiple types of 0-day attacks. Kishimoto et al. [4] addressed the challenges of preparing appropriate learning datasets for anomaly-based Intrusion Detection Systems (IDSs) due to the dynamic nature of Internet traffic. They proposed a method that combines multiple classifiers to follow traffic trends effectively. Evaluation was done using the Kyoto2006+ dataset. Lopez et al. [5] proposed an effective DDoS attack detection method using network traffic behavioural analytics. They emphasized on feature selection's significance, leveraging statistical analysis and information gain to identify relevant data and enhance detection accuracy and efficiency. Kokila et al. [6] proposed a comprehensive DDoS attack detection method for Software Defined Networking (SDN) controllers, emphasizing the importance of early detection using Support Vector Machines (SVM). They compared SVM's accuracy with other classifiers, highlighting its superiority. Additionally, they discussed integrating SVM-based traffic pattern detection with SDN controllers for online DDoS detection and suggested improving SVM performance by combining it with AVL trees to reduce testing time. Xu et al. [7] proposed SAiForest, an enhanced anomaly detection method, improving accuracy and efficiency compared to traditional Isolation Forest and LOF algorithms. SAiForest leverages selective integration and simulated annealing for optimized isolation tree selection. Validation was through experiments on large-scale datasets, demonstrating superior performance and simplicity on high-dimensional data. Thang et al. [8] proposed DBSCAN-MP, an innovative approach for anomaly detection in network traffic using DBSCAN with varying parameters for different clusters. Their method dynamically updates the normal behaviour and outperforms other clustering algorithms on the KDD Cup 1999 dataset. However, it faced challenges in detecting DOS attacks and experienced a higher false positive rate with changing network conditions. Cheng et al. [9] proposed a two-layer progressive ensemble method for outlier detection, integrating Isolation Forest (iForest) and Local Outlier Factor (LOF) algorithms. Their work efficiently performed outlier detection in complex datasets by utilizing iForest for quick scanning and LOF for accurate outlier identification.

Radford et al. [10] evaluated unsupervised anomaly detection methods on computer network traffic data, treating flow as a language and employing sequence aggregation rules and running them over LSTM models. Chen et al. [11] introduced an Autoencoder-based network anomaly detection method to address the challenges posed by cyberattacks in wireless communications. Leveraging Autoencoders, they captured non-linear feature correlations to enhance detection accuracy. They proposed two systems: one utilizing conventional Autoencoders and another employing Convolutional Autoencoders (CAE) for dimensionality reduction, resulting in shorter training times. Evaluation was performed using NSL-KDD dataset which demonstrated the superiority of the CAE-based method over other detection techniques. An et al. [12] proposed a novel approach that uses reconstruction probability derived from the variational autoencoder (VAE). By incorporating the probabilistic characteristics of the VAE, their method provided a more objective and principled anomaly score. Experimental results demonstrated its superiority over traditional autoencoder and principal component analysis (PCA) based methods. Utkin et al. [13] proposed ABIForest, an attention-based modification of the isolation forest for anomaly detection. By integrating attention weights via the Nadaraya-Watson regression, ABIForest outperformed other non-gradient-based algorithms for the supervised learning format. Ikram et al. [14] introduced an ensemble of deep neural networks integrated by XGBoost for intrusion detection. Results on UNSW-NB15 and VIT_SPARC20 datasets demonstrated high accuracy, highlighting the role neural networks can play in anomaly detection to minimize false alarms. It also projected the prospects of using IDS for online deployment.

2.1 Outcome of Literature Survey

Below are the outcomes of the literature survey:

- **Recent datasets were not used**—The use of recent datasets helps understand the evolving attack landscape. Also, network patterns and system configurations evolve over time which the older dataset may not reflect accurately in their distribution of normal and malicious traffic.
- **Comparison of performance using balanced and imbalanced dataset is not performed**—When a dataset is balanced, the anomalous and benign data points are roughly equal. This helps generate estimates with unbiased distinction patterns. It also contributes to the reduction of the problem of model overfitting. However, in the real world, anomalous data are very rare and thus, evaluating the model on an unbalanced dataset reveals its ability to handle the class imbalance and identify anomalies effectively despite their scarcity.

In this paper, we propose to use two recent datasets "ACI-IOT-2023" and "Mirai-based multi-class dataset," presented in papers [15, 16], respectively. We use them in balanced and imbalanced formats on various machine learning and deep learning techniques and present our analysis.

3 Proposed Methodology

In this section, we propose the methodology that we have considered in order to perform the comprehensive analysis of machine learning and deep learning models for anomaly detection. Figure 1 depicts the proposed methodology in chronological order. The first step discusses about the dataset. Using two datasets helps us compare the performance of the models more efficiently and generate a holistic analysis of the results. The next step involves data preprocessing, wherein the two datasets are processed in formats that are acceptable to the machine learning models. This step undertakes procedures such as data-cleaning and removal of redundant attributes, to name a few. Also, the formation of balanced and imbalanced datasets is performed in this step, which are the next action items after Data preprocessing in the proposed methodology. More details on the dataset will be discussed in the Sect. 3.1.

3.1 Dataset and Preprocessing

In this section, we will first discuss about the datasets and then present the various preprocessing techniques that we have considered to used in order to enhance the quality of the data.

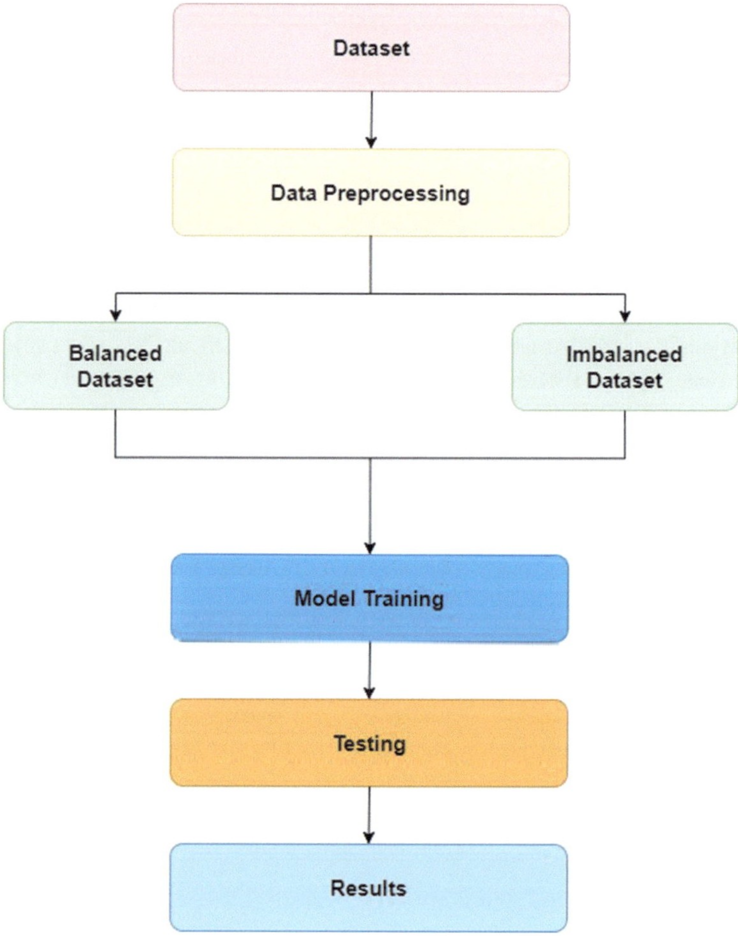

Fig. 1 Proposed methodology

Dataset-01

The first dataset that we have considered is the "ACI-IOT-2023" dataset presented by authors Nathaniel et al. in [15]. The features of the dataset are presented in Table 1. There are 85 in the dataset. This dataset was generated in the research conducted by the Internet of Things Research Lab (IoTRL) at the Army Cyber Institute (ACI). The collection was done from 30th October to 03rd November 2023. The dataset includes a dynamic mix of IoT devices and regular network devices to synthesize a near-realistic scenario. The attacks considered by the author were Denial-of-Service attacks, Bruteforce attacks, spoofing attempts, and reconnaissance activities such as Host Discovery, Port Scan, Ping Sweep, OS Scan, and Vulnerability Scans.

Table 1 Attributes of dataset "ACI-IOT-2023" [15]

Attributes	Attributes	Attributes	Attributes	Attributes
Flow ID	Src IP	Src port	Dst IP	Dst port
Protocol	Timestamp	Flow duration	Total fwd packet	Total bwd packets
Total length of fwd packet	Total length of bwd packet	Fwd packet length max	Fwd packet length min	Fwd packet length mean
Fwd packet length std	Bwd packet length max	Bwd packet length min	Bwd packet length mean	Bwd packet length std
Flow bytes/s	Flow packets/s	Flow IAT mean	Flow IAT std	Flow IAT max
Flow IAT min	Fwd IAT total	Fwd IAT mean	Fwd IAT std	Fwd IAT max
Fwd IAT min	Bwd IAT total	Bwd IAT mean	Bwd IAT std	Bwd IAT max
Bwd IAT min	Fwd PSH flags	Bwd PSH flags	Fwd URG flags	Bwd URG flags
Fwd header length	Bwd header length	Fwd packets/s	Bwd packets/s	Packet length min
Packet length max	Packet length mean	Packet length std	Packet length variance	FIN flag count
SYN flag count	RST flag count	PSH flag count	ACK flag count	URG Flag count
CWR flag Count	ECE flag count	Down/up ratio	Average packet size	Fwd segment size avg
Bwd segment size avg	Fwd bytes/bulk avg	Fwd packet/bulk avg	Fwd bulk rate avg	Bwd bytes/bulk avg
Bwd packet/bulk avg	Bwd bulk rate avg	Subflow fwd packets	Subflow fwd bytes	Subflow bwd packets
Subflow bwd bytes	FWD init win bytes	Bwd init win bytes	Fwd act data pkts	Fwd seg size min
Active mean	Active std	Active max	Active min	Idle mean
Idle std	Idle max	Idle min	Label	Connection type

From the dataset, we have shortlisted eight key features which stand out for their significant contribution to the detection and classification of anomalies. These are:

1. **Flow Duration**: It enables us to have deeper insights into the temporal characteristics of network flows. Variations observed in flow can serve as a crucial indicator of anomalous behaviour.
2. **Flow Packets**: It provides a measure of the intensity of packet transmission within flows. Abrupt fluctuations in packet rates can be a good indicator of anomaly.
3. **Flow IAT Mean**: Flow IAT Mean reflects the average time interval between consecutive flows.
4. **Forward IAT Mean**: Represents the regularity of packet transmission within flows.
5. **Forward Packets**: It discusses the intensity of packet transmission within flows. Anomalies characterized by abnormal packet rates can be effectively identified through this metric.

Table 2 Attributes of dataset "Mirai-based multi-class dataset" [16]

Attributes	Attributes	Attributes	Attributes
IP source address	IP destination address	IP flags	TCP flags
Source port number	Destination port number	Packet fragment	IP time-to-live (TTL)
IP checksum	Packet length	Packet sequence	Acknowledgement
Data offset values	TCP window size	TCP checksum	Label category

6. **Packet Length Mean**:They provide details on the packet size. It serves as a fundamental characteristic of network traffic, offering insights into the size and structure of packets within flows.
7. **Average Packet Mean**: Technically, Packet Length Mean and Average Packet Mean represent a similar value. Hence, to finalize a better candidate among the two, we have selected both to perform further analysis of data preprocessing
8. **Label**: Indicates classification of traffic as either malicious or benign.

From these candidate attributes, only some will be chosen to be used over the models. These finalized attributes are presented in Sect. 3.2.

Dataset-02

The second dataset that we have considered is the 'Mirai-based multi-class dataset' produced by the authors Hayelom et al. [16]. It is a curated collection of network traffic data files sourced and adapted from an existing IoT network intrusion dataset. It encompasses both benign (normal) traffic and malicious traffic associated with prevalent Mirai botnet attack types, such as SYN-Flooding, ACK-Flooding, and HTTP-Flooding. Each network traffic sample is represented by 16 features, which are presented in Table 2. From these sixteen, only few attributes will be considered for training the models as described in Sect. 3.2.

3.2 *Attribute Selection*

In order to select the relevant attributes from the shortlisted candidate attributes, we have used three techniques namely, Pearson's correlation [1] and Spearman correlation [2] and generated their heatmaps. Based on these heatmaps, the final attributes are selected. Also, there no missing data entries were found in the datasets.

The Pearson Correlation Heatmap of the Dataset-01 in Fig. 2 reveals: Flow Packets/s and Fwd Packets/s exhibit a correlation of 0.99, Fwd IAT Mean and Flow IAT Mean by 0.94, while Packet Length Mean and Average Packet Size display a perfect

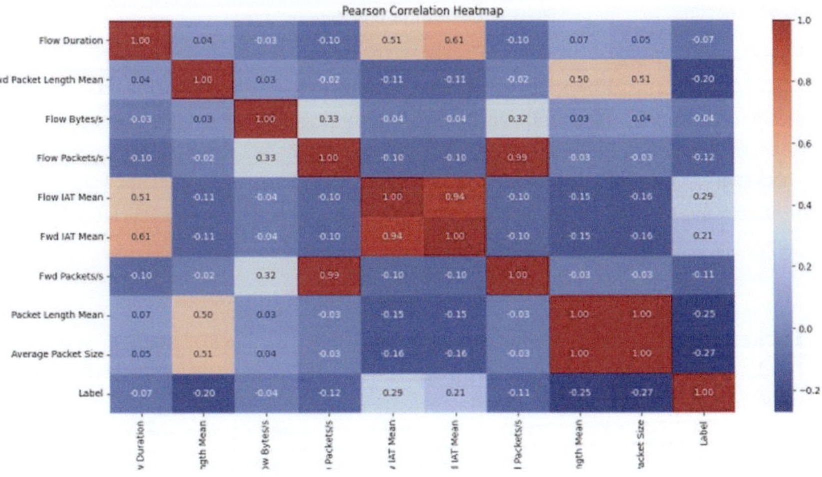

Fig. 2 Pearson correlation of Dataset-01

correlation of 1.00. These correlation value indicates how strong is the linear relationship and the direction between the respective pairs of variables. A correlation coefficient of 0.99 suggests an extremely strong positive linear relationship between Flow Packets/s and Fwd Packets/s. Similarly, a correlation coefficient of 0.94 indicates a very strong positive linear relationship between Fwd IAT Mean and Flow IAT Mean. Lastly, a correlation coefficient of 1.00 signifies a perfect positive linear relationship between Packet Length Mean and Average Packet Size, suggesting that these variables move perfectly in tandem.

From Fig. 3, the Spearman Correlation analysis of Dataset-01 reveals: correlations of Flow Packets/s and Fwd Packets/s exhibit at 0.99. Similarly, Fwd IAT Mean and Flow IAT Mean display a correlation of 0.95, Packet Length Mean and Average Packet Size show a perfect correlation of 1.00, while Packet Length Mean and Fwd Packet Length Mean, as well as Average Packet Size and Fwd Packet Length Mean, have correlations of 0.97. These correlation values indicates how strong is the monotonic relationship and the direction between the respective pairs of variables. For instance, a correlation coefficient of 0.99 suggests an extremely strong positive monotonic relationship between Flow Packets/s and Fwd Packets/s. Similarly, a coefficient of 0.95 indicates a very strong positive monotonic relationship between Fwd IAT Mean and Flow IAT Mean. A coefficient of 1.00 denotes a perfect positive monotonic relationship between Packet Length Mean and Average Packet Size, while correlations of 0.97 highlight significant positive monotonic relationships between Packet Length Mean and Fwd Packet Length Mean, as well as between Average Packet Size and Fwd Packet Length Mean.

Thus, based upon their analysis the selected attributes are : 'Flow Duration, Flow Packets/s, Flow IAT Mean, Packet Length Mean, and Label'.

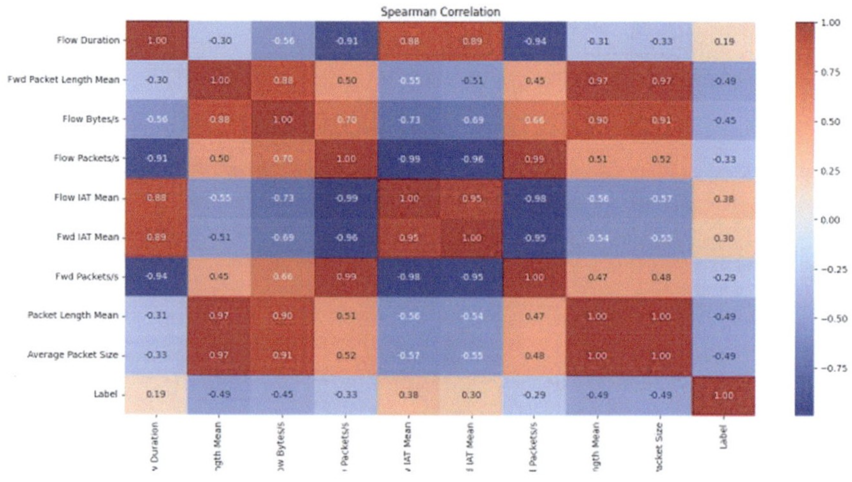

Fig. 3 Spearman correlation of Dataset-01

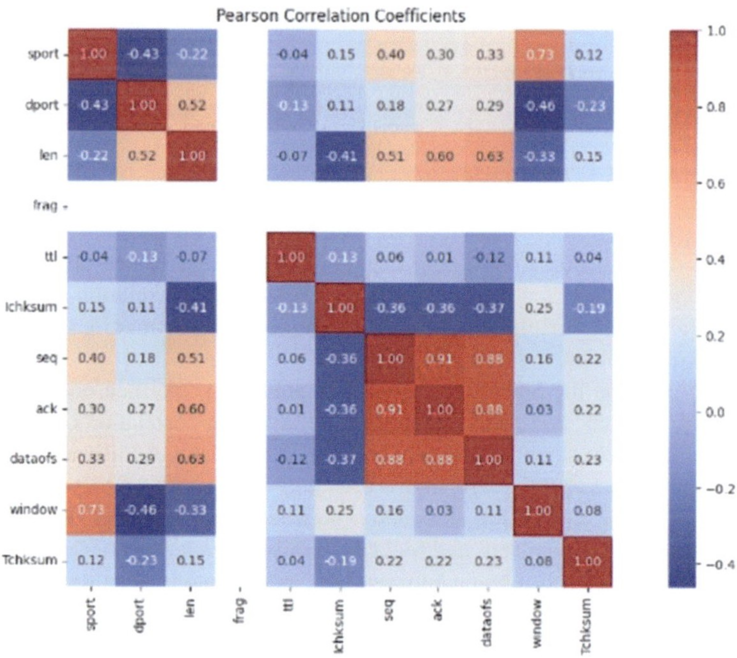

Fig. 4 Pearson correlation of Dataset-02

Figures 4 and 5 depict the heatmap for the dataset-02. From Fig. 4 the Pearson correlation analysis conducted on the dataset-02 reveals notable correlations between certain features. Specifically, the 'seq' and 'ack' features exhibit a strong correlation

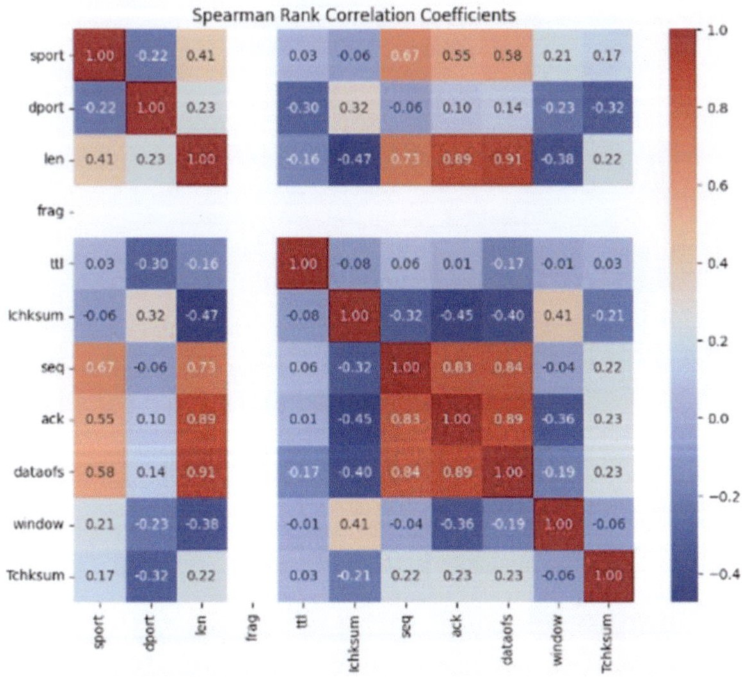

Fig. 5 Spearman correlation of Dataset-02

coefficient of 0.91, indicating a high degree of linear dependence between them. Similarly, the 'dataofs' feature demonstrates a substantial correlation of 0.88 with both 'seq' and 'ack' features, suggesting a significant linear relationship between these pairs of variables.

Figure 5 presents the Spearman Correlation analysis performed on the dataset-02. Notably, the 'len' and 'ack' features exhibit a strong correlation coefficient of 0.81, indicating a robust monotonic relationship between them. Additionally, the 'len' feature demonstrates a particularly high correlation of 0.91 with the 'dataofs' feature, suggesting a strong monotonic association between these variables. Furthermore, 'seq' and 'ack' as well as 'seq' and 'dataofs' exhibit correlations of 0.83 and 0.84, respectively, while the correlation between 'ack' and 'dataofs' stands at 0.89.

Thus, the selected features include Packet Length (len), Fragmentation Details (frag), Time-to-Live (TTL), IP Checksum (Ichksum), Sequence Number (seq), TCP Window Size (window), TCP Checksum (Tchksum), and Label. These features were chosen based on their relevance and minimal redundancy, determined through Pearson and Spearman correlation analysis. Though 'Fragmentation Details' attribute is constant, it is fundamental attribute of a packet and cannot be ignored. Hence, it is considered as an input for the model. Since it is constant, no change with respect to other attributes is observed in the heatmap.

Table 3 Balanced dataset for Dataset-01

S. No.	Scenario	Description
S1	Scenario-01	Training encompassed with 50% benign, Denial of Service (DoS), and brute force data, with testing conducted using benign and spoofing data.
S2	Scenario-02	Training with benign, spoofing, and brute force data, and testing with benign and DoS data
S3	Scenario-03	Training data from benign, spoofing, and DoS classes, testing against benign and brute force data
S4	Scenario-04	Training with benign, spoofing, and DoS data, with an 80–20 split for testing benign data
S5	Scenario-05	trained with benign and spoofing data and tested with benign and DoS data
S6	Scenario-06	training with benign and DoS data and testing with benign and spoofing data
S7	Scenario-07	Utilized all classes (benign, DoS, brute force, and spoofing) for training and testing, with an 80–20 split

Table 4 Balanced dataset for Dataset - 02

C. No.	Combination	Description
C1	Combination-01	The training dataset consists of normal data, AckFlooding and SynFlooding classes. The test data consists of normal data and HttpFooding classes
C2	Combination-02	All the classes were considered and then split of 80–20 split for train and test, respectively

Creation of Balanced Dataset

Table 3 depicts the different combinations that were used in order to balance dataset-01. Here, we can observe that in scenarios 01, 02, 03, 05 and 06, training is performed on a different subset of dataset-01 and testing is on another subset of dataset-01. This is performed to observe the response of the models when they encounter unknown attack (zero-day attack). Hence, Training data is prepared by using 50% of Benign class data while, rest 50% is prepared using combination of various attacks classes as described in Table 3. For the remaining scenarios, 80–20 split is performed. A similar strategy is used for the combinations generated in Table 4.

Thus from Tables 3 and 4, we can observe that Scenarios 01 to 05 and Combination C1, training set is different from that of the testing set. This is done to understand the ability of the models to identify those classes that are unknown to them. Furthermore, to maintain balance in the training data, the percentage of different class data points are kept same across all the scenarios.

Table 5 Strategy to create imbalanced data for both datasets

IS. No.	Scenario	Description
IS1	Imbalanced scenario-01	50% Benign, 50% Anomalous
IS2	Imbalanced scenario-02	75% Benign, 25% Anomalous
IS3	Imbalanced scenario-03	80% Benign, 20% Anomalous
IS4	Imbalanced scenario-04	85% Benign, 15% Anomalous
IS5	Imbalanced scenario-05	90% Benign, 10% Anomalous
IS6	Imbalanced scenario-06	95% Benign, 5% Anomalous

Table 6 Observation of different models on balanced data generated from Dataset-01 [15]

Models	S1 (%)	S2 (%)	S3 (%)	S4 (%)	S5 (%)	S6 (%)	S7 (%)
Vanilla autoencoder [8]	50	48	50	48.76	48.05	49.53	49
Variational autoencoder [9]	50	49	49	48.69	47.85	49.47	50
XGBoost [17]	53	53	53	86.49	53.68	50.34	76
SVM [6]	47	41	53	68.28	41.78	50.08	56
Unary SVM [18]	50	52	51	45.79	41.96	49.04	52
Isolation forest [10]	52.22	51.8	50.23	46.21	51.76	54.42	51.35
Attention based isolation forest [11]	50.89	47.65	50.6	48.52	45.24	50.55	48.63
Attention based vanilla autoencoder [19]	50.27	50.26	48.8	48.76	48.42	50.29	49.97
Attention based variational autoencoder [20]	50.28	50.2	48.7	48.96	48.18	50.26	50
DBScan [13]	62.47	45.63	49.33	89.62	93.90	79.45	76.28
Local outlier factor [21]	47.07	48.25	48.06	47.92	49.09	47.03	49.09

Creation of Imbalanced Dataset

In real world, occurrence of anomalous data is rare and thus data available about them is also limited. To replicate the same scenario of the real world, we have created imbalanced data for both the datasets. Table 5 highlights the creation process.

4 Implementation and Results

In this section, we present our implementation and discuss the results obtained. To perform analysis over the datasets, we have considered various Machine Learning and Deep Learning models as discussed in Sect. 3 and implemented the same in Python. The following Tables 6, 7, 8 and 9 present our observations. We have considered same hyperparameters for the models that are used in the reference papers.

From these tables, the below inferences were made:

Table 7 Different models on balanced data from Dataset-02 [16]

Models	C1 (%)	C2 (%)
Vanilla autoencoder [8]	55.73	49
Variational autoencoder [9]	52.65	50
XGBoost [17]	66.89	74.99
SVM [6]	70.44	99.52
Unary SVM [18]	50	74.88
Isolation forest [10]	90.54	49.88
Attention based isolation forest [11]	60.68	54.48
Attention based vanilla autoencoder [19]	48.24	49.52
Attention based variational autoencoder [20]	55.84	50.08
DBScan [13]	51.06	58.98
Local outlier factor [21]	45.08	49.78

- Models modified with the Attention layer did not contribute much to the prediction accuracy compared to their basic model versions.
- SVM methods performance was overall better on Balanced Dataset rather than Imbalanced Dataset. This must be so because SVM might struggle to find a suitable hyperplane when anomalies are severely outnumbered by normal instances. This is because SVM tries to maximize the margin between classes, and in imbalanced datasets, the minority class (anomalies) might not contribute sufficiently to determine a clear separation.
- DBScan performance was better on the imbalanced dataset. This must be because, when the dataset is balanced, the density distribution of normal and anomalous points might be more similar. Anomalies might not stand out as distinctly in terms of density differences compared to normal points. This could reduce the effectiveness of DBScan since it relies on density contrasts to identify outliers.
- Among the autoencoders, Vanilla Autoencoder was found to perform better compared to other autoencoders. However, it was observed that machine learning models were overall outperforming the Deep learning based autoencoder models.

Thus, from our analysis, it can be clearly seen that preparing a machine learning model for anomaly detection based on imbalanced data is very different from that based on balanced data. This highlights the importance of finalizing models based on analysis done using balanced as well as imbalanced datasets.

Table 8 Observation of different models on imbalanced data generated from Dataset-01 [15]

Model and scenarios	IS1		IS1		IS3		IS4		IS5		IS6	
Model name	Accuracy (%)	F1 score	Accuracy (%)	F1 score	Accuracy (%)	F1 score	Accuracy (%)	F1 score	Accuracy (%)	F1 score	Accuracy (%)	F1 score
Vanilla autoencoder [8]	48.83	0.33	73.91	0.64	78.80	0.71	78.23	0.75	88.78	0.84	94.96	0.92
Variational Autoencoder [9]	48.65	0.33	73.21	0.64	78.17	0.71	82.69	0.77	87.35	0.84	91.71	0.90
XGBoost [17]	49.68	0.34	75.10	0.66	80	0.73	84.46	0.79	88.78	0.86	93.08	0.92
Unary SVM [18]	47.44	0.33	72.12	0.63	76.87	0.70	81.47	0.76	86.35	0.83	90.57	0.90
Isolation Forest [10]	46.75	0.34	70.47	0.64	75.44	0.71	79.95	0.77	84.34	0.84	89.48	0.90
Attention based Isolation Forest [11]	45.89	0.33	70.47	0.64	74.17	0.70	79.27	0.77	84.30	0.84	89.40	0.90
DBScan [13]	71.52	0.68	82.66	0.77	86.18	0.81	89.53	0.85	93.25	0.9	96.76	0.95
Local Outlier Factor [21]	34.41	0.29	42.81	0.45	41.16	0.49	44.57	0.53	46.11	0.57	47.86	0.61

Table 9 Observation of different models on imbalanced data generated from Dataset-02 [16]

Model and scenarios	IS1		IS1		IS3		IS4		IS5		IS6	
Model name	Accuracy (%)	F1 score	Accuracy (%)	F1 score	Accuracy (%)	F1 score	Accuracy (%)	F1 score	Accuracy (%)	F1 score	Accuracy (%)	F1 score
Vanilla autoencoder [8]	50.27	0.34	74.09	0.64	79.41	0.71	85.01	0.[illegible]2	90.07	0.85	90.23	0.92
Variational autoencoder [9]	49.65	0.33	74.23	0.63	80.06	0.71	84.12	0.78	88.18	0.85	91.75	0.92
XGBoost [17]	49.86	0.34	75.22	0.66	80.25	0.73	85.39	0.[illegible]	90.09	0.87	97.27	0.96
Unary SVM [18]	48.19	0.33	72.78	0.64	77.84	0.71	82.58	0.78	87.2	0.84	91.4	0.91
Isolation forest [10]	43.78	0.32	64.61	0.62	68.77	0.69	71.99	0.77	76.92	0.82	80.54	0.86
Attention based isolation forest [11]	43.23	0.31	64.68	0.62	68.02	0.68	72.25	0.77	75.83	0.81	79.72	0.86
DBScan [13]	46.46	0.21	23.16	0.22	23.27	0.25	23.41	0.28	90	0.85	95	0.92
Local outlier factor [21]	45.02	0.39	45.77	0.42	46.3	0.48	47.09	0.51	48.12	0.54	49.19	0.61

5 Discussion

Anomaly detection plays a vital role in the security of a IoT network. The best illustration to understand this is the In-Vehicle Network (IVN) of a vehicle connected to the Internet (commonly known as Internet-of-Vehicles). IVN handles the communication of various internal units of vehicles for seamless coordination. These communications include data from sensors, engines, headlights, etc. However, due to the critical nature of the data in IVN, deployment of data encryption mechanisms is not feasible. Here, only IDS can be used. Thus, whenever a zero-day attack occurs, the anomaly detection module can help to recognize the attack and prevent it from occurring. Well-trained anomaly detection models (using balanced and unbalanced data), can help to detect the zero-day attacks with high accuracy and with fewer false positive alarms.

6 Conclusion and Future Work

In this paper, we attempted to perform anomaly detection using machine learning and deep learning models to generate a performance comparison between them. For our analysis, we first considered two latest (not benchmarked for anomaly detection) datasets and then selected those models that were commonly used for anomaly detection. In our analysis, we found that it is crucial to consider imbalanced data in order to generate an anomaly detection model, even if it is trained on balanced datasets. We also found that deep learning models such as autoencoders did not provide significant improvement over machine learning models. In future, we would like to perform a thorough analysis for the underperformance of the discussed methods to specific attacks. Also, we will explore the contributions of deep learning models and how they can contribute to anomaly detection.

References

1. Sharafaldin I, Lashkari AH, Ghorbani AA et al (2018) Toward generating a new intrusion detection dataset and intrusion traffic characterization. ICISSp 1:108–116
2. Sato M, Yamaki H, Takakura H (2012) Unknown attacks detection using feature extraction from anomaly-based ids alerts. In: 2012 IEEE/IPSJ 12th international symposium on applications and the internet. IEEE, pp 273–277
3. Song J, Takakura H, Kwon Y (2008) A generalized feature extraction scheme to detect 0-day attacks via ids alerts. In: 2008 international symposium on applications and the internet. IEEE, pp 55–61
4. Kishimoto K, Yamaki H, Takakura H (2011) Improving performance of anomaly-based ids by combining multiple classifiers. In: 2011 IEEE/IPSJ international symposium on applications and the internet. IEEE, pp 366–371
5. Lopez AD, Mohan AP, Nair S (2019) Network traffic behavioral analytics for detection of DDoS attacks. SMU Data Sci Rev 2(1):14

6. Kokila R, Selvi ST, Govindarajan K (2014) DDoS detection and analysis in SDN-based environment using support vector machine classifier. In: 2014 sixth international conference on advanced computing (ICoAC). IEEE
7. Xu D, Wang Y, Meng Y, Zhang Z (2017) An improved data anomaly detection method based on isolation forest. In: 2017 10th international symposium on computational intelligence and design (ISCID), vol 2. IEEE, pp 287–291
8. Thang TM, Kim J (2011) The anomaly detection by using DBSCAN clustering with multiple parameters. In: 2011 international conference on information science and applications. IEEE, pp 1–5
9. Cheng Z, Zou C, Dong J (2019) Outlier detection using isolation forest and local outlier factor. In: Proceedings of the conference on research in adaptive and convergent systems, pp 161–168
10. Radford BJ, Richardson BD, Davis SE (2018) Sequence aggregation rules for anomaly detection in computer network traffic. arXiv preprint arXiv:1805.03735
11. Chen Z, Yeo CK, Lee BS, Lau CT (2018) Autoencoder-based network anomaly detection. In: 2018 wireless telecommunications symposium (WTS). IEEE, pp 1–5
12. An J, Cho S (2015) Variational autoencoder based anomaly detection using reconstruction probability. Spec Lect IE 2(1):1–18
13. Utkin L, Ageev A, Konstantinov A, Muliukha V (2022) Improved anomaly detection by using the attention based isolation forest. Algorithms 16(1):19
14. Ikram ST, Cherukuri AK, Poorva B, Ushasree PS, Zhang Y, Liu X, Li G (2021) Anomaly detection using XGBoost ensemble of deep neural network models. Cybern Inf Technol 21(3):175–188
15. Bastian N, Bierbrauer D, McKenzie M, Nack E (2023) ACI IoT network traffic dataset 2023. https://doi.org/10.21227/qacj-3x32
16. Gebrye H (2023) Mirai-based multi-class dataset. https://doi.org/10.21227/h4ac-wr38
17. Dhaliwal SS, Nahid AA, Abbas R (2018) Effective intrusion detection system using XGBoost. Information 9(7):149
18. Amer M, Goldstein M, Abdennadher S (2013) Enhancing one-class support vector machines for unsupervised anomaly detection. In: Proceedings of the ACM SIGKDD workshop on outlier detection and description
19. Kundu A, Sahu A, Serpedin E, Davis K (2020) A3d: attention-based auto-encoder anomaly detector for false data injection attacks. Electr Power Syst Res 189:106795
20. Aslam N, Kolekar MH (2023) A-VAE: Attention based variational autoencoder for traffic video anomaly detection. In: 2023 IEEE 8th international conference for convergence in technology (I2CT). IEEE,pp 1–7
21. Paulauskas N, Bagdonas ĄF (2015) Local outlier factor use for the network flow anomaly detection. Secur Commun Netw 8(18):4203–4212
22. Li J, Cheng K, Wang S, Morstatter F, Trevino RP, Tang J, Liu H (2017) Feature selection: a data perspective. ACM Comput Surv (CSUR) 50(6):1–45

Deep Learning-Based Brain Tumor Detection Using Xception Model on MRI Images

Sunil Kumar Agarwal, Yogesh Kumar Gupta, and Sandeep Kumar Bothra

Abstract Brain tumor diagnosis and treatment create critical challenges due to the complexity of their nature. Deep learning techniques have appeared as reliable tools for the accurate identification and classification of brain tumors from MR images. This paper presents a novel approach utilizing the pretrained deep learning Xception model, to classify MRI images of brain tumors. The proposed method achieves an accuracy of 99%, which is far better than existing methods. Through comprehensive evaluation and experimentation, the research presents the effectiveness of the proposed CNN model. Our proposed approach can serve as an example for conducting and modeling other research. Furthermore, the paper explains the potential implications of deep learning in medical imaging and highlights avenues for future research.

Keywords Brain tumor detection · Deep learning · MRI images · Pretrained Xception deep learning model

1 Introduction

Abnormal cell development in the brain is known as a brain tumor and can create serious medical problems if not detected and treated within time [1]. Traditional methods of brain tumor detection can produce impressive results on brain images, but they have some limitations. This has led to an increased interest in using deep learning

S. K. Agarwal · Y. K. Gupta
Department of Computer Science, Banasthali Vidyapith, Jaipur, Rajasthan, India
e-mail: sun.aipc@gmail.com

Y. K. Gupta
e-mail: gyogesh@banasthali.in

S. K. Bothra (✉)
Department of Computer Science, S.S. Jain Subodh P.G. College, Jaipur, Rajasthan, India
e-mail: bothrajain@gmail.com

S. Kumar et al. (eds.), *Artificial Intelligence for Next Generation Computing, Volume 1*, Lecture Notes in Networks and Systems 1549,
https://doi.org/10.1007/978-981-95-0701-6_20

for brain tumor detection and classification [2]. This technique works perfectly on medical imaging applications.

Convolutional neural networks (CNNs) [3] are the most important and popular deep learning algorithms that are used to detect brain tumor. Transfer learning is another recently developed and popular deep learning method that performed very well in categorizing brain cancers [4]. Using a pretrained deep learning model, this method is compelling when working with a limited amount of training data [5]. One of the challenges of applying deep learning is the variability in the appearance of tumors in different patients. To overcome these challenges, researchers have developed some innovative models. Furthermore, another challenge is ensuring the strength and reliability of the proposed model. The suggested models must provide compatible and accurate results. To overcome these problems, researchers have developed several other methods for evaluating how well deep learning models can identify brain tumors. These techniques contain a cross-validation mechanism, to verify the performance of the newly created model [6]. Despite all these challenges, deep learning has shown great promise. Research has shown that deep learning techniques can yield better results compared to traditional tumor detection methods. Furthermore, deep learning algorithms may be automated, which could result in a more economical and effective method of brain tumor diagnosis [7].

This study implements and recognizes the functional principles of image processing and image classification using CNN. According to research, various types of tumors have been identified by medical practitioners, but mainly, three categories of tumors (glioma, meningioma, and pituitary) are detected and categorized by the CNN applications. Therefore, the primary issue that is emphasized in this study is how the created deep learning architecture is capable of categorizing MRI pictures of brain tumors with precision. This research introduced an integrated CNN, Xception [8] pretrained architecture. Through appropriate parameter settings of the neural network, we can create a CNN that effectively classifies the various types of tumors. The rest part of the manuscript is planned as follows: Part 2 covers the significance of earlier studies. Part 3 explains the materials and methods for analyzing MRI images. Part 4 describes the experimental results and detailed discussion, and Part 5 provides the conclusion and future directions of the paper.

2 Literature Review

In the field of medical image processing, deep learning and related techniques are broadly used to detect and classify the images. In the past, several types of research have been conducted in this area using MRI images.

In the paper [9], the authors proposed an approach for automated segmenting of brain cancer using three-dimensional magnetic resonance imaging. The authors employed an effective unsupervised technique known as the "FBB technique" to automate tumor identification. To accurately detect tumor boundaries regardless of their shape and volume, the method utilizes a geodesic level set-based 3D deformable model. The authors report an average time of around five minutes to detect and segment tumors with an accuracy of 38.04% and sensitivity of 89.01%. Although the study suggests good segmentation accuracy, the metrics may not capture all aspects of segmentation quality.

Other authors in their research paper [10] present a systematic investigation technique for extracting lesion information from MRI data. K-means methods are employed by the authors to successfully segment and classify data using a two-layer approach. The authors used the MATLAB R2013a platform to implement their algorithm and reviewed how well their approach worked using precision and sensibility expressions as analytical measures. The experimental results provide 85% accuracy and 100% sensitivity. However, the paper does not discuss specific methodologies, which may limit understanding of practical applications. In the paper [11], authors presented a data mining approach regarding MRI image analysis. The author uses denoising and imaging techniques in preprocessing to improve the precision and speed of the analysis. In the next step, the complex clustering method is used for segmentation. In the final stage, researchers using support vector machines (SVM) achieved high accuracy. Although the proposed research work provides various advantages, it has some limitations, like data constraints, scalability issues, and parameter sensitivity.

The authors of the research [12], present a method for dimension reduction and tissue classification to create a comprehensive architecture to define operative margins. Their proposed method performs tissue classification in two stages. Their method achieved 81.90% accuracy and 80.91% sensitivity. The research may have a limited data size, which can affect the generalizability of the results. An approachable algorithm for modeling glioblastoma is presented in the paper [13]. Their novel model evaluates cancer progression over time using an algorithm that was based on cellular automata and a fast-marching technique. This technique does not need extensive training and has an optimized latency. The suggested method can detect tumors in real time and then use these to estimate growth with the use of CRM. This research also has some potential limitations, such as a limited dataset and preprocessing dependence. Preprocessing, segmentation, pattern extraction, and pattern identification are the four stages of MRI image processing that were suggested in research [14]. The planned study compares two approaches for detecting brain cancer. The first technique separates brain cancer from images using an active contour nonparametric deformation model. The K-segmentation approach is the second technique. At the classification stage, SVM is utilized. The final result achieved 94.12% and 82.35% accuracy, respectively.

A brain tumor diagnosis and classification approach employed histogram analysis, resampling, KNN algorithm, and a distance matrix in the paper [15]. The pixel dissemination of the image is determined in the first step using histogram analysis. The KNN algorithm is then applied to brain tumor classification and identification after tuning the K parameter. This research achieved a classification score of around 95% for all tested images. Using an artificial neural network, the paper [16] suggested a multi-step strategy that included picture preprocessing, segmenting, feature collecting, SVM categorization, and tumor stage classification. The accuracy

Table 1 Summary of related work

Year	Reference number	Main findings	Limitations
2022	[17]	Investigation of deep learning methods for the classification of brain tumors	Lack of experimental validation on external datasets
2022	[18]	Enhancement of several tumor detection models which were based on the transfer learning method	Limited exploration of model interpretability and generalizability
2022	[19]	Categorization of brain tumors using an improved CNN model	Limited discussion on model interpretability
2020	[14]	MRI-based brain tumor detection and division	Lack of comparison with state-of-the-art methods
2019	[13]	Moving toward an automated prognostic system for estimating the growth of brain glioblastoma tumors	Limited validation on external datasets
2019	[12]	Hyperspectral brain imaging for intraoperative guidance using manifold embedding and semantic segmentation	Possible over-fitting due to limited dataset size
2019	[10]	Classifying brain tumors using an adaptive segmentation approach and a two-tier classifier	Limited details on the segmentation and classification techniques used

of the suggested approach was 97.44%. The important related study is summarized in Table 1. All the above-suggested research work provides an important contribution in the area of brain tumor identification and classification, but they are having some limitations, such as data and model limitations, lack of comparisons with other methods, and generalization across imaging modalities, which may be overcome in the proposed work.

3 Materials and Methods

The following section covers the materials and methods used in this research in more detail. This research work suggests the utilization of a CNN in integration with the Xception model to generate a structure capable of identifying and categorizing MRI images of brain tumors.

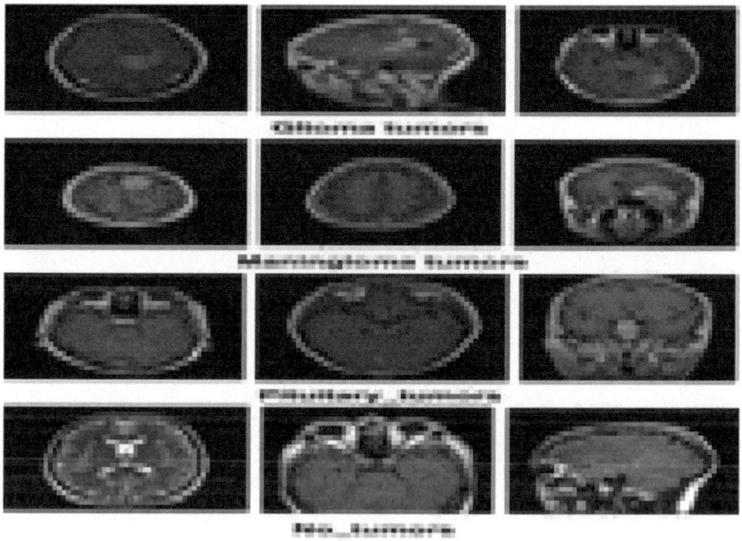

Fig. 1 Sample brain tumor MRI dataset

3.1 Dataset

System efficiency is determined using the datasets. The worldwide popular online repository Kaggle is the source of the dataset. It includes 3264 tumor pictures, which are categorized into four different types; dataset contains 926 pictures of glioma tumors, 937 pictures of meningioma tumors, 500 pictures of nontumor, and 901 pictures of pituitary tumor. All MRI pictures have the same size of 512 × 512 pixels. Figure 1 depicts the samples of the brain tumor MRI dataset.

3.2 Preprocessing

Data preprocessing is a crucial step that changes unprocessed data into a comprehensible format. Real-world data are often inconsistent, incomplete, and inaccurate, making it unsuitable for direct use in models. Preprocessing the data is essential to ensure that the model receives accurate and reliable data, thereby reducing errors and improving the quality of the results. The process of data preprocessing contains various steps [19, 20].

3.3 Feature Extraction

In image processing steps after augmenting the dataset, a vast amount of the dataset is converted into trained data, and respective attributes that signify each tumor are extracted. Convolutional filters are used to extract various features from the training dataset in CNN. In the proposed study, the data are analyzed on the Xception model to categorize three types of brain tumors [17].

3.4 Xception Model

The Xception model (Fig. 2) is a deep CNN that was developed by Google and is available as an open-source project [18]. Xception's convolution layer combines pointwise and depthwise convolutions, which have been shown to improve the accuracy of image classification tasks. The trained Xception model contains 36 convolution layers that build the base of the feature extraction network. These convolution layers are arranged in 14 different sections; each section contains linear residual connections, except the beginning and ending sections. Xception's architecture is designed to allow users to easily modify its parameters.

The heart of deep learning for brain tumor detection is a type of neural network that is trained on a vast number of MRI scans. This neural network is trained to differentiate normal and abnormal tissue. Neural networks perform this task using the supervised learning method, where a network is presented with a large number of labeled images and learns to divide the images based on their specific attributes.

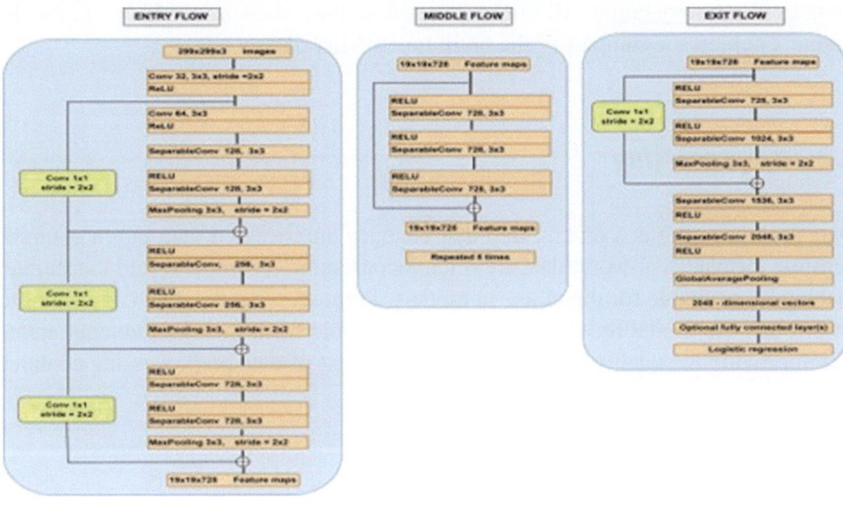

Fig. 2 The Xception architecture

Algorithm 1: Performance evaluation of Xception model

```
loadImage ();
dataPreprocessing ();
dataAugmentation ();
split Data();
loadModel();
for each epoch in epochNumber do
    for each batch in batchSize do
        ê = model(features);
        loss = crossEntropy(e, ê);
        optimization(loss);
        accuracy ();
        bestAccuracy = max(bestAcccuracy, accuracy);
return
```

Fig. 3 Performance evaluation algorithm

After several repetitive processes, the neural network can accurately classify brain images (Fig. 3).

Keynotes on the implementation methods are as follows:

1. Data preparation: Acquisition of a large labeled dataset of MRI images is necessary for preparing a reliable deep learning model.
2. Image preprocessing: Several types of preprocessing techniques are applied to the dataset to increase the diversity of the dataset and improve the model's efficiency.
3. Model selection: Model selection must depend on the complexity and size of the dataset and the task requirements. Recently developed transfer learning techniques which are used from various famous pretrained models such as Xception can speed up the training procedure and achieve better accuracy.
4. Training and Validation: Dividing the image dataset into two different categories that is training and validation is necessary to check the model's performance and avoid over-fitting. Several callbacks such as ReduceLROnPlateau, TensorBoard, and ModelCheckpoint can be used to monitor the training process and save the best model.
5. Evaluation metrics: Performance metrics are used to evaluate model performance.
6. Deployment: After training, the model can be deployed in a real-life scenario to automate the identification of brain tumors and support medical professionals in making clinical decisions.
7. Continuous Improvement: Regularly fine-tuning the model and updating the dataset with new images can improve the efficiency of the model over time

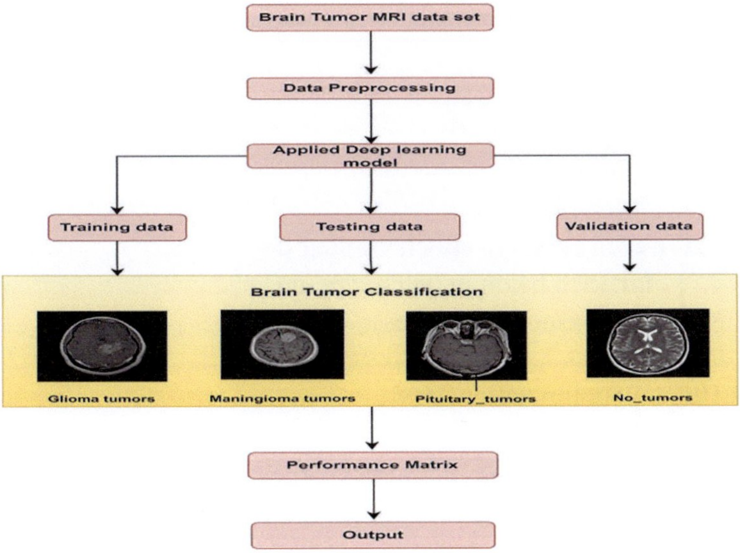

Fig. 4 Brain tumor classification flow graph

and keep it up to date with new advances in the field. A general flow graph of brain tumor classification is presented in Fig. 4.

In the proposed brain tumor identification method, the first step is to import the necessary libraries including cv2, os, sklearn, NumPy, Seaborn, TensorFlow, and several modules from the Keras libraries. The next step is to load the dataset by reading the training and test images from the folders, appending them to the *x*_train and *y*_train lists, and resizing the pictures to a constant size of 175 × 175 pixels. The dataset is then shuffled using the sklearn.utils.shuffle function. Image augmentation is applied to the data using the ImageDataGenerator class which belongs to the TensorFlow framework and the preprocessing image module which belongs to the Keras library. The train test split function belonging to sklearn library is then used to separate the dataset into training dataset and test dataset.

The model is built using a pretrained Xception model and is fine-tuned by unfreezing some layers and training them. Callbacks such as ReduceLROnPlateau, TensorBoard, and ModelCheckpoint are used to track the training procedure and save the best model. The precision of the trained structure is then determined by using it to predict things on the testing dataset. Using the function confusion matrix belonging to the sklearn package, the confusion matrix is produced, and several performance parameters are computed based on the available data in the performance matrix. Finally, performance metrics are printed.

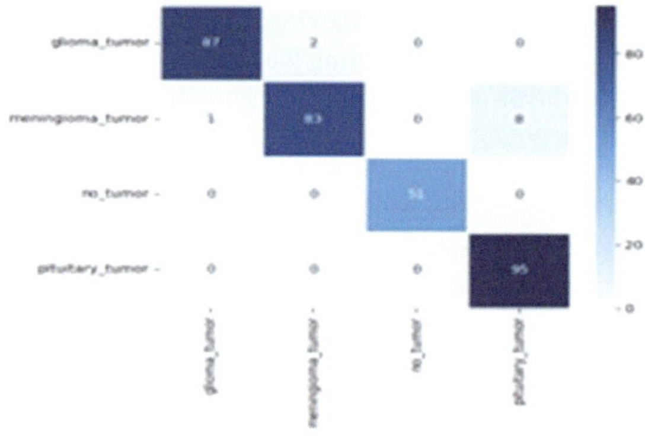

Fig. 5 Confusion matrix of brain tumor MRI dataset

4 Results and Discussion

4.1 Confusion Matrix Result

To examine the accuracy of a classification model, one can summaries the output using a confusion or performance matrix (Fig. 5) [21]. The matrix gives a complete summary of the model's efficiency by comparing its true and predicted values. It provides the error types generated in the model during execution and provides insights in the form of various parameters like accuracy, recall, precision, and specificity. As shown in the result, the confusion matrix presents an in-depth analysis of how well the model classified tumor images. The accuracy of the model is shown in the values along the diagonal of the matrix, which shows the number of true positive predictions. The output indicates that the proposed structure is capable of dividing the types of brain tumors accurately.

The number of accurate and inaccurate forecasts that the model generates is calculated using the matrix structure. The generated matrix structure represents four different values: t true positives (TP), which show the frequency of structure accurately anticipated a positive class; false positives (FP), which represent the frequency of structure erroneously anticipated a positive class; true negatives (TN), which represent the frequency of structure appropriately assessed the negative class; and false negatives (FN), which represent the frequency of structure falsely assessed the negative class.

4.2 Accuracy Results

The results generated by the Python script for a brain tumor identification using the Keras, TensorFlow, and Seaborn libraries for a frame size of 150 are quite impressive. The script achieved 99% accuracy, meaning it correctly classified 99% of the brain tumor pictures as either having a brain tumor or not. The neural network was fed training data 12 times during the script's training, or 12 epochs. The network weights are modified at each epoch to minimize a loss function that assesses the variance between predicted and real results. The network may learn more complicated patterns and increase its accuracy by training for longer epochs.

The fact that each epoch lasted more than 11 min for 195 iterations indicates that the dataset is quite large and the neural network had to process a large amount of data. The quantity of repetitions is equal to the quantity of batches of data processed during each epoch; in this case, it was 195. The training time of the dataset may depend on the size of the dataset, the complexity of the neural network architecture, and the available computing power.

Overall, the high accuracy achieved by the model suggests that it is a promising tool for a brain tumor or cancer identification from medical images. The accuracy and loss graphs of the Xception model are presented in Fig. 8. However, it is important to note that the script should be validated on independent test data to make sure that it is not over-fitted to the training data. The sample of the overall results of tumor classification is represented in Fig. 6. Lastly, to evaluate the efficiency of the model, we compared our findings with those results, obtained from some of the other research conducted in the same field.

Hemanth et al. [22] achieved 91% accuracy with LinkNet. Siddique et al. [23] presented an innovative DCNN model in which VGG16 was used to detect brain tumor and achieved 96% accuracy. After that, Hossain et al. [24] suggested an improved CNN to identify brain tumors, and they achieved 97.8% accuracy as a result. The performance of our suggested model in comparison with other well-known algorithms is displayed in Table 2 and Fig. 7 (Fig. 8).

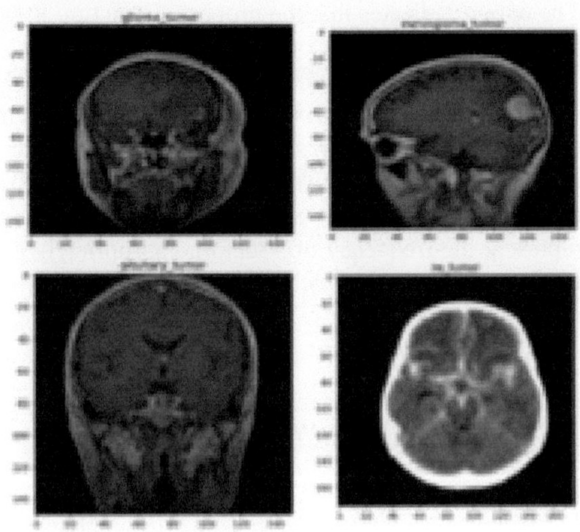

Fig. 6 Classifications of brain tumors

Table 2 Performance comparison with the existing methods

Name of researchers	Name of method and algorithms	Results in terms of Accuracy (%)
Hemanth et al. [22]	LinkNet	91
Siddique et al. [23]	DCNN (VGG16)	96
Hossain et al. [24]	CNN	97.87
Proposed model	DCCN (Xception)	99

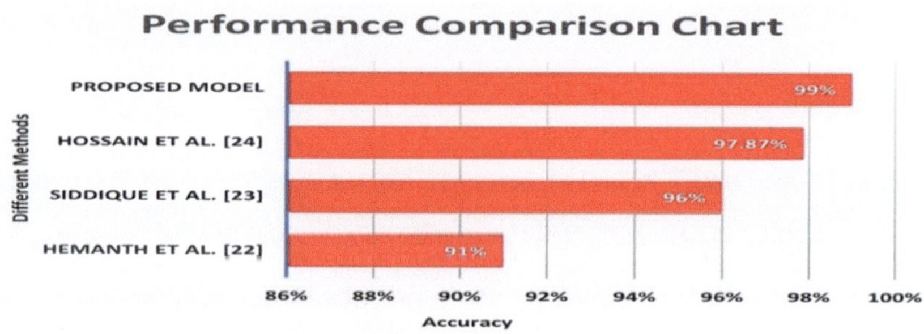

Fig. 7 Performance comparison chart with existing methods

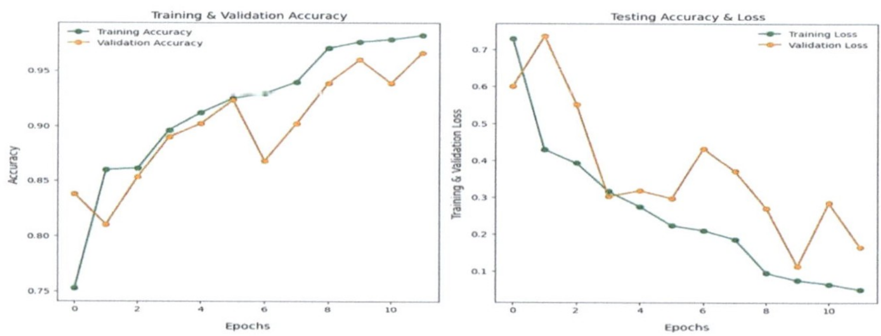

Fig. 8 Accuracy and loss graphs of the Xception model

5 Conclusion and Future Scope

This research employed the pretrained Xception model to categorize MRI pictures of brain tumors. We have achieved 99% accuracy using this model. Our neural network arrangement and parameter settings have shown classification results with high accuracy, without any over-fitting or under-fitting of the data. The results are presented in a performance matrix, which demonstrates the effectiveness of our approach. Our proposed approach can serve as an example for conducting and modeling other research. In conclusion, the observations indicate that deep learning is a promising tool for the analysis of medical images. Overall, deep learning can be a useful technique for medical professionals in detecting and diagnosing brain tumors, but further testing and validation are necessary before it can be used in a clinical setting [25–28].

There are several potential future directions for improving and extending the brain tumor identification technique. Some possible areas for future development include enhancing the dataset, improving the neural network architecture, incorporating other imaging modalities, and expanding to other medical applications. All these will expand the usefulness of deep learning and make it more versatile in clinical practice.

References

1. Biratu ES, Schwenker F, Ayano YM, Debelee TG (2021) A survey of brain tumor segmentation and classification algorithms. J Imaging 7(9):179. https://doi.org/10.3390/jimaging7090179
2. Al-Zoghby AM, Ebada AI (2023) Dual deep CNN for tumor brain classification. Diagnostics 13(12):2050. https://doi.org/10.3390/diagnostics13122050
3. Ayadi W, Elhamzi W, Charfi I, Atri M (2021) Deep CNN for brain tumor classification. Neural Process Lett 53(1). https://doi.org/10.1007/s11063-020-10398-2
4. Tan C, Sun F, Kong T, Zhang W, Yang C, Liu C (2018) A survey on deep transfer learning [Online]. Available: http://arxiv.org/abs/1808.01974
5. Alshalali T, Josyula D (2018) Fine-tuning of pre-trained deep learning models with extreme learning machine. In: Proceedings—2018 international conference on computational science and computational intelligence, CSCI 2018. Institute of Electrical and Electronics Engineers Inc., pp 469–473. https://doi.org/10.1109/CSCI46756.2018.00096
6. Xie Y et al (2022) Convolutional neural network techniques for brain tumor classification (from 2015 to 2022): review, challenges, and future perspectives. Diagnostics 12(8):1850. https://doi.org/10.3390/diagnostics12081850
7. Díaz-Pernas FJ, Martínez-Zarzuela M, González-Ortega D, Antón-Rodríguez M (2021) A deep learning approach for brain tumor classification and segmentation using a multiscale convolutional neural network. Healthcare (Switzerland) 9(2):153. https://doi.org/10.3390/healthcare9020153
8. Chollet F (2016) Xception: deep learning with depthwise separable convolutions [Online]. Available: http://arxiv.org/abs/1610.02357
9. Kermi A, Andjouh K, Zidane F (2018) Fully automated brain tumour segmentation system in 3D-MRI using symmetry analysis of brain and level sets. IET Image Process 12(11):1964–1971. https://doi.org/10.1049/iet-ipr.2017.1124
10. Zaw HT, Maneerat N, Win KY (2019) Brain tumor detection based on Naïve Bayes classification. In: 2019 5th international conference on engineering, applied sciences and technology (ICEAST). IEEE, pp 1–4
11. Snehal D (2022) Data analysis using data mining techniques as K Means and Apriori algorithms. Int J Adv Res Sci Commun Technol. https://doi.org/10.48175/ijarsct-5734
12. Shakeel PM, Tobely TEE, Al-Feel H, Manogaran G, Baskar S (2019) Neural network based brain tumor detection using wireless infrared imaging sensor. IEEE Access 7:5577–5588
13. Hossain T, Shishir FS, Ashraf M, Al Nasim MA, Shah FM (2019) Brain tumor detection using convolutional neural network. In 2019 1st international conference on advances in science, engineering and robotics technology (ICASERT). IEEE, pp 1–6
14. Amin J, Sharif M, Yasmin M, Fernandes SL (2020) A distinctive approach in brain tumor detection and classification using MRI. Pattern Recogn Lett 139:118–127
15. Siar M, Teshnehlab M (2019) Brain tumor detection using deep neural network and machine learning algorithm. In: 2019 9th international conference on computer and knowledge engineering (ICCKE). IEEE, pp 363–368

16. Hemanth G, Janardhan M, Sujihelen L (2019) Design and implementing brain tumor detection using machine learning approach. In: 2019 3rd international conference on trends in electronics and informatics (ICOEI). IEEE, pp 1289–1294
17. Padmapriya K, Divya S, Ezhumalai P (2022) A scrutiny on brain tumour classification using deep learning technique. Mater Today Proc. https://doi.org/10.1016/j.matpr.2020.11.561
18. Asif S, Yi W, Ain QU, Hou J, Yi T, Si J (2022) Improving effectiveness of different deep transfer learning-based models for detecting brain tumors from MR images. IEEE Access 10:34716–34730. https://doi.org/10.1109/ACCESS.2022.3153306
19. Mukkapati N, Anbarasi MS (2022) Brain tumor classification based on enhanced CNN model. Revue d'Intelligence Artificielle 36(1):125. https://doi.org/10.18280/ria.360114
20. Sadad T, Rehman A, Munir A, Saba T, Tariq U, Ayesha N, Abbasi R (2021) Brain tumor detection and multi-classification using advanced deep learning techniques. Microsc Res Tech 84(6):1296–1308
21. Abd El-Wahab BS, Nasr ME, Khamis S, Ashour AS (2023) BTC-fCNN: fast convolution neural network for multi-class brain tumor classification. Health Inf Sci Syst 11(1). https://doi.org/10.1007/s13755-022-00203-w
22. Hemanth G, Janardhan M, Sujihelen L (2019) Design and implementing brain tumor detection using machine learning approach. In: 2019 3rd international conference on trends in electronics and informatics (ICOEI), pp 1289–1294
23. Bakr Siddique MA, Sakib S, Rahman Khan MM, Tanzeem AK, Chowdhury M, Yasmin N (2020) Deep convolutional neural networks model-based brain tumor detection in brain MRI images. In: 2020 fourth international conference on I-SMAC (IoT in Social, Mobile, Analytics and Cloud) (I-SMAC), pp 909–914. https://doi.org/10.1109/ISMAC49090.2020.9243461
24. Hossain T, Shishir FS, Ashraf M, Al Nasim MA, Shah FM (2019) Brain tumor detection using convolutional neural network. In: 2019 1st international conference on advances in science, engineering and robotics technology (ICASERT). IEEE, pp 1–6
25. Bothra SK, Singhal S, Goyal H (2023) Hybrid JAYA algorithm for workflow scheduling in cloud. J Sci Tech Info Technol Mech Opt 23(2):313–322. https://doi.org/10.17586/2226-1494-2023-23-2-313-322
26. Bothra SK, Singhal S, Goyal H (2021) Deadline-constrained cost-effective load-balanced improved genetic algorithm for workflow scheduling. Int J Inf Technol Web Eng (IJITWE) 16(4):1–34. https://doi.org/10.4018/ijitwe.2021100101
27. Bothra SK, Singhal S, Goyal H (2022) Cost effective hybrid genetic algorithm for workflow scheduling in cloud. Sys res info technol (3):121–138. https://doi.org/10.20535/srit.2308-8893.2022.3.08
28. Bhatt C, Singhal S (2024) Multi-objective reinforcement learning for virtual machines placement in cloud computing. Int J Adv Comput Sci Appl 15(3). https://doi.org/10.14569/ijacsa.2024.01503105

Threshold Leaf Segmentation with K-Means Clustering and Logistic Particle Swarm Optimization

V. B. Prijimol and Sarada Jayan

Abstract Leaf segmentation is a crucial step in agronomy applications such as leaf counting, plant disease detection, phenotyping, and growth analysis, which helps farmers in plant breeding, plant health monitoring, horticulture, and improving crop yield. The problem of poor segmentation results often pose a significant challenge in the context of early detection of plant diseases. Traditional methods suffer from poor segmentation results and lack of accuracy. This work presents leaf segmentation using image processing techniques, segmenting the leaf from its background with K-means clustering, and optimizing the threshold with Logistic particle swarm optimization (PSO) which enhances the local search efficiency to address these challenges. The methodology entails taking 10 images of different classes from the PlantVillage dataset, applying contrast adjustment, and then reshaping. The preprocessed images are converted from RGB to $L^*\ a^*\ b^*$ color space, since it is device-independent and efficient in distinguishing subtle color differences, which enhances the segmentation accuracy. The leaf picture is first segmented into two clusters using K-means clustering, and these are subsequently optimized using Logistic Particle Swarm Optimization. The proposed method is compared with three algorithms to validate the performance. The results show significant change in terms of segmentation accuracy which demonstrate the effectiveness of the proposed method in addressing the inefficiencies in leaf segmentation.

Keywords Segmentation · K-Means · PSO · Fitness function · Logistic map · Chaotic optimization

V. B. Prijimol (✉) · S. Jayan
Department of Mathematics, Amrita School of Engineering, Amrita Vishwa Vidyapeetham, Bengaluru, India
e-mail: bl.sc.r4mat22004@bl.students.amrita.edu

S. Jayan
e-mail: j_sarada@blr.amrita.edu

S. Kumar et al. (eds.), *Artificial Intelligence for Next Generation Computing, Volume 1*, Lecture Notes in Networks and Systems 1549,
https://doi.org/10.1007/978-981-95-0701-6_21

1 Introduction

Agriculture plays a crucial role in the Indian economy, contributing significantly to the country's GDP, employment, and overall socioeconomic development. Indian agricultural statistical records show soil degradation from pesticide overuse and intensive agriculture. Farmers are migrating to other occupations due to uncertainty, low salaries, weather variations, and a lack of advanced farming facilities. To reduce farmer attrition, long-term agricultural sustainability must be improved. The strategy focuses on promoting development, advancing technology, and implementing comprehensive policy reforms in the agriculture sector [1]. For farmers to expand, stringent steps must be taken, including increasing crop yield, decreasing post-harvest losses, producing climate-tolerant seeds, and addressing pest infestations promptly [2]. Linking rural and urban areas is crucial for comprehending urbanization's impact on agrifood systems [3]. Agrobots can weed, monitor insects, and collect specialty crops. Weeding, harvesting, pesticide use, and productivity losses due to late pest and disease detection are minimized [4]. Timely intervention of disease in the early stages through precision farming can help in the extensive use of pesticides and promote sustainable agricultural practices.

In [5], a CNN-based AlexaNet model was introduced and compared with VGG-16 and Lenet-5 on a dataset consisting of 7000 images. The author also deployed machine learning techniques like SVM and KNN for the study. The author, in [6], performed a comparative analysis for detecting leaf lesions with a real-time image classification and disease detection models SVM, KNN and CNN on a Soyabean leaf disease dataset. CNN performed well with an accuracy of 96% extracting leaf pixels through leaf segmentation by performing K-means clustering on b-channel in CIELAB color space.

Image classification with deep learning systems is often challenging. Traditional image classification relies on raw pixel data to break down an image. Problems arise when two images of the same object have complex backgrounds and perspectives. Incorporating multiple traits in the deep CNN architecture could be beneficial for learning more reliable and discriminative information than using only one trait [7]. A context-aware 3D CNN model is constructed in [8], by extracting optimal features using the Lasso feature selection approach, to segment and categorize plant lesions. The plant life expectancy and overall survival rate are predicted using the linear regression model. Plant Village dataset is used to build prediction models for diseases in potatoes, tomatoes and peppers.

Color components alone cannot validate clustering results, hence, parameterization, texture, or gradient feature space extension is desirable. K-means clustering for image segmentation performs better in Lab Color space rather than RGB space [9]. A K-means or 3-means clustering algorithm based on color features is presented to detect rice leaf lesions. Color and spatial information are used to cluster pixels in CIELAB space. Only a^* and b^* are used for color classification because all color information is in those two levels. The experiments are tested on rice crops with diseases such as brown spots and leaf scald disease. Three-means clustering shows the background, infected part, and healthy section [10].

Hybrid K-means clustering segments images in YCbCr color space to address lighting and shadow issues, while PSO balances global solution search and convergence rate. It fixes poor segmentation in hyper green RGB photos due to reduced green information. Gray scale RGB photos feature color and saturation in addition to brightness, making segmentation harder [11].

PSO determines the best k value and initial cluster centers using the Structural Similarity Index (SSIM) fitness function. The method clusters image pixels in a 5-dimensional space using the K-means algorithm. Increasing dimensions can boost the accuracy of this procedure. Image quality assessment is better, and computing complexity is lower using SSIM [12].

The study presents a PSO-based K-means clustering used to overcome the drawback to selecting initial cluster centroids with K-means. Clustering is based on the minimum distance to the cluster centroid using three distance measures. Choice of distance measure pertains to priorities. Manhattan speeds up processing while Euclidean improves accuracy [13].

The main goal of this study is to improve the segmentation results of leaves for botanical analysis that help in disease detection, bio-fortification, weed identification, plant phenotyping, and yield production. In this work, a methodology to tackle the challenges associated with leaf segmentation as a result of poor segmentation results is highlighted. A combination of advanced image processing methods, including contrast enhancement, median filtering to mitigate noise, and segmentation of leaves in the CIELAB color space with only the a^* and b^* components, are utilized. Segmentation is performed using K-Means clustering on the combined components of the CIELAB color space. The segmentation findings are subsequently optimized using Logistic PSO to achieve desired results.The present article consists of five sections. The Sect. 2 outlines the rationale behind the methodologies employed in our study, drawing from previous research conducted by other researchers in the similar field. The proposed technique is illustrated in Sect. 3. The dataset, experimental findings and analysis are reported in Sect. 4. The conclusion and future directions are discussed in Sect. 5.

2 Related Work

Image segmentation and analysis can be done more efficiently with the help of color space transformations. The advantage of the Lab color space (CIELAB) is that it distinguishes between color (a^* and b^*) and luminosity (L^*). The a^* component of the $L^*\,a^*\,b^*$ color space distinguishes leaf spot diseased areas in the proposed method. PSO uses the Euclidean distance between particles and pixels in the fitness function to determine the initial centroids for the K-means clustering. Additionally, to improve the global search, the algorithm makes use of dynamic inertia [14]. The tabular analysis of other existing technologies used in earlier works relative to our work is presented in Table 1.

Natural environmental conditions present hurdles for plant disease identification, including noise and decreased contrast in lesion photos [15]. Haramaya University researchers randomly selected 198 contaminated tomato plant pictures to diagnose early blight tomato disease. Digital camera images are preprocessed with Median filtering to decrease noise and preserve essential details. Image contrast is increased by histogram equalization and contrast enhancement, with a contrast limit of 0 to 1. Transforming enhanced color images to $L^* a^* b^*$ color space and storing them in two color channels a^* and b^* reduces processing time, while L^* detects the contaminated region. K-means clustering, using Otsu's thresholding divides the image into two clusters. With CIELAB color space, leaf veins were not mistaken for damage. Otsu's thresholding is applied on the a* channel because humans can readily discern white foreground and black background. The entire tomato leaf's area is useful in determining the disease's severity [16]. Our technique is analogous to the cited papers by using similar color channels.

The article introduces Dynamic PSO and k-means clustering (DPSOK) to increase image segmentation quality and efficiency. Since PSO has strong global exploration by altering control parameters and k-means offers good local search, a hybrid alternative is suggested. The key principle is that K-means initializes all particles and clustering is implemented with PSO. Determining dynamic inertia weight enhances PSO performance, avoiding local optima traps and missing optimal points caused by linear decreasing inertia weight[17]. In [18], PSO parameters are configured by setting inertia, w to 0.9, cognitive and social coefficients, c_1, c_2 to 1.2, respectively. Median filtering, image clipping, optimization algorithms PSO, feature extraction with color co-occurrence, and classification with Minimum Distance Classifier are used to accurately identify and categorize leaf diseases in digital camera images. The proposed approach detects sunflower leaf diseases like white rust, bacterial leaf spot, Downy and Powdery mildew, Septoria leaf blight, and Sun rust. This method is better than others because it requires no segment count information and less processing to reach optimal results. The average accuracy is 98%. In our methodology, we establish identical parameters and apply median filtering to remove unwanted noise in the images.

[19] proposes three strategies for image segmentation. K-means clustering, K-means optimized with PSO, and K-means optimized with firefly algorithm. Segmentation distortions are removed during post processing. The experiments are performed on 50 images from each class taken from Plant village dataset. The results are computed for 11 sample images of different classes from the dataset of 20 classes. K-means with threshold optimized by Firefly Algorithm achieved 96.8% segmentation accuracy, outperforming other techniques.

Table 1 Tabular analysis of existing technologies used in earlier works

Study	Application	Methodology	Dataset	Segmentation	Key Features	Strengths	Limitations	Performance
[16]	Leaf disease detection	Otsu's thresholding	Haramaya University research site	K-Means clustering	Median filtering, CIELAB	Determine severity of disease in less time	Work is limited only to K-Means clustering	Only degree of infection is evaluated
[19]	Leaf segmentation	K-Means, K-Means PSO and K-Means Firefly	PlantVillage Dataset	K-Means clustering	CIELAB color space, Contrast adjustment	Fast convergence with Firefly optimization, addresses Complex backgrounds	Severity of disease is not addressed	96.8% accuracy
[12]	Image segmentation	PSO and K-Means	The Berkeley Segmentation Dataset and Benchmark	K-Means clustering	Structural Similarity Index(SSIM), converts pixels to 5-dimensional space	Accuracy increased on considering images in 5d space, SSIM has lower computational complexity	Comparison with other state of the art methods not addressed	Region Non-Uniformity used as an evaluation criteria
[14]	Leaf Segmentation	Initial centroids found by PSO and K-Means clustering in second stage for segmentation	Leaf spot disease in Apple plants	K-Means clustering	$a*$ component from CIELAB color space, dynamic inertia is used in PSO	$a*$ component helps in getting accurate results, balances local and global search	Figures are used for testing the performance	PSOK is better as compared to K-Means
[18]	Leaf disease detection	PSO	Not addressed	PSO	Median Filtering, co-occurrence method, Euclidean distance, Minimum distance classifier	Tuning PSO parameters $w = 0.9$ and c_1 $c_2 = 1.2$	Lacks in accuracy in some cases, Speed of detection is slow, database extension, few diseases are covered	98% accuracy
[13]	Image Segmentation	PSO-K-Means	UCI Machine Learning repository	K-Means	Three distance measure (Euclidean, Manhattan and Canberra)	PSO to overcome drawback to selection of initial centroid	Choose Manhaattan for fast process and Euclidean for accuracy	Performance metric based on RMSE and processing time with different distance measures
[10]	Leaf disease detection	2-Means clustering, 3-Means clustering	Not available	K-Means clustering	Ab channel of Lab color space used in K-Means, contrast enhancement image processing techniques	3 means clustering gives better segmentation healthy, infected and background	For some cases 3 clusters were not sufficient to recognize the infected part	Percentage of the infected area is used as performance metric against number of clusters used as a comparison
[17]	Image segmentation	PSO and K-Means	Random outdoor and indoor images taken from digital camera	PSO in the first stage and K-Means in the second stage	Dynamic inertia weight	Improve image segmentation quality and efficiency. PSO has strong global exploration and K-Means exploitation, fast convergence	Segmentation results compared from output images (Qualitative analysis)	DPSOK compared with PSOK and outperformed the results. It is tested on benchmark functions

3 K-Means Logistic Particle Swarm Optimization

The proposed work is an implementation of the work suggested in [19], with a novelty introduced to the image processing techniques and fitness function given by the Eq. 1.

$$ObjF = \begin{cases} \frac{2*abs(C_p+abs(V_p))}{e^2} & C_p > M_p \\ \frac{2*abs(M_p+abs(V_p))}{e^2} & C_p \leq M_p \end{cases} \tag{1}$$

where C_p is the current pixel, M_p is the mean pixel and V_p is the velocity of the current pixel. $ObjF$ is the fitness function to be used in PSO as the objective function.

The proposed work integrates chaos to tune PSO parameters. Chaos, besides of their random behavior are governed by a set of mathematical rules. Using chaotic variables to handle the optimization search helps rather than using random variables for a random search because of the properties like ergodicity and regularity [20]. In this article, the PSO parameters are dynamically altered with the aid of logistic chaotic map to adapt more flexibility to the optimization search space. The random numbers of the PSO algorithm are replaced with chaotic numbers generated from Logistic chaotic map. The continuum of the Logistic chaotic numbers is provided by the Eq. 2

$$x_{n+1} = l * x_n * (1 - x_n) \tag{2}$$

where l is the chaotic parameter for logistic map with value 4 and initial value x_0 set to 0.1 [21–23].

The segmentation process is done in two phases. The block diagram for the suggested technique is depicted in Figs. 1 and 2. In the first phase, the algorithm focuses on applying image processing techniques like reshaping, denoising with median fil-

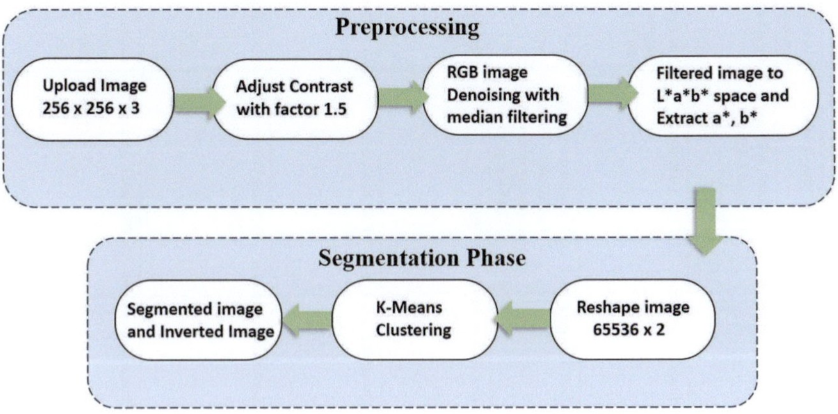

Fig. 1 Block diagram of phase I: preprocessing and K-means segmentation

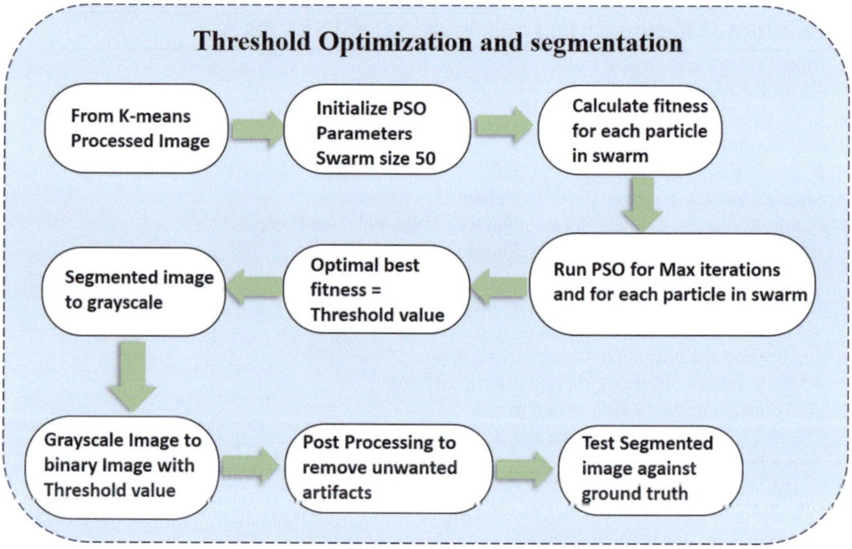

Fig. 2 Block diagram of phase II: threshold optimization with logistic PSO

tering [15] and transforming the images from RGB space to CIELAB in order to enhance the image quality [19]. The a^* and b^* components of the CIELAB space are extracted to do the clustering by K-means algorithm [16] with two clusters, two replicates and 10 iterations. The conversion from grayscale to binary scale is achieved by using a threshold of 0.5. Post-processing methods are utilized to eliminate artifacts that occur after the segmentation process. In phase two, the processed image is prepared for PSO. The threshold value is optimized with Logistic Particle swarm optimization to improve the overall segmentation accuracy. Parametric tuning for the inertia, cognitive and social coefficients are set as 0.9 and 1.2, respectively [18]. The algorithm 1 is run for a swarm of size 50 and over 100 epochs. The Eqs. 3 and 4 serve to update the PSO particle velocity and particle position update equations for the current swarm.

$$v(i+1) = 0.9 * v(i) + 1.2 * LC(i) * (p(\text{best}) - p(i)) + 1.2 * LC(i) * (g(\text{best}) - p(i)) \quad (3)$$

where $v(i+1)$ is the velocity of the particle for next iteration, $LC(i)$ is the chaotic value corresponding to the ith iteration generated using the Eq. 2, $p(\text{best})$ is the particle best position and $g(\text{best})$ is the global best position.

$$p(i+1) = p(i) + v(i+1) \quad (4)$$

Algorithm 1: K-means with Logistic chaotic PSO(KLPSO)

```
Input: Image and ground truth
Output: Segmented Image
R, G, B=split(image)
c=1.5 (contrast adjustment factor)
R' = c * R, G' = c * G, B' = c * B
adjusted image= = merge R', G', B' channels
Denoised image= median blur on adjusted image with kernelsize 5
Transform Denoised image to Lab image
L*, a*, b*= split(Lab image)
ab_image = extract and merge a*, b* channels
normalized image=ab_image/255
pixels= reshape normalized image to 2D array
Apply K-means clustering on pixels to find labels
Find cluster corresponding to leaf pixels
Create a leaf and non leaf mask based on leaf cluster
Apply leaf mask to original image to get segmented image and non leaf mask to get inverted image
Find the gray image from the segmented image and convert it to binary with a threshold of 0.5
processed image = apply post processing to remove unwanted artifacts
part segmented image = image with min( mean of inverted image, mean of segmented image)
Define the fitness function and function to generate chaotic numbers by Eqs. 1 and 2
Initialization for PSO include initial velocity set as e^1, particle best fitness and swarm best fitness as infinity, particle position as any value between 0 and 255, swarm best position is set as None
Update the velocity and particle position using Eqs. 3 and 4
Find the fitness with the updated particle and velocity
Update PBest and GBest if best fitness is found
PSO returns swarm best fitness and velocity
Repeat the PSO algorithm for a fixed number of rows and columns of the part segmented image
```

where $p(i)$ and $p(i+1)$ is the position of the particle in the current iteration and next iteration.

4 Experimental Results and Discussions

The suggested technique is tested on 10 images selected at random from PlantVillage dataset against the ground truth images from the segmented folder. PlantVillage dataset is available at [24]. In this study, K-means clustering and K-means-PSO [19] are compared to our suggested method, all of which incorporates novel image processing techniques as median filtering and histogram equalization. Also a comparison with the K-means-PSO technique in the article[19] is made to highlight the importance of better image processing. Nevertheless, our technique incorporates chaotic numbers as a substitute for random numbers during the Particle Swarm Optimization

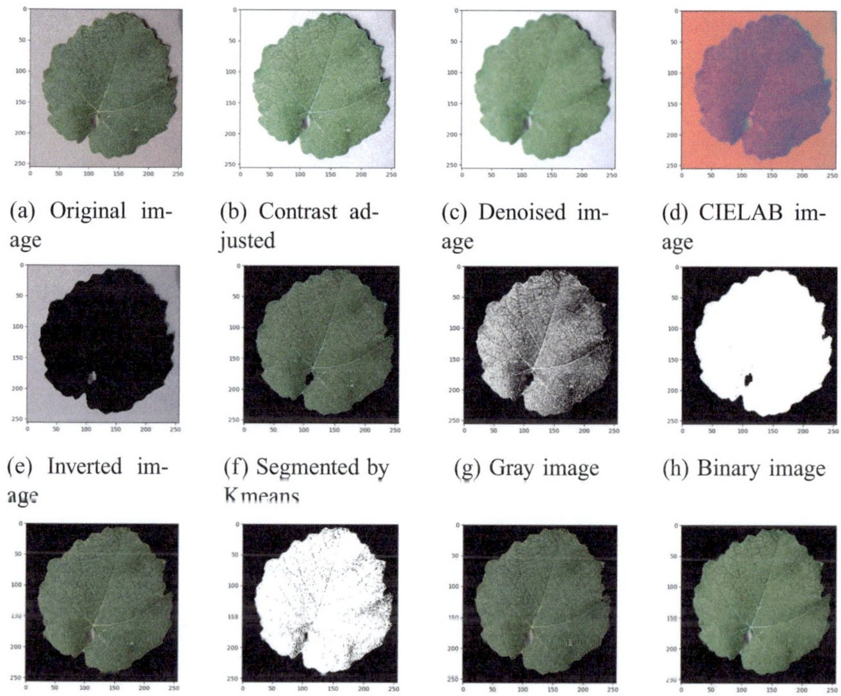

(a) Original image (b) Contrast adjusted (c) Denoised image (d) CIELAB image

(e) Inverted image (f) Segmented by Kmeans (g) Gray image (h) Binary image

(i) Processed image after Kmeans (j) Binary image after PSO threshold (k) Processed image after PSO (l) Ground truth

Fig. 3 Plots of the sequential steps involved in the proposed algorithm on Grape leaf

(PSO) phase. Refer to Fig. 3 to observe the sequential actions carried out on the grape image throughout the execution of the proposed algorithm. The performance metrics of the suggested approach used are accuracy, precision, recall, Jaccard index, Dice (F-measure), specificity and Mathew's Correlation Coefficient (MCC).

The evaluation metrics are computed using the provided equations, which can be found in Eqs. 5–11.

$$\text{Accuracy} = \frac{(\text{TP}_l + \text{TN}_l)}{(\text{TP}_l + \text{TN}_l + \text{FP}_l + \text{FN}_l)} \tag{5}$$

$$\text{Precision} = \frac{(\text{TP}_l)}{(\text{TP}_l + \text{FP}_l)} \tag{6}$$

$$\text{Recall} = \frac{\text{TP}_l}{(\text{TP}_l + \text{FN}_l)} \tag{7}$$

Table 2 Simulation results of three algorithms tested on randomly selected 10 images from the dataset

Leaf	Technique	Accuracy	Precision	Recall	Jaccard Index	Dice	Specificity	MCC
	Only K-means	0.8963	1.0	0.8505	0.8505	0.9192	1.0	0.7971
	K-means PSO as in article[19]	0.6604	0.8395	0.6309	0.5630	0.7204	0.7270	0.3303
	K-means PSO [19] with proposed prepro-cessing	0.9000	1.0000	0.8558	0.8558	0.9223	1.0000	0.8032
	Proposed algorithm	0.9689	1.0	0.9501	0.9501	0.9744	1.0	0.9369
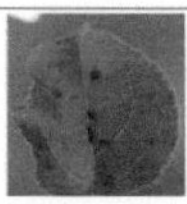	Only K-means	0.9198	0.9990	0.8934	0.8926	0.9433	0.9974	0.8226
	K-means PSO as in article[19]	0.9055	0.9999	0.8735	0.8734	0.9324	0.9998	0.7976
	K-means PSO [19] with proposed prepro-cessing	0.9209	0.9990	0.8949	0.8941	0.9441	0.9974	0.8247
	Proposed algorithm	0.9784	0.9952	0.9730	0.9685	0.9840	0.9900	0.9515
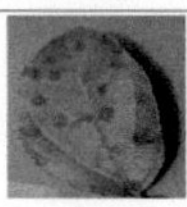	Only K-means	0.9257	0.9998	0.8910	0.8908	0.9423	0.9997	0.8500
	K-means PSO as in article[19]	0.6785	0.7029	0.9138	0.6591	0.7946	0.1772	0.1329
	K-means PSO [19] with proposed prepro-cessing	0.9258	0.9999	0.8910	0.8909	0.9423	0.9998	0.8502
	Proposed algorithm	0.9775	0.9999	0.9635	0.9634	0.9814	0.9999	0.9540

(continued)

Table 2 (continued)

Leaf	Technique	Accuracy	Precision	Recall	Jaccard Index	Dice	Specificity	MCC
	Only K-means	0.9143	1.0000	0.7796	0.7796	0.8761	1.0000	0.8269
	K-means PSO as in article[19]	0.8843	0.9988	0.7031	0.7025	0.8253	0.9995	0.7682
	K-means PSO [19] with proposed prepro-cessing	0.9208	1.0000	0.7961	0.7961	0.8865	1.0000	0.8395
	Proposed algorithm	0.9760	0.9981	0.9290	0.9274	0.9623	0.9991	0.9460
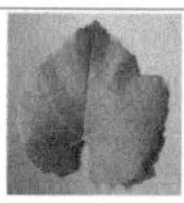	Only K-means	0.9127	1.0000	0.8604	0.8604	0.9250	1.0000	0.8354
	K-means PSO as in article[19]	0.7847	0.8873	0.7509	0.6855	0.8134	0.8410	0.5740
	K-means PSO [19] with proposed prepro-cessing	0.9140	1.0000	0.8625	0.8625	0.9262	1.0000	0.8376
	Proposed algorithm	0.9851	0.9983	0.9744	0.9728	0.9862	0.9980	0.9704
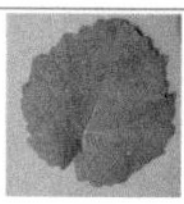	Only K-means	0.9115	0.9999	0.8704	0.8704	0.9307	0.9999	0.8251
	K-means PSO as in article[19]	0.9170	0.9927	0.8849	0.8792	0.9357	0.9861	0.8307
	K-means PSO [19] with proposed prepro-cessing	0.9170	0.9927	0.8849	0.8792	0.9357	0.9861	0.8307
	Proposed algorithm	0.9855	0.9992	0.9769	0.9761	0.9879	0.9988	0.9703

(continued)

Table 2 (continued)

Leaf	Technique	Accuracy	Precision	Recall	Jaccard Index	Dice	Specificity	MCC
	Only K-means	0.8856	1.0000	0.7623	0.7623	0.8651	1.0000	0.7903
	K-means PSO as in article[19]	0.8909	0.9472	0.8189	0.7831	0.8784	0.9576	0.7872
	K-means PSO [19] with proposed prepro-cessing	0.9089	1.0000	0.8108	0.8108	0.8955	1.0000	0.8305
	Proposed algorithm	0.9835	0.9998	0.9595	0.9592	0.9792	0.9998	0.9660
	Only K-means	0.9463	0.9990	0.8168	0.8161	0.8988	0.9997	0.8708
	K-means PSO as in article[19]	0.9022	0.8684	0.7837	0.7005	0.8239	0.9510	0.7583
	K-means PSO [19] with proposed prepro-cessing	0.9465	0.9990	0.8175	0.8168	0.8992	0.9997	0.8713
	Proposed algorithm	0.9818	0.9831	0.9425	0.9275	0.9624	0.9947	0.9507
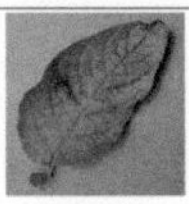	Only K-means	0.9052	1.0000	0.8155	0.8155	0.8984	1.0000	0.8260
	K-means PSO as in article[19]	0.9257	0.9526	0.9003	0.8617	0.9257	0.9526	0.8529
	K-means PSO [19] with proposed prepro-cessing	0.9251	1.0000	0.8543	0.8543	0.9214	1.0000	0.8604
	Proposed algorithm	0.9854	0.9988	0.9688	0.9677	0.9836	0.9990	0.9709

(continued)

Table 2 (continued)

Leaf	Technique	Accuracy	Precision	Recall	Jaccard Index	Dice	Specificity	MCC
	Only K-means	0.8669	1.0000	0.7919	0.7919	0.8839	1.0000	0.7605
	K-means PSO as in article[19]	0.7565	0.9851	0.6289	0.6230	0.7677	0.9830	0.5978
	K-means PSO [19] with proposed prepro-cessing	0.9031	1.0000	0.8484	0.8484	0.9180	1.0000	0.8177
	Proposed algorithm	0.9708	0.9997	0.9491	0.9489	0.9738	0.9997	0.9425

$$\text{Jaccard} = \frac{\text{TP}_l}{(\text{TP}_l + \text{FP}_l + \text{FN}_l)} \tag{8}$$

The simulation results are tabulated in Table 2. The results provide compelling evidence that our suggested algorithm functioned well in both healthy leaves and those with lesions.

In our proposed study, the rate of declassification between an object and its background was reduced. However, the precision score and specificity for the first image in the table in our proposed algorithm are 1.0 because of the texture and lighting conditions of the image, justifying the imbalance in the data.

Refer to Fig. 4 to observe the confusion matrices for the three methods that were evaluated after applying new image processing techniques. However, the confusion matrix for K-Means-PSO [19] didn't change for Grape leaf, supporting the tabular findings.

$$\text{Dice} = \frac{2 * (\text{Precision} * \text{Recall})}{(\text{Precision} + \text{Recall})} \tag{9}$$

$$\text{Specificity} = \frac{\text{TN}_l}{(\text{TN}_l + \text{FP}_l)} \tag{10}$$

$$\text{MCC} = \frac{(\text{TP}_l * \text{TN}_l) - (\text{FP}_l * \text{FN}_l)}{\sqrt{(\text{TP}_l + \text{FP}_l) * (\text{TP}_l + \text{FN}_l) * (\text{TN}_l + \text{FP}_l) * (\text{TN}_l + \text{FN}_l)}} \tag{11}$$

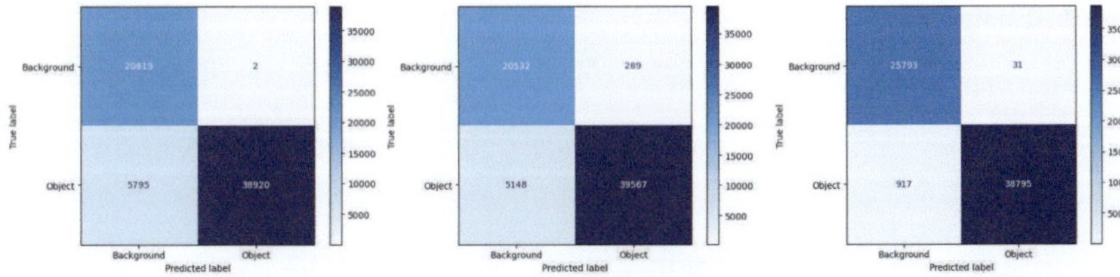

(a) Confusion matrix Kmeans

(b) Confusion matrix of k-means PSO [19] with and without new image processing

(c) Confusion Matrix of proposed work

Fig. 4 Confusion matrices for all three algorithms tested on Grape leaf

5 Conclusion and Future Directions

This work effectively applied a hybrid strategy, using the K-means and PSO algorithm incorporating chaos, to segment leaves. This approach aids farmers in plant phenotyping and enables early identification and categorization of illnesses. The objective is to reduce harvest losses and enhance the quality of farming. Enhancing the accuracy of leaf segmentation has significant implications for multiple facets of agricultural practices, including optimizing decision-making processes regarding treatment strategies to mitigate losses, enhancing agrifood systems, mitigating potential harm caused by excessive pesticide usage, and preventing the spread of diseases from infected areas. Precise segmentation can assist agrobots in monitoring the growth of plants, thereby decreasing the requirement for physical inspection. The proposed method gave promising results in terms of segmentation accuracy and outperformed when compared with other algorithms. The effectiveness of our approach may be limited when applied to datasets that are diverse in terms of lighting conditions and complicated backdrops. In the future, researchers may enhance these models by combining diverse datasets and applying various augmentation approaches. They can also utilize edge detection methods to improve the accuracy of segmenting regions by precisely defining their borders.

References

1. Chand R (2017) Doubling farmers' income rationale, srategy. Prospects and Action Plan, Niti Ayog National Institution for Transforming India
2. Chand R, Joshi P, Khadka S (2022) Indian agriculture towards 2030: pathways for enhancing farmers' income, nutritional security and sustainable food and farm systems. Springer Nature
3. World Health Organization (2023) The state of food security and nutrition in the world 2023: urbanization, agrifood systems transformation and healthy diets across the rural–urban continuum. Food Agric Org

4. Santos Valle S, Josef Kienzle (2020) Agriculture 4.0–agricultural robotics and automated equipment for sustainable crop production. Integr Crop Manag 24
5. Pushpa BR, Ashok A, Shree Hari AV (2021) Plant disease detection and classification using deep learning model. In: 2021 Third international conference on inventive research in computing applications (ICIRCA). IEEE
6. Nikith BV, Keerthan NKS, Praneeth MS, Amrita T (2023) Leaf disease detection and classification. Procedia Comput Sci 218
7. Fan X, Zhou R, Tjahjadi T, Das Choudhury S, Ye Q (2022) A segmentation-guided deep learning framework for leaf counting. Front Plant Sci 13
8. Shoaib M, Shah B, Hussain T, Ali A, Ullah A, Alenezi F, Syed I (2022) A deep learning-based model for plant lesion segmentation, subtype identification, and survival probability estimation. Front Plant Sci 13
9. Burney SA, Tariq (2014) K-means cluster analysis for image segmentation. Int J Comput Appl 96
10. Sethy PK, Negi B, Bhoi N (2017) Detection of healthy and defected diseased leaf of rice crop using K-means clustering technique. Int J Comput Appl 157
11. Zhang H, Peng Q (2022) PSO and K-means-based semantic segmentation toward agricultural products. Futur Gener Comput Syst 126
12. Zarei MA, Nickfarjam AM (2023) PSO-based procedure to find number of clusters and better initial centroids for K-means algorithm: Image segmentation as case study. In: 2023 6th International conference on pattern recognition and image analysis (IPRIA). IEEE
13. Pambudi EA, Andono PN, Pramunendar RA (2018) Image segmentation analysis based on K-means PSO by using three distance measures. ICTACT J Image Video Process 9
14. Anam S (2020) Segmentation of leaf spots disease in apple plants using particle swarm optimization and K-means algorithm. J Phys Conf Ser 1562
15. Liu J, Wang X (2021) Plant diseases and pests detection based on deep learning: a review. Plant Methods 17
16. Lijalem T, Asefa SA (2023) Detection of early blight Tomato leaf using k-means clustering. Int J Diabetes Metab Disord 8
17. Li H, He H, Wen Y (2015) Dynamic particle swarm optimization and K-means clustering algorithm for image segmentation. Optik 126
18. Singh V (2019) Sunflower leaf diseases detection using image segmentation based on particle swarm optimization. Artif Intell Agric 3
19. Kumar A, Sachar S (2023) Swarm intelligence for segmentation of leaf images. Natl Acad Sci Lett 46
20. Feng J, Zhang J, Zhu X, Lian W (2017) A novel chaos optimization algorithm. Multimed Tools Appl 76
21. Vallabhaneni M, Maddula B, Jayan S, Subramani R (2020) Chaotic Hooke-Jeeves algorithm using cubic map with MATLAB code. In: 2020 IEEE international conference for innovation in technology (INOCON). https://doi.org/10.1109/INOCON50539.2020.9298244
22. Maddula B, Vallabhaneni M, Jayan S, Subramani R (2020) Chaotic evolutionary algorithm. In: 2020 IEEE international conference for innovation in technology (INOCONhttps://doi.org/10.1109/INOCON50539.2020.9298233
23. Rani GS, Jayan S, Alatas B (2023) Analysis of chaotic maps for global optimization and a hybrid chaotic pattern search algorithm for optimizing the reliability of a bank. IEEE Access 11. https://doi.org/10.1109/ACCESS.2023.3253512
24. https://www.kaggle.com/datasets/abdallahalidev/plantvillage-dataset/code

Assessment of Collagen Morphology in Pancreatic Tumor Using Computational Translation of Histological Images

Garima Singhal, Nisha, Gavish Uppal, Manjit Kaur, Ruchi Sinha, and Rajesh Kumar

Abstract Collagen remodeling is a hallmark of cancer progression and is critical for diagnosis and treatment planning. Second harmonic generation (SHG) microscopy, while effective for collagen imaging, is limited by high costs and the need for specialized equipment. In this study, we implemented a deep learning-based architecture to translate bright-field (BF) images of H&E-stained pancreatic tissue microarrays (TMAs) into the "collagen" images. The similarity analysis demonstrated that translated collagen images were equivalent to the SHG images. These translated images were subsequently used to extract collagen measures using first-order intensity-based texture analysis, which enabled the evaluation of morphological differences between tumor and normal pancreatic tissues. Statistical analysis was conducted to evaluate the significance of these collagen measures in distinguishing the classes. Furthermore, machine learning (ML) models were compared to identify the optimal model for differentiating the classes based on collagen measures. First-order intensity-based collagen measure analysis indicated "variance" as the most significant feature in classifying the two groups, which was in agreement with ROC-AUC (receiver operating characteristics–area under curve) analysis. The study indicated that the collagen features extracted from the "translated image" can be potentially used to understand the role of collagen remodeling in the progression of pancreatic tumors and can facilitate the analysis of stromal biomarkers.

G. Singhal · Nisha · G. Uppal · R. Kumar (✉)
Department of Biomedical Engineering, Indian Institute of Technology Ropar, Ropar, Punjab, India
e-mail: rajeshkumar@iitrpr.ac.in

M. Kaur
Department of Pathology, All Indian Institute of Medical Sciences Bathinda, Bathinda, Punjab, India

R. Sinha
Department of Pathology, All Indian Institute of Medical Sciences Patna, Patna, Bihar, India

R. Kumar
Department of Physics, Indian Institute of Technology Ropar, Ropar, Punjab, India

S. Kumar et al. (eds.), *Artificial Intelligence for Next Generation Computing, Volume 1*, Lecture Notes in Networks and Systems 1549,
https://doi.org/10.1007/978-981-95-0701-6_22

Keywords Image translation · Collagen remodeling · Second harmonic generation (SHG) · Microscopy · CNN · Pancreatic tumor

1 Introduction

Pancreatic cancer is a highly aggressive and often fatal disease, characterized by its late detection and rapid progression [1]. Despite advances in treatment, it remains one of the most challenging cancers to treat, with a low survival rate. Pancreatic ductal adenocarcinoma (PDAC) is among the most devastating human malignancies and might become the fourth leading cause of cancer-related death by 2030 [2]. Due to its nonspecific symptomology and vague early symptoms, the disease is often challenging to detect and hence requires better techniques for accurate and early diagnosis [3].

Collagen is an essential constituent of the extracellular matrix (ECM), which under goes remodeling during cancer progression and could serve as a hallmark for disease advancement [4]. This remodeling is often characterized by the development of fibrosis and abnormalities in the ECM, contributing to the metastasis of cancer cells. Hence, understanding the organizational attributes of collagen is essential for enhancing our comprehension of disease evolution and progression, which can consequently assist in achieving more accurate diagnoses and developing individualized treatment strategies for various diseases, including pancreatic lesions.

In traditional histopathology workflows, examining collagen fibers in bright-field (BF) images of hematoxylin and eosin (H&E)-stained slides is the gold-standard practice [5]. However, over the year various analytical instruments and methods have been designed to characterize collagen fibers. Imaging methods for visualizing stromal collagen extensively used by researchers include polarized light microscopy, scanning electron microscopy (SEM), external dyes or antibody staining, and second harmonic generation (SHG) microscopy. SHG being a label-free and coherent optical process allows high-resolution and deep imaging of tissues, and is exploited for ultrastructural collagen visualization by researchers worldwide [6]. Due to high collagen specificity, this technique is used to identify and differentiate many diseases like nonalcoholic fatty liver [7], atherosclerosis [8], cartilage-related diseases [9], pancreatitis [10], and adenocarcinoma [1].

Although SHG microscopy is highly effective for imaging collagen fibers, its use is limited by high costs and the requirement for specialized equipment and the knowledge to operate it. Image-to-image translation involves mapping from one imaging modality to another and has been employed to develop synthetic images computationally that replicate the characteristics of one modality into another [11]. This process is achieved using machine learning (ML) or deep learning (DL) models that are trained to learn the relationship between the two modalities. Such techniques have the potential to translate the conventional histopathological H&E-stained BF images into the new collagen-based images, which has potential to accurately characterize collagen remodeling and, maybe incorporated in clinical settings without the need of specialized equipment and technical expertise. In a study conducted by Keikhosravi et al., convolutional neural network (CNN) has shown promise over other algorithms in translating HE-stained BF images across modalities while replicating its high-resolution imaging capabilities and demonstrated the capability to generate SHG-equivalent "collagen" images from BF images, enabling detailed and quantitative collagen analysis [12]. Similarly, in another study by Li et al., co-registered SHG and H&E-stained BF images were employed to distinguish between PDAC and chronic pancreatitis (CP) by utilizing graph neural networks (GNNs) to enable multimodal analysis [1].

In this study, BF images of H&E-stained histopathological tissue microarrays (TMAs) were translated into the collagen images using the CNN architecture. The translated collagen images were then used to extract and quantify statistical measures of collagen fibers to determine the significant changes in collagen morphologies in pancreatic tumors. Further, multiple commonly used ML approaches were compared to classify the tumors from normal pancreatic tissues using the core images of TMAs. This study demonstrates the feasibility of generating collagen-based images through a computational approach, which can be further utilized to study collagen remodeling and stromal analysis in pancreatic cancer progression and other related conditions.

2 Materials and Methods

The H&E-stained BF images and the reference SHG images of pancreatic TMA cores were used to implement the CNN architecture for translating BF images into collagen images. Image registration techniques were employed to align the images from the two different modalities during preprocessing. The collagen images generated by the CNN were then validated by comparing them to real SHG images using multiple analytical parameters. A schematic overview of the processes involved is provided in Fig. 1. Finally, the translated collagen images were examined and utilized to extract collagen features, which were then used to distinguish between control (normal) and pancreatic tumor (PT) tissues.

2.1 Samples

For our study, H&E-stained BF images from TMA slides provided by US Biomax, Inc. were utilized. These were collected from the WSISR-TMA image dataset with prior consent. The dataset included 63 TMA core images from 21 patients, comprising 28 core images from 17 patients with pancreatic tumors and 7 images from 4 control (normal) cases. A total of 28 augmented TMA core images of control cases were also included, which were generated by applying common techniques like basic changes (orientational), brightness and contrast (BC) changes, morphological operations (morph), and both BC and morph changes. The dataset comprised of 19 male

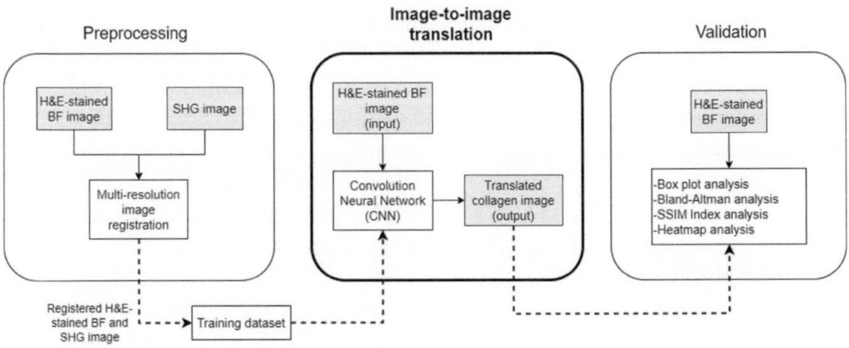

Fig. 1 Workflow for image-to-image translation utilizing TMAs of H&E-stained BF images of pancreatic tissues as input for collagen remodeling analysis. The process starts with preprocessing of input images, including the co-registration of SHG and H&E images. These images are then input into the CNN model to generate translated collagen images, which are subsequently validated against reference SHG images

and 16 female candidates, with ages ranging from 36 to 76 with a median of 52 years. The slides were kept at a temperature of 4 °C prior to being scanned utilizing a Digital Pathological Scanner with magnification of 20 × (0.504 μm/pixel). Moreover, for CNN validation, 36 SHG images of TMA cores were sourced from a GitHub repository [13], which were not used in the test dataset. Since the purpose was to validate the CNN model that can translate the H&E-stained BF images into the collagen-based images, the diagnostic and demographic details of these images were not of the prime concern. The SHG images were acquired utilizing a custom-built system equipped with a 890-nm mode-locked Titanium-sapphire laser and a 445 ± 20 nm narrow band pass filter at a magnification of 20 × in backward mode.

2.2 Image Preprocessing

Before image translation, H&E-stained BF images and the SHG images were registered to match their resolution and dimensions. SHG images highlight collagen fibers, while BF image show a broader range of tissue components, making manual alignment challenging. Consequently, an automated two-step intensity-based registration method was implemented to accurately align SHG and H&E images, ensuring that images from the two modalities were co-registered [14].

The registration transformation involved translation, rotation, and scaling matrices that operated on the H&E whole-core image to segment the ECM image from the H&E-stained image. Preprocessing of the H&E BF image included a segmentation step to separate stromal regions before registration, involving color space thresholding and segmentation techniques such as histogram modification and Otsu's thresholding to create a logical mask. This mask was then smoothed using Gaussian filtering to improve segmentation accuracy.

During the registration process, optimal transformation parameters were identified to quantify the shared information between the H&E and SHG image. This process involved estimating joint and marginal probability distributions using Parzen Windows and maximizing mutual information (MI) through an evolutionary optimizer. A multi-resolution strategy with Gaussian pyramids improved efficiency and minimized the possibility of local optima, ensuring convergence toward global MI optima.

2.3 Image Translation

After image registration, the aligned SHG and H&E images were input into the CNN architecture to translate them into collagen images from the BF images. The study by Keikhosravi et al., assessed multiple networks proposed for image translation using the HE-stained BF images and concluded that most among these either generated inaccurate outcomes or failed to converge [12]. Further, it was suggested to use an encoder–decoder network comprising of residual blocks and evaluated its efficacy to state-of-the-art Generative Adversarial Networks (GANs) [15, 16]. This model demonstrated superior performance in generating translated collagen images from H&E-stained BF images that could be utilized for analysis of collagen remodeling.

The proposed network architecture (Fig. 2) included encoder and decoder blocks with residual modules that were designed to address the vanishing gradient issue and facilitate optimization. The encoder reduced the spatial dimensions of the image while increasing the number of feature maps, and the decoder performed the inverse operation. A pixel shuffle layer was employed to enhance spatial resolution while decreasing channel depth, upscaling the translated collagen image from a lower-resolution feature space. The network was pretrained on over a million pairs of SHG and H&E images from human pancreatic and breast cancer samples [12]. It received input H&E-stained images sized $128 \times 128 \times 3$ pixels. The model integrated residual blocks with varying filter sizes and skip connections, producing high-quality translated "collagen" images that were equivalent to the reference SHG images. This approach effectively preserved the edges in the translated collagen images, ensuring accurate and reliable image translation.

Network training specifics. The network was pretrained using 1388 pairs of pancreatic TMA H&E-stained and SHG images, as well as 489 pairs of H&E-stained breast biopsy whole slide images [12]. The architecture produced the translated collagen images using the stain normalized H&E images. The input images were resized to suitable dimensions with a 10% overlap between neighboring windows to enhance performance, prevent information loss, and facilitate comprehensive analysis. The sub-pixel convolution technique, which involves using both group convolution and periodic shuffling, was employed to reconstruct the output image from the feature

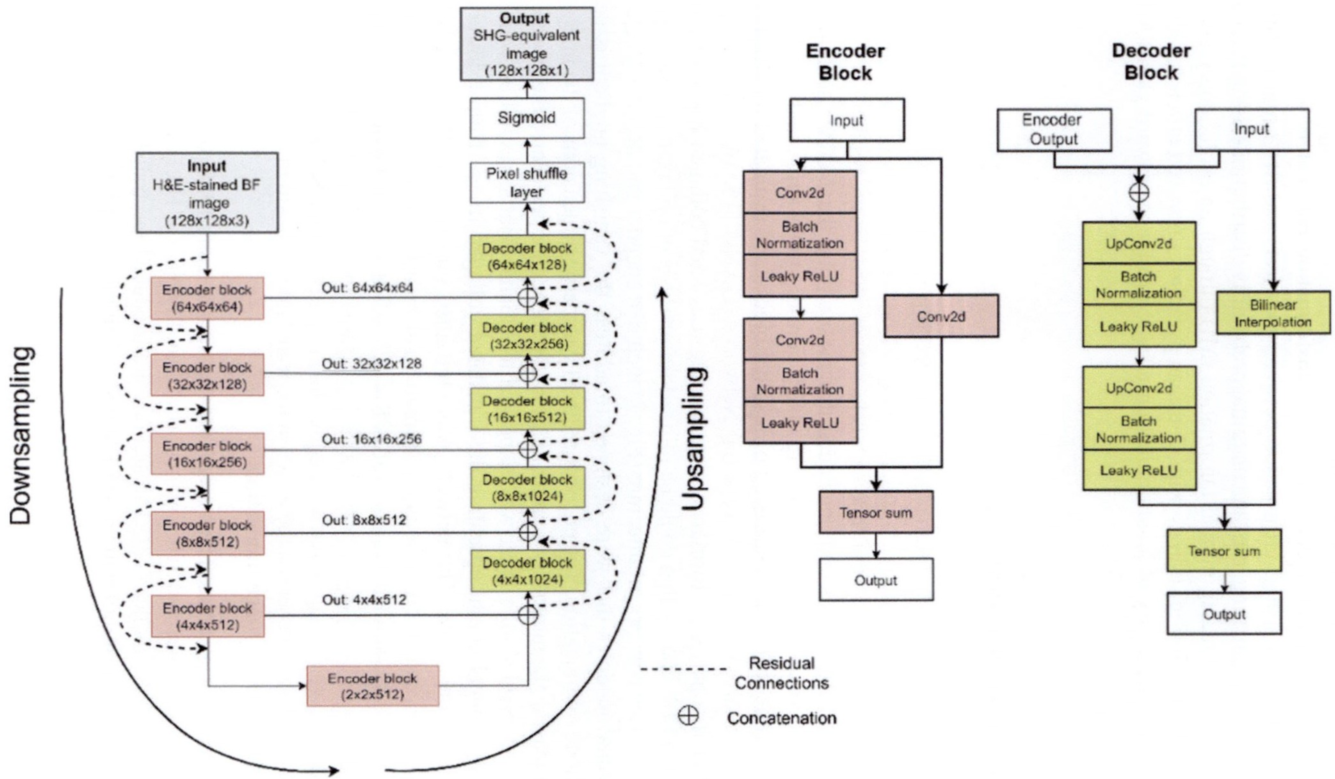

Fig. 2 CNN architecture for image-to-image translation

maps of the last layer of the decoder. This reconstruction was performed in such a way that the size of the output patch matches that of the input image. The resulting images were subsequently merged using the stitching plugin in the Fiji software.

2.4 *Assessment of Collagen Remodeling in Pancreatic Tumor*

The translated images generated by the CNN model were analyzed to extract various collagen measures for distinguishing normal and pancreatic tumor tissues. Collagen measures were extracted through the first-order statistics (FOS), which utilizes histogram of pixels intensities without considering their spatial relationships with neighboring pixels [17]. Six FOS measures (i.e., mean, variance, skewness, kurtosis, energy, and entropy) were calculated from the translated collagen images using ImageJ software. The mean represents the overall density of collagen in the pancreatic tissue, higher values of which could indicate fibrosis [17]. Variance gives the measure of intensity variability around the mean, suggesting whether the collagen distribution is heterogeneous or uniform [18]. Skewness indicates the degree of asymmetry, while kurtosis reflects the sharpness of the intensity distribution [19]. Energy represents the sum of squared intensity values, and entropy measures the randomness of the distribution [18, 20]. Statistical analysis was conducted on these collagen measures to determine the differences between the two groups. These collagen measures were subsequently utilized in the ML models. Seven commonly used ML models i.e., logistic regression, K-nearest neighbors (KNN), support vector machine (SVM), Random forest, Gaussian Naïve Bayes (Gaussian NB), decision tree, and gradient boosting were evaluated for classification accuracy using Jupyter Notebook. To identify the most significant collagen features responsible for predicting the class, feature importance and ROC-AUC (receiver operating characteristic–area under the curve) analyses were performed.

3 Results

3.1 *Validation of Image Translation*

The translated collagen images using the CNN model were compared with the SHG images (reference image), which represents the true collagen. Qualitatively, the translated collagen image and the reference SHG image were observed to be similar (Fig. 3). The Structural Similarity Index (SSIM) was calculated for translated image (output) and its corresponding reference SHG image. The "SSIM Index plugin" was utilized to record changes in various aspects like contrast and luminance, which serve as a qualitative metrics of image quality. The SSIM index showed a value of 0.578, indicating good similarity between the translated and SHG images.

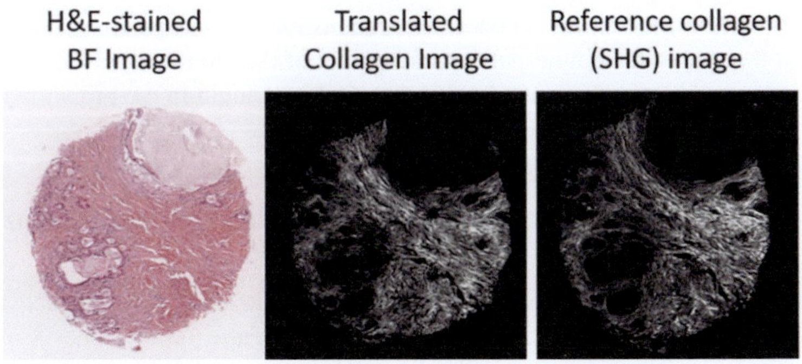

Fig. 3 Qualitatively, no significant differences were observed between the translated collagen image and the SHG image

The Curve Align program was employed to examine the fiber orientations and alignments for the translated and reference collagen image pairs using the curvelet approach [21]. A statistically significant relationship was observed using boxplots of the alignment (Fig. 4a) and orientation (Fig. 4b). The calculated Pearson's correlation coefficient (0.95) indicated a strong agreement between the alignment values. The Bland–Altman plot illustrating differences in alignment values (Fig. 4c) and sine of orientation values (Fig. 4d) were also shown. Heatmaps depicting the fiber orientations for both the translated image (Fig. 4e) and the reference SHG image (Fig. 4f) were produced.

3.2 Statistical Analysis of Collagen Measures in Pancreatic Tumor

Statistical analysis was implemented using GraphPad Prism v9.4.1 software, where the Mann–Whitney U-Test was executed to differentiate pancreatic tumor from control (normal). The mean $\pm$ standard deviation (SD) corresponding to each group was plotted using bar graphs (Fig. 5).

A statistically significant difference was observed between the two classes in all (mean, variance, skewness, kurtosis, energy, and entropy) collagen measures. These measures provide unique insights into the distribution and alignment of collagen, making them valuable for quantifying and describing the textural features of stromal collagen in pancreatic tumor.

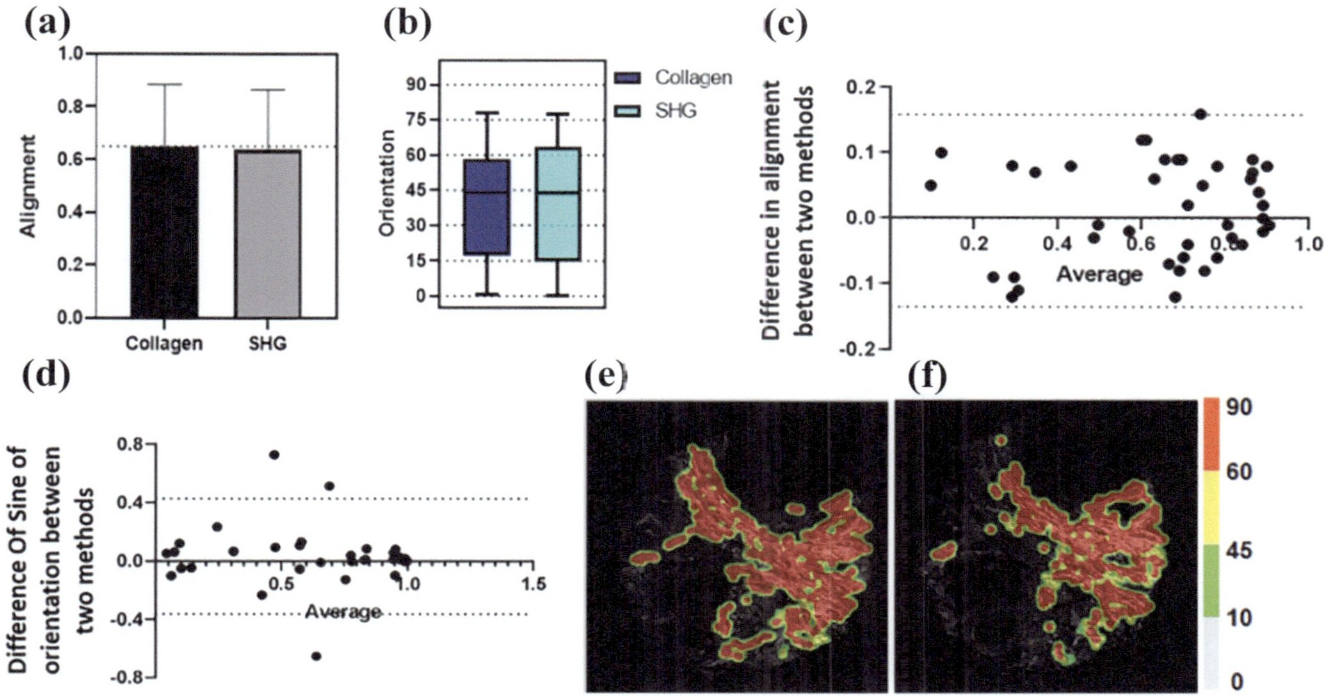

Fig. 4 Validation of image-to-image translation: **a** mean ± SD of alignment values and, **b** box plot of orientation values for translated and SHG images. Bland–Altman plots for difference in alignment, and **c** sine values of orientation, **d** between two methods. Heatmaps of the fiber orientations for both the translated collagen, **e** and **f** the reference SHG image

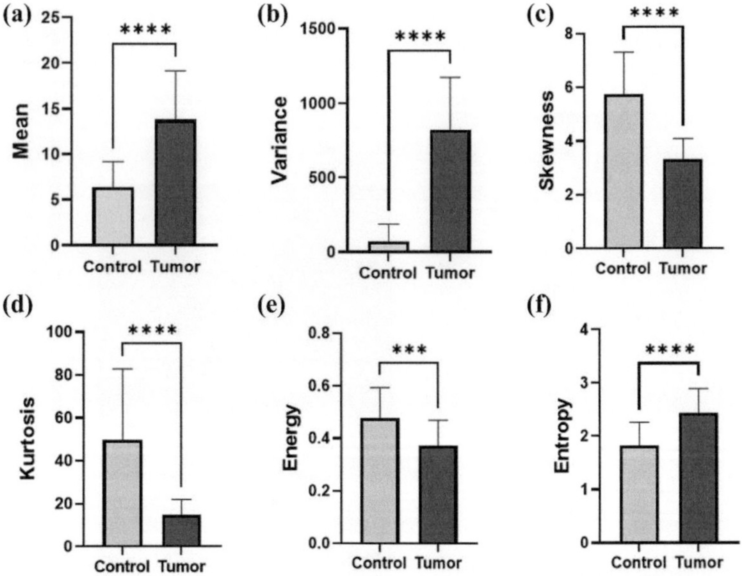

Fig. 5 Statistical analysis ($n = 63$, $*p < 0.05$, $**p < 0.01$, $***p < 0.001$, $****p < 0.0001$) of each collagen measure for classifying pancreatic tumor and control groups. The results were plotted as mean ± standard deviation for the collagen measures **a** mean, **b** variance, **c** skewness, **d** kurtosis, **e** energy, and **f** entropy

3.3 Performance of ML Models

Seven machine learning models for classifying control and pancreatic tumors were compared, and the performance of each network was assessed. Five-fold stratified cross-validation was employed to divide the data into training and testing datasets. The SVM classifier demonstrated the highest accuracy among all the models analyzed (Fig. 6). Furthermore, feature importance was evaluated for each model, which identified "variance" as the most significant feature consistently across all the models, including the SVM classifier (Fig. 7). Subsequently, the ROC-AUC curve analysis for each collagen measure was performed using the SVM model (Fig. 8, Table 1). Among all the collagen measures evaluated, the "variance" emerged as the most effective classifier, achieving the highest AUC value of 0.99. The highest AUC value indicates that the "variance" measure could be a key factor in differentiating between the classes, highlighting its potential significance in the classification process.

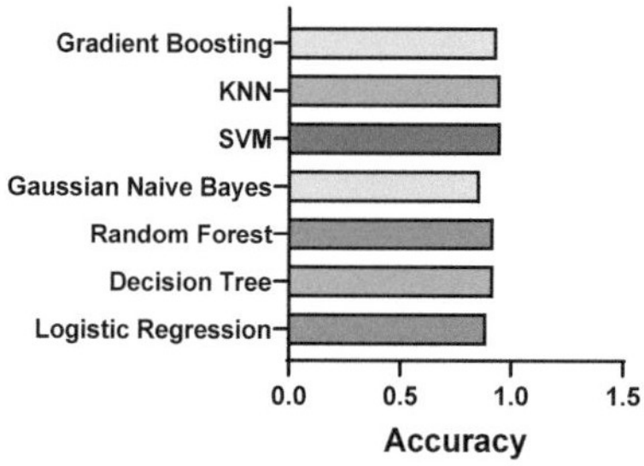

Model	Accuracy
Logistic Regression	0.8910
Decision Tree	0.9218
Random Forest	0.9218
Gaussian NB	0.8603
SVM	0.9526
KNN	0.9520
Gradient Boosting	0.9370

Fig. 6 Model performance comparison and accuracy for machine learning (ML) models

4 Discussion and Conclusion

In this study, we employed CNN to translate BF images of H&E-stained TMA core into the collagen images that are equivalent to SHG images. This approach offers an accessible way for detailed and quantitative collagen analysis in histological setting, potentially enhancing research and diagnostic capabilities on collagen remodeling in cancer progression and other related conditions. The SSIM index of 0.578 indicated that the CNN model captured the essential structural details for accurate collagen representation. Curve Align software further validated this by showing a statistical similarity in fiber orientations and alignments between the translated and SHG images. The boxplots and Bland–Altman plot of fiber orientation values showed a significant correlation, supporting the efficacy of the image-to-image cross-modality translational approach. The statistical results of first-order intensity-based texture analysis demonstrated significant differences between pancreatic tumor samples and control samples across all collagen measures, including mean, variance, skewness, kurtosis, energy, and entropy. The results of this texture analysis demonstrated its potential to serve as quantitative descriptors, offering deeper insights into collagen remodeling.

Since collagen measurements have demonstrated statistically significant differences in distinguishing pancreatic tumors from normal samples, these measurements were subsequently utilized in the ML models. The results of ML models showed that these measures were able to classify the tumor and the normal pancreatic tissues accurately with a maximum accuracy of 0.9526. The feature importance analysis of collagen fibers revealed that "variance" outperformed other features, highlighting its major role in the prediction process. Moreover, the ROC-AUC curve analysis for collagen measures showed a similar result, with "variance" exhibiting the highest value. This finding further supports the result of the feature importance analysis. While the study has shown the feasibility of image-to-image translation for analysis of collagen remodeling in pancreatic lesions, several limitations require further investigation. The SSIM index demonstrated a fair degree of similarity between the translated and SHG images, but there is potential to improve the neural network's ability to capture the subtler details by using a larger dataset. Furthermore, distinguishing

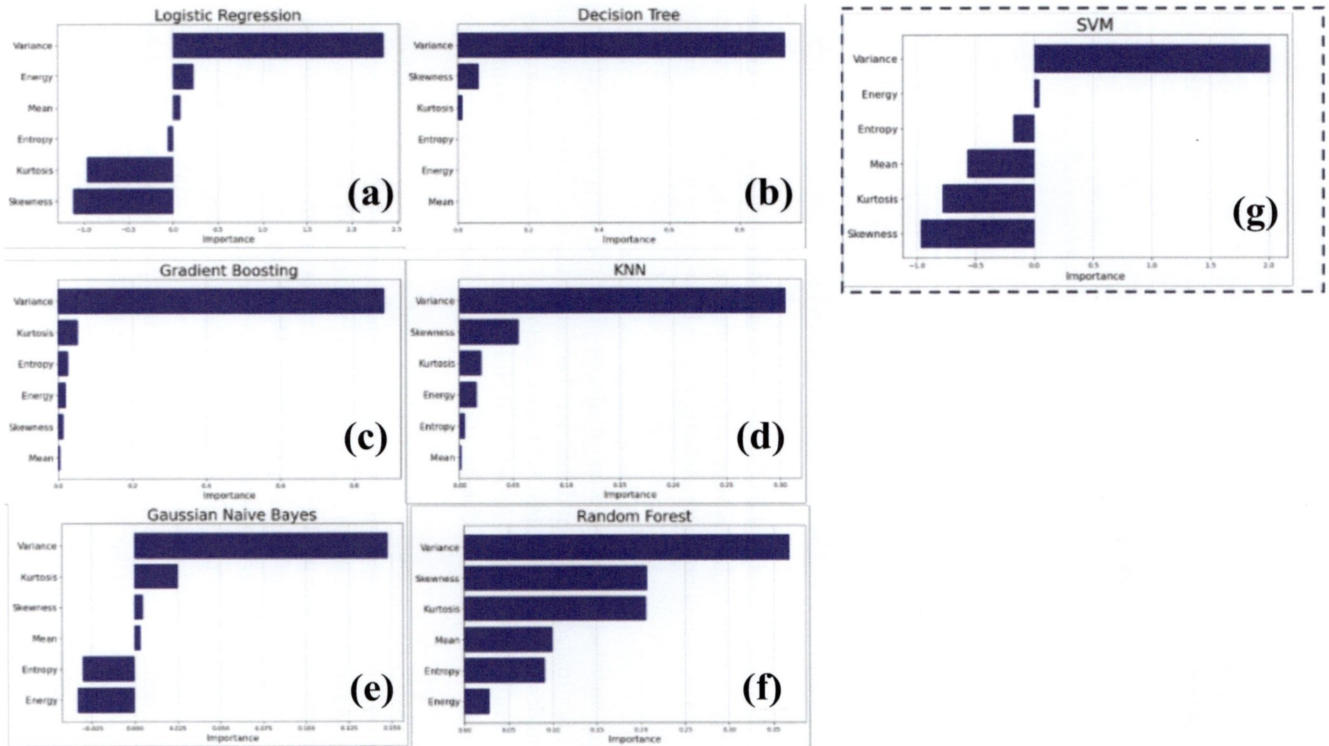

Fig. 7 Feature importance plots: **a** logistic regression, **b** decision tree, **c** gradient boosting, **d** KNN, **e** Gaussian NB, **f** random forest, **g** SVM

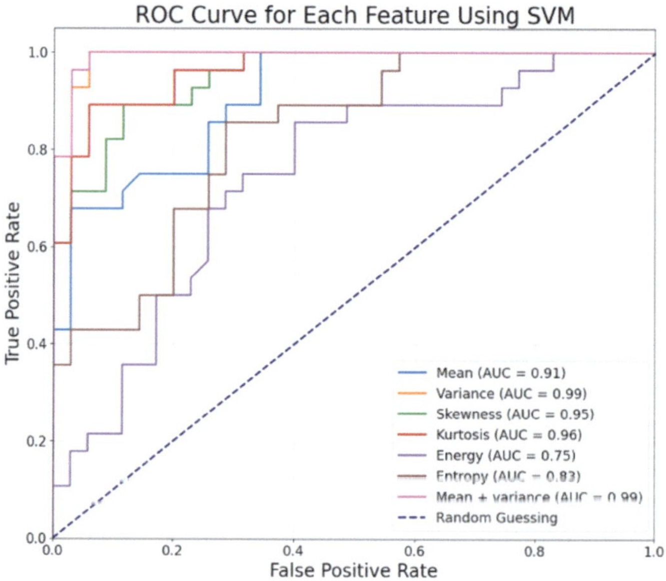

Fig. 8 Receiver operating characteristic (ROC) curve analysis for collagen measures: mean, variance, skewness, kurtosis, energy, and entropy

Table 1 ROC curve analysis summarizes the results for each collagen measure in classifying pancreatic tumors and normal tissues

Feature	Accuracy	Sensitivity	Specificity	Youden's index	AUC
Mean	0.779487	0.726667	0.828571	0.555238	0.91
Variance	0.921795	0.933333	0.914286	0.847619	0.99
Skewness	0.875641	0.866667	0.885714	0.752381	0.95
Kurtosis	0.857692	0.900000	0.828571	0.728571	0.96
Energy	0.688462	0.700000	0.685714	0.385714	0.75
Entropy	0.751282	0.700000	0.800000	0.500000	0.83

between various pancreatic conditions, such as chronic pancreatitis, reactive chronic pancreatitis, intraepithelial neoplasia, and pancreatic adenocarcinomas, by utilizing stromal collagen features requires additional research. Such research could have direct relevance in clinical settings, where the morphological features are often overlapped and confusing among these conditions. Moreover, studies on different tissue types could further broaden the analysis and characterization of collagen remodeling across a wider range of pathological conditions.

In conclusion, the study demonstrated the feasibility and effectiveness of employing CNNs to translate whole-core H&E-stained BF microscopy images into collagen images, which are equivalent to SHG images. The approach of image translation offers a cost-effective and accessible alternative for detailed collagen analysis in histopathological slides, including TMAs, demonstrating its potential in clinical settings. Future advancements in deep learning algorithms and image translation techniques are expected to refine and broaden the applicability of the computational methods, paving the way for more widespread adoption of collagen measures in stromal analysis.

References

1. Li B, Nelson MS, Savari O, Loeffler AG, Eliceiri KW (2022) Differentiation of pancreatic ductal adenocarcinoma and chronic pancreatitis using graph neural networks on histopathology and collagen fiber features. J Pathol Inform 13:100158. https://doi.org/10.1016/j.jpi.2022.100158
2. Rahib L, Smith BD, Aizenberg R, Rosenzweig AB, Fleshman JM, Matrisian LM (2014) Projecting cancer incidence and deaths to 2030: the unexpected burden of thyroid, liver, and pancreas cancers in the United States. Cancer Res 74(11):2913–2921. https://doi.org/10.1158/0008-5472.CAN-14-0155
3. Drifka CR, Eliceiri KW, Weber SM, Kao WJ (2013) A bioengineered heterotypic stroma–cancer microenvironment model to study pancreatic ductal adenocarcinoma. Lab Chip 13(19):3965. https://doi.org/10.1039/c3lc50487e
4. Song K, Yu Z, Zu X, Li G, Hu Z, Xue Y (2022) Collagen remodeling along cancer progression providing a novel opportunity for cancer diagnosis and treatment. Int J Mol Sci 23(18):10509. https://doi.org/10.3390/ijms231810509
5. Hristu R et al (2021) Influence of hematoxylin and eosin staining on the quantitative analysis of second harmonic generation imaging of fixed tissue sections. Biomed Opt Express 12(9):5829. https://doi.org/10.1364/boe.428701
6. Kular JK, Basu S, Sharma RI (2014) The extracellular matrix: structure, composition, age-related differences, tools for analysis and applications for tissue engineering. J Tissue Eng 5:2041731414557112. https://doi.org/10.1177/2041731414557112
7. Chang PE, Goh GBB, Leow WQ, Shen L, Lim KH, Tan CK (2018) Second harmonic generation microscopy provides accurate automated staging of liver fibrosis in patients with non-alcoholic fatty liver disease. PLoS ONE 13(6):e0199166. https://doi.org/10.1371/journal.pone.0199166
8. Le TT, Langohr IM, Locker MJ, Sturek M, Cheng J-X (2007) Label-free molecular imaging of atherosclerotic lesions using multimodal nonlinear optical microscopy. J Biomed Opt 12(5):054007. https://doi.org/10.1117/1.2795437
9. Makkithaya KN et al (2024) Investigating cartilage-related diseases by polarization-resolved second harmonic generation (P-SHG) imaging. APL Bioeng 8(2). https://doi.org/10.1063/5.0196676
10. Tokarz D et al (2019) Characterization of pancreatic cancer tissue using multiphoton excitation fluorescence and polarization-sensitive harmonic generation microscopy. Front Oncol 9:272. https://doi.org/10.3389/fonc.2019.00272
11. Vemulapalli R, Van Nguyen H, Zhou SK (2015) Unsupervised cross-modal synthesis of subject-specific scans. In: 2015 IEEE international conference on computer vision (ICCV). IEEE, pp 630–638. https://doi.org/10.1109/ICCV.2015.79
12. Keikhosravi A, Li B, Liu Y, Conklin MW, Loeffler AG, Eliceiri KW (2020) Non-disruptive collagen characterization in clinical histopathology using cross-modality image synthesis. Commun Biol 3(1):414. https://doi.org/10.1038/s42003-020-01151-5
13. Li B, Keikhosravi A, Loeffler AG, Eliceiri KW (2021) Single image super-resolution for whole slide image using convolutional neural networks and self-supervised color normalization. Med Image Anal 68:101938. https://doi.org/10.1016/j.media.2020.101938
14. Keikhosravi A, Li B, Liu Y, Eliceiri KW (2020) Intensity-based registration of bright-field and second-harmonic generation images of histopathology tissue sections. Biomed Opt Express 11(1):160. https://doi.org/10.1364/BOE.11.000160

15. Zhu JY, Park T, Isola P, Efros AA (2017) Unpaired image-to-image translation using cycle-consistent adversarial networks. In: Proceedings of the IEEE international conference on computer vision. Institute of Electrical and Electronics Engineers Inc., pp 2242–2251. https://doi.org/10.1109/ICCV.2017.244
16. Isola P, Zhu JY, Zhou T, Efros AA (2017) Image-to-image translation with conditional adversarial networks. In: Proceedings—30th IEEE conference on computer vision and pattern recognition, CVPR 2017. Institute of Electrical and Electronics Engineers Inc., pp 5967–5976. https://doi.org/10.1109/CVPR.2017.632
17. Mostaço-Guidolin LB et al (2014) Evaluating collagen morphology and pathological lipid deposition using multiphoton image statistics, Proc. SPIE 8948, Multiphoton Microscopy in the Biomed Sci XIV, 89482A. https://doi.org/10.1117/12.2038232
18. Birk JW et al (2014) Second harmonic generation imaging distinguishes both high-grade dysplasia and cancer from normal colonic mucosa. Dig Dis Sci 59(7):1529–1534. https://doi.org/10.1007/s10620-014-3121-7
19. Mostaço-Guidolin LB et al (2013) Collagen morphology and texture analysis: From statistics to classification. Sci Rep 3(1):2190. https://doi.org/10.1038/srep02190
20. Hristu R et al (2018) Quantitative second harmonic generation microscopy for the structural characterization of capsular collagen in thyroid neoplasms. Biomed Opt Express 9(8):3923. https://doi.org/10.1364/BOE.9.003923
21. Candès E, Demanet L, Donoho D, Ying L (2006) Fast discrete curvelet transforms. Multiscale Model Simul 5(3):861–899. https://doi.org/10.1137/05064182X

Deep Learning for Detecting Common Spine Disorder

Kumari Bhawana, Amritanjali, and I. Mukherjee

Abstract The prevalence of disorders affecting the vertebral column has increased recently as a result of injuries, sedentary lifestyle. Scoliosis and spondylolisthesis anomalies can start affecting kids early on and, if not addressed, can lead to terrible pain. Severe scoliosis might also result in heart and lung issues. Early diagnosis can therefore facilitate the application of treatments or interventions and stop the progression of the disease. Currently used diagnosis techniques rely on medical experts by visually inspecting radiographs. The observations are used by conventional AI-based diagnosis systems to carry out automated classification, enabling quick and simple diagnosis support tools. In this work, we leverage deep learning advancements in transfer learning to diagnose scoliosis and spondylolisthesis directly from X-ray pictures, without requiring any measurement. We gathered unprocessed data from authentic X-ray pictures showing spondylolisthesis, normal, and scoliosis. We also used data augmentation techniques to generalize the training process. Deep learning models, including DenseNet-121, and ResNet-152, were evaluated. For three-class classification, ResNet-152 performed significantly better than DenseNet-121 (94.42%), with a mean accuracy of 98.95%. The capacity to accurately diagnose the participants' vertebral column disorders from routine X-ray pictures is demonstrated by the measured performance indicators. With minimal effort and errors, the current study offers a useful supporting technique to assist in early diagnosis and lowering the necessity for surgical procedures.

Keywords Spinal disorders · Scoliosis · Spondylolisthesis · Medical imaging · Deep learning

K. Bhawana (✉) · Amritanjali · I. Mukherjee
Department of Computer Science and Engineering, BIT Mesra, Ranchi, India
e-mail: phdcs10002.20@bitmesra.ac.in

S. Kumar et al. (eds.), *Artificial Intelligence for Next Generation Computing, Volume 1*, Lecture Notes in Networks and Systems 1549,
https://doi.org/10.1007/978-981-95-0701-6_23

1 Introduction

The 33 little bones that make up the vertebrae that make up the spinal column are divided into five different regions: the cervical, thoracic, lumbar, sacrum, and coccygeal. It is necessary for the stability and motion of the human body. More significantly, the spinal column offers defense. The spinal cord is a component of the central nervous system (CNS) and is in charge of sending information about movement and senses to and from the brain. It can result in numbness or weakness in one or both legs, as well as difficulty walking and standing [1]. Thus, degeneration of the spine lowers overall quality of life and causes a variety of conditions (such as pain, numbness, and restricted motion) [2]. Spinal cord disorders significantly reduce the quality of human life and encompass various conditions, including scoliosis, spondylitis, spondylolisthesis, oncology, osteoarthritis, infections, degenerative diseases, and adult spinal deformity [3]. There are numerous diseases that can impact the spinal column. Since back pain is a significant social, medical, and economic issue in our society, it is critical to identify spine abnormalities as soon as possible.

This research investigates two categories of degenerative pathologies: spondylolisthesis and scoliosis. A lumbar or thoracic spine curvature in the coronal plane, or sideways, is known as scoliosis. The physician makes the diagnosis based on spine X-ray pictures and may also use magnetic resonance imaging (MRI) to rule out tumors. The prevalence of scoliosis, a common spine condition, varies by nation and ranges from 0.47 to 5.2% [4]. For instance, the number of Americans reported to have scoliosis ranges from 6 to 9 million. Depending on how much slippage there is, this is usually graded differently (e.g., low grade vs high grade) [5]. A damaged vertebral slipping forward on the vertebrae right below it might result in spondylolisthesis [6]. The prevalence of spondylolisthesis in adult populations is 6%. A physical examination is the first step in the diagnosis procedure for diseases of the spinal column [7]. Figure 1 shows benchmark dataset of spine diseases [1].

2 Related Works

In recent year, deep learning has been increasingly popular in medical imaging analysis, and it holds promise for more precise and effective diagnosis of a range of spinal illnesses. Masood et al. [8] present a system for lumbar assessment that combines innovative illness categorization techniques with autonomous spinal measurement extraction. Their paradigm has the potential to simplify diagnostic decision-making because it incorporates semantic segmentation of vertebral bodies and achieves excellent accuracy in categorizing illnesses such as lumbar lordosis and spondylolisthesis. In the meantime, Ma et al. [9] investigate the use of MRI images and deep learning algorithms to identify lesions in cervical spinal cord illnesses. Their research shows

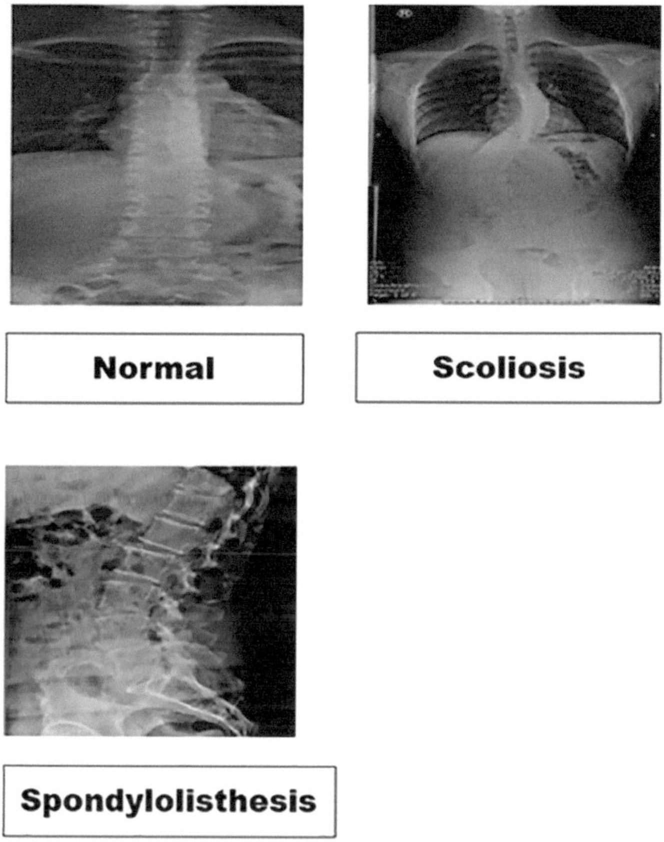

Fig. 1 Spine dataset: spine image benchmark dataset

that using a faster-region convolutional neural network (Faster R-CNN) may effectively locate and recognize lesions, which can be very helpful for radiologists and spine surgeons when making diagnoses. Bezabh et al. [10] concentrate on cervical spondylosis myelopathy (CSM), stressing the value of an accurate and timely diagnosis by categorizing X-ray pictures. Their suggested CNN concatenation feature extraction method successfully identifies and categorizes cervical spine illnesses with high accuracy rates, demonstrating the potential of deep learning to enhance diagnostic procedures. Mohammad et al. [1] collected raw X-ray data from 338 subjects and developed models for three-class classification (scoliosis, spondylolisthesis, and healthy) and pair-wise binary classifications by using different deep learning models. The results showed high accuracy rates, with the best model achieving a mean accuracy of 96.73% for three-class classification and over 98% for binary classifications. Islam et al. [11] the document reports that Random Forest with feature selection achieves the highest accuracy of 94% among the compared methods, and identifies the supporting and contradicting attributes for the abnormality of the lumbar spine.

Saravagi et al. [6] the study successfully demonstrates the application of two deep learning models, VGG16 and InceptionV3, for diagnosing lumbar spondylolisthesis using X-ray radiographs. Data augmentation techniques were applied to address the limited dataset, and both models were optimized using TFLite for efficient deployment on smaller devices. The VGG16 model achieved a higher accuracy of 98%, surpassing InceptionV3's 96%, and was further compressed without significant loss in performance. Singh et al. [12] it involves collecting real-time and benchmark MRI datasets, validated by medical experts, and applying preprocessing techniques like histogram equalization and median filtering. Achieves high classification accuracy with multiple algorithms, with the best results from multi-support vector machine (96%), followed by random forest (94%), decision tree (93.5%), and Naïve Bayes (91%). Mane et al. [13] the effectiveness of deep learning models, particularly CNNs, for detecting osteoporosis from spine X-ray images. Among the models tested, the VGG16 combined with a random forest classifier exhibited the highest accuracy of 95%.

3 Deep Learning-Based Spondylolisthesis and Scoliosis Classification

The current research is limited by the availability of datasets and their scope of improvement for better diagnosis of the problem. Our goal is the early detection of two categories of degenerative pathologies: spondylolisthesis and scoliosis. Proposed model is given in Fig. 2.

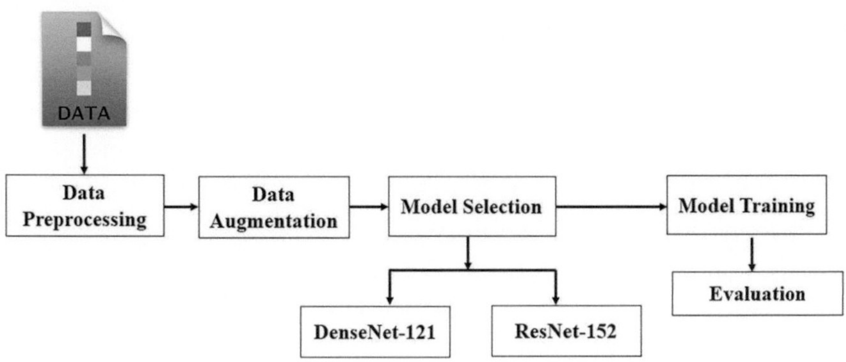

Fig. 2 Proposed methodology

3.1 Problem Formulation

Process clinical dataset in real-time and predicting common spinal cord disorders with an early-stage diagnosis.

3.2 Data Collection and Preparation

In data preparation phase, the benchmark X-ray dataset (Kaggle) was used for normal, scoliosis, and spondylolisthesis spine disorders. The benchmark dataset was used for initial training, validation and testing of the selected deep learning models. Later, dataset was collected from hospital and merged with benchmark dataset. Then, dataset augmentation techniques (rescale, shear, zoom, horizontal flip) were used to prepare the dataset for further training and testing

3.3 Model Selection and Configuration

We have selected two deep learning models: DenseNet-121 and ResNet-152. DenseNet-121 is a densely connected convolutional neural network that addresses the vanishing gradient problem by connecting each layer directly with every other layer. ResNet-152—Highly effective deep learning model for data classification across many domains, leveraging its ability to learn very deep representations and capture discriminative features in complex data. Its skip connections enable training of very deep networks without the vanishing gradient problem.

Deep Learning Models

(a) DenseNet-121: Densely connected convolutional neural network that addresses the vanishing gradient problem by connecting each layer directly with every other layer. This direct connection allows each layer to receive input from all preceding layers and pass its own feature maps to all subsequent layers, promoting feature reuse and enhancing information flow throughout the network. DenseNet-121's image classification process involves dense connections between layers, feature reuse, consistent feature map sizes within dense blocks, bottleneck layers for model optimization, and a global average pooling layer for classification.
(b) ResNet-152: Highly effective deep learning model for data classification across many domains, leveraging its ability to learn very deep representations and capture discriminative features in complex data. Its skip connections enable training of very deep networks without the vanishing gradient problem. It stands out for its optimization ease, high accuracy, ability to handle very deep networks,

effective feature learning, and computational efficiency, making it a powerful choice for data classification tasks.

3.4 Training Setup

The total dataset, including hospital and Kaggle data, comprises 728 samples and spilt the data into training and testing. In training dataset include 441 samples and in testing dataset include 287 samples. We performed fine tuning of pretrained models DenseNet-121 and ResNet-152 using cross-entropy loss function along with Adam optimizer. Activation function is ReLU and output function is SoftMax. The hyper-parameters were initialized with learning rate as 0.001, batch size 35 and number of epochs 60.

3.5 Evaluation Parameters

Accuracy: Estimates the model's overall accuracy by dividing the number of successfully predicted occurrences by the total number of instances.

$$\text{Accuracy} = \frac{\text{Number of Correct Predictions}}{\text{Total number of Predictions}} \tag{1}$$

Precision: Demonstrates the percentage of optimistic predictions that come true. A high degree of precision reduces false positives.

$$\text{Precision} = \frac{\text{True Positive}}{\text{True Positive} + \text{False Positive}} \tag{2}$$

Recall: Calculates the percentage of true positives that are properly identified. Recall is high, which reduces false negatives.

$$\text{Recall} = \frac{\text{True Positive}}{\text{True Positive} + \text{False Negative}} \tag{3}$$

F1-Score: The harmonic mean of recall and precision yields a single measure that strikes a balance between the two. Helpful in cases of unequal class distribution.

$$\text{F1 Score} = 2 \times \frac{\text{Precision} \times \text{Recall}}{\text{Precision} + \text{Recall}} \tag{4}$$

where FP (false positives) is the number of images incorrectly categorized as another class, FN (false negatives) is the number of images overlooked by the classifier, and TP (true positives) is the number of correctly classified images (i.e., for each of the classes).

4 Evaluation Results

4.1 *Dense Net-121 Model*

Figure 3 shows confusion matrix of the DenseNet-121 model. Figures 4 and 5 show the accuracy graph and loss graph of model with mean accuracy of 94.42%, a maximum accuracy of 100%, and a minimum accuracy of 50.0%, the Dense Net-121 model showed strong performance The excellent diagnostic accuracy and dependability of DenseNet-121 can be attributed to its architecture, which encourages feature reuse and effective information flow through dense connections. This illustrates how sophisticated architectures may effectively diagnose spinal cord abnormalities using medical imaging data.

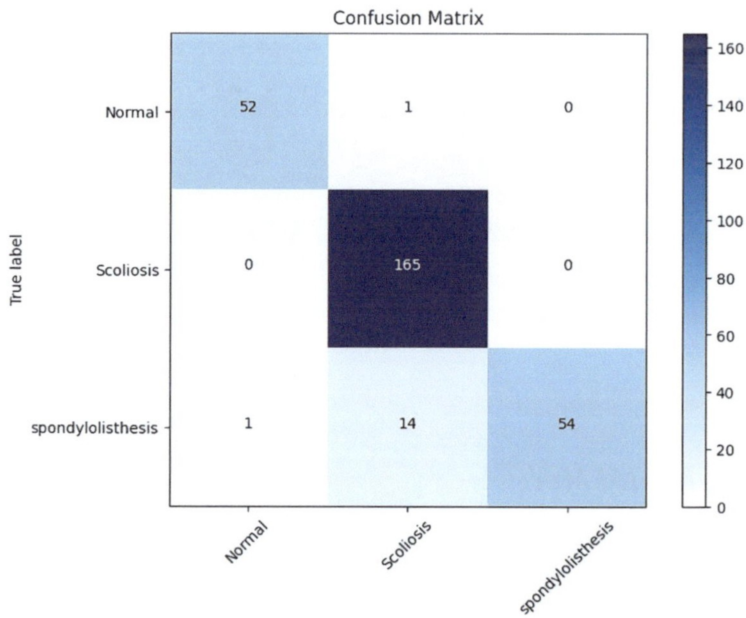

Fig. 3 Confusion matrix

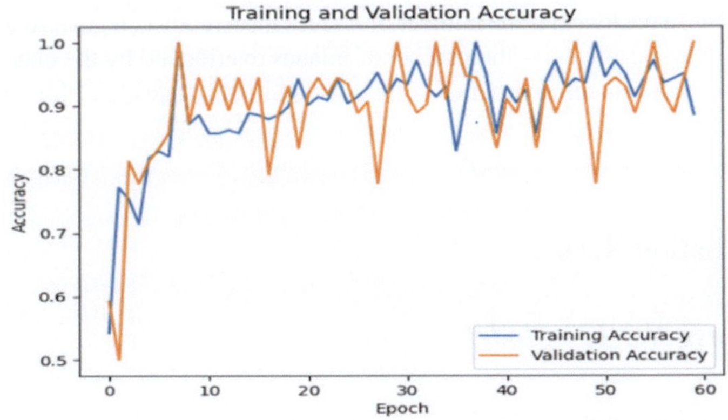

Fig. 4 Accuracy graph

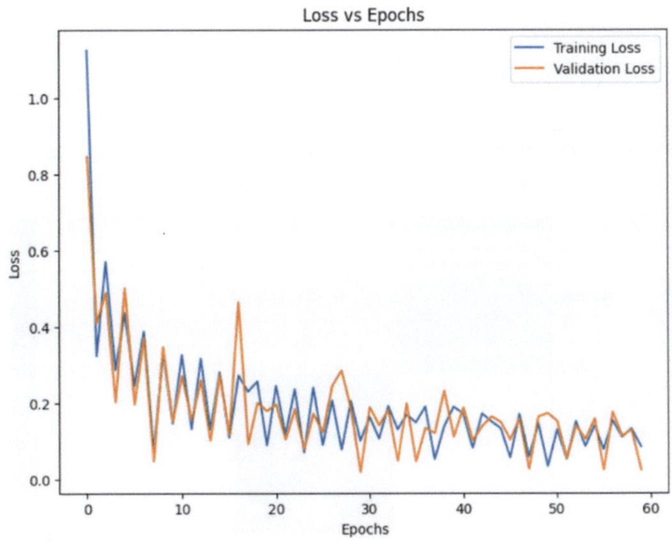

Fig. 5 Loss graph

4.2 *ResNet-152 Model*

In Fig. 6, confusion matrix of ResNet-152 is shown. Figs. 7 and 8 illustrate the accuracy graph and loss graph of model with a mean accuracy of 98.95%, a maximum accuracy of 100%, and a minimum accuracy of 72.22%, ResNet-152 produced exceptional results. Its accuracy in identifying spinal diseases is demonstrated by its low standard deviation of 10.17. When combined with skip connections, its deep architecture efficiently captures complex information found in medical imaging data.

Because of ResNet-152's outstanding performance, which demonstrates its superiority over DenseNet-121, it was chosen as the model of choice for this study's accurate and reliable classification of spinal cord illnesses.

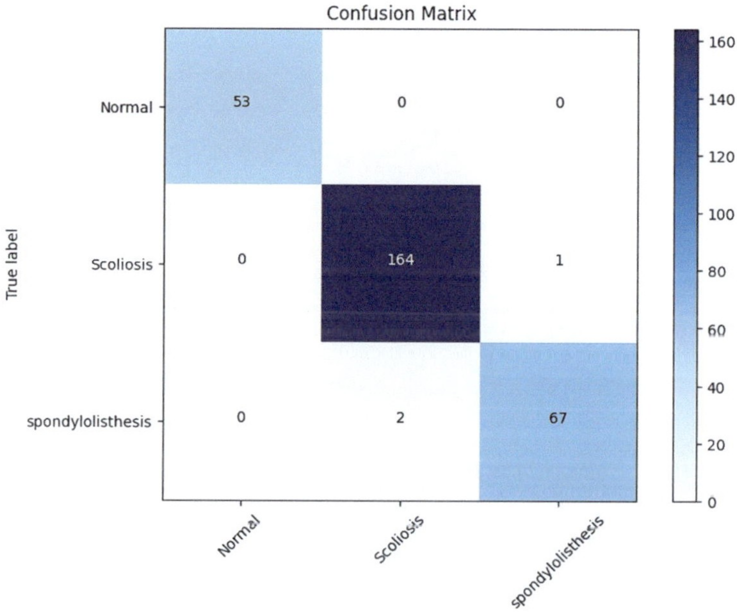

Fig. 6 Confusion matrix

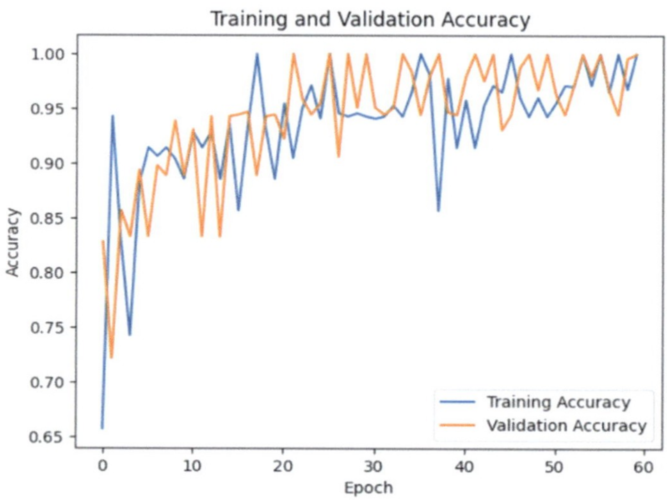

Fig. 7 Accuracy graph

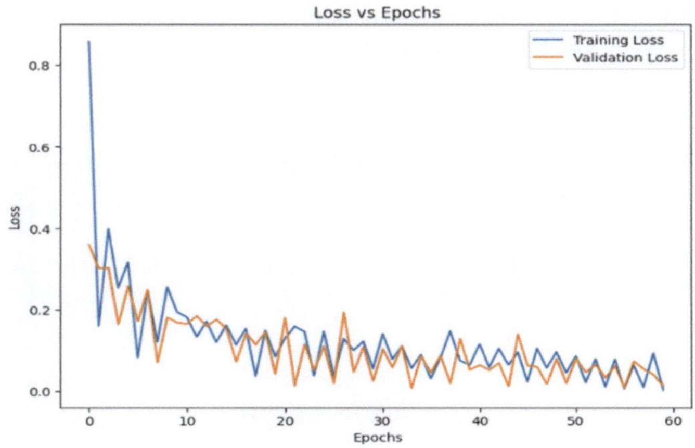

Fig. 8 Loss graph

Table 1 Result of disorder detection: normal, scoliosis, spondylolisthesis, epoch-60 and batch size-35

Model name	Mean accuracy	Max. accuracy	Min. accuracy	SD
DenseNet-121	94.42	100	50.0	22.94
ResNet-152	98.95	100	72.22	10.17

Table 2 Result of F1-Score, precision, recall of disorder detection

Model name	F1-Score	Precision	Recall
DenseNet-121	94	96.66	92
ResNet-152	97.33	97.66	97

The training and validation accuracy curve line grows, indicating that the model is accurately classifying data. The small gap between the two curves indicates that the model is well-generalizing to previously unknown data. It is good to see that the training loss and validation loss curves often decrease as training progresses. The model is a better generalization since there is less of a gap between the two curves. Tables 1 and 2 display the results of both the models.

5 Comparative Analysis

In previous work, a maximum classification accuracy of 96% was achieved for three-class classification using DenseNet-201. By incorporating both benchmark and hospital datasets in the current study, we achieved an improved accuracy of 98%

using ResNet-152. In difficult tasks like medical image classification, modern architectures like ResNet and DenseNet have greatly outperformed traditional models. By using dense connections to solve the vanishing gradient problem, DenseNet-121 was able to attain a higher accuracy of 94%. ResNet-152 achieved the greatest mean accuracy of 98% for three-class classification with its skip connections, which enable very deep network training without vanishing gradients. This indicates that its effectiveness due to its greater capacity to acquire deep representations and capture discriminative characteristics.

6 Conclusion, Limitation, and Future Work

A comprehensive evaluation was conducted on various deep learning models, including DenseNet-121, and ResNet-152. Following a rigorous assessment, ResNet-152 emerged as the most effective architecture for spine disease classification. To optimize model performance and mitigate potential overfitting, transfer learning from pretrained architectures was employed. This technique capitalizes on the existing knowledge of these models, accelerating the training process and improving overall accuracy. Training on combined dataset resulted in better performance in terms of correctly identifying normal subjects from those suffering from scoliosis, Spondylolisthesis. Data augmentation techniques were used to further enhance model robustness and minimize susceptibility to distortions in future predictions.

Due to lack of availability of annotated data, more detailed analysis of the disorders was not done, such as disorder localization and determining the severity of the disorders.

In order to improve the diagnostic capabilities, next study will measure Cobb's angle for scoliosis assessment and incorporate vertebrae segmentation. Furthermore, in order to increase diagnostic precision, we want to pinpoint the affected areas in spondylitis and spondylolisthesis instances while removing localization mistakes. The development of a decision support system and the monitoring of treatment outcomes will help medical personnel make well-informed judgments, which will eventually improve patient care and treatment efficacy. These developments will guarantee improved tracking of the course of the disease, more precise diagnosis, and tailored treatment plans.

References

1. Fraiwan M et al (2022) Using deep transfer learning to detect scoliosis and spondylolisthesis from X-ray images. Plos one 17(5):e0267851
2. Alafeef M et al (2020) Shannon entropy and fuzzy C-means weighting for AI-based diagnosis of vertebral column diseases. J Ambient Intell Humanized Comput 11(6):2557–2566
3. Haldeman S et al (2012) Advancements in the management of spine disorders. Best Pract Res Clin Rheumatol 26(2):263–280

4. Konieczny MR, Senyurt H, Krauspe R (2013) Epidemiology of adolescent idiopathic scoliosis. J Child Orthop 7(1):3–9
5. American Association of Neurological Surgeons. Scoliosis (2021). Available from: https://www.aans.org/Patients/Neurosurgical-Conditions-and-Treatments/Scoliosis. [cited 15 Jan 2022]
6. Saravagi D et al (2022) Diagnosis of lumbar spondylolisthesis using optimized pretrained CNN models. Comput Intell Neurosci (1):7459260
7. Morales H (2018) Infectious spondylitis mimics: mechanisms of disease and imaging findings. In: Seminars in ultrasound, CT and MRI, vol 39, no 6, pp 587–604
8. Masood RF et al (2022) Deep learning based vertebral body segmentation with extraction of spinal measurements and disorder disease classification. Biomed Signal Process Control 71:103230
9. Ma S, Huang Y, Che X, Gu R (2020) Faster RCNN-based detection of cervical spinal cord injury and disc degeneration. J Appl Clin Med Phys 21(9):235–243
10. Bezabh YA et al (2024) Classification of cervical spine disease using convolutional neural network. Multimed Tools Appl 1–17
11. Islam MS, Asaduzzaman M, Rahman MM (2019) Feature selection and classification of spinal abnormalities to detect low back pain disorder using machine learning approaches. In: 2019 1st international conference on advances in science, engineering and robotics technology (ICASERT). IEEE, pp 1–4
12. Singh D, Singla J, Rahmani MKI, Ahmad S, ur Rehman M, Jha S, Prashar D, Nazeer J (2023) Lumbar spine disease detection: enhanced CNN model with improved classification accuracy. IEEE Access 11:141889–141901
13. Mane PR, Vemulapalli J, Reddy NS, Anudeep N, Prabhu G (2016) Osteoporosis detection using deep learning on X-ray images of human spine. J Phys: Conf Ser 2571(1):012017

CaKnow: Connecting Images by Representing Captions as Knowledge Graph

Sourav Pandey, Shreya Gupta, and Ashish Singh Patel

Abstract Due to the unstructured nature of images, representing salient information present in images is challenging. Knowledge graphs are promising solutions for representing unstructured information. However, constructing a knowledge graph requires an ontology, which is a complex task due to the semantic gap (between object and their context-dependent relationship). Moreover, several visual datasets, such as MS-COCO, have vast domain coverage; however, accessing the attributes and information present in such datasets is limited to a single image only. To address this challenge, we have developed a domain-independent ontology (*CaKnow*) based on the MS-COCO dataset capable of representing information in visual datasets. The proposed ontology helps understand the dataset and connect the information present in images with each other using semantic links. In addition, *CaKnow* is also helpful in building the knowledge graph where the common objects and context can be linked to each other and semantics present in the image can be queried. The ontology presented in the paper defines a vocabulary for representing the various elements in the MS-COCO dataset along with other visual datasets, including the image attributes, object classes, relationships, and metadata. CaKnow ontology can be used to construct the knowledge graph of images present in the visual datasets where each image is added as an individual, providing a structured and comprehensive representation of the salient information present in an image and linking it with other images, resulting in a knowledge graph. Furthermore, the entities and relations of the knowledge can be connected to the Linked Open Data Cloud, becoming the Web of Data due to common and domain-independent context.

Keywords Knowledge graph · Microsoft COCO · Ontology · Connecting images

---

S. Pandey · A. S. Patel (✉)
National Institute of Technology Mizoram, Aizawl, India
e-mail: ashish.cse@nitmz.ac.in

S. Pandey
e-mail: dt23cs011@nitmz.ac.in

S. Gupta
National Institute of Technology Manipur, Imphal, India

S. Kumar et al. (eds.), *Artificial Intelligence for Next Generation Computing, Volume 1*,
Lecture Notes in Networks and Systems 1549,
https://doi.org/10.1007/978-981-95-0701-6_24

1 Introduction

In today's data-driven world, visual data, which includes images, videos, and other multimedia, is rapidly expanding. However, visual data is often unstructured, presenting a significant challenge in representing and extracting meaningful semantics [20]. The complexity arises due to the sheer volume and diverse nature of visual content, making it challenging to derive standardized and interpretable information [18]. As a result, sharing and integrating visual data across various platforms and systems becomes increasingly complex, especially when these systems are independently developed. Ontologies provide a promising solution for addressing these challenges by promoting data interoperability. An ontology is a controlled, structured vocabulary representing entities within a specific domain. Visual data can be efficiently represented using semantic technology as entities and relations to create a knowledge base (knowledge graph) which can be queried, reasoned, and analyzed [17].

The current approach to image analysis (identifying events, objects, descriptions, etc.) often involves relying on existing and benchmark datasets to tackle complex problems. However, each dataset has several characteristics and millions of images that are not linked within, such that each image is depicted as an isolated but related (part of the same dataset) entity. It is crucial to establish technology for linking images present to create comprehensive datasets and extend the scope beyond the basic recognition task. Ontologies can enhance contextual understanding by linking objects in images with their semantic relationships and relevant contextual information [1]. However, constructing and integrating ontologies pose particular challenges. Visual data often needs standardized annotations, making it easier to represent the diverse range of objects and relationships semantically.

In this study, we develop an ontology using the W3C OWL 2 Web Ontology Language and Protégé software for visual datasets to integrate visual representations and descriptions of all the images of the dataset for reasoning and analysis beyond a single image as a dataset. This approach will help bridge the gap between visual content and semantic representation, ultimately supporting advanced AI applications [22]. As a use case, the ontology is built to represent the image and its captions present in the Microsoft COCO dataset [12], which can be extended to other visual datasets. To address the challenges in visual data representation and semantic extraction, this study makes the following contributions:

1. A visual ontology *CaKnow*, based on image captions to support integrating images with common attributes that can express real-world entities.
2. A knowledge graph is built by adding an image as an individual for retrieval and facilitating advanced querying and reasoning capabilities regarding the relationships within the dataset, building links between different images (individuals) and real-world entities.
3. Constructing a structured knowledge base using a knowledge graph to interconnect objects and their descriptions across multiple images, thereby enabling more substantive data retrieval and relation.

4. The proposed *CaKnow* ontology will facilitate linking real-world entities with existing entities in Linked Open Data Cloud, enabling a structured and integrated web of data.

The rest of the paper is structured as follows: Sect. 2 provides the background on the need for knowledge in the vast domain of ontology creation. Section 3 describes the ontology development, elaborating on their properties and structure. Section 4 discusses the *CaKnow* ontology, explaining how it can be used to describe images using captions and annotations. Finally, Sect. 5 concludes the paper with a discussion of future work.

2 Background

The unstructured nature of images makes them challenging to organize and retrieve effectively. However, knowledge graphs offer a powerful tool to represent salient information and interconnect common entities, objects, and descriptions.

By modeling the semantic content of images, researchers can better visualize and locate images for various applications, from academic to commercial. Knowledge graphs offer a structured representation of knowledge, capturing relationships between entities, concepts, and attributes. They can expand queries, provide context, and improve image retrieval based on semantics. Machine learning, particularly deep learning, is often used with knowledge graphs to classify objects, scenes, and semantics. These models can enhance image analysis by learning from visual and semantic data. However, ensuring model explainability while maintaining accuracy remains challenging, as many operate as black boxes.

KG4Vis [11] uses a knowledge graph to recommend visualizations by learning entity and relation embedding through a TransE-based technique. The knowledge graph, constructed from dataset-visualization pairs, is used to model mapping rules between data features and visualization design choices. However, *KG4Vis* has limitations in supporting only a limited number of visualizations and performing slightly lower than neural network-based methods in capturing non-linear relationships.

Similarly, *VisualSem* [2] is a multimodal knowledge graph aimed at vision and language research. It integrates textual descriptions with curated image sources and uses pre-trained neural networks to retrieve entities through k-nearest neighbor search. Despite its benefits, *VisualSem* faces challenges like maintaining image quality for knowledge graph nodes and ensuring that multilingual glosses provide complementary, not redundant, information. Roopak et al. [8] proposed the *OntoKnowNHS* which is the Domain Ontology-based query term enrichment and Knowledge enrichment of the images with the help of Wikidata and Google's Knowledge Graph API using the Flickr30k dataset. The main drawback of this ontology was not able to capture the text and caption of the image using their semantic relation. However, Le et al. [10] approach to semantic-based image retrieval by integrating KD-Tree structures through developing ontology using datasets like MS-COCO and Flickr. But the ontology framework may require further enrichment to handle more complex multi object data. Muscetti et al. [16] proposed an ontology that supports automated processes such as ontology population, where new instances are added to an existing ontology automation reduces the time and effort for the MS-COCO dataset. This ontology lacks in complexity can lead to high development costs and time-consuming processes

Several previous works, by Guo et al. [7] and Ma et al. [13] have attempted to store visual data as a knowledge base using the images and videos. These knowledge graphs can capture the semantic relationships between visual elements, enabling more effective retrieval and exploration of image content. However, challenges remain in ensuring the accuracy, explainability, and scalability of these approaches, particularly when dealing with the complexity and diversity of visual data. The proposed *CaKnow* ontology aims to go beyond object classes by expanding the relationships and attributes to discover connections between images, enabling reasoning over the relationships within the knowledge graph [13]. This broader approach allows the ontology to describe properties, entities, and their relationships, enhancing the ability to reason over the graph and providing a more comprehensive understanding of the semantic content within images.

3 Ontology Development and Knowledge Graph Construction

In this section, we discuss the development of an ontology-based framework for representing captions from selected images of the MS-COCO dataset. The primary goal of this work is to enhance object and relationship detection methods by employing a structured, semantically enriched representation of image data. This approach leverages ontology concepts to refine image analysis tasks, improving the accuracy and the interpretability of results.

The *CaKnow* ontology framework is a domain-independent ontology that captures the relationships between various entities, actions, and attributes within image captions. The approach begins by parsing the image captions, extracting subject-predicate-object triples, and mapping them to corresponding concepts in the ontology using the MS-COCO dataset as shown in Fig. 1. These concepts include objects, actions, and attributes. The relationships between the extracted entities are then inferred based on the ontology's structure and the contextual information present in the captions [5, 9] .

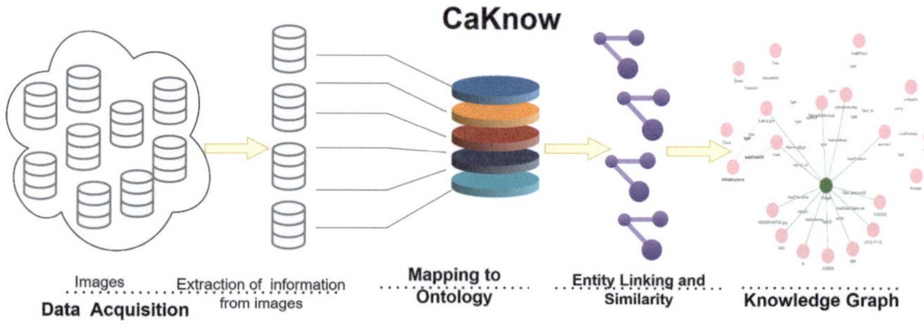

Fig. 1 Overview of the knowledge graph development using *CaKnow* ontology

Table 1 Comparison of ImageNet, MS-COCO, and cityscapes datasets

Dataset	Focus	Categories	Contextual information
ImageNet	Object classification	1000 object categories	Limited context, images often cropped around object
MS-COCO	Object detection, segmentation, captioning	80 object categories	Emphasizes objects' context in natural scenes
Cityscapes	Urban scene understanding, semantic segmentation	30 urban related classes	Focused on road scenes with specific urban objects

3.1 Image Datasets

MS-COCO is currently the preeminent object detection dataset and benchmark. The data captures complex real-world scenes containing common objects in their natural context. Objects are annotated using per-instance segmentation to facilitate precise localization. The dataset comprises 886,284 labeled instances across 122,266 images, spanning various scenarios from dining table close-ups to complex traffic scenes [12]. The MS-COCO dataset, particularly the 2017 version, is one of the most widely adopted and comprehensive resources for computer vision tasks, including object detection, segmentation, keypoint estimation, and caption generation. It is renowned for its large-scale, high-quality annotations and diverse categories, providing a broad representation of real-world scenarios [14]. Compared to earlier benchmarks such as PASCAL VOC, MS-COCO is significantly larger in scale, encompassing over 2.5 million labeled instances across 328,000 images (Table 1).

The dataset's contextual information allows for deeper semantic understanding, which is essential for ontology-based systems that integrate visual data with reasoning and inference tasks. This enables enhanced decision-making processes in

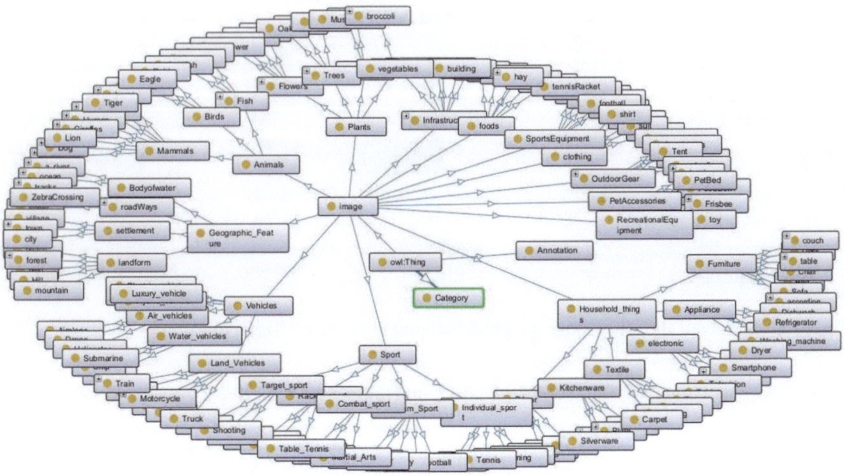

Fig. 2 Ontograph of the *CaKnow* ontology

domains like autonomous driving, medical imaging, and AI research by comprehensively representing the real world and thus used in this work [15, 21].

3.2 *CaKnow* Ontology Construction

In this section, we present *CaKnow* ontology designed to represent and integrate image captions as a knowledge graph with the cognitive information about the image, enabling more powerful and flexible image understanding and retrieval tasks [4] The *CaKnow* ontology is constructed by image captions of the MS-COCO dataset. The selected image captions are processed to extract subject-predicate-object (SPO) triples, which serve as the foundational building blocks of our ontology. The extracted Subject and Object form the Class of the ontology while predicate forms the object properties of the ontology. These triples encapsulate relationships between entities in an image, such as objects and their interactions. For example, a caption like *a man riding a horse* would be parsed into the triple (man - riding - horse). The ontograph of the *CaKnow* ontology is shown in Fig. 2.

The ontology also contains the data properties and object properties. These properties allow us to assign specific attributes to individuals (data properties) and define relationships between different individuals (object properties). For example, the object property riding connects man and horse, while data properties might include the color or size of the objects. These properties are integral to the knowledge base and facilitate more advanced reasoning tasks. The complete set of data and object properties used in the ontology is shown in Figs. 3 and 4. Thus, the ontology contains Class, Object properties, based on these triplets.

Fig. 3 Data properties to represent visual relations present in images

The *CaKnow* Ontology can be accessed by researchers and developers at github[1]

3.3 Representation

The proposed ontology, named *CaKnow*, is designed as a domain-independent framework that can be used to construct knowledge graphs for images in visual datasets. Each image in the dataset is represented as an individual, offering a structured and comprehensive representation of its concepts and their interconnections. This framework enables semantic querying and reasoning over image content, making it a powerful and valuable tool for improving object detection, pattern recognition, and image retrieval tasks across various domains. The *CaKnow* ontology is designed to generate a knowledge graph for each image in a dataset. Python 3.0 and *Protégé* version 5.6.3 were used to build and visualize the ontology and knowledge graph. The *rdflib* library in Python was employed to construct the knowledge graph, rep-

[1] https://github.com/Sourav544/CaKnow-COCO-ontolgoy.

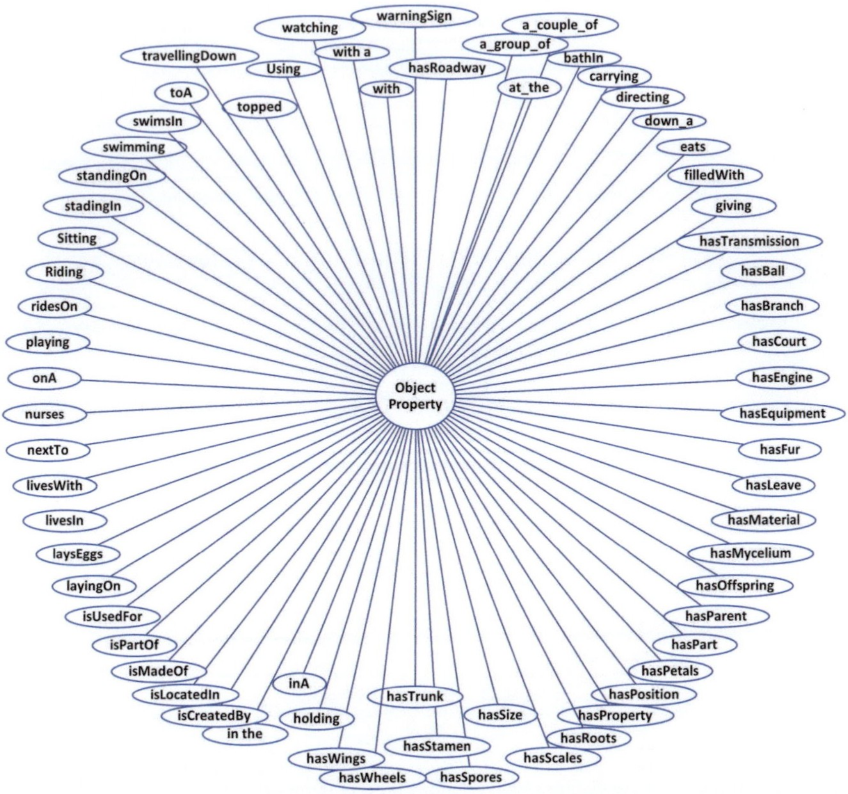

Fig. 4 Object properties to represent visual relations present in images

resenting image captions as a formal set of interconnected concepts in a structured, machine-readable format [5]. For visualizing the knowledge graph, the *pyvis* Python library, designed for network graph visualization, was utilized [19].

The critical contribution of *CaKnow* lies in its ability to represent complex relationships between entities, actions, and attributes present in image captions, as well as the cognitive information about the images [9]. By mapping image captions to a knowledge graph, *CaKnow* enables effective querying and retrieval of images based on semantic content. The graph structure also allows for the deduction of additional facts and relationships between entities based on ontology rules and the spatial context of the image. This method provides a rich and structured representation of the entities, relationships, and attributes present in the MS-COCO dataset, with each image represented as an individual in the knowledge graph, enabling a range of downstream applications, such as image retrieval, visual question-answering, and scene understanding, as well as further research directions in multimodal learning and knowledge representation [6, 20].

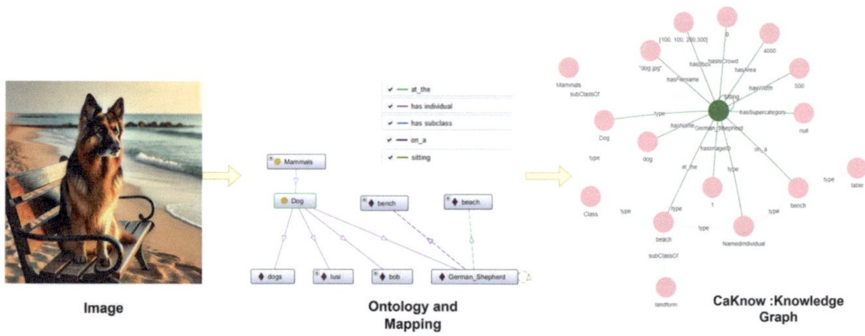

Fig. 5 Knowledge graph from the MS-COCO dataset of a dog image

The proposed ontology identifies relationships between various entities by leveraging image captions as a knowledge source. It facilitates the retrieval of information from the dataset through targeted searches of annotations, IDs, and semantic concepts. This approach captures the inherent associations between objects, actions, and attributes commonly depicted in images. The ontology framework encompasses classes for these fundamental elements, and object properties are defined to represent the interactions between them. Figure 5 illustrates an example where the mammals class has a subclass Dog, which contains an individual German Shepherd. This German Shepherd individual is connected to two other individuals, bench and beach, through the defined object properties, thereby representing the relationships among these entities.

4 Ontology Evaluation

The *CaKnow* ontology is evaluated on its ability to represent the visual information present in MS-COCO dataset. In addition, the capability is demonstrated by linking images through semantic similarity. In addition *CaKnow* ontology is also evaluated by syntactic, semantic, and pragmatic analysis [3].

4.1 *CaKnow Characteristics*

The richness of *CaKnow* to represent information present in images is presented in Table 2 that provides an overview of key characteristics of the ontology and its associated data. It includes the counts of object properties, data properties, and classes. Additionally, it highlights the estimated number of triples and the number of linkages for the whole dataset, offering insights into the structure and complexity of the dataset. And a pie chart is used to visualize the content of table in Fig. 6.

Table 2 Ontology and triples metrics of the MS-COCO dataset

Metrics	Count
Number of object properties	66
Number of data properties	26
Number of classes	174
Total number of individuals	122,266
Total number of triplets	9,027,585
Total number of linkages	4,259,902

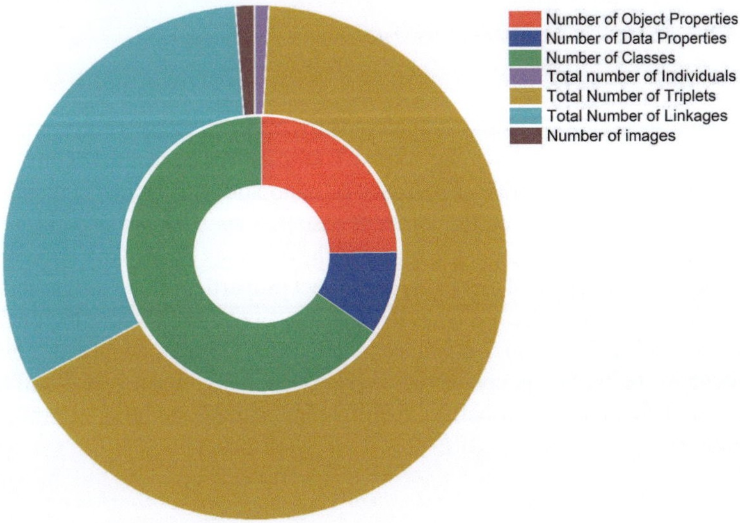

Fig. 6 Distribution of class, object, data properties, individulas, triplets, and links by linking two images of the MS-COCO dataset

4.2 Linking Images Through Semantic Similarity

In creating the ontology and knowledge graph for the two MS-COCO images (ID: 89078 and ID: 121673), each image's captions are broken down into subject-predicate-object (SPO) triples. For the first image, which features a woman snowboarding, the ontology captures details such as (Woman, wearing, Pink jacket) and (Woman, performing, Snowboarding). Similarly, the second image, showing a man surfing, generates triples like (Man, wearing, Red t-shirt) and (Man, performing, Surfing). These triples form the basis of the knowledge graph, where entities (e.g., Woman, Man, Snowboard, Surfboard) are represented as nodes, and their relationships (e.g., wearing, riding, performing) as edges.

The two images are linked in the knowledge graph through common concepts, such as the shared action of riding an object (snowboard or surfboard), the depic-

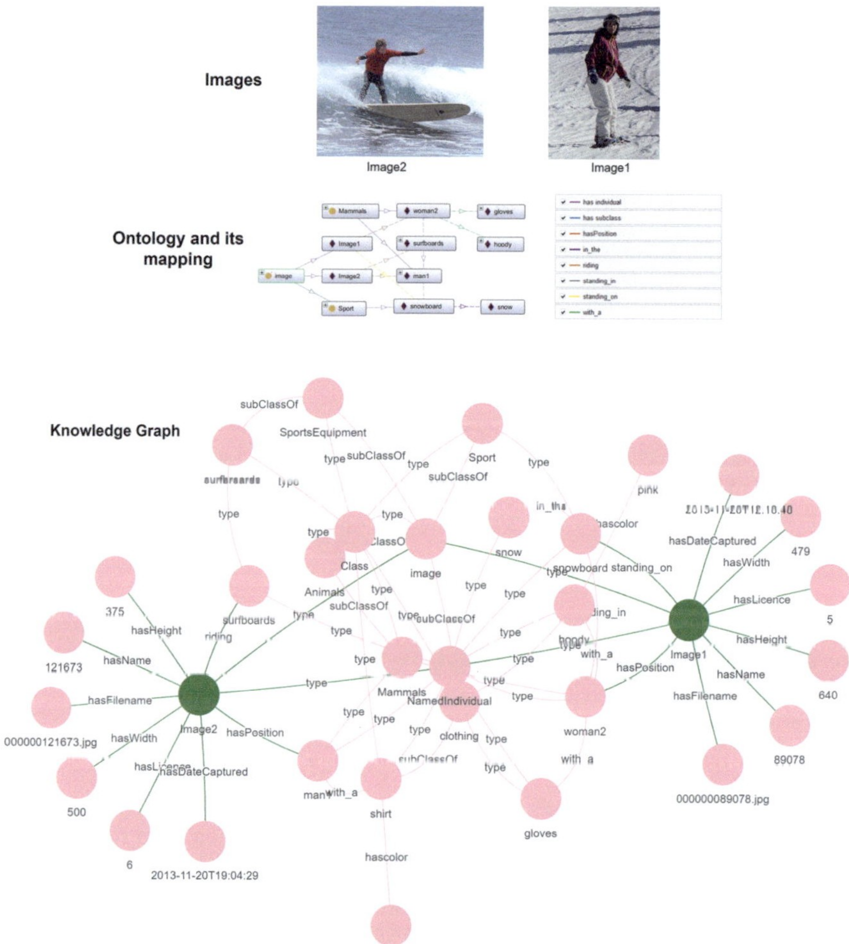

Fig. 7 Knowledge graph linking two MS-COCO dataset images

tion of human subjects engaged in outdoor sports, and the adaptation to specific environments (snow vs. waves). The knowledge graph linking these shared concepts illustrates how seemingly unrelated activities like snowboarding and surfing can be semantically connected. As shown in Fig. 7, we can visualize these connections, where both images are represented as clusters linked by shared concepts such as riding, human subject, and outdoor sports, highlighting the semantic relationships between them.

Table 3 Ontology evaluation: syntactic, semantic, and pragmatic

Metric	Semiotic metric score
Syntactic	0.52
Semantic	0.73
Pragmatic	0.17

4.3 Caknow Analysis

Evaluating ontologies from various perspectives is essential to ensure their quality before use or reuse.

Ontology evaluation is a fundamental phase in the ontology development process. In this work, we have utilized the *OntoKeeper* tool [3] to evaluate the *CaKnow* Ontology, and based on the calculations, we have assessed the Syntactic, Semantic, and Pragmatic aspects of the ontology using the equations provided in the paper. The results of these evaluations are presented in Table 3.

5 Conclusion and Future Work

In this paper, we introduced *CaKnow*, a caption-based ontology derived from the MS-COCO dataset, designed for use in web-based retrieval systems such as Richpedia and similar platforms. The ontology provides a structured and semantic representation of the MS-COCO dataset, enabling more intelligent and efficient data retrieval and analysis.

However, the primary limitation of *CaKnow* is the complexity and scalability of its ontology, which encompasses a wide range of concepts and relationships from the diverse MS-COCO dataset. This complexity can make the ontology challenging to maintain and update. While the ontology offers a structured framework for knowledge representation, it may not significantly enhance image recognition performance without advanced integration techniques.

Future work can focus on further enhancing the ontology by incorporating it into various advanced systems. These include:

1. Question Answering Systems: Developing the ontology to enable more accurate and context-aware query responses.
2. Image Retrieval Systems: Leveraging the ontology to improve semantic-based image searches, making retrieving relevant images based on captions and relationships between objects easier.
3. Building Complex Knowledge Systems: Expanding the ontology to integrate with other datasets creates a more comprehensive knowledge graph that can support complex reasoning and decision-making tasks.

References

1. Al Azemi A, Nefti S, Manzoor U, Rezgui Y (2011) Building a bilingual bio-ontology platform for knowledge discovery. Int J Innov Comput Inform Control 7(12):7067–7075
2. Alberts H, Huang T, Deshpande Y, Liu Y, Cho K, Vania C, Calixto I (2020) Visualsem: a high-quality knowledge graph for vision and language. arXiv preprint arXiv:2008.09150
3. Amith M, Manion F, Liang C, Harris M, Wang D, He Y, Tao C (2019) Architecture and usability of ontokeeper, an ontology evaluation tool. BMC Med Inform Decis Mak 19:1–18
4. Baclawski K, Bennett M, Berg-Cross G, Schneider T, Sharma R, Singer J, Sriram RD (2021) Ontology summit 2020 communiqué: knowledge graphs. Appl Ontol 16(2):229–247
5. Bonatti PA, Decker S, Polleres A, Presutti V (2019) Knowledge graphs: new directions for knowledge representation on the semantic web (dagstuhl seminar 18371). In: Dagstuhl reports. vol 8. Schloss Dagstuhl-Leibniz-Zentrum fuer Informatik
6. Ehrlinger L, Wöß W (2016) Towards a definition of knowledge graphs. SEMANTiCS (Posters Demos SuCCESS) 48(1–4):2
7. Guo Y, Yin F, Li XH, Yan X, Xue T, Mei S, Liu CL (2023) Visual traffic knowledge graph generation from scene images. In: Proceedings of the IEEE/CVF international conference on computer vision, pp 21604–21613
8. Hybridised KCN (2021) Ontoknownhs: ontology driven knowledge centric novel hybridised semantic scheme for image recommendation using knowledge graph. In: Knowledge graphs and semantic web: third tberoamerican conference and second Indo-American conference, KGSWC 2021, Kingsville, Texas, USA, 22–24 Nov 2021, Proceedings. Springer, p 138
9. Lang C, Braun A, Valada A (2022) Robust object detection using knowledge graph embeddings. In: DAGM German conference on pattern recognition. Springer, pp 445–461
10. Le TM, Dinh NT, Van TT (2023) Developing a model semantic-based image retrieval by combining KD-tree structure with ontology. Expert Syst e13396
11. Li H, Wang Y, Zhang S, Song Y, Qu H (2021) Kg4vis: a knowledge graph-based approach for visualization recommendation. IEEE Trans Visual Comput Graph 28(1):195–205
12. Lin TY, Maire M, Belongie S, Hays J, Perona P, Ramanan D, Dollár P, Zitnick CL (2014) Microsoft coco: common objects in context. In: Computer vision—ECCV 2014: 13th European conference, Zurich, Switzerland, 6–12 Sept 2014, Proceedings, Part V 13. Springer, pp 740–755
13. Ma Y, Wang Y, Wu Y, Lyu Z, Chen S, Li X, Qiao Y (2022) Visual knowledge graph for human action reasoning in videos. In: Proceedings of the 30th ACM international conference on multimedia, pp 4132–4141
14. Marino K, Salakhutdinov R, Gupta A (2016) The more you know: using knowledge graphs for image classification. arXiv preprint arXiv:1612.04844
15. Menglong C, Detao J, Ting Z, Dehai Z, Cheng X, Zhibo C, Xiaoqiang X (2019) Image classification based on image knowledge graph and semantics. In: 2019 IEEE 23rd International conference on computer supported cooperative work in design (CSCWD). IEEE, pp 81–86
16. Muscetti M, Rinaldi AM, Russo C, Tommasino C (2022) Multimedia ontology population through semantic analysis and hierarchical deep features extraction techniques. Knowl Inf Syst 64(5):1283–1303
17. Patel AS, Merlino G, Bruneo D, Puliafito A, Vyas O, Ojha M (2021) Video representation and suspicious event detection using semantic technologies. Semantic web 12(3):467–491
18. Patel AS, Vyas R, Vyas O, Ojha M (2022) A study on video semantics; overview, challenges, and applications. Multimedia Tools Appl 81(5):6849–6897
19. Sheth A, Padhee S, Gyrard A (2019) Knowledge graphs and knowledge networks: the story in brief. IEEE Internet Comput 23(4):67–75
20. Vinyals O, Toshev A, Bengio S, Erhan D (2015) Show and tell: a neural image caption generator. In: Proceedings of the IEEE conference on computer vision and pattern recognition, pp 3156–3164

21. Wajid MS, Terashima-Marin H, Najafirad P, Wajid MA (2024) Deep learning and knowledge graph for image/video captioning: a review of datasets, evaluation metrics, and methods. Eng Rep 6(1):e12785
22. Wang C, Zhuo X, Li P, Chen N, Wang W, Chen Z (2020) An ontology-based framework for integrating remote sensing imagery, image products, and in situ observations. J Sens 2020(1):6912820

Enhancing Deep Fake Image Generation and Detection Through Transfer Learning

Ronak Bhalgami, Vasu Bhut, Shivam Ardeshna, Trusha Patel, and Rikita Chokshi

Abstract The emergence of deep fake technologies has prompted serious concerns regarding the authenticity and trustworthiness of digital media. This paper addresses the challenge of deep fake image detection using a combination of generative adversarial networks (GANs) for synthetic image generation and convolutional neural networks (CNNs) for classification. A custom dataset comprising 5000 real-face images and synthetic images generated using GANs is used for training and evaluation. Our approach leverages transfer learning by fine-tuning a DenseNet121 architecture pre-trained on ImageNet for improved feature extraction and classification performance. The model is trained and validated using a split of 70% training, 10% validation, and 20% testing datasets. The training process is evaluated through metrics including accuracy, loss, and receiver operating characteristic (ROC) curve analysis. The proposed methodology demonstrates promising results in discerning between real and fake images, achieving high accuracy in distinguishing manipulated images from genuine ones. This research contributes to the ongoing efforts in combating the adverse effects of deep fake technologies on various domains, including media, politics, and cybersecurity.

Keywords Deep fake · Machine learning · Dense121

1 Introduction

The integrity of visual content in the digital age is a concern due to the deep fake technologies' rapid advancement. These artificial media, which are frequently driven by generative adversarial networks (GANs) [6], cast doubt on the veracity of internet media by allowing the production of incredibly lifelike but artificial images. This

R. Bhalgami · V. Bhut · S. Ardeshna · T. Patel (✉) · R. Chokshi
Computer Engineering Department, CSPIT, CHARUSAT, Changa, Gujarat, India
e-mail: trushapatel.ce@gmail.com

R. Bhalgami
e-mail: 21ce009@charusat.edu.in

S. Kumar et al. (eds.), *Artificial Intelligence for Next Generation Computing, Volume 1*, Lecture Notes in Networks and Systems 1549,
https://doi.org/10.1007/978-981-95-0701-6_25

technology poses multifaceted challenges across various sectors, including media, politics, entertainment, and cyber security. The pervasive spread of deep fakes undermines public trust in visual information [5]. In journalism, manipulated content can distort factual reporting and weaken audience trust in the media. In politics, deep fakes threaten democratic processes by spreading misinformation and influencing public opinion. Additionally, the entertainment industry confronts ethical and legal challenges with the unauthorized use of celebrities' likenesses in deceptive contexts [14].

Addressing the widespread threat of deep fakes demands innovative detection and mitigation strategies [10]. This research leverages machine learning, specifically convolutional neural networks (CNNs), to detect deep fake images [19]. Our methodology involves training a robust classification model on a curated dataset of 5,000 real-face images and GAN-generated synthetic images.

This research is crucial for combating the harmful impacts of deep fakes and restoring trust in visual media [13]. Through transfer learning and fine-tuning pre-trained models like DenseNet121 [1], we aim to improve the effectiveness of deep fake detection. Our approach is inspired by recent advancements in the field, including techniques proposed by leading researchers in deep fake detection technologies.

Our research aims to reduce fraud and misinformation caused by deep fakes, preserving the integrity of visual information in the digital age. The following sections will detail our experimental approach, results, and analysis, highlighting the effectiveness of our method in countering AI-generated media deception.

2 Related Work

The field of deep fake detection has seen rapid growth, with research focused on addressing the rise of manipulated media. Key contributions and methods have been explored in detecting deep fake images, leveraging generative adversarial networks (GANs) for data generation and the VGG16 model for classification.

GANs [7] serve as a foundation for training generative models, consisting of a generator and discriminator in a competitive dynamic to create increasingly convincing fake images. This innovation has significantly advanced image generation, which now poses challenges for deep fake detection.

EfficientNet [18], known for its scalability and efficiency, highlights the power of transfer learning, a technique where a pre-trained model aids the training of new models. EfficientNet's architecture is particularly effective for optimizing performance in complex detection tasks.

Simonyan and Zisserman (2014) introduced VGG16, a deep CNN used for image classification and detection. Its simplicity and effectiveness make it ideal for distinguishing between real and fake images in deep fake detection systems. Auto Augment [2] is a notable automated data augmentation method that tailors augmentation strategies to specific datasets, enhancing model performance and generalization. In deep fake detection, Auto Augment provides crucial diversity in training data, improving models' ability to detect subtle anomalies in fake images.

The integration of multimodal fusion [8] strategies boosts deep fake detection systems by incorporating visual and audio data, improving detection accuracy. This approach is vital for keeping pace with increasingly sophisticated fake content, enhancing the robustness of detection systems.

Transformers [20] utilize self-attention mechanisms to efficiently capture dependencies, proving successful in tasks like machine translation and image captioning. In deep fake detection, attention mechanisms focus on critical regions of an image, identifying subtle inconsistencies or artifacts from the generation process, crucial for identifying manipulated media.

3 Proposed Approach

In our comparative analysis, existing deep fake detection systems typically rely on predefined datasets of real and fake images generated by standard GANs, which often lack diversity in characteristics like lighting and texture. As a result, classification models such as ResNet and VGG16 struggle to detect subtle differences, limiting their generalization and robustness in real-world applications.

In contrast, our approach utilizes a fine-tuned GAN to generate a wider variety of fake images with diverse lighting, textures, and complexities. This enhanced dataset improves detection across various manipulations. We also employ DenseNet121 for classification, benefiting from its densely connected layers that facilitate better feature propagation. This combination allows our model to effectively distinguish subtle differences and generalize better, making it more accurate in real-world deep fake detection scenarios (Fig. 1).

Generative adversarial networks (GANs) are employed to produce realistic fake images. A GAN consists of a generator and a discriminator, which are trained together through an adversarial process [7]. The CIFAR-10 dataset, which contains 60,000 color images across 10 different classes, is used as the source of real images. However, the images are resized to 256 $\times$ 256 pixels for compatibility with the deep learning models used in this study.

The training process of the GAN involves the generator creating fake images from noise inputs and the discriminator attempts to discriminate between authentic and fraudulent pictures. The generator's goal is to create pictures that are indistinguishable from genuine ones in order to fool the discriminator. Simultaneously, the discriminator is trained to reliably identify real and fraudulent pictures. This adversarial training continues until the generator produces highly realistic images.

Upon completing the GAN training, the generator is used to create a new set of fake images. These generated images, which are 256 $\times$ 256 pixels in size, are then combined with real images from the CIFAR-10 dataset to form a new dataset. This dataset includes an equal number of real and fake images, ensuring a balanced distribution for training the classification model.

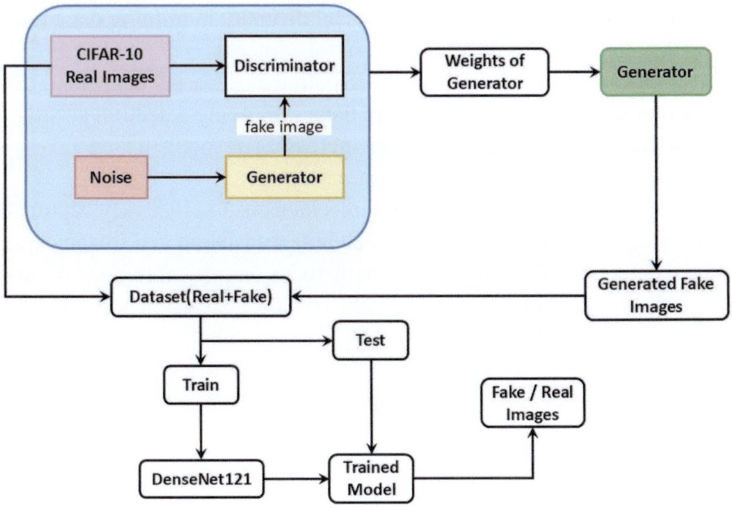

Fig. 1 The proposed approach

The DenseNet121 model, a densely connected convolutional neural network, is utilized for classifying images as real or fake. DenseNet121 connects each layer to every other layer, which improves efficiency and performance by facilitating feature reuse [9]. The combined dataset of real and fake images is split into training and testing sets. Typically, 80% of the data is used for training and 20% for testing, though this ratio can be adjusted based on specific requirements.

The DenseNet121 model is trained on the training set to learn the distinguishing features between real and fake images. The model's parameters are optimized using a suitable loss function, such as binary cross-entropy, and an optimizer like Adam [12]. A validation set is used during training to tune hyperparameters and prevent overfitting [11].

After training, the DenseNet121 model is evaluated on the test set to assess its performance in detecting fake images. Metrics including accuracy, precision, recall, and F1-Score are used to evaluate the model's efficacy [17]. Visualizations like confusion matrices and ROC curves are also created to give deeper insights into the model's performance [16].

This methodology demonstrates a comprehensive approach to generating and detecting fake images using GANs and DenseNet121. The combination of these techniques provides a robust framework for identifying synthetic images, which is crucial for applications in digital forensics, media integrity, and cybersecurity.

4 Methodology Review

4.1 Generating Fake Images Using GANs

In our study, we utilized generative adversarial networks (GANs), an influential generative model [7], as a cornerstone for synthesizing realistic fake images. GANs operate through an extraordinary structure comprising two neural networks: a generator and a discriminator. These networks are trained in a contentious fashion, competing against each other to improve their abilities. The generator is tasked with convincing that images are fake to remove random noise inputs, while the discriminator's role is to distinguish between real and fake images. This adversarial training process leads to the style of both networks, ultimately allowing the generator to produce high-quality images that closely resemble real samples.

Figure 2 provides a visual representation of the GAN architecture, offering insight into how the generator operates within the network. The figure illustrates the iterative process where the generator learns to synthesize diverse fake images by transforming latent noise vectors into complex visual patterns. The discriminator, on the other hand, refines its ability to differentiate between generated images and authentic data, driving the overall improvement of the GAN model.

The effectiveness of GANs lies in their capacity to capture intricate data distributions and generate novel, realistic content. This capability has profound implications for various applications, including image generation, data augmentation, and, notably, the creation of synthetic data for training deep learning models. In our research, leveraging GANs enabled us to generate a large dataset of fake images, which served as valuable input for training our deep fake detection model.

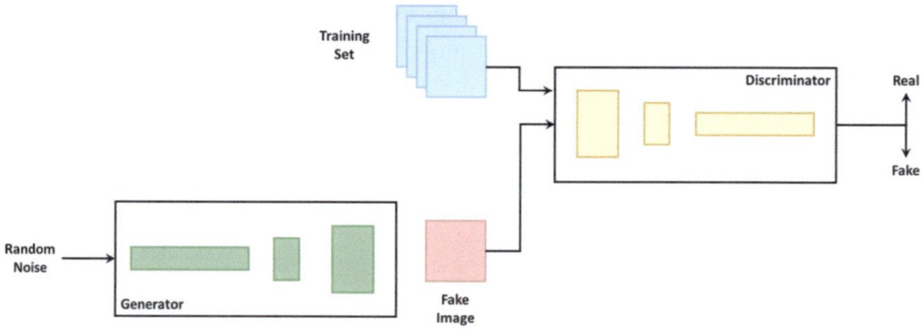

Fig. 2 GAN detecting image as real or fake

4.2 Dataset Description and Image Augmentation

We employed a diverse dataset of real images showcasing various objects, people, and scenes, which were used to generate fake versions through a fine-tuned GAN model. The generated fake images include a wide range of variations in lighting, texture, background, and object features, closely simulating authentic manipulations and capturing a broad spectrum of potential forgeries.

This combined dataset of real and fake images was then used for classification with DenseNet121, a deep learning model pre-trained on ImageNet. With images of varying resolutions, lighting conditions, and angles, the classifier learns to detect subtle differences between real and fake images. This variability is crucial for ensuring the model generalizes well and accurately identifies forgeries across different types of images, regardless of content or context.

In our study, enhancing the robustness and generalization of our deep fake detection model was pivotal, and to achieve this, we employed sophisticated data augmentation techniques, specifically Auto Augment represents a cutting-edge approach to data augmentation by dynamically optimizing augmentation policies tailored to the characteristics of the dataset. By doing so, the technique introduces diverse variations into the training data, which in turn enriches the model's exposure to different image transformations and variations.

Auto Augment operates by dynamically adjusting augmentation policies during the training process, effectively diversifying the dataset without the need for manual intervention. This dynamic optimization process ensures that the model learns to detect subtle features indicative of fake images, improving its overall capability to generalize to unseen data and variations. By leveraging Auto Augment, we aimed to equip our deep fake detection model with the resilience and adaptability necessary to accurately discern between real and manipulated images in diverse scenarios.

The integration of Auto Augment into our methodology underscores the importance of data augmentation in enhancing model performance and combating issues related to overfitting and limited dataset diversity. By systematically expanding the training dataset through tailored augmentation strategies, we aimed to mitigate potential biases and improve the model's ability to capture nuanced patterns indicative of deep fake images.

4.3 Deep Fake Image Detection

In our research, we focused on developing a robust deep learning model tailored for binary classification, specifically distinguishing between real and fake images, using a transfer learning approach with a DenseNet121 [9]. Transfer learning enabled us to leverage the foundational training DenseNet121 model, which had been trained on a big scale data like ImageNet, and adapt it to our specific task of deep fake image detection. This approach allowed us to benefit from the feature extraction capabilities

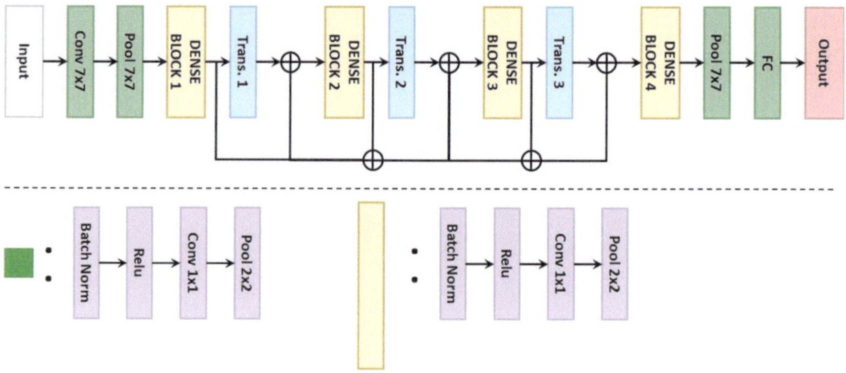

Fig. 3 Architecture of DenseNet121

of DenseNet121, which is renowned for its effectiveness in capturing complex visual patterns.

The deep learning model was trained using a dataset comprising both real and fake images generated by the generative adversarial network (GAN). This dataset served as crucial input for training the model to differentiate between authentic and manipulated images. By incorporating images generated by GANs, we ensured that the model was exposed to a diverse range of synthetic data, thereby enhancing its ability to generalize to unseen variations and subtle manipulations commonly found in deep fake images.

Figure 2 in our study illustrates the architecture of the deep learning model, highlighting key components such as the features that are extracted from hidden layers of DenseNet121 and the subsequent classification layers. The hidden layers used for feature extraction are responsible for capturing relevant visual features from the input images, while the classification layers leverage these features to make predictions (i.e., classifying images as either real or fake). This architecture showcases the intricate design of our model, emphasizing its capability to effectively discern between authentic and manipulated content (Figs. 3 and 4).

Through the utilization of transfer learning with DenseNet121 and training on a various dataset of real and fake images, our deep learning model was equipped with the necessary knowledge and discriminative power to combat the challenges posed by deep fake media in modern digital landscapes.

4.4 Experimental Results

We have directed comprehensive tests to estimate the performance of our deep learning model on a dedicated test dataset, showcasing its effectiveness in detecting

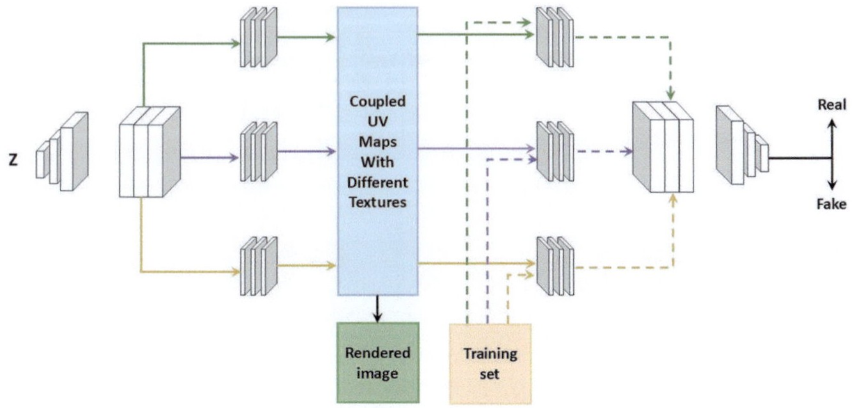

Fig. 4 The figure depicts how and image is categorized

fake images with high accuracy and reliability. Leveraging the integration of GAN-based image generation, data augmentation techniques, and deep learning-based detection, our methodology represents a holistic approach toward combating the challenges posed by manipulated media and advancing the field of deep fake detection.

Figure 5 highlights sample predictions generated by our model, demonstrating its ability to discern between real and fake images effectively. These visualizations provide concrete examples of the model's discriminative capabilities, showcasing its capacity to accurately classify images based on subtle manipulations indicative of deep fake content.

By integrating cutting-edge methodologies and harnessing advancements in generative adversarial networks (GANs), data augmentation techniques, and transfer learning, our study makes a substantial contribution to the continuous endeavor of combating manipulated media and safeguarding the authenticity of digital content within today's dynamic and rapidly evolving landscape. The utilization of GANs, known for their ability to generate highly realistic synthetic data, allows us to create diverse and authentic-looking fake images representative of the challenges posed by deep fake content. By leveraging these advancements, our study enables the development of robust detection models capable of discerning between genuine and manipulated media with heightened accuracy and reliability.

Furthermore, the integration of sophisticated data augmentation techniques enriches the diversity and complexity of our training dataset, enhancing the model's capacity to generalize and recognize subtle patterns indicative of deep fake alterations. This approach ensures that our detection system is adept at identifying manipulated media across various scenarios and contexts (Fig. 6).

Furthermore, we assessed the performance of our model quantitatively, achieving notable results with a test loss of 22.87% and a test accuracy of 91.77%. These

Fig. 5 The figure shows the predictions of an image which is fake or real

metrics underscore the robustness and reliability of our deep fake detection system in real-world scenarios (Fig. 7).

The performance of the DenseNet121 model in distinguishing between real and fake images is evaluated using typical classification criteria like as accuracy, recall, and F1-Score (Table 1).

Precision, at 99.50%, shows the model's ability to minimize false positives, while recall, at 99.86%, highlights its effectiveness in reducing false negatives. The F1-Score of 99.68% demonstrates a balanced performance in detecting fake images. These metrics validate DenseNet121's robustness in fake image recognition with low false positives and strong fake detection capability. Its performance makes it suitable for applications in digital forensics, media integrity, cyber security, and social media monitoring. In forensics, it can identify manipulated images used in investigations, while in media, it helps verify image authenticity to combat misinformation. For cybersecurity, it detects deep fakes used in phishing or fraud, and social

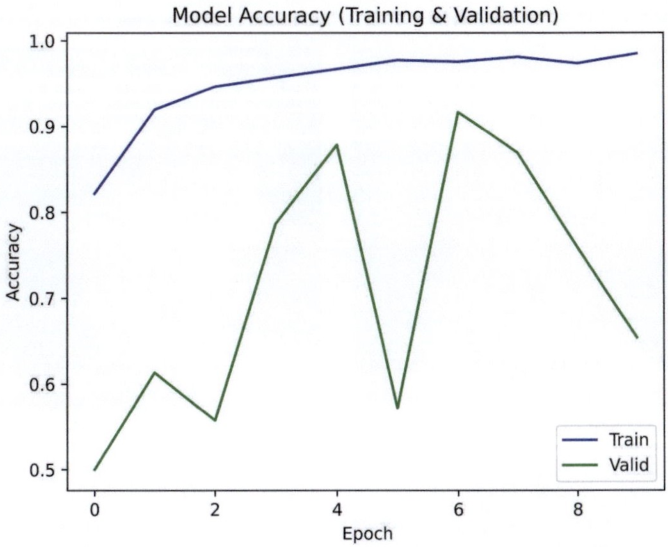

Fig. 6 Accuracy graph

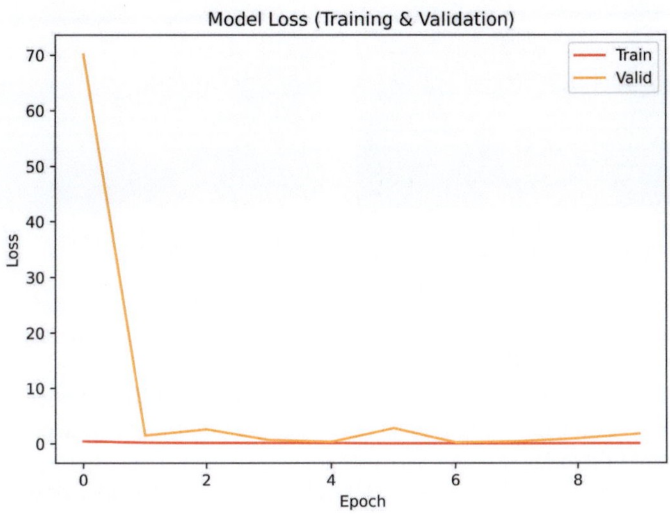

Fig. 7 Loss graph

Table 1 Classification report

Precision (%)	99.50
Recall (%)	99.86
F1-Score (%)	99.68

media platforms can use it to flag fake content. DenseNet121's integration with GAN-generated datasets offers a comprehensive solution for detecting synthetic media in an evolving landscape. A diagram of the system's workflow, showcasing image input and classification as real or fake, further illustrates the detection process [3] (Fig. 8).

Moreover, the strategic application of transfer learning, leveraging pre-trained models like DenseNet121, empowers our study to capitalize on existing knowledge and feature representations learned from large-scale datasets. This facilitates the growth of deep learning models specifically tailored for deep fake detection, achieving notable performance metrics such as high accuracy and reliability (Fig. 9).

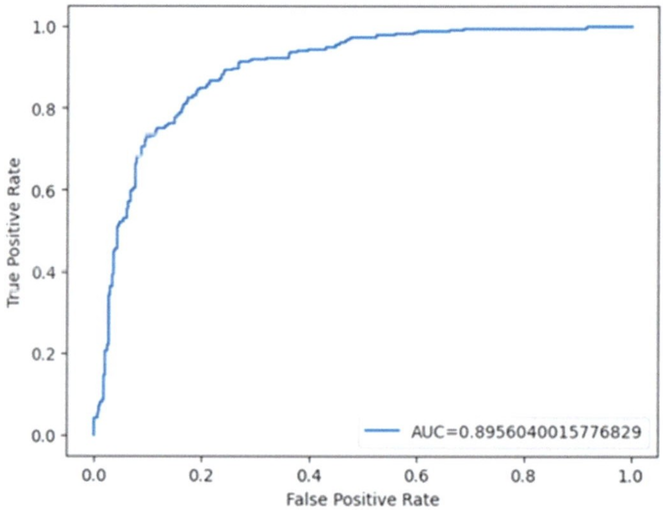

Fig. 8 ROC and AUC curve

Fig. 9 Heatmap

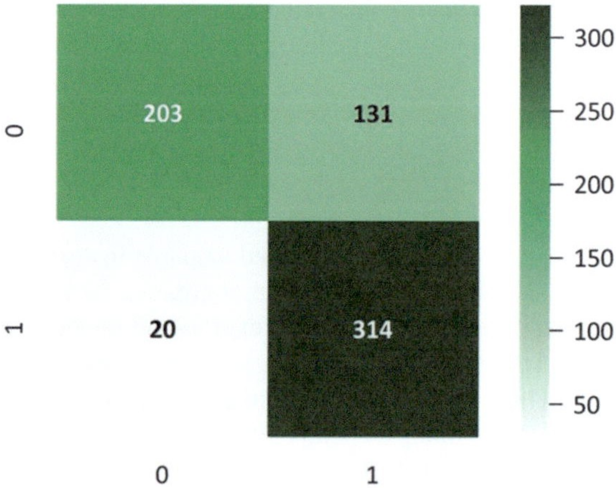

The integration of GANs, data augmentation techniques, and deep learning-based detection represents a synergistic approach toward enhancing model performance and addressing the challenges posed by deep fake media. Through extensive experimentation and validation, our methodology demonstrates promising results and opens avenues for further advancements in deep fake detection research.

5 Conclusion

In conclusion, our research presents a robust methodology for addressing the challenges posed by deep fake image detection using advanced machine learning techniques. Leveraging the power of generative adversarial networks (GANs) for synthetic image generation and transfer learning with DenseNet121 for feature extraction, our approach demonstrates promising results in discerning between real and fake images with high accuracy and reliability.

Through comprehensive experimentation and evaluation on a custom dataset, we achieved notable performance metrics, including a test accuracy of 91.77% and a test loss of 22.87%. These results underscore the effectiveness and robustness of our deep fake detection system in real-world scenarios.

Our study contributes significantly to the ongoing efforts aimed at combating the adverse effects of deep fake technologies on various domains, including media, politics, entertainment, and cybersecurity. By integrating cutting-edge methodologies and leveraging advancements in machine learning, we aim to restore trust in visual media and mitigate the prevalence of fraud and misinformation facilitated by deep fakes in today's digital age.

Moving forward, our work opens avenues for further advancements in deep fake detection research, emphasizing the importance of interdisciplinary approaches and technological innovations in addressing the pervasive challenges posed by manipulated media. Through continued exploration and refinement of our methodology, we strive to enhance the resilience and effectiveness of deep fake detection systems, thereby safeguarding the integrity of digital content in rapidly evolving landscapes.

6 Future Work

While our research has provided valuable insights and achieved notable success in deep fake image detection, several avenues for future exploration and improvement remain. Future research could explore the integration of multimodal features, such as audio and text, in addition to visual cues, to enhance the robustness of deep fake detection systems [4]. Another promising direction involves the development of adversarial training techniques and defense mechanisms tailored specifically for deep fake detection [15]. By focusing on these advanced research areas, we can further advance the capabilities of deep fake detection systems and contribute to the ongoing efforts to combat the proliferation of manipulated media in digital environments.

References

1. Awotunde JB, Jimoh RG, Imoize AL, Abdulrazaq AT, Li CT, Lee CC (2023) An enhanced deep learning-based deepfake video detection and classification system. Electronics 12(1). https://doi.org/10.3390/electronics12010087. https://www.mdpi.com/2079-9292/12/1/87
2. Cubuk ED, Zoph B, Mane D, Vasudevan V, Le QV (2019) Autoaugment: learning augmentation policies from data
3. Fawcett T (2006) Introduction to ROC analysis. Pattern Recognit Lett 27:861–874. https://doi.org/10.1016/j.patrec.2005.10.010
4. Festus Ayetiran E (2024) Özgöbek: a review of deep learning techniques for multimodal fake news and harmful languages detection. IEEE Access 12:76133–76153. https://doi.org/10.1109/ACCESS.2024.3406258
5. Gambín AF, Yazidi A, Vasilakos A, Haugerud H, Djenouri Y (2024) Deep-fakes: current and future trends. Artif Intell Rev 57(3). https://doi.org/10.1007/s10462-023-10679-x
6. Gandhi A, Jain S (2020) Adversarial perturbations fool deepfake detectors. In: 2020 international joint conference on neural networks (IJCNN), pp 1–8. https://doi.org/10.1109/IJCNN48605.2020.9207034
7. Goodfellow IJ, Pouget Abadie J, Mirza M, Xu B, Warde-Farley D, Ozair S, Courville A, Bengio Y (2014) Generative adversarial networks
8. Guo Y, Liu Y, Oerlemans AAJ, Lao S, Wu S, Lew MS (2016) Deep learning for visual understanding: a review. Neurocomputing 187:27–48. https://api.semanticscholar.org/CorpusID:207111722
9. Huang G, Liu Z, van der Maaten L, Weinberger KQ (2018) Densely connected convolutional networks
10. Khan S, Chen JC, Liao WH, Chen CS (2024) Adversarially robust deep-fake detection via adversarial feature similarity learning. Springer Nature Switzerland, pp 503–516. https://doi.org/10.1007/978-3-031-53311-237
11. Kingma DP, Ba J (2017) Adam: a method for stochastic optimization
12. Krizhevsky A (2012) Learning multiple layers of features from tiny images. University of Toronto
13. Salman S, Shamsi J, Qureshi R (2023) Deep fake generation and detection: Issues, challenges, and solutions. IT Prof 25(01):52–59. https://doi.org/10.1109/MITP.2022.3230353
14. Salvi D, Liu H, Mandelli S, Bestagini P, Zhou W, Zhang W, Tubaro S (2023) A robust approach to multimodal deepfake detection. J Imaging 9(6):122. https://doi.org/10.3390/jimaging9060122. https://www.mdpi.com/2313-433X/9/6/122
15. Seow JW, Lim MK, Phan RC, Liu JK (2022) A comprehensive overview of deep-fake: generation, detection, datasets, and opportunities. Neurocomputing 513:351–371
16. Sokolova M, Lapalme G (2009) A systematic analysis of performance measures for classification tasks. Inf Process Manage 45:427–437. https://doi.org/10.1016/j.ipm.2009.03.002
17. Srivastava N, Hinton G, Krizhevsky A, Sutskever I, Salakhutdinov R (2014) Dropout: a simple way to prevent neural networks from overfitting. J Mach Learn Res 15(1):1929–1958
18. Tan M, Le QV (2020) Efficientnet: rethinking model scaling for convolutional neural networks
19. Tolosana R, Vera-Rodriguez R, Fierrez J, Morales A, Ortega-Garcia J (2020) Deepfakes and beyond: a survey of face manipulation and fake detection. Inf Fusion 64:131–148. https://doi.org/10.1016/j.inffus.2020.06.014. https://www.sciencedirect.com/science/article/pii/S1566253520303110
20. Vaswani A, Shazeer N, Parmar N, Uszkoreit J, Jones L, Gomez AN, Kaiser L, Polosukhin I (2023) Attention is all you need

Spectral Analysis of Water Ice in Erlanger Crater, Lunar North Pole's Permanently Shadowed Area

Chandani Sahu, Himanshu Govil, Rajesh Doriya, and Gaurav Mishra

Abstract Permanently shadowed regions (PSRs) on the Moon, located near its north and south poles, are characterized by extreme cold and a lack of sunlight, creating conditions conducive to the trapping of volatiles such as water ice. The Moon's axial tilt, which is nearly perpendicular to sunlight, results in these areas remaining perpetually dark and cold, allowing water ice to accumulate. This ice is crucial for future lunar exploration and potential human habitation. Research efforts, including the Chandrayaan-1 mission, have focused on detecting water ice within these PSRs. Chandrayaan-1, launched by the Indian Space Research Organization, utilized its Moon mineralogy mapper (M3) hyperspectral data to analyze the lunar surface. The M3 instrument was particularly effective in identifying spectral signatures indicative of hydration feature and ice molecules at the Moon's north pole. The M3 hyperspectral images facilitated the analysis of water ice absorption features, and different techniques of image processing were applied to enhance the accuracy of the data. The spectral angle mapper (SAM) classification method was employed to map water ice deposits, and the Target Detection Wizard tool in ENVI software was used to validate the findings against reference spectra. Results from these studies indicate significant amounts of water ice may exist in the PSRs. The study effectively identified and classified water ice in the permanently shadowed regions of Erlanger Crater at the lunar North Pole using spectral angle mapper (SAM) and Thor Target Detection techniques. Spectral analysis from the Moon mineralogy mapper (M3) revealed water ice absorption features at 1.5 and 2.9 μm, validated by comparisons with the

C. Sahu (✉) · H. Govil · G. Mishra
Department of Applied Geology, National Institute of Technology, Raipur, Chhattisgarh, India
e-mail: csahu.phd2023.geo@nitrr.ac.in

H. Govil
e-mail: hgovil.geo@nitrr.ac.in

G. Mishra
e-mail: mshgaurav04@gmail.com

R. Doriya
Department of Information Technology, National Institute of Technology, Raipur, Chhattisgarh, India
e-mail: rajeshdoriya.it@nitrr.ac.in

S. Kumar et al. (eds.), *Artificial Intelligence for Next Generation Computing, Volume 1*, Lecture Notes in Networks and Systems 1549,
https://doi.org/10.1007/978-981-95-0701-6_26

John Hopkins University spectral library. SAM classified water ice pixels with an angular value of 0.33 radians (19.26°), primarily around the crater's rim and inner walls, while Thor Target Detection identified additional signatures across the region. This discovery not only enhances our understanding of the polar regions of the moon but also holds implications for future exploration missions.

Keywords Moon mineralogy mapper · Spectral angle mapper · Target detection · Hyperspectral data

1 Introduction

Permanently shadowed regions are areas that never receive sunlight, resulting in unique characteristics on planetary bodies that distinguish them from other locations [1]. Their darkness is primarily due to the Moon's axis that is perpendicular to the sunlight direction, which facilitates the freezing of water and volatiles in the polar region of moon. Since sunlight never reaches these areas, the temperature in the shaded regions remains consistently low. The permanently shadowed regions of the moon have very low and cold temperatures ~ 120 which act as traps for volatile materials as a result [1].

The issue of the existence of H_2O and other volatiles on the Moon is crucial in terms of science and exploration [2]. Various factors are responsible to deliver the water molecules on the lunar surface, i.e., bombardment of water containing impactors like comets, asteroids, meteorites, the reaction of protons from solar winds and endogenic water on Moon. Extensive cometary volatility at the lunar poles would give fresh possibilities over geological time to tackle significant questions around the sources and fluxes of impacting cometary items on the Moon [2]. Exploration of planets and planetary bodies is essential to understand their formation and study of their evolution responsible for their current geological state [3]. Various studies, such as those by [4], have utilized data from the Lunar Orbiter Laser Altimeter to pinpoint the locations of permanently shadowed regions (PSRs), along with thermal maps using the Diviner Lunar Radiometer Experiment, water abundance measurements from the Lunar Prospector Neutron Spectrometer, and slope maps from Lunar Orbiter Laser Altimeter. The objective of these endeavors is to pinpoint regions that possess the greatest scientific capacity for examining volatiles, specifically with the identification of volatile ice in the lunar poles [5]. Among these initiatives, the Chandrayaan-1 mission was specifically launched to map the lunar surface and detect water deposits near the Moon's poles. For information on the lunar surface and permanently shadowed regions, various lunar remote sensing images are there, i.e., multi and hyperspectral datasets in the visible and near infrared, and shortwave infrared, which adds to the information about the mineralogical composition of the lunar surfaces as a result of spectral diagnostic absorption features of major minerals and elements. Electronic and vibrational are two important types of absorption processes that take place in the material based on which material and other elements can be easily identified. The electronic process comprising the UV, visible and NIR region occurs in lower wavelength regions. Whereas the vibrational absorption process occurs in visible to the NIR region at wavelengths of around 1–30 μm. They produce complex spectra. NIR reflectance spectroscopy is very useful in measuring the molecular vibrations and can use to discriminate water ice from OH- and OH- bearing minerals. The best spectral and spatial resolution data of the lunar surface, including the polar regions, are obtained from NIR reflectance measurements taken with the Moon mineralogy mapper instrument on board the Chandrayaan-1 mission. The purpose of the M3 imaging spectrometer is to map and identify the mineral and lithology compositions of the lunar surface [6].

The permanently shaded area of the moon is completely dark and highly cratered with the possibility of ice traps. These shadowed regions are indirectly illuminated by sunlight, reflected from neighboring topography that was directly illuminated. The search for ice focused on these shaded regions. Therefore, getting information about this region with Chandrayaan-1 M3 are preferable choice. The goal for the utilization of M3 data is to map the surface chemistry of moon and retrieve the lunar mineral resources on the moon.

In the present study, the spectra collected from Erlanger Crater are compared to JHU (JOHN HOPKINS UNIVERSITY) reference spectra, and the spectral angle mapper (SAM) classification is utilized to identify target water ice pixels. The primary focus of the present study is on detecting the diagnostic absorption feature, which occur near wavelengths of 1.5 and 2.9 μm. For validation purposes, the Thor Target Detection Wizard is employed to map the targeted water ice pixels.

1.1 Volatiles on the Lunar Surface

Prior to the year 2000, the Moon was generally regarded as a dry celestial body which has no water and volatiles within it and only the permanently shadowed areas have the possible exception of volatile deposits [7]. The study of moon volatiles began as soon as the Apollo Sample were returned, then it became the primary motivation to study the lunar volatiles [8]. The sample returned from Apollo contained hydrous minerals and other weathering products indicating water, but the bulk study of Apollo basalt indicated that they were dry [9]. Over the past decade, proofs of the existence of water molecules have been uncovered using various laboratory researches and satellite data. These findings have demonstrated that water exists on the bright surface of the Moon as well as within the lunar interior and in certain permanently shaded areas. Water was discovered in volcanic glass, olivine-hosted melt inclusions and Apatite in the Apollo sample through the use of analytical methods [10, 11]. The idea that the lunar surface is dry has been called into question by these observations. Further implications of this data for the genesis of the Earth and Moon system as well as the inside of the solar system include that the isotopic composition and amount of water may vary throughout the lunar interior [12]. Due to the rapid removal of components from the lunar atmosphere through solar wind, solar radiation, and gravitational escape, the volatiles and water particles on the lunar surface are thought to be unstable. The evaporation rate which is expected to be obtained at the coolest region of the moon surface is the factor that limits this process [13]. The composition of lunar volatiles is determined by variety of nuclear mass and optical stereoscopic technique [14]. Before the arrival of the Lunar Reconnaissance Orbiter (LRO) and LCROSS at the Moon in 2009, researches conducted with the Lunar Prospector Neutron Spectrometer (LPNS) provided evidence of elevated subsurface volatile concentrations at polar region of moon, which is associated with icy traps [15]. The volatile distribution on lunar polar regions is limited, neutron data indicates that amount of hydrogen is highest at the higher latitude both inside and outside of the permanently shadowed regions [14]. These lunar volatiles are compound that vaporized when it comes in contact with the air and other environment for example hydrogen (H), water (H_20), helium (He), carbon dioxide (Co_2), and carbon monoxide (Co). These volatiles play vital role in the study of sustainable presence of human in lunar surface. The existence of surface water in the permanently shadowed regions is supported by ultraviolet measurements, and near-infrared spectra and laser reflectance measurements [15–17], which look at the upper nanometers to millimeters of the lunar surface. The exosphere and lunar surface are closely related. Every volatile species in the solar system provides insights on the origins and development of surface exospheres. Only a small number of neutral organisms had been directly detected in the lunar exosphere up to the past ten years, and most others had only been given upper limits [18]. Exospheric volatiles have been the subject of numerous investigations in an effort to understand the interaction of the solar wind with airless bodies throughout the solar system. Recent studies suggest that a significant portion of carbon ions may leave the lunar surface in the form of methane (CH_4), with estimates ranging from 25 to 76% [18]. Similarly, 10–50% of solar wind protons (H+) coming from the sun rays may be reacted to the regolith and converted into molecular hydrogen (H_2) upon interaction with the lunar surface [14]. The results of this analysis, however, show significant degrees of uncertainty, which is indicative of underlying knowledge gaps about the lunar carbon and hydrogen cycles generated by the solar wind. The wide ranges in these percentages highlight the need for further research to refine our knowledge of the complex processes governing the interaction between solar wind and the lunar surface. Improving our understanding of these cycles is crucial for accurately modeling the formation and evolution of the lunar exosphere, as well as for interpreting remote sensing data and in-situ measurements of volatile species on the Moon.

2 Study Area and Data Used

The Erlanger Crater, located at 86.94° N 28.62° E with a diameter of 10.94 km, was chosen as the study area for this work. This crater is situated between the much larger craters of Peary and Byrd near the Moon's north pole. Given that a significant portion of Erlanger falls within permanently shadowed regions, there exists a higher probability of finding water ice deposits within this crater compared to other areas. According to [17]'s analysis of the water–ice signature from the Chandrayaan-1 M3 dataset, 3.8% of cold traps have ice exposures, one of which being the Erlanger crater region. In this study Chandrayaan-1 Moon mineralogy mapper (M3) data was used to detect the pure water ice pixels. The M3 is an imaging spectrometer in a push broom mode that worked from the visible into near infrared (0.42–3.0) where highly diagnostic mineral absorption bands occur. The M3 sensor is an image spectrometer that was designed to map and identify the mineral compositions of the lunar surface. Using a push broom mode and a wavelength range of 0.46–2.97 μm in 85 consecutive spectral bands, this hyperspectral sensor records the radiation reflected from the Moon's surface. Its 20–40 nm spectral sampling provides 140–280 m of spatial resolution [19]. For the present study, one strip of M3 dataset M3G20090604T160320_V01_RFL downloaded from the Lunar Orbital Data Explorer (ODE). This dataset is Level 2 data product which is thermally and photometry corrected reflectance dataset captured during OP2C optical period [20] (Fig. 1).

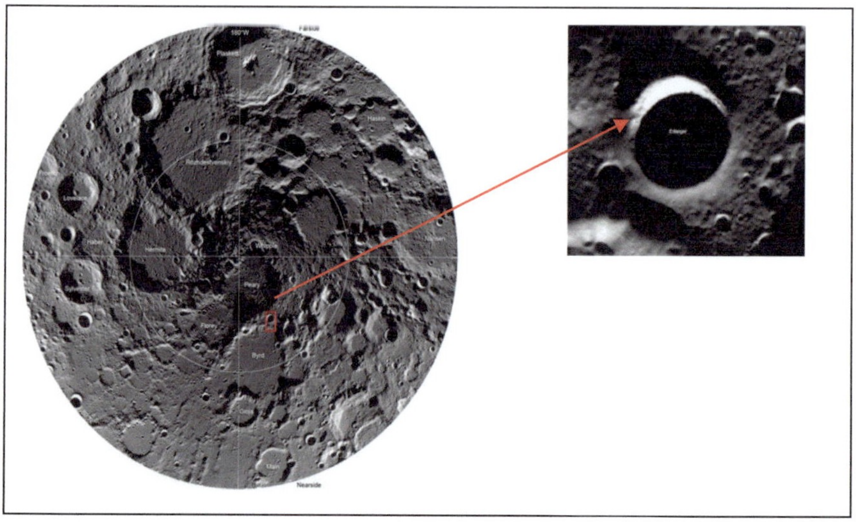

Fig. 1 LROC WAC mosaic of Lunar North Pole. The Erlanger crater is highlighted by a red rectangle

3 Methodology

The flowchart of the proposed work is given in Fig. 2. For the detection of water in the study area, M3 hyperspectral reflectance data was used. The preprocessing of the image has been performed to reduce the noise from the dataset. The reflectance M3 data were subjected to data reductions techniques including de-stripping. In this work, THOR de-striping tool has been used to remove noise stripes from image. ENVI THOR de-striping calculates the mean of every *N*th line and normalizes each one to its respective mean. After that smoothing is performed to remove the noise from the spectra, since the spectra collected from M3 data is noisy because they are collected from permanently darker regions which are not directly illuminated by sunlight but getting reflected sunlight from crater wall and other topographic features. The chosen criteria for this investigation are the identification and characterization of the 1.5- and 2.9-micron diagnostic near-infrared absorption feature of water ice obtained by the Moon mineralogy mapper carried by Chandrayaan-1. After getting the spectra of permanently shadowed regions, the spectra have been compared to the reference JHU (JOHN HOPKINS UNIVERSITY) spectra and SAM angle was calculated between collected spectra and reference spectra using spectral analyst. To further constrain the water ice pixels, SAM classification has been performed which calculates the angle between known spectra and unknown spectra. To classify pure water ice pixels, the minimum noise fraction (MNF) transformation is employed to reduce the dimensionality of hyperspectral data while minimizing noise. The MNF transform is a linear transformation that consists of two successive principal components analysis (PCA) transformations. Once the dimensionality of the data has been reduced using MNF, the next step involves applying the Pixel Purity Index (PPI). The MNF bands serve as input for the PPI, which is used to extract pure pixels from the hyperspectral image. Pixel purity index allows for spatial data reduction. After MNF transformation, a pixel purity score is calculated in the image by randomly generated line in the *n*-D space. After PPI, the carefully chosen pure pixels are used in *n*-D scatter plots. Every axis of this *n*-dimensional plots represents the individual bands.

In this work, this scatter plot is used to extract the pure end member of water ice deposits. After getting endmembers of water ice deposits, its spectra were compared to reference JHU (JOHN HOPKINS UNIVERSITY) spectra for spectral validation. After the validation, the pure end members were selected to perform SAM classification. For this study, the certain maximum angle is set apart in the algorithm that is 0.1 by default, in which small angle denotes a nearer match to the known spectrum. To substantiate the result obtained from SAM target detection method was performed to detect the pure water ice pixel in study area. The target detection wizard provides the step-by-step procedure through the process and gives accurate results. It detects the target on the basis of matched filtering, spectral angle mapper, spectral similarity match, etc. In this study, only spectral angle mapper (SAM) has been considered to detect the target water ice pixel in Erlanger crater.

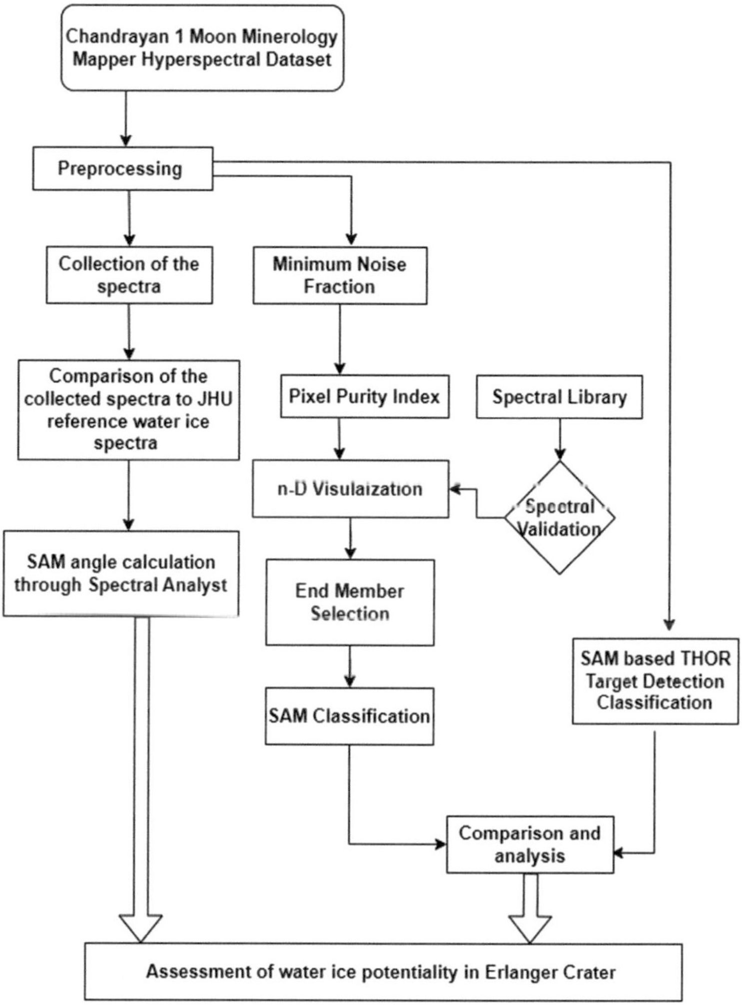

Fig. 2 Flow chart of the methodology, in which input Moon mineralogy mapper datasets

4 Results and Discussions

In this study, spectra were collected from different regions, were analyzed based on their band positions related directly to OH and H_2O abundance. The results obtained from M3 shows specific water ice absorption bands near 1.5 and 2.9 μm in Fig. 3 (Left). The inadequate light source from continually darkened locations causes the water ice pixels' retrieved spectra to be noisy, but the absorption bands depth is reliable with the water ice pixels. The hydration absorption feature has been seen in IR range and the spectra is compared with the JHU (JOHN HOPKINS UNIVERSITY) spectral library reference spectra as shown in Fig. 4a. The collected spectra and the reference spectra are showing somewhat similarity as in both the spectra there are two prominent absorption depth near 5 and 2.9 μm regions. Further, SAM angle is calculated between the collected spectra and JHU (JOHN HOPKINS UNIVERSITY) reference water ice spectra. The SAM value calculated for this pure collected pixel spectra is 0.33 radians which is equal to 19.26°. This value is less than the threshold value (30°) based on literature approach [17]. The potential ice-bearing pixels close to the Erlanger Crater rim are depicted in Fig. 4b's SAM classified image of the research area. Red pixels indicate the classed water ice pixels. The water ice pixels are spread around crater wall and rim. Such absorption feature on the lunar surface is generally associated with OH or other OH bearing minerals like plagioclase, pyroxene, but may indicate H_2O and H_2O ice. SAM is performed for image derived spectra. The certain maximum angle is set apart in the algorithm that is 0.1 by default, in which small angle denotes a nearer match to the known spectrum.

To validate the result obtained from SAM classification, Thor Target Detection tool was used to search for pure water ice pixels. In this target detection method, only SAM has been considered to detect the target water ice pixel in Erlanger crater. The

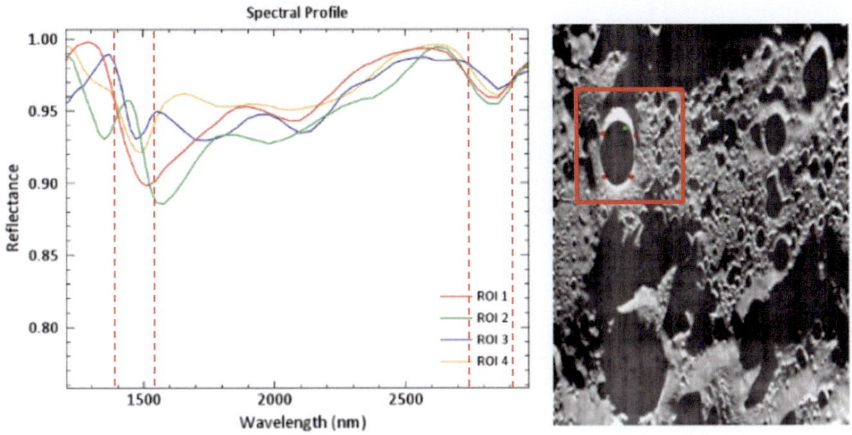

Fig. 3 Lunar spectra collected from Erlanger crater showing significant hydration feature near 1.5 and 2.9 μm (Left). Red and green points under the red rectangle showing the regions from spectra were extracted (Right)

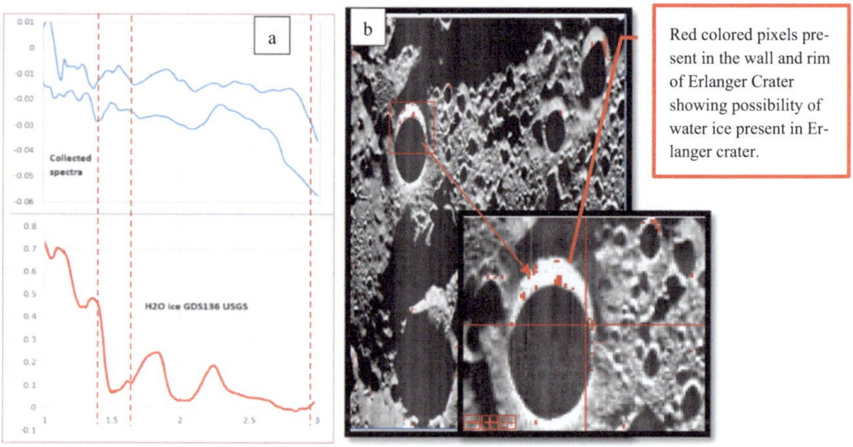

Fig. 4 **a** Comparison of collected spectra with reference reflectance spectra of water ice from USGS library. **b** SAM classified image of Erlanger crater showing water ice pixels, also showing zoomed in view of the crater

result obtained from Thor Target Detection is shown in Fig. 5, in which red color pixels represents the presence of possible water ice molecules near crater rim and inside the crater. In this target detection method, the value of spectral angle mapper (SAM) was considered 0.33 radians according to literature-based approach.

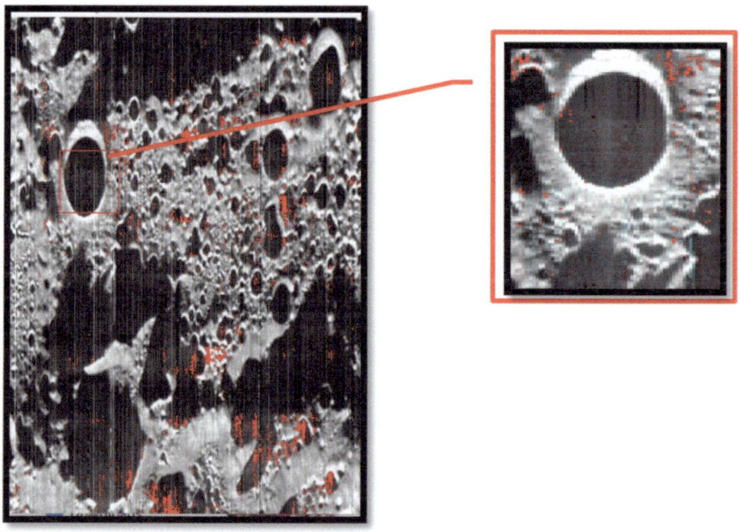

Fig. 5 Thor Target Detection classified image showing red color targeted water ice pixels in crater rim area and inside the crater

As per Fig. 5, the target detection classified image displays a wider spread of potential water ice signatures across the lunar surface, with numerous red-highlighted regions. In contrast, the SSA classified image is more selective, with fewer but more distinct regions marked in red as shown in Fig. 4b. The target detection approach provided a broad identification of potential water ice regions, useful for initial exploration, whereas SSA offered a more precise classification, filtering out false positives and honing in on areas with a higher probability of containing water ice. Utilizing both methods together enhanced the effectiveness of water ice detection on the lunar surface, with target detection giving a preliminary overview and SSA refining the results for more accurate classification.

5 Conclusion

The present work employed spectral angle mapper (SAM) and Thor Target Detection techniques to locate and classify water ice in the permanently shadowed portions of Erlanger Crater at the lunar North Pole. A notable water ice absorption characteristic was identified by spectral analysis of data from the Moon mineralogy mapper (M3) at the 1.5 and 2.9 μm region of NIR bands, despite the challenges posed by weak light sources in the shadowed regions. Comparisons with the JHU (JOHN HOPKINS UNIVERSITY) spectral library further validated these findings, showing strong similarity in the absorption features, particularly near 2.9 and 1.5 μm. The SAM classification yielded reliable water ice pixels, with an angular value of 0.33 radians (19.26°), which is below the literature threshold for accurate water ice identification. The water ice pixels identified by SAM were concentrated around the rim and inner walls of Erlanger Crater. Thor Target Detection further corroborated these findings, identifying additional water ice signatures across the region. The combination of SAM and Thor Target Detection provided a broad and refined classification, with SSA offering more precise identification by minimizing false positives. Overall, this study demonstrates the effectiveness of using both SAM and Thor Target Detection in identifying water ice and volatiles in permanently shadowed regions of the moon. In addition to the findings, it is important to acknowledge the limitations faced in this study, particularly regarding the noisy spectra caused by low light conditions in the permanently shadowed regions of Erlanger Crater. The weak illumination in these areas led to challenges in acquiring clear spectral data, which could affect the precision of water ice identification. This noise may result in potential misclassifications or incomplete detection of water ice and other volatiles.

The integration of these two methods enables a more comprehensive detection strategy, with target detection giving an initial broad mapping and SSA refining the accuracy. These results are crucial for future lunar exploration, where water ice plays a key role in understanding lunar geology and supporting potential in situ resource utilization.

Acknowledgements The authors would like to express their sincere gratitude to ISRO for providing the Chandrayaan-1 M3 dataset and for funding this project through Announcement of Opportunity (AO) grant number ISRO/SPO/CH-2AO/II/3/2022.

References

1. Lawrence DJ (2017) A tale of two poles: toward understanding the presence, distribution, and origin of volatiles at the polar regions of the Moon and Mercury. J Geophys Res: Planets 122(1):21–52. https://doi.org/10.1002/2016JE005167
2. Matsuda S, Iio H, Bajo K, Nagao K (2009) 40th lunar and planetary science conference concentration isotopic ratio He (10 cm/g) Solar Ne/Ne, pp 7–8
3. Chauhan P, Kaur P, Srivastava N, Sinha RK, Jain N, Murty SVS (2015) Hyperspectral remote sensing of planetary surfaces: an insight into composition of inner planets and small bodies in the solar system. Curr Sci 108(5):915–924
4. Lucey PG et al (2014) Journal of geophysical research: planets the global albedo of the Moon at 1064 nm. J Geophys Res: Planets 119:1665–1679. https://doi.org/10.1002/2013JE004592. Received
5. Lemelin M, Li S, Mazarico E, Siegler MA, Kring DA, Paige DA (2021) Framework for coordinated efforts in the exploration of volatiles in the south polar region of the moon. Planet Sci J 2(3):103. https://doi.org/10.3847/PSJ/abf3c5
6. Pieters CM et al (2009) The Moon mineralogy mapper (M3) on Chandrayaan-1. Curr Sci 96(4):500–505
7. Arnold JR (1979) Ar, vol 84, no 9
8. McCubbin FM et al (2021) Endogenous lunar volatiles. Rev Mineral Geochem (New Views of the Moon-2)
9. Maxwell JA, Peck LC, Wiik HB (2016) Chemical composition of Apollo 11 lunar samples 10017, 10020, 10072 and 10084. Geochimica et Cosmochimica Acta Supplement, Volume 1. In: Levinson AA (ed) Proceedings of the Apollo 11 Lunar Science Conference held 5–8 January, 1970 in Houston, TX. Volume 2: chemical and isotope analyses, vol 6, no August. Pergammon Press, New York, p 1369
10. Saal AE, Hauri EH, Lo Cascio M, Van Orman JA, Rutherford MC, Cooper RF (2008) Volatile content of lunar volcanic glasses and the presence of water in the Moon's interior. Nature 454(7201):192–195. https://doi.org/10.1038/nature07047
11. McCubbin FM, Steele A, Hauri EH, Nekvasil H, Yamashita S, Hemley RJ (2010) Nominally hydrous magmatism on the Moon. Proc Natl Acad Sci USA 107(25):11223–11228. https://doi.org/10.1073/pnas.1006677107
12. Robinson KL, Taylor GJ (2014) Heterogeneous distribution of water in the Moon. Nat Geosci 7(6):401–408. https://doi.org/10.1038/ngeo2173
13. Watson K (1961) The behavior of volatiles on the lunar surface, vol 66, no 9, pp 3–4
14. Hurley DM et al (2017) Contributions of solar wind and micrometeoroids to molecular hydrogen in the lunar exosphere. Icarus 283:31–37. https://doi.org/10.1016/j.icarus.2016.04.019
15. Feldman WC et al (2000) Polar hydrogen deposits on the Moon. J Geophys Res: Planets 105(E2):4175–4195. https://doi.org/10.1029/1999JE001129
16. Fisher EA et al (2017) Evidence for surface water ice in the lunar polar regions using reflectance measurements from the lunar orbiter laser altimeter and temperature measurements from the diviner lunar radiometer experiment. Icarus 292:74–85. https://doi.org/10.1016/j.icarus.2017.03.023
17. Li S et al (2018) Direct evidence of surface exposed water ice in the lunar polar regions. Proc Natl Acad Sci USA 115(36):8907–8912. https://doi.org/10.1073/pnas.1802345115

18. Prem P et al (2021) Lunar volatiles and solar system science. Bull AAS 53(4). https://doi.org/10.3847/25c2cfeb.f62324b8
19. Sivakumar V, Neelakantan R (2015) Mineral mapping of lunar highland region using Moon mineralogy mapper (M3) hyperspectral data. J Geol Soc India 86(5):513–518. https://doi.org/10.1007/s12594-015-0341-1
20. Kumaresan PR, Saravanavel J, Palanivel K (2018) Lithological mapping of Eratosthenes crater region using Moon mineralogy mapper of Chandrayaan-1. Planet Space Sci 182(August):2020. https://doi.org/10.1016/j.pss.2019.104817

Beyond Singular Emotions: A Deep Learning-Based Approach for Detecting Multi-Emotion Intensities in Text

Divya Gupta and Shampa Chakraverty

Abstract Human emotions are complex and cannot occur in singularity. They exist in blends and mixes of more than one emotion. Traditionally, the datasets and computational models were built up either for multi-class, multi-label classification, or simple linear regression for emotion prediction. However, we found a scarcity of computational resources that can aid in detecting the presence of more than one emotion and their strength in text. Therefore, a computational approach needs to be devised for quantifying multiple emotions simultaneously. For devising such an approach, we utilize available textual resources, *Tweet Emotion Intensity* (*EmoInt*) and *NRC Affect Intensity Lexicon* (*AIL*), for training individual Bidirectional Encoder Representation from Transformer (BERT) regressors for predicting intensities of *target emotions*, namely, *anger*, *fear*, *joy*, and *sadness*. All the regressors achieved a Pearson Correlation Coefficient r above 0.7, indicating high correlation. Using these regressors, we annotate a multi-label dataset *GoEmotions* to generate a *Multi-Target GoEmotions* (*MTGoEmotions*) dataset. With *MTGoEmotions*, we fine-tune a BERT model to perform multi-output regression and output a vector of *target emotions* intensities. The multi-output regressor performed with a low mean square error MSE of 0.014 and an r of 0.88, indicating very high correlation. Using the intensity vector, we also quantify secondary emotions: *pride*, *guilt*, *envy*, and *despair*.

Keywords Affective computing · Blended emotion · Emotion analysis · Mixed emotion · Multi-output regression · Secondary emotion · Sentiment analysis

D. Gupta (✉)
Department of Computer Science and Engineering, Netaji Subhas University of Technology, New Delhi, India
e-mail: divya.gupta.cs19@nsut.ac.in

S. Chakraverty
Prof., Department of Computer Science and Engineering, Netaji Subhas University of Technology, New Delhi, India
e-mail: shampa@nsut.ac.in

S. Kumar et al. (eds.), *Artificial Intelligence for Next Generation Computing, Volume 1*,
Lecture Notes in Networks and Systems 1549,
https://doi.org/10.1007/978-981-95-0701-6_27

1 Introduction

Human emotions are complex and multifaceted, often manifesting as intricate blends of emotions rather than singular feelings. These blends of emotions, generally termed secondary emotions, exist in two forms: blended and mixed. Secondary blended emotions are combinations of one or more primary emotions, such as anger, fear, and joy. Secondary mixed emotion is a form of blended emotion in which two conflicting emotions exist simultaneously, such as anger-joy, sadness-joy. In today's age, the intersection of human primary and secondary emotions and social media has become increasingly prevalent. The digital realm provides a platform for individuals to express their spectrum of emotions through text. Our continuously evolving online presence has created a unique dynamic where individuals experience a mix of emotions simultaneously. Thus, capturing emotions from our digital interactions can benefit applications based on social media monitoring.

Traditionally, text-based emotion analysis has relied on sentiment analysis, which categorizes emotions as positive, negative, or neutral. However, such simple categorization may not efficiently capture the expressed emotion. Progressively, research endeavors have led to advancements in detecting multiple primary emotions such as anger, fear, and joy [5, 13]. Yet, these methods often overlook the intensities of emotions, missing out on the varying degrees and depths of emotions a text can exhibit. Emotion intensity, a crucial dimension of emotional expression, transcends simple binary labeling to include the potency of emotions, ranging from mild to extremely strong. Therefore, quantifying emotion intensity from text is necessary to unlocking a comprehensive understanding of human emotions.

Different research works have contributed textual resources to the field of emotion quantification from text. These datasets feature text samples for different emotions, which are manually annotated with intensity (ranging from 0–1, 0–100, low to high) [2, 4, 9]. These datasets can be used to train model to predict emotions in singularity. Nevertheless, there is a need of an approach that can identify and predict the intensity of multiple emotions present in a text simultaneously. Therefore, we design a computational approach that predicts the intensity of four *target emotions*, namely, *anger*, *fear*, *joy*, and *sadness*. We use these four emotions because they are fundamental and commonly experienced by people. The major contributions of our work are:

1. Construct a corpus by combining two textual resources *Tweet Emotion Intensity* (*EmoInt*) [9] and *NRC Affect Intensity Lexicon* (*AIL*) [11] to train regression models for *target emotions*, with an enhanced affective understanding.
2. Generate a *Multi-Target GoEmotions* (*MTGoEmotions*) dataset from an existing multi-label text dataset *GoEmotions* [5]. Using this multi-target dataset, fine-tune a computational model to output a vector of *target emotions* intensities for an input text. We also explore inferring secondary emotion intensities from these fundamental emotions.

2 Related Work

Identifying emotions in text has been primarily been treated as either a task of multi-class classification or multi-label classification. Different manually annotated datasets have been proposed to support training computational models for such tasks. Table 1 lists the different datasets proposed for emotion detection in text. These datasets are applicable for *Ekman* emotions (*anger*, *disgust*, *fear*, *joy*, *sadness*, and *surprise*) [6], *Plutchik* emotions (*anger*, *disgust*, *fear*, *joy*, *sadness*, *surprise*, *anticipation*, and *trust*) [12], mixed emotion classification, and valence. Based on the type of machine learning task feasible on a dataset, they can be categorized into multi-class, multi-label, and simple regression datasets. Multi-class and multi-label datasets are used for classification and simple regression datasets for conducting linear regression. In multi-class dataset, an instance is assigned a single class label whereas a multi-label dataset instance is labeled with one or more than one emotion. In simple regression datasets, a numeric value for an emotion can be predicted.

Emotion datasets: Multi-class dataset available for research are *ISEAR*, *Blogs*, *Affective Text*, *GoodNewsEveryone*, and *Data for Emotion Detection* [2, 4, 8, 13, 14]. Scherer and Wallbott proposed *ISEAR* dataset which contained tweets labeled for *Ekman* emotions along with *shame* and *guilt* [13]. A blog-based dataset, *Blogs*, prepared by Aman and Szpakowicz, contained blogs annotated with *Ekman* emotions, *mixed* emotion marking the presence of more than one *Ekman* emotion, and *noemotion* indicating absence of any *Ekman* emotion. Emotion detection datasets such as *Affective Text* and *GoodNewsEveryone*, have also been curated from news headlines

Table 1 List of corpora in English language for emotion detection

Dataset	Size of Dataset	Content	Emotion	Annotation
ISEAR [1994] [13]	7666	Tweet	Ekman + {shame, guilt}	Labels
Blogs [2007] [2]	5025	Blogs	Ekman + {mixed, noemotion}	Intensity (Low, Medium, High)
Affective Text [2007] [14]	1250	News title	Ekman and Valence	Labels for emotions and valence (− 1/0/1)
EmoInt [2017] [9]	7097	Tweets	Anger, fear, joy, sadness	Intensity ([0, 1])
GoodNews Everyone [2020] [4]	5000	News headline	Plutchik	Intensity (Low, Medium, High)
GoEmotions [2020] [5]	58,000	Reddit comments	Ekman	Labels
Data for Emotion Detection [2023] [8]	20,000	Social posts	Ekman	Labels

or titles. *Affective Text* garnered by Strapparava and Mihalcea, contained news titles labeled with *Ekman* emotions and Valence ($-1/0/1$). Bostan et al. compiled *GoodNewsEveryone* from news headlines annotated with *Plutchik* emotions. An emotion value is assigned either of intensity labels *Low*, *Medium*, or *High*. Besides these collections of tweets, news, and blogs, a work by Machova et al. prepared a dataset *Data for Emotion Detection* from social conversation posts on *Kaggle* labeled with one of *Ekman* emotions.

Apart from multi-class datasets, *GoEmotions* dataset proposed by Demszky et al. contained multi-labeled Reddit comments annotated with one or more of *Ekman* emotions [5]. Within simple regression dataset category falls *EmoInt* dataset proposed by Mohammad and Bravo-Marquez, which contained tweets annotated with an intensity ranging [0, 1] of emotions, *anger*, *fear*, *joy*, and *sadness* [9].

Computational models for emotion prediction using EmoInt: As *EmoInt* is the only dataset suitable for emotion regression, we review research works experimenting on it. A regression system by Mohammad and Bravo-Marquez achieved a maximum correlation of 0.71 for any emotion in *target emotions* [9]. For experimenting to build better models, the dataset was experimented under *WASSA-2017 shared task* [10]. Under *WASSA-2017*, Akhtar et al. built an ensemble using Support Vector Regression (SVR), Convolutional neural network (CNN), Long Short-Term Memory (LSTM) network, and Gated Recurrent Unit (GRU) network, achieving a maximum Pearson correlation of 0.748 for any emotion [1]. Another work by Koper et al. under the *task* built a CNN-LSTM regression system achieving 0.72 correlation [7]. Nevertheless, due to the single emotion annotated text samples in *EmoInt*, these studies are limited to quantifying them in singularity.

Noticeable from the above review, available datasets limit the research to emotion classification and linear regression, leading to a dearth of studies predicting the presence of multiple emotions and their strength. Therefore, to bridge this gap, we aim to:

1. Utilize existing textual corpora *Tweet Emotion Intensity* [9] and *NRC Affect Intensity Lexicon (AIL)* [11] to train a deep learning language model for univariate regression for each of our *target emotions*: *anger*, *fear*, *joy*, and *sadness*. Each model predicts an intensity score of their respective emotion in range [0, 1] where 0 is the minimum and 1 is the maximum intensity.
2. Apply emotion regressors on *GoEmotions* corpora for emotion intensity annotation, thus, generating a multi-target dataset *MTGoEmotions*.
3. Employ *MTGoEmotions* to train a deep learning language model to perform multi-target regression on a text input. The model outputs a vector of four real-value scores ranging [0, 1], each representing the intensities of the *target emotions* predicted in a text.
4. Demonstrate an application of the multi-target regression to quantify mixed and blended secondary emotion in a text.

In the next section, we describe in detail the existing textual resources used in our work.

3 Datasets Overview

In this section, we discuss about the three existing textual resources: *Tweet Emotion Intensity*, *NRC Affect Intensity Lexicon*, and *GoEmotions*. We select these datasets to serve our purpose of preparing *MTGoEmotions* for multiple emotion intensity regression. In our approach, *Tweet Emotion Intensity* dataset supplies the text samples annotated with the intensity of our *target emotions*. We augment the volume of annotated samples in this dataset with *NRC Affect Intensity Lexicon* dataset because it accounts for the same set of emotions. This aids in offering a more nuanced comprehension of the affective potency of words while training a model. Another dataset utilized is *GoEmotions*, which encompasses multi-labeled text. It includes text samples that not only express a single emotion but multiple emotions simultaneously. Next, we discuss about these datasets in detail.

Tweet Emotion Intensity (EmoInt): Mohammad and Bravo-Marquez proposed a dataset that contains a total of 7097 English tweets for four primary emotions: *anger*, *fear*, *joy*, and *sadness* [9]. Each tweet is annotated with an intensity score of an emotion, ranging [0, 1]. The dataset contains non-overlapping four subsets of tweets containing train, validation, and test sets, for each emotion. *anger* emotion subset contains 1701 tweets with train set of size 857, validation set of size 84 and test set of 760 tweets, which we represent as 857-84-760. *fear* emotion subset contains 2252 tweets (1147-110-995), *joy* emotion subset contains 1611 tweets (823-74-714), and *sadness* emotion subset contains 1533 tweets (786-74-673).

NRC Affect Intensity Lexicon (AIL): Mohammad designed a dataset that contained nearly 6000 English words entries associated with four basic emotions: *anger*, *fear*, *joy*, and *sadness* [11]. Each word has an associated emotion and intensity score. The intensity score is a real-valued score indicating the intensity of the associated emotion on a scale from 0 to 1. A score of 1 indicates that a word conveys the highest amount of emotion. A score of 0 denotes the lowest amount of emotion in a word. For example, lexicon *Outrage* has associated *Emotion*: *Anger* and *Intensity*: 0.95. This means that the word *Outrage* is strongly associated with the emotion *anger*, with a high strength of 0.95.

GoEmotions: Demszky et al. introduced *GoEmotions* dataset of 58,000 Reddit English comments split into train (43,410), validation (5426), and test (5427) sets [5]. Each comment is multi-labeled with 27 different emotion types and a *Neutral* type. We used the hierarchical clustering of emotion taxonomy at Ekman's level of the dataset. The Ekman level of the taxonomy maps 27 emotion types to 6 *Ekman* emotion categories; the *Neutral* type maps to *Neutral* emotion category. To maintain the consistency with the above two datasets *EmoInt* and *AIL*, we also exclude the text instances containing *disgust* and *surprise* emotion and *Neutral*. After removal of such instances, the dataset contained 23,395 instances in train, 2981 in validation, and 2896 in test set.

In next section, we discuss our methodology in detail.

4 Methodology

In this section, we discuss the workflow of our approach which includes preprocessing text datasets, multi-target emotion dataset generation, and fine-tuning BERT-based models for univariate and multi-output regression.

Dataset preprocessing: Prior to employing the datasets described in the above section, we preprocessed all the text samples except *AIL* dataset. *AIL* dataset contains lexicons and do not require any text preprocessing. Text preprocessing of the two datasets, *EmoInt* and *GoEmotions*, includes removing hashtags, user mentions, punctuation, emoticons, emojis, symbols, pictographs, transport and map symbols, and stopwords.

Hereafter, the datasets *EmoInt* and *GoEmotions* refer to their preprocessed version for any of their mentions in the methodology outlined next.

Outline: Fig. 1 displays the workflow of our proposed methodology for predicting intensities of *target emotions* exhibited in a text. The approach works in three phases, as illustrated in Fig. 1: *Phase I*, highlighted in green; *Phase II*, highlighted in red; and *Phase III*, highlighted in blue. In the first phase *Phase-I*, we construct a training corpus *TC* by combining lexicons from *AIL* dataset and tweets from *EmoInt* dataset. Using *TC*, we train regressors for each of *target emotions*, which combinedly form *EmoAnnotator*, as shown in Container *A*. *Phase-II* uses *EmoAnnotator* to augment *GoEmotions* dataset with intensities of *target emotions*, thus generating a multi-target emotion dataset, *MTGoEmotions*. Further, *Phase-III* uses *MTGoEmotions* to fine-tune a multi-output regressor for predicting an emotion intensity vector for an input text.

We explain the detailed working of each phase in the following subsections.

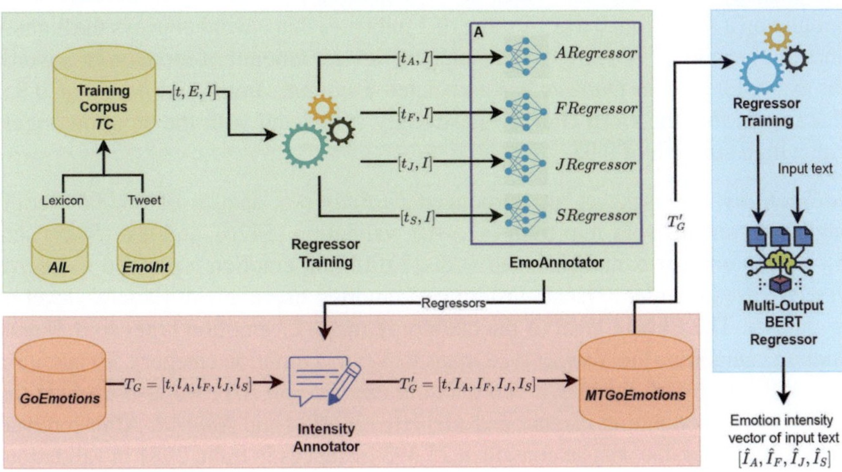

Fig. 1 Workflow of the proposed methodology for predicting multiple emotion intensities expressed in a text

4.1 Phase I-TC Construction and Emotion Regressor Training

In this phase, we prepare a training corpus *TC* by merging two datasets: *AIL*[11] and *EmoInt* [9]. The constructed *TC* contains instances $[t, E, I]$, where t is a text (tweet/lexicon), E is the emotion in t, and I is the intensity of E in t. Over *TC* corpus, we train four regression models: *ARegressor* for *anger*, *FRegressor* for *fear*, *JRegressor* for *joy*, and *SRegressor* for *sadness*. Each regressor is a Bidirectional Encoded Representation from Transformers (BERT)-based model. We use BERT model for its capacity to process text efficiently. It stores the context of a text in classification token $[CLS]$, a pooled output embedding vector of size 768. Further, this CLS token is used for regression task.

For training a BERT-based regression model for an emotion E, we pass only those text instances from *TC* corpus which exhibit E. Thus, we use $[t_A, I]$ for training *ARegressor*, where t_A is the text containing *anger*, and I is the intensity of *anger* in t_A. Similarly, we use *fear* exhibiting instances $[t_F, I]$ to model *FRegressor*, *joy* samples $[t_J, I]$ to train *JRegressor*, and *sadness* exhibiting text $[t_S, I]$ to model *SRegressor*. Next, for fine-tuning a BERT-based regressor, we forward $[CLS]$ token to a dense layer of dimension 768X1 with sigmoid activation. This layer produces a score in range $[0, 1]$, which represents intensity of an emotion. 0 represents lowest intensity and 1 represents highest intensity of an emotion. Some of the hyperparameters each regressor uses, include model optimizer $optimizer$, learning rate lr, batch size b, maximum sequence length len, and dropout rate $dropout$.

After fine-tuning, these four regressors *ARegressor*, *FRegressor*, *JRegressor*, and *SRegressor* combine to form *EmoAnnotator*. A regressor in *EmoAnnotator* independently predicts the intensity of respective emotion in a text. Next, we employ *EmoAnnotator* in *Phase II* to generate *MTGoEmotions* dataset as described in next subsection.

4.2 Phase II-MTGoEmotions generation

In this phase, Module *Intensity Annotator* applies *EmoAnnotator* to transform *GoEmotions* to *MTGoEmotions* dataset. Each text entry $T_G = [t, l_A, l_F, l_J, l_S]$ is transformed to $T'_G = [t, I_A, I_F, I_J, I_S]$, where l_e indicates the binary label of an emotion e and I_e is the intensity score of emotion e in text t. The suffix e denotes *anger* with A, *fear* with F, *joy* with J, and *sadness* with S.

For applying regressors of *EmoAnnotator*, *Intensity Annotator* module utilizes labels (0/1) of *target emotions* in *GoEmotions*. 0 indicates absence and 1 indicates presence of an emotion. For a text annotation, it activates only those emotion regressors for intensity prediction whose respective emotion label is 1. For example, an input text has labels $[0, 1, 0, 1]$ in order of *target emotions*. Since *fear* and *sadness* have labels 1, regressors for *fear* and *sadness* emotion, i.e., *FRegressor* and *SRegres-*

sor will make predictions, while *ARegressor* and *JRegressor* will remain inactive. Being inactive means that the intensities for *anger* and *joy* will be assigned 0. After annotation, we forward the new augmented dataset, *MTGoEmotions*, for training multi-output regressor in *Phase III* described next.

4.3 Phase III-Multi-output Emotion Intensity Prediction

In this phase, we fine-tune a multi-output BERT-based regression model on *MTGoEmotions*. *MTGoEmotions* dataset split is the same as in *GoEmotions* for training, validation, and testing. Each input T'_G includes a text feature t and target vector $[I_A, I_F, I_J, I_S]$. Similar to BERT in *Phase-I*, the CLS token of t is processed to a 768-dimension vector. This vector is next forwarded to a dense layer of dimension 768 X 4 with sigmoid activation. This layer produces a vector $[\hat{I}_A, \hat{I}_F, \hat{I}_J, \hat{I}_S]$ of size 4 where each value is a real-value score ranging [0, 1]. Each score represents the predicted intensity scores for *anger*, *fear*, *joy*, and *sadness*, respectively. Similar to *Phase-I*, 0 represents the lowest and 1 represents the highest intensity of an emotion. For fine-tuning, multi-output regressor uses following hyperparameters: model optimizer $optimizer$, learning rate lr, batch size b, maximum sequence length len, and dropout rate $dropout$.

For model training and testing both in *Phase-I* and *Phase-II*, we use appropriate metrics described in next subsection.

4.4 Metrics Used

In this subsection, we describe the metrics used to measure performance of the regression models trained in our work.

Mean Square Error (MSE): *MSE* often named as *L2* loss, measures the error between the predicted and target values by averaging the squared difference. Squaring the difference heavily penalizes the predictions deviating significantly from target values. Averaging the squared differences normalizes the total error by the number of samples. It can be calculated as:

$$\text{MSE} = \frac{1}{N} * \sum (y_i - \hat{y}_i)^2 \tag{1}$$

where N is the number of samples, y_i is the target value of i-th sample, and $\hat{y}_i$ is the predicted value of i-th sample.

Pearson Correlation Coefficient: Pearson correlation Coefficient, also known as Pearson correlation index, is a statistical measure for linear correlation between two variables, denoted by r. It quantifies the strength and direction of their correlation, ranging between −1 and +1. A value between 0 and 1 indicates a positive correlation between the two variables. A correlation value of 0 indicates no correlation. And, a value between −1 and 0 denotes a negative correlation between the two variables. It can be calculated as:

$$r(y, \hat{y}) = \frac{\sum_{i=0}^{N-1}((y_i - \text{mean}(y)) * (\hat{y}_i - \text{mean}(\hat{y})))}{\sqrt{\sum_{i=0}^{N-1}(y_i - \text{mean}(y))^2} * \sqrt{\sum_{i=0}^{N-1}(\hat{y}_i - \text{mean}(\hat{y}))^2}} \quad (2)$$

where N is the number of samples, y_i is the target value of i-th sample, $\hat{y}_i$ is the predicted value of i-th sample, and $mean(.)$ is the function which calculates mean of a variable.

4.5 *Calculate Secondary Emotions from Intensity Vector*

In this section, we discuss the method used for calculating the intensity of mixed and blended secondary emotion from primary emotions, *anger*, *fear*, *joy*, and *sadness*. For combinations of these four emotions, we refer to Plutchik's emotion dyads [12]. The combinations we measure are: mixed secondary emotion- *Pride* (Anger-Joy), *Guilt* (Fear-Joy), and blended secondary emotion- *Envy* (Anger-Sadness), *Despair* (Fear-Sadness). To quantify these emotions, we use the *Minimum Value method* described by Berrios et al. [3]. According to *Minimum Value method*, the intensity of a secondary emotion can be estimated by taking the lower value between two emotions. For example, if joy is 0.3 and fear is 0.6 in range [0, 1], the value of mixed emotion *Guilt* is 0.3 ($min(0.3, 0.6)$). If one emotion is not felt (e.g., joy is zero), the minimum value is zero, indicating no mixed emotion.

In next section, we present the implementation details and results of our experiment.

5 Experiment and Results

We implemented our system in Python 3.7.9 on a machine with Windows 11 64-bit Operating system, Intel(R) Core(TM) i5-10300H CPU @ 2.50GHz, 16 GB RAM. For building regressors of *EmoAnnotator* module in *Phase-I* and multi-output regressor in *Phase-III*, we used BERT model *bert-base-uncased*. Further subsections present the phase-wise results of the implementation.

Emotion Regressor training in *Phase-I*: Table 2 displays the results of the regressors trained in *Phase-I*. The models mentioned under column *Regression Model* of Table 2 uses following hyperparameter values: $optimizer$=AdamW, lr=5e-5, b=16, len=40, and $dropout$=0.3. The model *FRegressor* marked with * uses the same set of values except for lr of 3e-5. Each regressor was trained for different number of

Table 2 Results of training emotion regressors of *EmoAnnotator* in *Phase-I*

S. No.	Regression model	Training			Testing	
		Number of Epochs	Training Loss	Validation Loss	Mean squaredError (MSE)	Pearson correlation coefficient (r)
1.	ARegressor	1	0.02	0.34	0.01	0.73
2.	FRegressor*	2	0.03	0.4	0.02	0.74
3.	JRegressor	5	0.02	0.22	0.02	0.75
4.	SRegressor	7	0.003	0.32	0.02	0.79

Table 3 Annotated text samples of *anger* emotion in *MTGoEmotions*

S.No.	Text	Intensity
1	Goddamn it...	0.81
2	This is so infuriating	0.74
3	And there's good reason for that. The SD is absolute shit	0.54
4	Shit, I guess I accidentally bought a Pay-Per-View boxing match	0.54
5	If you want nuance, deep thought and constructive dialogue maybe try a topic not as asinine as the cultural value of a reality TV star	0.29

epochs as mentioned under column *Number of Epochs*. *ARegressor* achieved a performance of r=0.73, *FRegressor* of r=0.74, *JRegressor* of r=0.75, and *SRegressor* of r=0.79.

Annotation results in *Phase-II*: Table 3 displays a few text samples of *MTGoEmotions* dataset which exhibit *anger* emotion in varying intensity. Column *Text* list the unprocessed version of text. Column *Intensity* lists the intensity of emotion. As shown in Table 3, there are instances of *anger* with varying intensity in *MTGoEmotions*.

Multi-Output Regressor Training in *Phase-III*: Fine-tuning BERT model for multi-output regression uses the following hyperparameter values: $optimizer$ =AdamW, lr=5e-5, b=16, len=40, and $dropout$=0.3. Figure 2 displays the epoch-wise results of training phase. Figure 2a shows the training and validation loss for each epoch. Figure 2b displays the change in r for each epoch. As shown in Figs. 2a and 2b, the model was appropriately fine-tuned in 2 epochs with a training loss of 0.16, a validation loss of 0.19, and r of 0.89. The trained regressor achieved an r of 0.88 with *MSE* of 0.014 in testing phase.

Demonstration of secondary emotion intensity quantification: To demonstrate an application of the multi-target emotion model, we infer the intensities of mixed and blended secondary emotions based on the predicted intensities of primary emotions. Table 4 displays text sample for each secondary emotion. For example, entry in the first row of Table 4 displays *Pride* emotion, which is a combination of *anger* and *joy*.

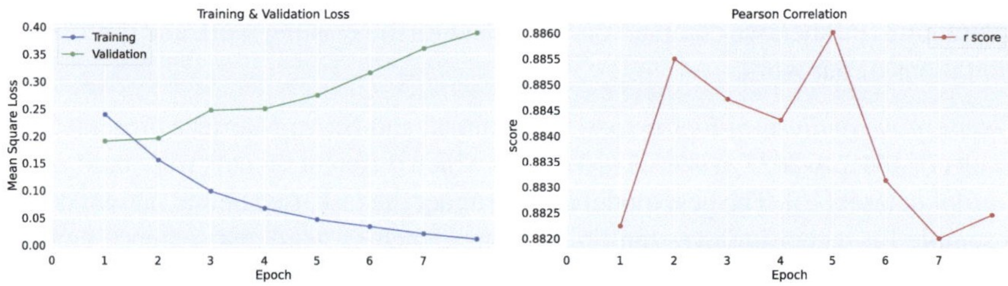

(a) Training and validation loss (b) Pearson Correlation Coefficient

Fig. 2 Results from training phase of multi-output BERT regressor

Table 4 Sample text of *MTGoEmotions* exhibiting secondary emotions

Secondary emotion	Type	Combination	Text	Emotion score
Pride	Mixed	Anger-joy	And they don't give any actual details, just a string of baseless claims. Wonderful.	min (0.32, 0.64) = 0.32
Guilt	Mixed	Fear-joy	Indeed. This episode is for me the best one officially. I'm scared for the finale	min (0.77, 0.41) = 0.41
Envy	Blended	Anger-sadness	Unfortunately, I don't. Sorry	min (0.38, 0.67) = 0.38
Despair	Blended	Fear-sadness	Please no hate it's my first post bro please	min (0.27, 0.43) = 0.27

The text sample for *Pride* exhibits *anger* of intensity 0.32 and *joy* of intensity 0.64. Therefore, using the *Minimum Value* method, the intensity of *Pride* is $min(0.32, 0.64) = 0.32$.

6 Discussion

From the above results, it is evident that BERT-based regression model can be efficiently employed for building an acceptable quality of emotion predictors for annotating *GoEmotions*. Further from *MTGoEmotions*, we inferred that it contained text samples with emotion scores ranging from lower intensity to higher intensity. This denotes a good coverage of text samples, making it suitable for multi-output regressor training. Further, we compare our models with the existing work and discuss the limitations of our work.

Comparative analysis of Regressors: Mohammad and Bravo-Marquez presented the results of *WASSA-2017*, a shared task under which different models were built using *EmoInt* dataset [10]. The best model trained under the task used neural networks and ensemble for regression over multiple features, such as word and sentence embeddings, along with different emotion lexicon datasets. It achieved an r of 0.765 for *anger*, 0.732 for *fear*, 0.762 for *joy*, 0.732 for *sadness*, and an average performance of 0.747. Comparing it with the results of *EmoAnnotator* regressors in Table 2, our models achieved an average r of 0.752. While all models showed comparable performance to the existing models, the model for *sadness* emotion showed an improved performance (0.79 > 0.73). For comparing our BERT multi-output regressor, we could not locate any research work exploring detection of multiple emotions simultaneously. Nevertheless, with an r of 0.88, our final model performs rather well.

Limitation of our methodology: We identified two limitations during the implementation of our approach. One of the limitations was the availability of intensity annotated datasets for a full spectrum of human emotions. Secondly, our approach employs only the text part of the samples for training individual BERT regressors. Future studies may extend our approach to other primary emotions apart from our *target emotions*, provided the availability of appropriate datasets. Furthermore, the performance of the regressors could be improved by including text parts along with different features such as emojis and emoticons.

7 Conclusion

In our work, we successfully trained emotion regressors *ARegressor*, *FRegressor*, *JRegressor*, and *SRegressor* on *AIL* and *EmoInt*. Using these models, we transformed *GoEmotions* to a multi-target dataset, *MTGoEmotions*, annotated with intensities of *target emotions*, namely, *anger*, *fear*, *joy*, and *sadness*. Further, a multi-output BERT regressor was fine-tuned on *MTGoEmotions* dataset for predicting *target emotions* intensities in a text. We also demonstrated the use of multi-output emotion prediction for quantifying secondary emotions. This quantification approach may fill in the need of novel datasets specifically built for secondary emotion intensity prediction. Ability to capture secondary emotion intensities from text can aid in monitoring emotional shift through social media.

References

1. Akhtar MS, Ekbal A, Cambria E (2020) How intense are you? predicting intensities of emotions and sentiments using stacked ensemble. IEEE Comput Intell Mag
2. Aman S, Szpakowicz S (2007) Identifying expressions of emotion in text. Springer, Berlin Heidelberg
3. Berrios R, Totterdell P, Kellett S (2015) Eliciting mixed emotions: a meta-analysis comparing models, types, and measures. Front Psychol
4. Bostan LAM, Kim E, Klinger R (2019) Goodnewseveryone: a corpus of news headlines annotated with emotions, semantic roles, and reader perception. In: International Conference on Language Resources and Evaluation
5. Demszky D, Movshovitz-Attias D, Ko J, Cowen A, Nemade G, Ravi S (2020) Goemotions: a dataset of fine-grained emotions. In: Proceedings of the 58th annual meeting of the association for computational linguistics
6. Ekman P (1992) An argument for basic emotions. Cognition and Emotion
7. Köper M, Kim E, Klinger R (2017) Ims at emoint-2017: emotion intensity prediction with affective norms, automatically extended resources and deep learning. In: Proceedings of the 8th workshop on computational approaches to subjectivity. Sentiment and Social Media Analysis
8. Machová K, Szabóova M, Paralič J, Mičko J (2023) Detection of emotion by text analysis using machine learning. Front Psychol
9. Mohammad S, Bravo-Marquez F (2017) Emotion intensities in tweets. In: Proceedings of the 6th joint conference on lexical and computational semantics
10. Mohammad S, Bravo-Marquez F (2017) Wassa-2017 shared task on emotion intensity. In: Proceedings of the 8th workshop on computational approaches to subjectivity. Sentiment and Social Media Analysis
11. Mohammad SM (2018) Word affect intensities. In: Proceedings of the 11th edition of the language resources and evaluation conference
12. Plutchik R (2001) The nature of emotions: Human emotions have deep evolutionary roots, a fact that may explain their complexity and provide tools for clinical practice. Am Sci
13. Scherer KR, Wallbott HG (1994) Evidence for universality and cultural variation of differential emotion response patterning. J Pers Soc Psychol
14. Strapparava C, Mihalcea R (2007) SemEval-2007 task 14: affective text. In: Proceedings of the fourth international workshop on semantic evaluations

Pixel.ai: Image Enhancement Using Deep Learning Techniques

Soumya Parekh, Ritik Mody, Riya Tasgaonkar, Nikhil Namboodiri, and Kavita Kelkar

Abstract With the explosion in the number of selfies taken daily, the demand for more accurate and visually pleasing images is increasing. However, even images captured by modern cameras are prone to degradation by noise, blur, or loss of color. Although tools like Photoshop can aid in restoring such images, they require significant graphic editing skills and manual interaction. This study proposes an facial image enhancement pipeline to address these challenges through deep learning methods. We explore colorization methods in various color spaces along with denoising and deblurring techniques on images with Gaussian noise and blur. The CNN networks achieve high image quality (PSNR and SSIM) and are paired with appropriate metrics to evaluate colorfulness, noise, and focus measures. All models are combined into a single pipeline, minimizing manual interaction for image enhancement. This creates a seamless experience for users to improve the visual appeal of their photographs.

Keywords Image enhancement · Deep learning · Image colorization · Image deblurring · Image denoising · Image processing

1 Introduction

With the fast-paced digital era and its rapidly evolving technology, it has become easier than ever to capture memories in the form of digital photographs. Although they last longer than their analog counterparts, digital images are also subjected to degradation. There could be many reasons for degradation, including issues in formation (haze, noise, motion, lens blur), post-processing (compression, super-resolution), transmission, or storage. These degradations can occur in many forms, such as color loss or introducing distortions due to noise and blur.

S. Parekh (✉) · R. Mody · R. Tasgaonkar · N. Namboodiri · K. Kelkar
K. J. Somaiya College of Engineering, Mumbai, India
e-mail: remove@gmail.com

S. Kumar et al. (eds.), *Artificial Intelligence for Next Generation Computing, Volume 1*,
Lecture Notes in Networks and Systems 1549,
https://doi.org/10.1007/978-981-95-0701-6_28

Image enhancement refers to the process of correcting or removing certain information in an image according to specific needs. For example, it eliminates noise, reveals blurred details, and adjusts levels to highlight features of an image. However, restoring the original image from the ruined output often involves much human effort and interaction with photo editing software like Photoshop.

This study proposes an image enhancement approach for facial photographs based on deep learning techniques to recover a clean, non-degraded image from a distorted image. The implemented framework would perform enhancements on the image, such as colorization, denoising, and deblurring. To reduce user intervention, the pipeline would automatically detect the degraded aspects of the image and apply restoration techniques in the correct order. This will streamline the process to make facial images more visually appealing and suitable for various applications.

2 Previous Work

2.1 Color Detection and Image Colorization

Image colorization is the process of producing an output image that represents the semantic colors and tones of the greyscale image. Zhang et al. [1] devised a colorization model implemented as a feed-forward pass in a convolutional neural network (CNN). They introduced a new perceptual loss function to encourage the model to generate more saturated colors. This function was combined with traditional L1 loss to ensure that the generated colors were also accurate.

Another approach by Baldassarre et al. [2] implemented a colorization technique using the CIELAB (LAB) color space to represent the input images since it separates the color characteristics from the luminance that contains the main image features. The study trained a deep CNN from scratch using Inception ResNet v2 as a high-level feature extractor. Combining the luminance with the predicted color components ensured a good level of fidelity in the final reconstructed image. However, such colorization models failed to perform well in accurately colorizing more low-level information since they were trained on images with more high-level details. A similar method leveraging feature extraction with a pre-trained Mask-R CNN [3] leads to significantly better colorization results compared to previous approaches.

In 2021, Kumar et al. presented the ColTran [4], a novel approach for diverse high-fidelity image colorization based on self-attention by leveraging the conditional transformer layer. More recent work has also seen a shift toward generative adversarial networks (GANs) [5] using a generative prior network to guide a decoder toward more natural colors.

A color detection process is run to identify if the image is less colorful before applying colorization. For this purpose, we adopt a popular metric proposed by Hasler & Suesstrunk [6]. The colorfulness metric was calculated using the means and standard deviations between the red, green, and blue (RGB) channels of the image, and it achieved above 90% correlation between human evaluation and ground truth.

2.2 Noise Detection and Image Denoising

Noise refers to random artifacts introduced in the image due to variations in color and brightness [7]. Removing these unwanted distortions while preserving the information in the original image is known as denoising.

In 2017, Chuah et al. [8] devised a CNN model to recognize if Gaussian noise is present in an image, along with the extent of the noise. It uses a CNN with Pooling, ReLU activation layers, and Softmax activation function. The evaluation metrics used were training and testing accuracies.

The Convolutional Blind Denoising Network (CBDNet), proposed by Guo et al. [9], includes a CNN architecture designed to be robust on both synthetic and real noisy image pairs. It consists of a noise estimation subnetwork and a non-blind denoising subnetwork. The asymmetric loss function improves the model's generalization ability for real-world noise while performing interactive denoising conveniently.

More recent work by Cheng et al. [10] proposed the Noise Basis Learning Network (NBNet), a model that learns a set of reconstruction basis vectors in the feature space. The input image's features are projected onto a signal subspace, effectively removing noise components.

Niresi and Chi [11] suggests using the Huber Loss function applied to a novel, unsupervised denoising algorithm. It achieved robustness in denoising images with mixed types of noise without any regularization or pretraining.

2.3 Blur Detection and Image Deblurring

Blurring is usually caused by the acquisition of the same information from the scene on different receiver cells. Image deblurring aims to remove those blurring artifacts from images.

Bansal et al. [12] suggested using the Laplacian filter to detect blurred images, using the variance of Laplacian as an evaluation metric. A higher variance indicated a higher focus measure and a lesser blur, and vice versa. Other approaches [13] for blur detection include leveraging Fast Fourier transforms to detect higher-frequency components in the image, but this does not extend well to our case.

In 2016, Dong et al. [14] proposed the Super Resolution CNN (SRCNN) for image super-resolution using a three-step process involving patch extraction, non-linear mapping, and reconstruction. The Deblurring Super Resolution CNN (DBSR-CNN) introduced by Albluwi et al. [15] improved on the architecture using four convolutional layers for feature extraction, enhancement, concatenation, and reconstruction. Both approaches aimed to learn end-to-end mappings between low-resolution and high-resolution images. Nah et al. [16] proposed a blind deblurring neural network for estimating sharp images. It is trained in the multi-scale space to follow a course to fine approach for image super-resolution. These courser scale features aided deblurring while preserving finer information, outperforming many state-of-the-art benchmarks.

Other approaches [17] include predicting a blur kernel accurately reflecting the image blur before the deblurring process and uncertainty-aware methods [18] combining deep residual priors and handcrafted priors when the extent of blur is unknown. More recent work by Chen et al. [19] combines a compact diffusion model with a regression-based approach to generate a prior feature and integrates it from multiple scales for improved deblurring accuracy.

3 Methodology

To reduce the manual task of selecting the enhancement operations, a pipeline was developed to automatically determine which enhancement operation to apply for the given input image. It determines if the input image is grayscale, noisy, and blurred and then applies suitable image enhancement operations to it (Fig. 1).

Initially, the given image is checked for color saturation. If the colorfulness is below a certain threshold, the colorization model restores the image's hue. Post that, the noise detection process is performed. If the image is indicated to have Gaussian noise, it is passed through the denoising network. It was observed that in some instances of extremely noisy images, blurring artifacts were introduced in the image. Hence, blur detection is the last step, which appropriately removes blur if the original image had it or if it was introduced in the previous operations.

3.1 Dataset

This study uses the Flickr Faces HQ (FFHQ) Dataset [20] for training models and evaluating results. The dataset, created by Nvidia Labs [21], consists of 70,000 high-quality PNG images at 1024×1024 resolution and contains considerable variation in

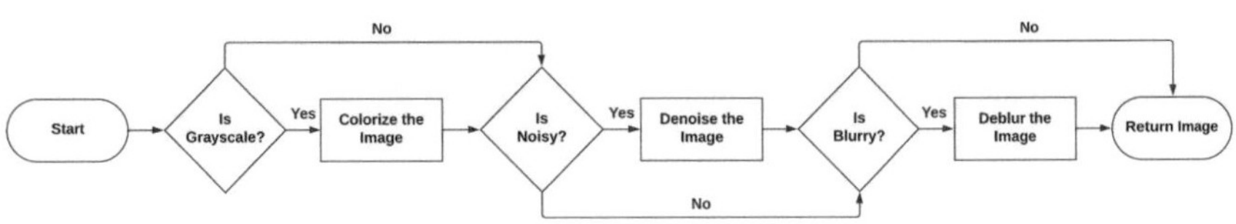

Fig. 1 Flowchart depicting steps in the image enhancement pipeline

terms of age, ethnicity, accessories, and image background. Due to a lack of resources to train models on the entire dataset, a subset of 6000 images at 128 × 128 resolution with an 80:20 train-test split was used.

The original FFHQ dataset is publicly available on the NVLabs website [21]. The image results generated in this study are included in an external repository here.

3.2 Image Colorization

Color Detection We adapt the approach followed by Hasler & Suesstrunk [6] to detect if the image is colored. rg and yb represent the opponent color space in the RGB color channels. After that, the standard deviation σ_{rgyb} and mean μ_{rgyb} are calculated to compute the final colorfulness metric, C.

$$rg = R - G \tag{1}$$

$$yb = \frac{1}{2}(R + G) - B \tag{2}$$

$$\sigma_{rgyb} = \sqrt{\sigma_{rg}^2 + \sigma_{yb}^2} \tag{3}$$

$$\mu_{rgyb} = \sqrt{\mu_{rg}^2 + \mu_{yb}^2} \tag{4}$$

$$C = \sigma_{rgyb} + 0.3 * \mu_{rgyb} \tag{5}$$

Human evaluators predicted the images in our dataset to be colorful when their colorfulness metric was above 5. Hence, $C = 5$ was selected as a threshold for detecting if image inputs were colored.

Color Spaces For the task of image colorization, the following color spaces were considered for implementation:

- RGB: The image is comprised of three color channels: red, green, and blue.
- YIQ: The Y channel represents the luma information, and the I and Q channels represent the chrominance information.
- LAB: The A (green to red) and B (blue to yellow) channels represent the color information, while the L (lightness) channel encodes the intensity of the color.

Model Architecture

Inception ResNet v2 Inception ResNet v2 is a CNN trained on more than one million images of the ImageNet database. The network is 164 layers deep and it can classify

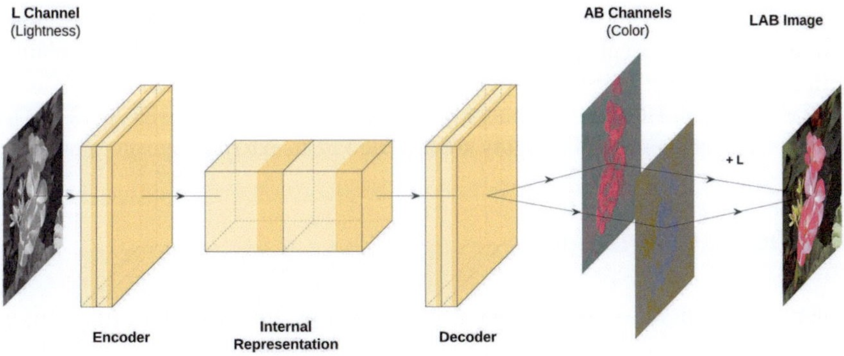

Fig. 2 Zhang et al.'s CNN for colorization based on LAB colorspace

images into several categories such as mouse, pencil, and animal. This knowledge can be transferred to the colorization problem by extrapolating the classification layer of the Inception ResNet and merging it with the CNN layer [2]. However, this approach does not work when applied to the task of human face colorization as it fails to adapt its knowledge to the problem. The majority of the pictures are just colored with a light brown tone, leading to very low accuracy.

Auto Encoder The model used to solve the task of image colorization is implemented using an architecture using an encoder and decoder [1]. The L channel is used as the grayscale input to the network, which the encoder passes through convolution and downsampling layers to create a feature map. The decoder passes this vector through convolution and upsampling layers to expand it back into its original shape and predict the A and B channels. Finally, these predictions are combined with the original L channel to output a LAB representation of the image, which can then be converted to RGB and downloaded after post-processing (Fig. 2).

Of the above-mentioned color spaces, the best results were achieved from the LAB color space, although similar experiments were performed on the RGB and YIQ color spaces as well. In the RGB color space, it was observed that images lost their resolution, and in the YIQ color space, images preserved resolution but were not as vivid. However, the LAB color space preserved the sharpness and vivid hues of the original image.

The model was trained for 30 epochs on the dataset with the Adam optimizer using mean squared error (MSE) as the loss function and accuracy as an evaluation metric. It achieved a training accuracy of 63.45% and a validation accuracy of 59.03%, which is better at colorizing lower-level details of the face compared to the ResNet (Table 1).

Table 1 Comparison of deep learning colorization model with other methods

Method	PSNR	SSIM
Instance aware [3]	27.56	0.93
Generate color prior [5]	21.81	0.88
Proposed approach	23.90	0.77

3.3 Image Denoising

Data Preparation A noisy image dataset was prepared for this task by synthetically adding Gaussian noise to images. Randomly varying the range of variance (1% to 10%) and noise factor (0.3 to 0.7) helped combat overfitting in the deep learning models.

Noise Detection Based on Chuah's [8] approach to detecting Gaussian noise in images, a shallow CNN was developed with max pooling and the sigmoid activation function to classify inputs as 0 (not noisy) or 1 (noisy). The model was trained for ten epochs with the Adam optimizer using binary cross entropy as a loss function. It achieved a training accuracy of 100% and a testing accuracy of 99%.

Model Architecture The image-denoising model proposed by Guo [9] includes a noise estimation network to detect the extent of noise in the image, which is fed to the auto-encoder model for denoising.

A similar architecture was adopted to build the noise estimation CNN with five layers of convolution filters applied to the input image. This output, concatenated to the original input, is transferred to the denoising model. The encoder passes it through convolution and downsampling layers to create a feature map. While creating feature maps, only the essential features of the input image are retained. Hence, the noisy elements of the image are discarded. The decoder passes this feature map through convolution and upsampling layers to expand it back into its original shape and predicts the clean image. To prevent blurring and degradation of images, skip connections [9] were used within the auto-encoder network (Fig. 3).

The model is trained for 30 epochs with the Adam optimizer using MSE as a loss function. While denoising images, it achieves 91.12% training accuracy and 91.24% testing accuracy. To evaluate the developed model, its performance was compared with other commonly used image-denoising algorithms, such as bilateral filtering, Principal Component Analysis (PCA), Total Variation (TV) Regularization, the non-local means algorithm, and wavelet transforms. Peak signal-to-noise ratio (PSNR), MSE, and structural similarity index measure (SSIM) were used as evaluation metrics. For the given task, the proposed model outperformed the standard techniques for all the metrics (Table 2).

Table 2 Comparison of deep learning denoising model with other methods

Method	PSNR	SSIM	MSE
Bilateral filtering	22.321	0.6	381.091
PCA	16.92	0.374	1321.511
TV regularization	24.369	0.713	237.162
NLM algorithm	25.435	0.736	186.014
Wavelet	22.816	0.633	339.974
Proposed approach	27.685	0.836	110.798

3.4 Image Deblurring

Data Preparation The dataset was created by adding random levels of Gaussian blur to the original data. Kernels of sizes 5 and 7 are used to vary the blur strength.

Blur Detection

Fast Fourier Transform Fast Fourier Transform (FFT) depicts the image in both real and imaginary components by separating it into spatial and frequency domains. By examining the frequency distribution, one can ideally classify the image as blurred [13] if there is a low amount of high-frequency components. However, deciding on a radius around the image's center point to zero out the FFT shift was much trickier since it depended on the size of the image in order to rely on a fixed threshold to classify images as blurred. Hence, this approach was not explored further to detect blur in the image.

Variance of Laplacian We explored Bansal's [12] technique to detect blur in the image through a focus measure by computing the variance of Laplacian of input images. Blurred images have a lower variance and focus measure. Human evaluation of blurred images resulted in using a value of 100 as a focus measure. Below this threshold, images are classified as blurred, and vice versa.

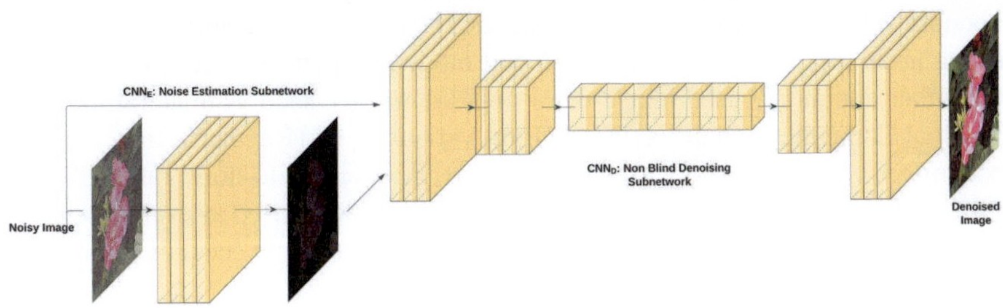

Fig. 3 Guo et al.'s architecture of convolutional blind denoising network

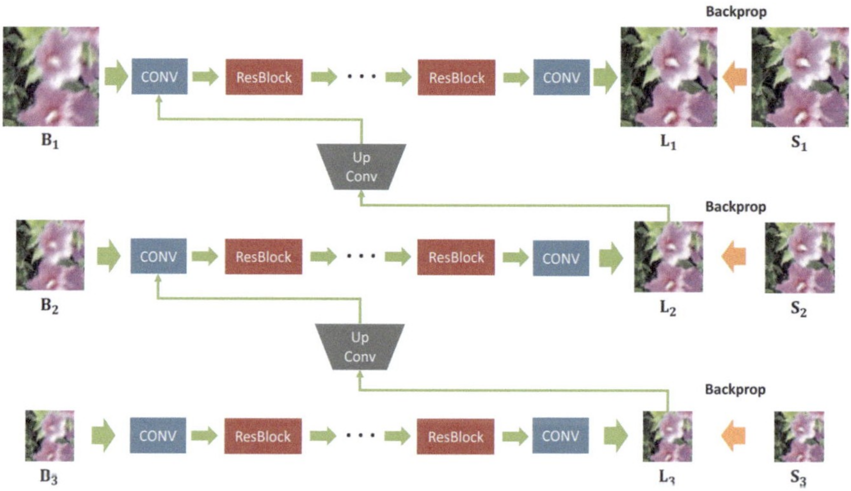

Fig. 4 Nah et al.'s architecture for the deblurring network

Table 3 Comparison of deep learning deblurring model with other methods

	PSNR	SSIM
SRCNN [14]	29	0.70 to 0.85
DBSRCNN [15]	30	0.75
Uncertainty aware [18]	28.64	0.892
Proposed approach	27.46	0.87

$$\text{Laplace}(f) = \frac{\partial^2 f}{\partial x^2} + \frac{\partial^2 f}{\partial y^2} \tag{6}$$

Model Architecture An initial model is constructed on the lines of the work by Dong [14] and Albluwi [15] using a 9-5-1 filter pattern as proposed by the authors. However, this resulted in issues for some images due to colored patches in the generated output. Nah [16] explored another approach to deblur the image using course features and preserve finer details through convolution and deconvolution networks. The encoder generates a feature map, which is a vector containing the features of the image. This feature map is passed through 9 residual blocks. The output of the residual blocks is passed through the deconvolution network to generate an image of the original size (Fig. 4 and Table 3).

4 Results

All deep learning models were combined into a single pipeline to automatically enhance images with multiple degradation issues. The processed images and experimental results are available in an external repository here. It includes experimental results of the image colorization, noise detection, image denoising, blur detection, and image deblurring models. The repository also contains results of images with multiple degradations passed through the composite image enhancement pipeline.

5 Conclusion

Image enhancement is a research problem with critical real-life applications. Exceptional success in deep learning approaches has resulted in the rapid growth of colorization, deblurring, and denoising techniques. Within this project's scope, we have adapted these techniques for facial image enhancement into an automated pipeline for detecting and applying the appropriate corrections in the right order. It creates a seamless and hassle-free experience for the users, unlike other image processing operations, which require human interactions to select appropriate operations and thresholds for tasks. The model can be further improved in future to achieve better results through hyperparameter tuning. It can also be fine-tuned to cater to the needs of specific domains like satellite images [22], CCTV footage [23], and medical sciences [19]. Additionally, the functionality can be extended to other image restoration techniques like inpainting [24] and video restoration as well [25].

References

1. Zhang R, Isola P, Efros AA (2016) Colorful image colorization. In: Leibe B, Matas J, Sebe N, Welling M (eds) Computer vision—ECCV 2016, vol 9907. Springer International Publishing, pp 649–666
2. Baldassarre F, Morín DG, Rodés-Guirao L (2017) Deep koalarization: image colorization using CNNs and Inception-ResNet-v2
3. Su JW, Chu HK, Huang JB (2020) Instance-aware image colorization, pp 7968–7977
4. Kumar M, Weissenborn D, Kalchbrenner N (2021) Colorization transformer. In: 9th International conference on learning representations, ICLR 2021, Virtual Event, Austria, 3–7 May 2021. OpenReview.net
5. Wu Y, Wang X, Li Y, Zhang H, Zhao X, Shan Y (2021) Towards vivid and diverse image colorization with generative color prior, pp 14377–14386
6. Hasler D, Suesstrunk SE (2003) Measuring colorfulness in natural images. In: Human vision and electronic imaging VIII, vol 5007. SPIE, pp 87–95
7. Kim JJ (2022) What is denoising?
8. Chuah JH, Khaw HY, Soon FC, Chow CO (2017) Detection of Gaussian noise and its level using deep convolutional neural network. In: TENCON 2017–2017 IEEE region 10 conference, pp 2447–2450

9. Guo S, Yan Z, Zhang K, Zuo W, Zhang L (2019) Toward convolutional blind denoising of real photographs. In: 2019 IEEE/CVF conference on computer vision and pattern recognition (CVPR), pp 1712–1722
10. Cheng S, Wang Y, Huang H, Liu D, Fan H, Liu S (2021) NBNet: noise basis learning for image denoising with subspace projection, pp 4896–4906
11. Niresi KF, Chi CY (2022) Unsupervised hyperspectral denoising based on deep image prior and least favorable. Distribution 15:5967–5983
12. Bansal R, Raj G, Choudhury T (2016) Blur image detection using Laplacian operator and Open-CV. In: 2016 International conference system modeling and advancement in research trends (SMART), pp 63–67
13. Rosebrock A (2025) Blur detection with OpenCV
14. Dong C, Loy CC, He K, Tang X (2016) Image super-resolution using deep convolutional networks. IEEE Trans Pattern Anal Mach Intell 38(2):295–307
15. Albluwi F, Krylov VA, Dahyot R (2018) Image deblurring and super-resolution using deep convolutional neural networks. In: 2018 IEEE 28th international workshop on machine learning for signal processing (MLSP), pp 1–6
16. Nah S, Kim TH, Lee KM (2017) Deep multi-scale convolutional neural network for dynamic scene deblurring. In: 2017 IEEE conference on computer vision and pattern recognition (CVPR), pp 257–265
17. Zhang K, Luo W, Zhong Y, Ma L, Stenger B, Liu W, Li H (2020) Deblurring by realistic blurring, pp 2737–2746
18. Tang X, Zhao X, Liu J, Wang J, Miao Y, Zeng T (2023) Uncertainty-aware unsupervised image deblurring with deep residual prior, pp 9883–9892
19. Chen Z, Pawar K, Ekanayake M, Pain C, Zhong S, Egan GF (2023) Deep Learning for image enhancement and correction in magnetic resonance imaging-state-of-the-art and challenges. J Digit Imaging 36(1):204–23
20. Karras T, Laine S, Aila T (2019) A style-based generator architecture for generative adversarial networks. In: 2019 IEEE/CVF conference on computer vision and pattern recognition (CVPR), pp 4396–4405
21. NVlabs/ffhq-dataset (2024)
22. Asokan A, Anitha J, Ciobanu M, Gabor A, Naaji A, Hemanth DJ (2020) Image processing techniques for analysis of satellite images for historical maps classification-an overview. Appl Sci 10(12):4207
23. Pillai MS, Chaudhary G, Khari M, Crespo RG (2021) Real-time image enhancement for an automatic automobile accident detection through CCTV using deep learning. Soft Comput-A Fusion Found, Methodol Appl 25(18):11929–11940
24. Xiang H, Zou Q, Nawaz MA, Huang X, Zhang F, Yu H (2023) Deep learning for image inpainting: a survey. Pattern Recognit 134:109046
25. Rota C, Buzzelli M, Bianco S, Schettini R (2023) Video restoration based on deep learning: a comprehensive survey. Artif Intell Rev 56(6):5317–5364

Privacy-Preserving Healthcare Data Analysis Using Blockchain and Homomorphic Encryption

Rudra Krishnasrija, Subhasish Ghosh, Solleti Krishna Chaitanya Subhash, D. Yogesh, and Amit Kr Mandal

Abstract The rapid growth of IoT in healthcare results in large volume of healthcare data, analyzing such data helps in determining the patient health condition, identifying the disease etc. These data are distributed across various places and private in nature. InterPlanetary File System (IPFS) is often used by healthcare providers to securely store the raw data and Content Identifier (CID) is used to share it across different blockchain network. However, if CID is shared then the stored plaintext data will be available which is undesirable for highly sensitive healthcare data. Again, if we analyze data from a single healthcare provider it may result in a biased learning model. Thus, it necessitates robust privacy-preserving solutions for sensitive data sharing without depriving the analytics aspects of it. Traditional techniques based on encryption and decryption are not suitable as it is not scalable and does not support analytics on encrypted data. For this purpose, we proposed a system integrating Homomorphic encryption with blockchain technology to address these challenges. It allows us to securely share the sensitive health data in encrypted format across various healthcare provider network. This heterogeneous data is utilized to create a robust machine learning model while maintaining data privacy. The system is implemented using Hyperledger Fabric and the TenSEAL library. The devised architecture supports multiple Health Service Provider (HSP) networks which utilizes chaincodes and homomorphic encryption for data logging and storing it in IPFS enabling

R. Krishnasrija (✉) · S. Ghosh · S. K. Chaitanya Subhash · D. Yogesh · A. K. Mandal
Department of Computer Science and Engineering, SRM University AP, Amaravati, Andhra Pradesh, India
e-mail: krishnasrija_rudra@srmap.edu.in

S. Ghosh
e-mail: subhasish_ghosh@srmap.edu.in

S. K. Chaitanya Subhash
e-mail: chaitanyasubhash_solleti@srmap.edu.in

D. Yogesh
e-mail: yogesh_d@srmap.edu.in

A. K. Mandal
e-mail: amitmandal@ieee.org

S. Kumar et al. (eds.), *Artificial Intelligence for Next Generation Computing, Volume 1*, Lecture Notes in Networks and Systems 1549,
https://doi.org/10.1007/978-981-95-0701-6_29

privacy preserving decentralized storage. The experimental results demonstrates that the proposed model exhibits high throughput and low latency and also maintains high accuracy while preserving the data privacy. This shows the effectiveness of the proposed approach in enhancing data security and privacy in digital healthcare environments.

Keywords Machine learning · Data privacy · Homomorphic encryption · Blockchain

1 Introduction

The widespread adoption of Internet of Things (IoT) devices in healthcare has resulted in the establishment of extensive sensor networks that continuously gather and upload large volumes of data. This data is crucial for analytics and decision-making processes across various sectors, including health monitoring [1], disease detection [2], governance [3, 4] etc. However, the massive influx of data may come from multiple Health Service Providers (HSP). Therefore, ensuring the integrity of the data collected at various locations is crucial [5]. Further, the stored data poses significant risks of data leakage during the processing. Thus, ensuring the privacy and security of health data is paramount, especially during analytics when data is most vulnerable [6]. In addition, the federal regulations such as PDP, GDPR [3] mandates patient records, especially those pertaining to chronic conditions, require stringent protection due to their sensitive nature [6]. Traditional encryption methods, while effective in securing data, but extremely expensive when it comes to performing computation on encrypted data, which is essential for modern data analytics and machine learning applications [7].

In recent Years, significant advancements have been made in analyzing and securing IoT-based healthcare applications, particularly through the integration blockchain technology and machine learning [1]. Researchers have also explored the use of InterPlanetary File System (IPFS) to enhance the security and efficiency of data storage. For instance, [8] developed a blockchain-IPFS hybrid framework for secure and decentralized data sharing in healthcare, addressing issues such as data integrity and access control. Despite these advancements, scalability remains a challenge, as noted by [3], which highlights that increasing size of the blockchain can result in higher storage and computational costs, potentially limiting its feasibility for some IoT applications. Additionally, the compromise of IPFS hash integrity poses a significant challenge, compromising the security of stored data. A breached hash can lead to unauthorized access or manipulation, threatening the privacy and security of healthcare information [9]. Also few authors developed frameworks that enable secure data sharing and computation while preserving patient privacy. For instance, [10] proposed a homomorphic encryption-based framework to facilitate secure data sharing within IoT healthcare systems and [6] introduced a privacy-preserving blockchain framework using partially homomorphic encryption for secure patient data handling. Similarly, [5] surveyed the potential of combining blockchain and homomorphic encryption to enhance data privacy and ensure secure computations on encrypted data [7]. This research has shown that achieving higher accuracy in homomorphic encryption is challenging, particularly due to the complexity of operations and the overhead introduced by encryption [11]. Furthermore, it is particularly challenging to achieve high accuracy for textual and image datasets [12] using homomorphic encryption due to the increased computational burden and potential for errors [7]. However, managing the encryption mechanism across various blockchain nodes is a challenging task.

To overcome the identified challenges, we have developed a secure system that utilizes homomorphic encryption for securing sensor data management within Hyperledger-based IoT networks. It ensures integrity of data generated at various location by using blockchain. The distributed ledger technology used in blockchain helps in preventing any infringement of crucial sensor data generated by various Health Service Providers. Whereas, privacy preservation can be attained by utilizing Homomorphic encryption [7] as it facilitates healthcare providers to perform necessary computations on encrypted data. Thus, the crucial health data generated by the sensor will be obfuscated using homomorphic encryption and then stored at IPFS using distributed ledger technology. This process helps in maintaining the confidentiality and integrity of patient information while still being able to perform critical analytics [5]. The experimental results indicate that the model achieves a training accuracy of 82% with a gradual reduction in loss over time. Additionally, the ROC AUC curves across multiple classes show strong predictive performance, with values ranging from 0.81 to 0.96, demonstrating effective classification. This demonstrates the effectiveness of the proposed approach in enhancing data security and privacy in digital healthcare environments.

2 Related Work

In recent years, sensitivity of healthcare data and the rise of IoT devices have heightened the focus on privacy in IoT healthcare applications. This section reviews recent research addressing privacy challenges in this domain.

In this direction, Zareen et al. [9] developed a service model that integrates blockchain and the InterPlanetary File System (IPFS) for Internet of Things (IoT) applications. The model uses IPFS to store credentials of nodes and data in a hashed format, while encrypted hashes are recorded on the blockchain to secure the data. Although blockchain storage is costly, it addresses data tampering and leakage issues that arise with centralized databases. The author proposed a service sharing mechanism and a consensus process with smart contracts to handle disputes effectively. Building on this Hasan et al. [8] proposed a blockchain-based solution to address the shortcomings of centralized systems in managing IoT streaming data. It incorporates proxy re-encryption to maintain data privacy and utilizes the InterPlanetary File System (IPFS) for decentralized storage, which effectively handles large data sizes. But, it still faces challenges related to the integration and interoperability with existing IoT infrastructure and protocols. Dorri et al. [13] used blockchain technology to enhance privacy in IoT information management systems through an integrity verification technique. While this decentralized ledger eliminates reliance on trusted third parties, it suffers from high computational overhead and latency due to intensive cryptographic operations. From the above literature, although IPFS and blockchain provide privacy benefits, vulnerabilities and practical challenges may still compromise the security of IPFS hashes.

To overcome above limitations few literatures employed encryption methods, particularly homomorphic encryption is utilized as a key strategy for safeguarding privacy. The research outlined in the paper by Lin et al. [2] introduced Privacy-Enhanced Data Fusion Strategy (PDFS) for managing sensitive healthcare data from IoT devices. PDFS incorporates a sensitive task classification mechanism, and reward-based task contract design using deep reinforcement learning, and homomorphic encryption-based data fusion. But computational overhead introduced by deep reinforcement learning may impact the system's efficiency and scalability in real-world applications. Liu et al. [10] proposed a homomorphic encryption-based framework designed for secure data sharing within IoT healthcare systems enabling data owners to delegate data computation tasks to untrusted entities. However, their method did not adequately address scalability and efficiency challenges associated with homomorphic encryption [14]. Few studies employed leveled homomorphic encryption to implement standard machine learning models. For example, Lou and Jiang et al. [12] applied this approach to ResNet-20 for the CIFAR-10 dataset and ResNet-18 for the ImageNet dataset, using a leveled variant of the Fully Homomorphic Encryption scheme.

Also few studies focused on secure data transmission and privacy in IoT networks, employing homomorphic encryption and blockchain technology. Ali et al. [5] developed a blockchain-based database with homomorphic encryption for secure searches and keyword-based access. It includes a secure key revocation mechanism and policy updates. But it does not address scalability challenges for deploying the blockchain data transmission method with many IoT devices, potentially impacting system efficiency. The other work proposed by Ali et al. [1] is a solution for securing healthcare data by leveraging blockchain technology for decentralized storage and transparent transaction records. Loukil et al. [6] introduced PrivDA, a novel approach for privacy-preserving data aggregation in IoT ecosystem. It leverages blockchain technology and employs homomorphic encryption to obfuscate IoT data, enhancing privacy protection. But it does not delve into the specific technical considerations and does not address the issue of interoperability with existing IoT devices and systems. Peng et al. [15] explored the use of blockchain technology and homomorphic encryption to secure data transfer in IoT systems. It emphasizes process of encrypting data and details various blockchain technologies and consensus algorithms, showcasing significant improvements in data transmission accuracy and efficiency. But it did not address deploying the proposed blockchain data secure transmission method in scenarios involving a massive number of IoT devices, which could impact the overall efficiency and performance of the system. Al et al. [4] proposed a permission-based blockchain combined with homomorphic encryption to enhance IoT security and privacy. It addresses the complexity and storage issues in current blockchain-based IoT systems by optimizing data storage and providing lightweight authentication. But this research limits to a framework that relies solely on a single approach, utilizing a permission-based blockchain to optimize data storage via smart contracts. Hence reviewed studies demonstrate advancements in utilizing blockchain and IPFS for enhancing privacy in IoT healthcare applications.

3 System Architecture

The system architecture integrates blockchain, IPFS, and homomorphic encryption to create a secure, decentralized framework for managing healthcare data, ensuring confidentiality, integrity, and accessibility of sensitive medical records. The key components of the architecture are outlined below.

1. **Health Service Provider (HSP)**: Each HSP node, representing entities like hospitals, clinics, or research facilities, is pivotal in generating, processing, and requesting healthcare data. These nodes interact with the blockchain for immutable operation logging and utilize IPFS for secure data storage, ensuring that all patient-related data remains confidential yet accessible.
2. **Blockchain Network**: It provides a secure, immutable platform for recording transactions and data access logs. Smart contracts (Chaincode) facilitate this process by executing the underlying functions required for data retrieval and other operations, ensuring that all actions adhere to predefined rules and permissions. It stores the Content Identifiers (CIDs) of IPFS file records. These CIDs act as unique references, enabling secure and efficient retrieval of the decentralized data stored on IPFS.
3. **InterPlanetary File System (IPFS)**: IPFS is employed to store data in a decentralized manner. Data on IPFS is indexed by a unique Content Identifier (CID), allowing efficient and secure data retrieval.
4. **Node API Server**: This server facilitates secure interactions between the HSP nodes and the blockchain, managed through Hyperledger Fabric APIs. It ensures secure data transactions, handles API calls to the network gateway for analytics and manages communications with IPFS for data storage.
5. **Gateways**: The system employs two distinct gateways to manage data flow and processing. The Fabric Gateway interfaces the blockchain network and the HSP nodes, ensuring seamless communication for data transactions and Chaincode executions. The Network Gateway connects to the analytics server, securely transferring encrypted data for analysis.
6. **Analytics server**: This server performs model analysis on the encrypted data. Leveraging logistic regression models integrated with homomorphic encryption, it enables the secure evaluation of encrypted input data, preserving the

confidentiality of sensitive information while enabling accurate multi-class classification.

The architecture shown in Fig. 1 includes multiple Health Service Provider (HSP) networks, each tailored to specific geographic or operational needs, and employs blockchain with smart contracts (Chaincode) for immutable transaction logging and decentralized data storage through IPFS. Data from HSP networks is encrypted using homomorphic encryption with specific context keys before being securely transmitted to IPFS, where it is stored. The blockchain records metadata associated with data transactions, such as access logs and details, ensuring all operations are traceable and verifiable. Chaincode manages transaction execution following established protocols and permissions, reinforcing data security. This robust framework ensures seamless interaction between components, allowing for secure data processing without decryption and preserving data privacy during analytics or data sharing across nodes, thus providing a comprehensive solution to the challenges in healthcare data management.

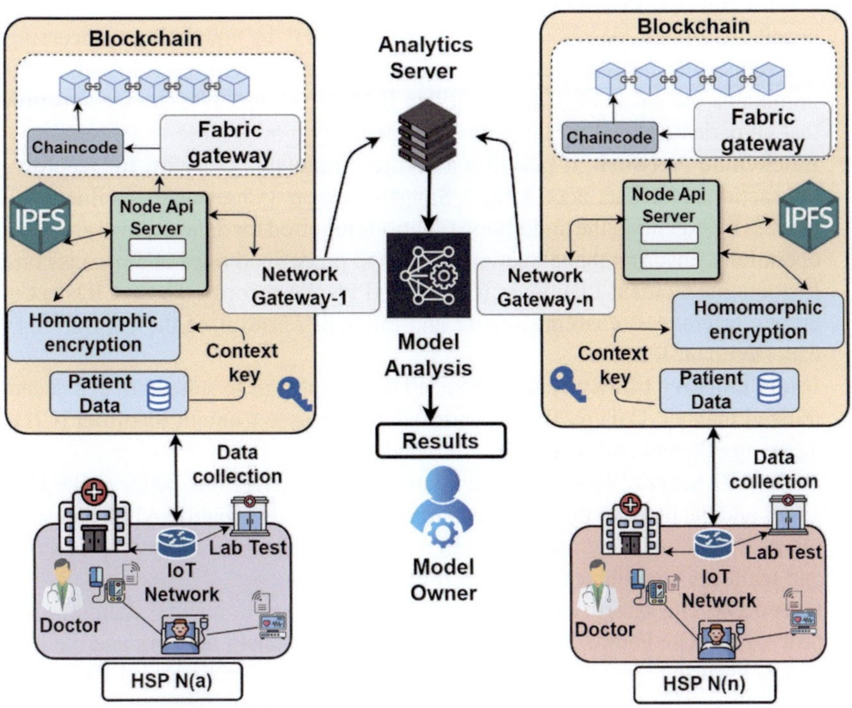

Fig. 1 Privacy-preserving healthcare using blockchain and homomorphic encryption

4 Methodology

This methodology develops a secure multi-class classification model using logistic regression integrated with homomorphic encryption. It involves preprocessing the Texas Hospital Discharge Dataset, training the logistic regression model, optimizing parameters, and applying homomorphic encryption to protect model weights and biases. Secure evaluations on encrypted data ensure privacy while enabling accurate classification. Additionally, the architecture leverages blockchain and decentralized storage solutions to securely manage and share encrypted data across multiple HSP. The detailed steps and integration process are described below.

4.1 Data Preprocessing

The preprocessing steps applied to the Texas Hospital Discharge Dataset [16]. The dataset is initially loaded and rows with missing values are removed. Unnecessary columns like RECORD_ID and PAT_STATUS are dropped. Categorical variables such as Sex Code, Admitting Diagnosis, Principal Diagnosis, and Other Diagnosis Codes are encoded into numerical formats. Occurrence Day and Code are converted into categorical codes, while HCFA MDC is encoded into numerical categories. Ethnicity, Source of Admission, and Race are transformed into categorical codes. Length of Stay, Type of Admission, and APR MDCDRG are also encoded categorically. Illness Severity and Risk Mortality are categorized numerically, with Risk Mortality values further mapped using RISK_MORTALITY_MAP for simplified classification.

These preprocessing steps prepare dataset for analysis, enabling numerical encoding that facilitates machine learning algorithms for predictive analytics.

Data augmentation in the context of the Texas hospital discharge dataset aims to address its highly imbalanced nature as shown in Fig. 2 and enhance its representativeness for analysis and modeling. We have utilized the Conditional Generative Adversarial Network (CTGAN) technique for synthesizing new data points based on the existing dataset structure. Metadata is inferred for column types and ranges. CTGANSynthesizer uses CTGAN to generate synthetic data resembling original data's statistical properties.

4.2 Integration of Homomorphic Encryption Scheme with Logistic Regression Model

The algorithm 1 presented combines the principles of homomorphic encryption with multi-class logistic regression. First, logistic regression model (LR) is initialized with parameters for the number of input features (*n_features*) and output classes

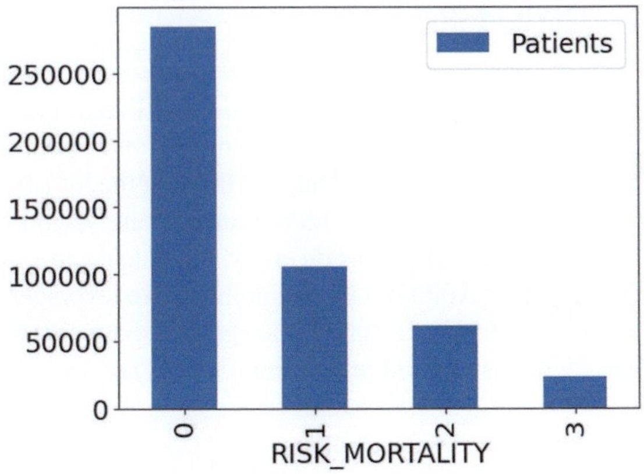

Fig. 2 Distribution of records for each of the four classes in the dataset

(*n_classes*).The LR model uses ADAM optimizer, softmax, and categorical cross-entropy for training iterations. During each epoch, the model computes predictions (*out*) on the training data (x) and calculates the loss (*loss*) against the true labels (y). Gradients of the loss are computed and propagated back through the model using backpropagation to update parameters for minimizing loss and improving predictions. For privacy, the LR model is encapsulated within an EncryptedLR class, initializing and updating weights and biases using homomorphic encryption techniques. Encryption context (*ctx_training*) is established with parameters such as polynomial modulus degree (*poly_mod_degree*) and coefficient modulus bit sizes (*coeff _mod_ bit_sizes*). The context generates necessary keys for secure computation. Before evaluation, the test data (*x_test*) is encrypted using the established encryption context. Each feature vector in *x_test* is transformed into encrypted CKKS vectors (*enc_x_ test*), ensuring that all operations on these vectors maintain their encrypted state. During evaluation, the encrypted LR model computes predictions (*enc_out*) on the encrypted test data (*enc_x_test*). The encrypted predictions are decrypted to plaintext and compared with true labels (*y_test*) to compute accuracy, ensuring that sensitive data remains secure throughout the evaluation process.

The logistic regression model predicts the probability $P(yi|X_i, w, b)$ for each class y_i given input features X_i, weights w, and bias b. This is computed as Eq. 1:

$$P(yi|X_i, w, b) = \text{softmax}(w^T x_i + b) \tag{1}$$

where w^T x_i represents the dot product of weights and input features, and softmax normalizes the outputs to a probability distribution.

Homomorphic encryption uses CKKS scheme for secure addition and multiplication on encrypted data, protecting privacy during machine learning tasks. To integrate logistic regression with homomorphic encryption, the model parameters $\{w_k, b_k\}K_{k=1}$ are initially encrypted into ciphertexts. During inference, given an encrypted input vector $\text{Enc}(x_i)$, the encrypted logistic regression model computes as Eq. 2:

$$\text{Enc}(P(y_i = k|\mathbf{x}_i; w_k, b_k)) = \frac{\exp\left(\text{Enc}(w_k)^T \text{Enc}(x_i) + \text{Enc}(b_k)\right)}{\sum_{j=1}^{k} \exp\left(\text{Enc}(w_j)^T \text{Enc}(x_i) + \text{Enc}\left(b_j\right)\right)} \tag{2}$$

The resulting ciphertext $\text{Enc}(P(y_i = k|\mathbf{x}_i; w_k, b_k))$ can then be decrypted to obtain the predicted probabilities for classification. This formulation ensures that the entire computation from input features to prediction remains encrypted, preserving data privacy throughout the logistic regression process.

4.3 Blockchain and IPFS Integration

To ensure the secure and decentralized sharing of sensitive healthcare data across multiple Health Service Providers (HSPs), the proposed model integrates blockchain technology, specifically Hyperledger Fabric, and IPFS. Each HSP encrypts its local dataset using homomorphic encryption before sharing it with other nodes. The Content Identifiers (CIDs) of these encrypted datasets are stored on IPFS, while the metadata and transaction logs, including data access permissions and operations, are immutably recorded on the blockchain using smart contracts (Chaincode). These smart contracts manage the retrieval of the encrypted data for analysis and ensure that only authorized nodes can access the relevant CIDs from IPFS. This decentralized architecture enables secure, traceable data exchanges and preserves privacy by keeping the data in encrypted form throughout the process, ensuring that both learning and transaction records remain confidential and immutable.

Algorithm 1 Training and Evaluating Encrypted Multi-Class Logistic Regression Model

```
procedure EncryptedLR(torch_lr)                    ▷ torch_lr is pre-trained LR model
    initialize weight and bias from torch_lr
end procedure
procedure FORWARD(enc_x)
    return softmax(enc_x.dot(weight) + bias)
end procedure
procedure BACKWARD(enc_x, enc_out, enc_y)
    Δw += enc_x · (enc_out − enc_y)
    Δb += (enc_out − enc_y)
    count += 1
end procedure
procedure UPDATE_PARAMETERS(η)
    weight −= η · (Δw/count + weight)
    bias −= η · (Δb/count)
    reset Δw, Δb, count
end procedure
procedure PLAIN_ACCURACY(x_test, y_test)
    out ← softmax(x_test.dot(weight) + bias)
    return mean(argmax(out, axis=1) == y_test)
end procedure
procedure ENCRYPT(context)
    weight ← ts.ckks_vector(context, weight)
    bias ← ts.ckks_vector(context, bias)
end procedure
procedure DECRYPT
    weight ← weight.decrypt()
    bias ← bias.decrypt()
end procedure
procedure INITIALIZE_ENCRYPTION_CONTEXT
    poly_mod_degree ← 8192
    coeff_mod_bit_sizes ← [40, 21, 21, 21, 21, 21, 21, 40]
    ctx_training ← ts.context(ts.SCHEME_TYPE.CKKS, poly_mod_degree, -1,
  coeff_mod_bit_sizes)
    ctx_training.global_scale ← 2^21
    ctx_training.generate_galois_keys()
end procedure
procedure ENCRYPT_TEST_SET(x_test)
    t_start ← time()
    enc_x_test ← [ts.ckks_vector(ctx_training, x.tolist()) for x in x_test]
    t_end ← time()
    return enc_x_test, t_end - t_start
end procedure
procedure ENCRYPTED_EVALUATION(model, enc_x_test, y_test)
    for enc_x, y in zip(enc_x_test, y_test) do
        enc_out ← model(enc_x)
        out ← softmax(torch.tensor(enc_out.decrypt()))
        correct += (argmax(out) == y)
    end for
    return correct / len(y_test)
end procedure
```

5 Results and Discussion

In our blockchain system, we utilized a batch size of 100 transactions to optimize the aggregation process before committing them into blocks. This approach was designed to balance latency and throughput, ensuring efficient block generation. A default batch timeout of 2 s was employed to facilitate timely block commits, maintaining system responsiveness. To simulate real-world conditions, we used ten Caliper workers to generate and manage client transactions, providing a robust assessment of the blockchain's performance under concurrent load demands.

Initially, we conducted end-to-end testing with a dataset, encrypting the data using homomorphic encryption and storing the IPFS CID in the blockchain. This initial test was successful, with the encrypted data securely stored on IPFS and, subsequently the CID in the respective blockchain network. However, when we scaled up to a larger dataset, the process demanded significantly more computational resources, making testing with such large-scale data impractical. Consequently, we did not proceed with end-to-end testing on the larger dataset. Instead, we tested the system without homomorphic encryption, evaluating the latency and throughput for READ and WRITE operations within the blockchain network. The throughput for WRITE operations showed an initial linear increase with the transaction send rate escalating from 100 to 400 TPS, but plateaued and slightly declined as the transaction rate reached 600 TPS, indicating system saturation as depicted in Fig. 3a. Despite this, the system efficiently processed a high volume of transactions under increasing loads. In contrast, READ operations exhibited a consistent linear increase up to 1400 TPS, followed by a slight decline in throughput, highlighting system saturation while maintaining high performance as shown in Fig. 3c. This underscores the system's efficiency in handling READ operations with lower computational demand compared to WRITE operations. The latency for WRITE operations revealed a significant increase as the transaction send rate rose, particularly between 400 and 600 TPS, with maximum latency reaching about 35 s at 600 TPS as shown in Fig. 3b. This trend suggests that higher transaction rates exacerbate processing delays, likely due to increased system congestion and limitations in processing capacity. In contrast, latency metrics for READ operations remained consistently low across all tested transaction send rates, with minimal increases observed. Even at the highest rate of 1800 TPS the maximum latency did not exceed 10 s as shown in Fig. 3d. Using a sample dataset of nearly 20,000 EHR records [16], this test was conducted without homomorphic encryption to evaluate the baseline performance of a single hospital blockchain network. From the above observations the blockchain network demonstrates high efficiency and scalability in handling READ operations, it encounters bottlenecks with WRITE operations at higher transaction rates.

Our experiments also assessed the performance of a machine learning (ML) model using homomorphic encryption to ensure privacy throughout both training and inference phases. We specifically trained a logistic regression model on encrypted healthcare data, preserving data confidentiality while evaluating its classification accuracy.

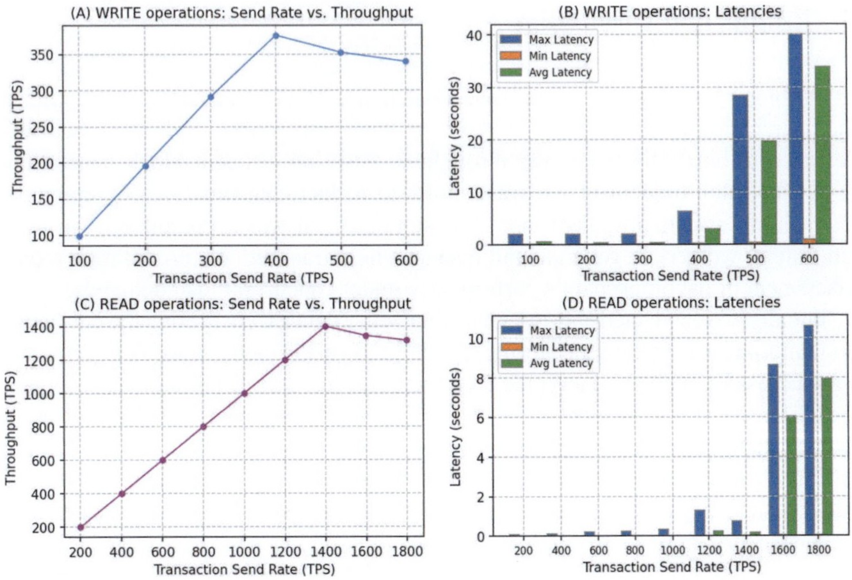

Fig. 3 Performance analysis of READ and WRITE operations in blockchain system

Multiple experiments were conducted across various epochs and classes, demonstrating the model's effectiveness and performance metrics in securely processing and predicting encrypted data. Initially, with homomorphic encryption, it is observed that the graph starts with a high loss, which rapidly decreases as the number of epochs increases, particularly sharp in the initial epochs. This steep descent reflects the model's quick learning from the data. As the training progresses, the loss levels off and reaches a stable low point, forming a plateau of around 20, 000 epochs. This plateau signifies the model's effective learning and convergence to an optimal state as shown in Fig. 4a. The early rapid improvement indicates good initial learning, and the stable plateau suggests the model has successfully learned the patterns in the data and is not overfitting as shown in Fig. 4b. The confusion matrix details the model's accuracy across classes, emphasizing correct predictions along the diagonal, such as 25,880 for class 0 and 21,877 for class 1, and misclassifications off-diagonal, for instance, 2853 instances of class 0 as class 1. It identifies class confusion patterns, essential for optimizing model performance and accuracy as shown in Fig. 4c. The ROC curve assesses the classification model's performance across various classes, where each curve represents a distinct class. AUC values indicate the model's capability to distinguish between positive and negative instances. Class 0 demonstrates excellent discrimination (AUC = 0.98), while Classes 1 and 2 show good discrimination (AUC = 0.93 and 0.91, respectively). Class 3 exhibits excellent discrimination with an AUC of 0.99 as shown in Fig. 4d. Higher AUC values closer to 1 indicate better performance in correctly predicting positive instances.

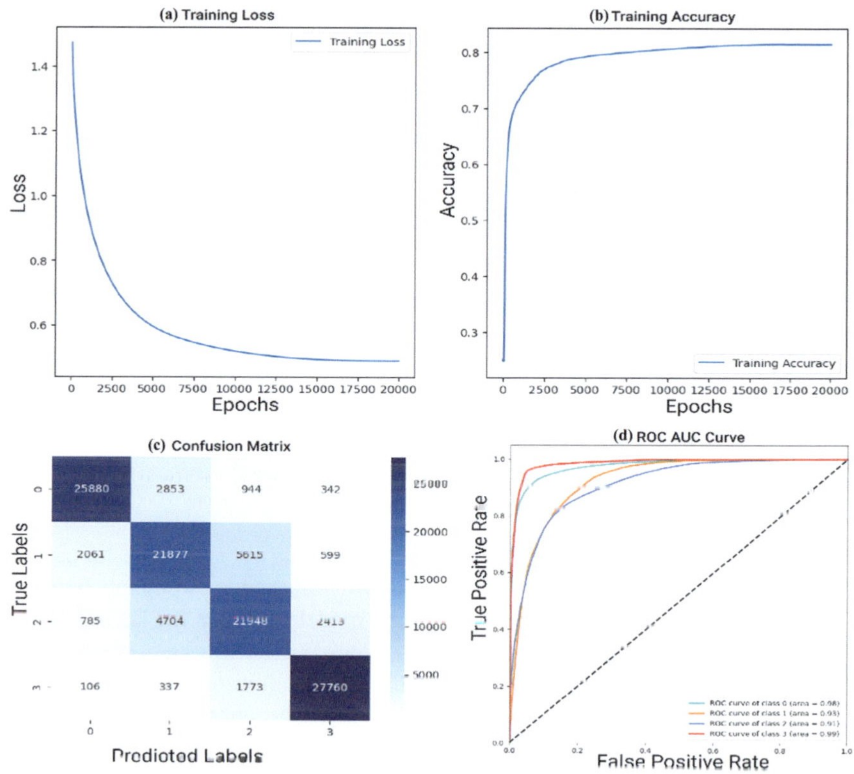

Fig. 4 Evaluation of model performance without homomorphic encryption

Form the observations with homomorphic encryption shows the confusion matrix for the encrypted model results good performance in correctly identifying class 0 and class 3, with high true positives (222 for class 0 and 257 for class 3). However, there are notable misclassifications, particularly class 1 frequently misclassified as class 2 (54 instances) and class 2 misclassified toward both class 1 and class 3 as shown in Fig. 5a. The ROC curve with homomorphic encryption shows the model's performance across classes: Class 0 (AUC = 0.92) demonstrates excellent discrimination, Class 1 (AUC = 0.81) and Class 2 (AUC = 0.83) shows moderate discrimination, and Class 3 (AUC = 0.96) achieves high discrimination as shown in Fig. 5b. These AUC values reveal the model's strong overall performance, with some variation in its ability to differentiate between specific classes.

From the above observations in terms of efficiency and performance, the model without homomorphic encryption shows higher accuracy, lower misclassification rates, and consistent discriminative power across classes. The encrypted model also demonstrates strong capabilities but with trade-offs in accuracy and misclassification rates, reflecting the challenge of balancing data security with model performance in sensitive applications.

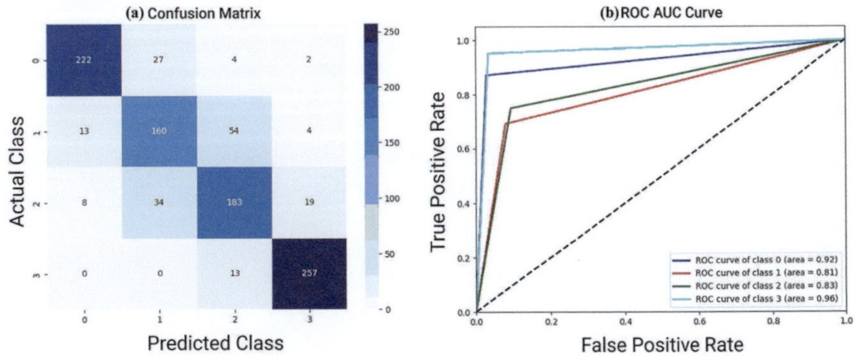

Fig. 5 Evaluation of model performance with homomorphic encryption

6 Conclusion and Future Work

The future work involves optimization of homomorphic encryption techniques to further reduce computational overhead and improve transaction throughput. Further, we aim for transitioning to a federated learning model to reduce the latency. Additionally, we wish to conduct a real-world implementation to assess the practical challenges and the system's performance in different healthcare environments.

The future work involves optimization of homomorphic encryption techniques to further reduce computational overhead and improve transaction throughput. Further, we aim for transitioning to a federated learning model to reduce the latency. Additionally, we wish to conduct a real-world implementation to assess the practical challenges and the system's performance in different healthcare environments.

References

1. Ali A, Al-Rimy BAS, Alsubaei FS, Almazroi AA, Almazroi AA (2023) HealthLock: blockchain-based privacy preservation using homomorphic encryption in internet of things healthcare applications. Sensors 23(15):6762
2. Lin H, Garg S, Hu J, Wang X, Piran MJ, Hossain MS (2020) Privacy-enhanced data fusion for COVID-19 applications in intelligent internet of medical things. IEEE Internet Things J 8(21):15683–15693
3. Bernabe JB, Canovas JL, Hernandez-Ramos JL, Moreno RT, Skarmeta A (2019) Privacy-preserving solutions for blockchain: review and challenges. IEEE Access 7:164908–164940
4. Al Hwaitat AK, Almaiah MA, Ali A, Al-Otaibi S, Shishakly R, Lutfi A, Alrawad M (2023) A new blockchain-based authentication framework for secure IoT networks. Electronics 12(17):3618
5. Ali A, Almaiah MA, Hajjej F, Pasha MF, Fang OH, Khan R, Teo J, Zakarya M (2022) An industrial IoT-based blockchain-enabled secure searchable encryption approach for healthcare systems using neural network. Sensors 22(2):572
6. Loukil F, Ghedira-Guegan C, Boukadi K, Benharkat A-N (2021) Privacy-preserving IoT data aggregation based on blockchain and homomorphic encryption. Sensors 21(7):2452

7. Acar A, Aksu H, Uluagac AS, Conti M (2018) A survey on homomorphic encryption schemes: theory and implementation. ACM Comput Surv (Csur) 51(4):1–35
8. Hasan HR, Salah K, Yaqoob I, Jayaraman R, Pesic S, Omar M (2022) Trustworthy IoT data streaming using blockchain and IPFS. IEEE Access 10:17707–17721
9. Zareen H, Awan S, Sajid ME, Baig SM, Faisal M, Javaid N (2021) Blockchain and IPFS based service model for the internet of things. In: Complex, intelligent and software intensive systems: proceedings of the 15th international conference on complex, intelligent and software intensive systems (CISIS-2021). Springer, Berlin, pp 259–270
10. Liu Y, Luo Y, Zhu Y, Liu Y, Li X (2018) Secure multi-label data classification in cloud by additionally homomorphic encryption. Information Sciences 468:89–102
11. Al Badawi A, Jin C, Lin J, Mun CF, Jie SJ, Tan BHM, Nan X, Aung KMM, Chandrasekhar VR (2020) Towards the AlexNet moment for homomorphic encryption: HCNN, the first homomorphic CNN on encrypted data with GPUs. IEEE Trans Emerg Topics Comput 9(3):1330–1343
12. Lou Q, Jiang L (2019) She: a fast and accurate deep neural network for encrypted data. Adv Neural Inform Process Syst 32
13. Dorri A, Kanhere SS, Jurdak R, Gauravaram P (2017) Blockchain for IoT security and privacy: the case study of a smart home. In: 2017 IEEE International conference on pervasive computing and communications workshops (PerCom Workshops), pp 618–623
14. Vizitiu A, Niţă CI, Puiu A, Suciu C, Itu LM (2020) Applying deep neural networks over homomorphic encrypted medical data. Comput Math Methods Med 2020(1):3910250
15. Peng S, Cai Z, Liu W, Wang W, Li G, Sun Y, Zhu L (2022) Blockchain data secure transmission method based on homomorphic encryption. Comput Intell Neurosci 2022(1):3406228
16. Texas Health. Texas inpatient public use data file (PUDF)|Texas DSHS. https://www.dshs.texas.gov/texas-health-care-information-collection/health-data-researcher-information/texas-inpatient-public-use. Last accessed 04 Feb 2024

An Efficient Hybrid Deep Learning-Based Automatic Feature Engineering and Classification Framework for Smartphone Sensor-Based Human Activity Recognition System

Nurul Amin Choudhury, Badal Soni, and Gyanda Kaushal

Abstract Human Activity Recognition (HAR) is a popular area in sensor technology and smart learning algorithms. HAR is the process of recognizing individual or group activities using a set of sensors and an effective learning algorithm. Most of the research on HAR systemsiseitherbasedondatasetscollectedinacontrolledenvironmentorhas minimal human activities associated with them. This paper proposes an efficient and lightweight hybrid deep learning neural network convolutional neural network—long short-term memory (CNN-LSTM) to recognize human activities in daily living using built-in smartphone sensors. A state-of-the-art wearable sensors-based HAR dataset is developed using the inbuilt accelerometer and gyroscope sensors in an unsimulated environment consisting of 20 user data and 8 daily living human activities. Multiple benchmark deep learning models and an intense data pre-processing pipeline are also designed for model validation and to make the dataset usable for classification. The proposed CNN-LSTM achieved an average performance accuracy of 97%, outperforming all the incorporated benchmark models in optimal computational time margin. Further, the training and validation loss was minimal compared to the benchmark models as they lack the combination of spatial and temporal feature handling. The proposed model attained a lower loss percentage of 0.0977% compared to 0.1235% and 0.2569% with LSTM and DNN benchmark models.

Keywords Human Activity Recognition (HAR) · Activities of daily living (ADLs) · Deep learning · Feature engineering · Signal processing

---

N. A. Choudhury (✉) · B. Soni · G. Kaushal
Department of Computer Science and Engineering, National Institute of Technology, Silchar, Assam, India
e-mail: nurul21rs@cse.nits.ac.in

B. Soni
e-mail: badal@cse.nits.ac.in

S. Kumar et al. (eds.), *Artificial Intelligence for Next Generation Computing, Volume 1*,
Lecture Notes in Networks and Systems 1549,
https://doi.org/10.1007/978-981-95-0701-6_30

1 Introduction

In this advanced technological era, Human Activity Recognition (HAR) plays a crucial role in identifying and classifying human activities in daily life using diverse sensor modalities and state-of-the-art learning algorithms [1, 2]. Human activities in daily living include walking, running, sitting, standing, jumping, and many more. To understand the complexities of our daily activities, we can distinguish these activities between two types of activities: static and dynamic. Static activities are recognized by limited body movements and low energy use, such as standing, sitting and sleeping. On the other hand, dynamic activities involve vigorous body movements and require more energy, and this includes activities like running, jogging, and stairclimbing. In the field of Human Activity Recognition (HAR) wide range of sensors are used, such as accelerometers, gyroscopes, magnetometers, GPS, ambient sensors, biometric sensors and light sensors. These sensors act as the eyes and ears of HAR systems, recording intricate details of human movements and behaviour with optimal precision [1, 3].

With the advancement in the hardware of smartphones after the integration of sensors such as accelerometers and gyroscopes, it has become a popular device used for data collection in the HAR and biometric authentication systems [4]. Inbuilt smartphone sensors such as accelerometers and gyroscopes have become very popular for data acquisition of multiple applications. Various types of learning algorithms are used, ranging from conventional machine learning to cutting-edge deep learning models, while developing HAR systems for precise activity recognition. These algorithms analyse sensor data to detect patterns and attributes related to different activities. This enables accurate prediction and classification. The field of Human Activity Recognition (HAR) has a wide range of applications, including healthcare and biometric authentication. By integrating HAR technology into wearable devices or smartphones, healthcare professionals can monitor patients' body activity and identify irregularities that may indicate health issues. Similarly, HAR plays a vital role in biometric authentication systems, where people's unique body movement patterns can be used as an additional layer of security.

In recent times, the HAR system has seen significant advancements in terms of its learning techniques and sensor modalities. Researchers around the globe incorporate multiple wearable and vision-based sensors for collecting raw sensor data and processing them to make them robust towards feature extraction and activity classification. Machine, ensemble and deep learning algorithms are frequently incorporated into sensor-based HAR systems for activity classification and real-time model generalization. Machine learning models are fast and can learn patterns optimally if adequate training instances are provided to them. Similarly, ensemble models work better in sensor-based HAR systems as they compute results based on result aggregation and maximal voting, which enhances the overall outcome of the HAR system utilizing their base learners. However, both machine and ensemble learning models require manual feature engineering to achieve optimal performance.

On the contrary, deep learning models perform automatic feature engineering using complex neural networks and regularization modules. They also perform effectively for sensor-based HAR systems as demonstrated in multiple state-of-the-artworks [1, 5]. However, we need to ensure the data availability and adequate training and validation instances for optimal pattern generalization. Numerous smartphone sensor-based HAR datasets are available for HAR system development and exploitation, but they are either collected in a simulated environment or have minimal activity count associated with them. Further, state-of-the-art deep learning-based HAR models are complex and take long training computational time for model generalization. To address these issues, this paper proposes an efficient hybrid deep learning-based automatic feature engineering and classification framework for a smartphone sensor-based HAR system. The main novelties and contributions of this paper are as follows-

1. An efficient, lightweight and custom hybrid deep learning neural network (CNN-LSTM) has been designed and developed to recognize daily living human activities using inbuilt smartphone sensors.
2. A state-of-the-art wearable sensors-based HAR dataset is developed using an inbuilt accelerometer and gyroscope sensors in an unsimulated environment consisting of 20 user data and 8 daily living human activities.
3. Multiple benchmark deep learning models and evaluation metrics have been incorporated for training and validation of the proposed CNN-LSTM model.
4. A detailed analytical discussion with a benchmark model comparison has been done for a comprehensive understanding.

2 Relatedworks

Researchers [1, 2, 6, 7] around the globe design and develop HAR systems using multiple ways using diverse learning algorithms and sensor mixes. Challa et al. [7] proposed a hybrid CNN-LSTM model capable of automatic feature extraction from raw sensor data with minimal pre-processing. The incorporation of CNN and LSTM facilitated extracting spatial and temporal information from long sequence data patterns. The proposed model achieved an average performance accuracy of 95% on three benchmark datasets, i.e. WISDM, UCI-HAR and PAMAP2. The authors in [8] proposed a novel Multi-Res-Net for recognizing daily human activities. They incorporated residual neural networks with attention to automatic feature extraction from the sensor data. While comparing the proposed model, the authors managed to outperform the benchmark models with optimal performance margin.

Muralidharan et al. [9] proposed a 1D-CNN along with some machine learning algorithms like Logistic regression, Linear and Kernel SVM, Decision Tree and Random Forest for recognizing daily living human activities. The incorporated models use a dataset retrieved from smartphone accelerometers and gyroscope sensors for the classification of ADLs like—sitting, standing, climbing, etc. Upon in-depth analysis, they intercept that the combination of the SVM classifier with the

1D-CNN feature extractor demonstrates the best performance results. The authors in [10] developed an adaptive batch-sized-based CNN-LSTM framework for recognizing six daily living human activities collected from inbuilt smartphone sensors. The proposed model uses dynamic batch sizes while training the CNN-LSTM model to efficiently capture both short and long-term dependencies between human activities. Incorporating the proposed model, they achieved superior performance in their own and two publicly available HAR datasets.

Jeyakumar et al. [11] presented an end-to-end concept-based explainable activity recognition system for recognizing both higher and lower-level daily living human activities. The proposed model incorporates multiple CNN and LSTM layers for extracting spatial and temporal features with a decoder module for recognition explanation. Likewise, the authors in [12] introduced a novel physique-based HAR system using smartphone sensors. Their methodology segregates the traditional HAR datasets based on the user physiques as they analysed the motion and orientation difference between diverse user sets. Incorporating multiple shallow and ensemble learning models, the proposed methodology achieved an average accuracy of 99% compared to the traditional HAR systems. In [13], the authors combined the evolutionary algorithm with deep learning to build a robust feature extractor for sensor-based HAR systems. They proposed an RCNN-BiGRU for extracting informative features from the raw input data and incorporated a procedure inspired by the behaviour of marine predators (MPA) for selecting sub-optimal features. After statistical comparison, they found that the basic MPA method outperforms the benchmark with optimal performance margin. Choudhury et al. [14] developed an efficient and light weight hybrid deep learning framework for recognizing daily living human activities using smartphone sensors. They minimized the computational time of convolution and LSTM layers for better efficiency and robust feature extraction without hampering the performance of the overall HAR model. Using the minimal hyperparameters and low-cost data segmentation technique, they achieved optimal performance with their own and public datasets.

Gupta et al. [5] proposed a novel hybrid deep neural network model named CNN-GRU that was successfully validated on the WISDM dataset and acquired better performance accuracy compared to the benchmark state-of-the-art deep neural network models. Mukherjee et al. [15] proposed EnsemConvNet, an ensemble of three classification models, namely CNN-Net, Encoded-Net, and CNN-LSTM, for effectively classifying daily living human activities. The incorporated models are built upon simple 1D-CNN but differ in terms of the number of dense layers and kernel size used in the architecture. The authors in [16] proposed a CSI-based HAR model that uses channel state information (CSI) for classifying seven different human activities collected by a Raspberry Pi 4 microcontroller. CSI data is then converted into images fed as inputs to a 2D-CNN classifier. The developed model retained an average performance accuracy of 95%.

3 Proposed Methodology

The proposed methodology of the experimental framework consists of different phases, from inbuilt smartphone sensor calibration to building the custom hybrid deep learning models with evaluation metrics settings as follows.

3.1 Data Collection

Numerous wearable and smartphone sensor-based HAR datasets are available for end-to-end HAR system development and research exploitation. Most of the smartphone sensor-based HAR datasets have multiple features associated with them and comprise multiple user data from diverse ADLs. However, only a handful of datasets have data on different human activities in uncontrolled, real-world environments as they were collected in controlled laboratory environments. To encounter this issue, we collected different ADL data from 20 individuals (aged between 18 $\pm$ 25 years) in a non-simulated real-world environment. As we collected the data in an uncontrolled environment, we managed to collect real-life raw samples of different ADLs performed by the users. Moreover, multiple smartphone modules are used as a data collection module. They were mounted in the front pocket of each participant to collect the data, as shown in Fig. 1.

Each person performed eight daily human activities, such as walking, running, proceeding upstairs, proceeding downstairs, sitting, standing, sleeping and jumping, which were the most frequent routines done by an individual in day-to-day life. Accelerometer and gyroscope are chosen as the inbuilt sensors since they can measure the proper motion acceleration and orientation of individual activities, respectively.

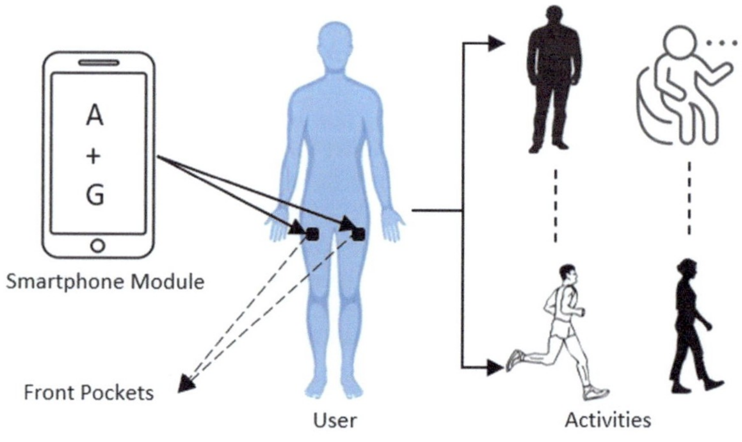

Fig. 1 Data collection module with sensor mounting location of the proposed framework

Table 1 Description of the formulated smart phone sensor-based HAR dataset

Activity	Description	#Samples
Walking	Person walks on a flat surface	63,000
Running	Person runs or jogs on a flat surface	63,000
Standing	Person stand sidle	63,000
Sleeping	Person sleeps or lies down	63,000
Going Upstairs	Person walks downstairs	62,000
Going downstairs	Person walks downstairs	63,000
Sitting	Person sits in one place	63,000
Jumping	Person jumps repeatedly	28,000
Total	–	46,800

The data is collected at a steady rate of 100 Hz, and each individual activity was performed at the user's convenience. The dataset consists of 468,000 samples, as described in Table 1, where each sample consists of 16 raw features, i.e. 3D—Rotational Rate, 3D—Rotational Vector, 3D—Acceleration due to Gravity, 3D—Linear Acceleration, 3D—Gravity in 3D and cos RV(x) gathered using the Sensor-Record android application. The activity-wise data patterns of the generated dataset are shown in Fig. 2.

3.2 Data Pre-processing

Once the data is collected and all the necessary ADLs data is in place, we pre- process it to make it robust towards corrupted artefacts and unwanted data instances. First, a check for any missing and corrupted values in the dataset as they affect dataset quality and make the dataset unable for classification and feature engineering. Upon analysis using pandas and numpy libraries, it was found that the dataset was free from missing and corrupted data instances. Noisy data generated meaningless data patterns and hampers model training while learning new rules for model generalization. We trimmed the dataset by removing the first and last 500 entries for each activity, as these were the noisy data that arose during the mounting and dismounting of the device.

Further, we checked for outliers and redundant data instances in the collected data. As we calibrated the sensor at a higher frequency for data collection, we found a good number of duplicated instances in the raw sensor data, and we dropped them using the panda's library function named drop duplicates. However, upon analysing the dataset with the local outlier factor (LOF), we did not encounter any form of artefacts present in our formulated dataset. Similarly, the issue of class imbalance was also absent in our dataset, except for the jumping class label, as described in the Table 1. We did not solve this minor issue of class imbalance with the jumping

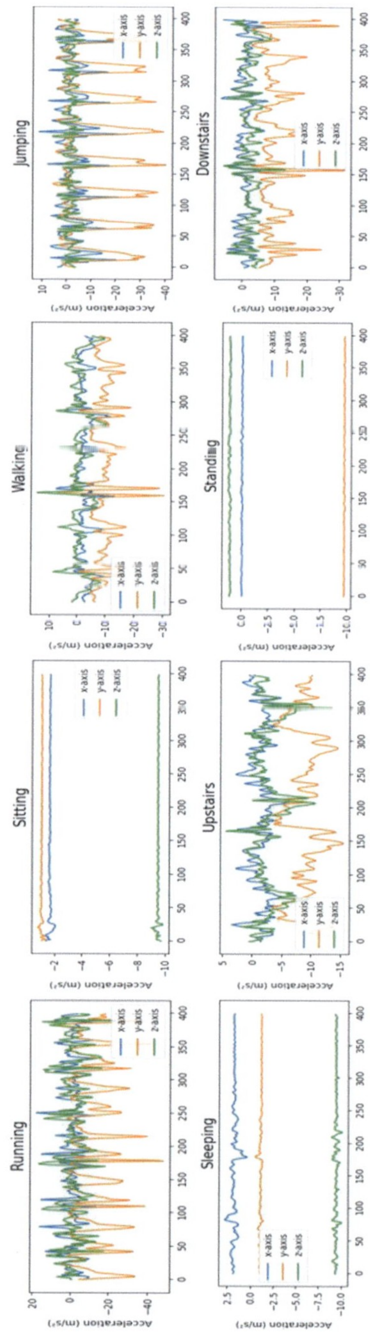

Fig. 2 Activity-wise acceleration due to gravity patterns of the generated dataset

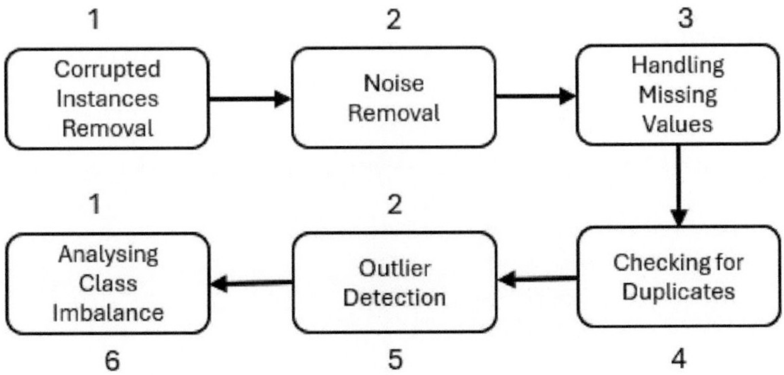

Fig. 3 Data Pre-processing pipeline

activity, as significant data patterns were associated with it. Each entry in the dataset was meticulously labelled to correspond to the specific activity it represented and the subject that performed the activity. Following the labelling process, the labelled data for each activity was aggregated into a single comma-separated value (CSV) file. The detailed data pre-processing pipeline is illustrated in Fig. 3.

3.3 Proposed Model

In this research, a novel hybrid CNN-LSTM framework has been designed and developed for automatic feature engineering and recognition of human activities. CNN and LSTM layers were incorporated to effectively capture spatial and temporal dependencies in between long sequential data. The CNN component comprises three stacked *Time-Distributed Conv-1D* layers with 64 filters, kernel size 3, and *ReLU* activation for adding non-linearity, followed by a *Dropout* layer to prevent overfitting, a *Max Pooling-1D* layer for down sampling the feature set, and a *Flatten* layer to prepare the data for the LSTM. Multiple stacked convolution layers help the model extract intricate features in the raw sensor data.

Further, we added an *LSTM* layer, with dynamic neurons and a *Dropout* layer to capture long-term dependencies in the sequential data. The *memory* unit in the LSTM layer helps the overall feature engineering pipeline by facilitating the powerofmemorizinglong-termdependenciesintherawsensordata.Moreover, dense layers with adjustable neurons and *ReLU* activation are included to add non-linearity and completeness. Then, for the result aggregation, another dense layer with *SoftMax* activation is incorporated into the classification tasks. The entire model was compiled using *Adam* optimizer and *categorical cross-entropy* loss function as illustrated in Fig. 4.

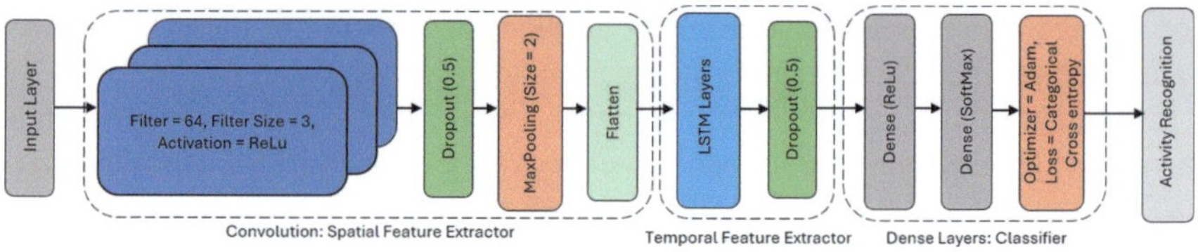

Fig. 4 Architecture of the proposed hybrid CNN-LSTM model

3.4 Benchmark Model and Evaluation Metrics

Along with the proposed CNN-LSTM model, we incorporated multiple benchmark deep learning models such as dense neural network (DNN) and LSTM for performance analysis. These models were chosen as they are frequently incorporated for classifying normal to complex human activities in state-of-the-art sensor based HAR systems [17]. Moreover, multiple evaluation metrics like—Accuracy (A), F1-Score (F1), Recall (R), Precision (P), Testing Loss and Computational Time (C_T) [18] have been incorporated for model effectiveness analysis and comparison.

4 Experimental Result and Discussion

To develop our proposed and benchmark models, we used Google Colab felicitated by Python (version—3.10.2) Scikit-Learn machine learning library, Tensor-Flow (version—2.15.0) and Keras (version—2.15.0) libraries. All the incorporated models were developed and executed on the Google Colab platform, supported by 12.7 GB RAM, 100 GB storage and Tesla T4GPU.

4.1 Performance Analysis of the Proposed CNN-LSTM Model

Totrain, validate and test the proposed and the benchmark models, we split the dataset instances into ratios of 70:10:20, respectively, using stratification of labels. The majority of the data was fed during the model training phase, as these models are data-hungry to facilitate the automatic feature engineering process. The proposed model was trained using the hyper parameters described in the Table 2. The same activation function, neuron count, dropout rate, dense function, and epochs are kept for the DNN and LSTM models to give the same behaviour as the proposed CNN-LSTM model. Upon classification, the proposed model achieved the highest performance accuracy of 97.12% and an average performanceaccuracyof97%. The benchmark models did not manage to acquire optimal teaching accuracy as they overfit during the model training phase. The average performance accuracy is illustrated in the Fig. 5.

DNN and LSTM secure higher testing loss on the test set as they overfit while training. Likewise, other performance metric scores were also lower for DNN and LSTM compared to the CNN-LSTM as they over fit and failed in generalizing input data patterns optimally. The proposed model generalizes the data patterns and achieves optimal performance accuracy of 97.12% with enhanced P, R and F1 scores compared to the benchmark DNN and LSTM models, as described in Table 3.

Table 2 Details of incorporated hyperparameters used in the proposed CNN-LSTM model

Stage	Hyperparameters	Used parameter
Architecture	Convolution1 Filters Kernel size	Conv1D 64 3
	Convolution2 Filters Kernel size	Conv1D 64 3
	Convolution3 Filters Kernel size	Conv1D 64 3
	Dropout1 Max Pooling CNN activation LSTM activation output activation LSTM neuron Dense	0.5 Max Pooling 1D(2) relu Relu softmax 200 200
Training	Epochs Batch size Loss function Optimizer	150 256 Categorical Cross-entropy Adam

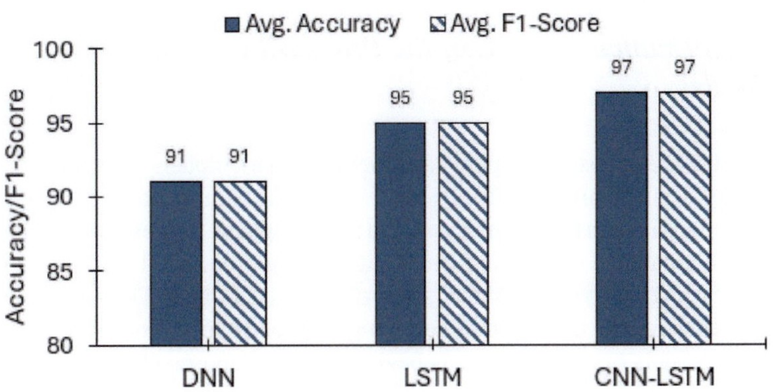

Fig. 5 Average accuracy and F1-score comparison of the proposed model with bench- mark models

Table 3 Detailed performance description of the proposed CNN-LSTM with other incorporated benchmark models

Model	A (%)	Train Loss	P	R	F1	*Ct*
DNN	91.44	0.2569	0.93	0.90	0.91	3150
LSTM	95.94	0.1235	0.95	0.95	0.95	4050
CNN-LSTM	97.12	0.0977	0.97	0.97	0.97	2250

4.2 Efficacy Analysis of the Proposed Model

The proposed model achieved significantly better performance than the bench- mark-modelsasithasbetterextractionabilities.Thestackedconvolutionand LSTM layers help the proposed CNN-LSTM model to extract spatial and temporal dependencies from the long input sequence data. On the other hand, bench- mark models like DNN lack the capability to optimally synchronize spatial and temporal information from the input sequence as they do not have convolution operation on the input data. DNN also lacks grasping long-term dependencies, and they lack memory units for data storage. Due to the above reasons, DNN possesses higher training loss and lower training accuracy than the LSTM and proposed CNN-LSTM model as illustrated in Fig. 6a.

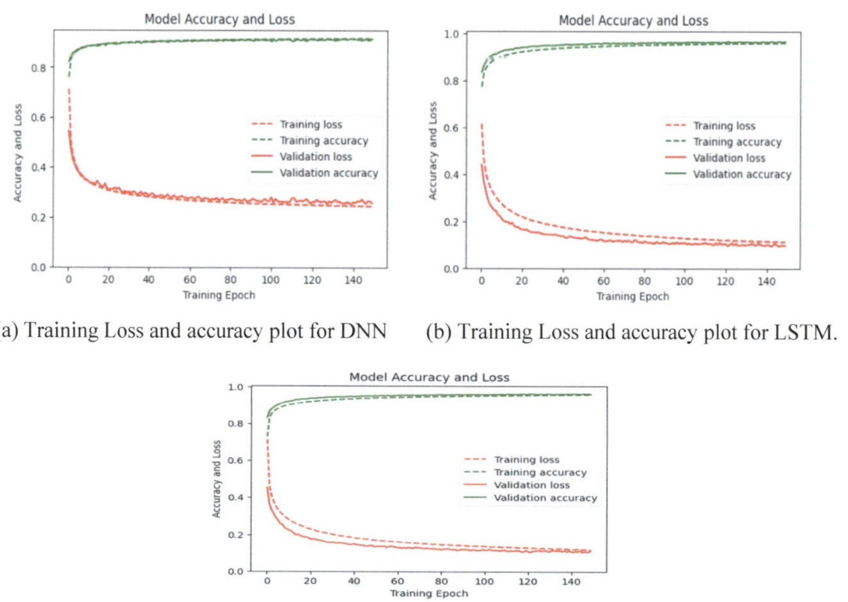

(a) Training Loss and accuracy plot for DNN (b) Training Loss and accuracy plot for LSTM.

(c) Training Loss and accuracy plot for proposed hybrid CNN-LSTM model.

Fig. 6 Accuracy cum loss plot for our proposed model with benchmark models

Further, LSTM successfully handled the long-term temporal dependencies on the raw sensor data but failed to extract intricate spatial information. The lack of a convolution layer degraded the model performance by introducing overfitting, and it managed to achieve an average performance accuracy of 95%, while the training accuracy was 98%. The loss while training was minimal, as shown in Fig. 6b. However, the presence of model overfitting degraded the overall model's performance. On the contrary, the proposed CNN-LSTM's convolution and memory layer optimally extract spatial and temporal information from the input raw sensor data and regularize the musing dropout and max pooling functionalities to prevent overfitting. Due to this, it achieves higher training scores and minimal training and validation loss as illustrated in Fig. 6c. Moreover, the loss on LSTM and DNN was significantly higher compared to the proposed CNN-LSTM, with lower P, R and F1 scores, as described in the Table 3.

5 Conclusion

In this paper, we experimented with the implementation of an efficient hybrid deep learning-based CNN-LSTM framework for automatic feature engineering and classification of sensor-based HAR systems. A state-of-the-art smartphone sensor-based HAR system is generated using an inbuilt accelerometer and gyroscope sensor in an unsimulated real-world environment to ensure good quality data. The dataset has 8 ADLs of 20 different users of different gender and physique types. To make the dataset usable, an intense data pre-processing pipeline has been designed to make the dataset robust towards noise, corrupted data, outliers and many more. The proposed CNN-LSTM model was trained and validated on the generated dataset, and it achieved the highest performance accuracy of 97.12% and an average performance accuracy of 97%. The proposed model also outperformed benchmark models with optimal performance and computational margin.

In future, we will design and develop an enhanced smartphone sensor-based HAR dataset consisting of more ADLs and will deploy the proposed model after further tuning for real-time activity classification.

References

1. Choudhury NA, Soni B (2023) In-depth analysis of design and development for sensor-based human activity recognition system. Multimedia Tools Appl 1–40
2. Ramanujam E, Perumal T, Padmavathi S (2021) Human activity recognition with smartphone and wearable sensors using deep learning techniques: a review. IEEE Sens J 21(12):13029–13040
3. Singh S, Choudhury NA, Soni B (2023) Gait recognition using activities of daily livings and ensemble learning models. In: International conference on advances in IoT and security with AI. Springer, pp 195–206

4. Choudhury NA, Singh S, Soni B (2024) An efficient ensemble framework for human gait recognition using CNN-LSTM with extra tree classifier and smartphone sensors in real-world environment. IEEE Sens Lett 8(9):1–4
5. Gupta S (2021) Deep learning based human activity recognition (HAR) using wearable sensor data. Int J Inf Manag Data Insights 1(2):100046
6. Choudhury NA, Soni B (2022) An efficient CNN-LSTM approach for smartphone sensor-based human activity recognition system. In: 2022 5th International conference on computational intelligence and networks (CINE), pp 01–06
7. Challa SK, Kumar A, Semwal VB (2022) A multibranch CNN-BILSTM model for human activity recognition using wearable sensor data. Vis Comput 38(12):4095–4109
8. Al-qaness MAA, Dahou A, Elaziz MA, Helmi AM (2023) Multi-resatt: multilevel residual network with attention for human activity recognition using wearable sensors. IEEE Trans Ind Inform 19(1):144–152
9. Muralidharan K, Ramesh A, Rithvik G, Prem S, Reghunaath A, Gopinath M (2021) 1d convolution approach to human activity recognition using sensor data and comparison with machine learning algorithms. Int J Cogn Comput Eng 2:130–143
10. Choudhury NA, Soni B (2023) An adaptive batch size-based-CNN-LSTM framework for human activity recognition in uncontrolled environment. IEEE Trans Industr Inf 19(10):10379–10387
11. Jeyakumar JV, Sarker A, Garcia LA, Srivastava M (2023) X-char: a concept- based explainable complex human activity recognition model. In: Proceedings of the ACM on interactive, mobile, wearable and ubiquitous technologies, vol 7, no 1
12. Choudhury NA, Moulik S, Roy DS (2021) Physique-based human activity recognition using ensemble learning and smartphone sensors. IEEE Sens J 21(15):16852–16860
13. Helmi AM, Al-qaness MA, Dahou A, Abd Elaziz M (2023) Human activity recognition using marine predators algorithm with deep learning. Futur Gener Comput Syst 142:340–350
14. Choudhury NA, Soni B (2023) An efficient and lightweight deep learning model for human activity recognition on raw sensor data in uncontrolled environment. IEEE Sens J 23(20):25579–25586
15. Mukherjee D, Mondal R, Singh PK, Sarkar R, Bhattacharjee D (2020) EnsemConvNet: a deep learning approach for human activity recognition using smartphone sensors for healthcare applications. Multimedia Tools Appl 79:31663–31690
16. Moshiri PF, Shahbazian R, Nabati M, Ghorashi SA (2021) A CSI-based human activity recognition using deep learning. Sensors 21(21):7225
17. Amin Choudhury N, Moulik S, Choudhury S (2020) Cloud-based real-time and remote human activity recognition system using wearable sensors. In: 2020 IEEE international conference on consumer electronics—Taiwan (ICCE-Taiwan), pp 1–2
18. Choudhury NA, Soni B (2022) Effect of shallow and ensemble learning models for human activity recognition in uncontrolled environment. In: 2022 IEEE 19th India council international conference (INDICON), pp 1–6

Temporal Price Trend Prediction for Cotton Crop Using Machine Learning

Ritu Garg, Jyoti Ohri, and Sarthak Mittal

Abstract India is a land of agriculture; we are the leading producer of many crops all over the world. It is a major producer of milk, pulses, and jute and is the second largest producer of cotton, sugarcane, vegetables, etc. In today's world where almost everything is digital, only the agricultural sector has no technological advances. India is still showing a slower pace to adapt to more advanced technologies when it comes to agriculture. Machine learning has emerged as a big data technology and is expected to grow rapidly shortly. With the help of machine learning, we can improve the current situation of the poor pricing model of farmers' crops and help them by getting the best Minimum Support Price (MSP) for their crops. One of the majorly produce commodities in India which is raw cotton, it has the most volatile price trend as compared to any other crop in India. Due to this volatile nature of the price trend, our farmers suffer the most and the government can also not decide on an appropriate MSP to support them. In this paper, we proposed an approach for time-series data of the market prices of raw cotton to predict the market price trend using different machine learning models. We have used the Facebook Prophet algorithm for a complete year trend that will help our industrialists as well to determine their future ideology and Random Forest regression for day-wise trends. We have considered absolute percentage error as the evaluation metric. By using both Facebook Prophet and Random Forest, we have got the mean absolute percentage error of 0.08 in Facebook Prophet and 0.04 in Random Forest that indicates the efficacy of the proposed models in price trend prediction for cotton crop.

Keywords Price trend · Time-series dataset · Prophet · Random forest

R. Garg (✉) · J. Ohri · S. Mittal
National Institute of Technology Kurukshetra, Kurukshetra, India
e-mail: ritu.59@nitkkr.ac.in

J. Ohri
e-mail: ohrijyoti@nitkkr.ac.in

S. Kumar et al. (eds.), *Artificial Intelligence for Next Generation Computing, Volume 1*,
Lecture Notes in Networks and Systems 1549,
https://doi.org/10.1007/978-981-95-0701-6_31

1 Introduction

Cotton is one of India's largest produced and cultivated crops which accounts for about 22% of the world's total production. Indian Economy depends on its production and export in a large number of ways including the use of raw cotton from the textile industry to the handloom industry to the medical sector to the cosmetics sector to also feeding the cotton seeds to the cattle as well. Over the years, there has been a significant increase in cotton production just because of increasing demand and greater export. The area which comes under the cultivation of cotton is 4.7% of the total cropped area in the country. It is grown mainly in semi-arid areas in the country. The rainfall level required for its proper growth is 50–100 cm. It requires a high temperature and bright sunshine for its growth and a clear sky during the flowering stage. India is the leading cotton-producing country by defeating China in the race.

Deepa et al. [1] emphasize the cotton cultivation distribution in India. The major contribution toward its cultivation comes from Northern zones including Punjab, Haryana, and Rajasthan. In the central zone, Gujarat, Maharashtra, and Madhya Pradesh are large provinces for cultivating cotton. Karnataka, Andhra Pradesh, and Tamil Nadu also constitute a good percentage of India's total cotton production. In India, the Maharashtra region is best known for its pro cotton shooter. The production amount provided by the state is more than 29.78% of the total cotton production in India. Gujarat is the second largest producer in India, constituting 19.33% of total production. Andhra Pradesh's contribution to production is 12.46% of the complete offering in India. Punjab's production of cotton accounts for 12.42% of complete India's production. Haryana accounts for about 11.91%, followed by Karnataka, Madhya Pradesh, and Tamil Nadu.

The Directorate of Economics & Statistics, Agriculture (DESAg), in the Ministry of Agriculture is the prime source of agricultural statistics. Other sources include the National Sample Survey Organization (NSSO), State's Directorate of Economics and Statistics (DES), and Central Statistical Organization. A major part of this data is time-series data, which measures agriculture prices over changing time frequencies—yearly, monthly, quarterly or daily. These series of prices show trends, along with the seasonality and cyclic nature of agricultural prices. These time-series data associated with agricultural prices can be used to forecast short-term and long-term price estimates.

References [2, 3] emphasize agricultural price forecasting as they are important for determining subsidy rates, minimum support prices, export prices, etc., therefore monitoring and controlling agricultural price volatility and preplanning policies to mitigate any adverse situations. Time series of crop prices can also help to understand, interpret, and explain instances of an economic crisis in the agricultural sector. Moreover, agriculture price forecasts are essential in determining risks in agri-commodity futures and options trading markets.

The current research work, proposed a solution to predict the future market price trend of cotton which will help the farmers, industrialists, and government as well on deciding MSP and selling this at the best price. Working over the time-series data which is a special use-case of temporal data since in time-series, data is collected on a fixed interval of time. Here, we have used the market prices of cotton crop in different markets places across the nation for a fixed period of the last five years. The irregularity in the price trend of the crop in India is due to different weather, diversities, and different sowing timing in the nation.

Also, every farmer wants good MSP from the government, and industrialists also need some price trend to buy their raw materials accordingly. Thus, we have overcome all these issues and provided a reliable future price trend with appreciable accuracy by considering all the factors like area sown, climate, Inflation rate, yield, and month-year. Since the price trend depends on the available data from trusted government sources regarding the complete anatomy of the cotton crop which is irregular, so the problem is much real and we need solutions for the same to help our farmers for their better living.

A lot of research has been done on seasonal time-series forecasts as described by the authors in [4] but due to the unavailability of datasets and low accuracy of models, none of them have predicted the expected results with good accuracy. Thus, we derived the dataset from various government statistics sources [5–7]. The dataset includes factors like fertility of the soil, weather changes, area is sown, and other climatic and historic factors as well. In the current research, we use the auto-regressive Facebook Prophet model along with the Random Forest and Linear Regression model for the future price trend prediction which has been trained on a dataset of more than 250K entries. Facebook Prophet is based on the additive model and works best for time-series data to capture the seasonality trend efficiently.

The outline of the paper is as follows: Sect. 2 shows the complete literature review comprising the learning from different references from other papers along with different attempts performed in the direction of cotton price prediction. Section 3 gives the details about the dataset, starting from collecting dataset to refining, pre-processing to decomposition. Section 4 describes the methodology of the paper including the proposed approach, workflow, and, implementation details of the algorithms used in the research. Section 5 provides an analysis of the observations drawn from the algorithms and Sect. 6 provided the conclusion of the research.

2 Literature Review

The authors in Refs. [3, 7] talked much about cotton cultivation and its usefulness. Cotton is one of the most valuable fiber crops in this world. It has been providing the cotton textile industry with the basic and most important raw material which is cotton fiber. Apart from the textile industry, raw cotton has been required in lot more sectors like cosmetics, medical, handloom, and much more. In all places, raw cotton is in high demand and that's why it has been cultivated on a large scale in India and majorly in tropical and sub-tropical areas. If we talk about India, a total of 80% of the total surplus comes from only three states: Haryana, Punjab, and Rajasthan. Cotton cultivation has been providing the livelihood and daily wages to a larger population of the nation including direct employment to more than sixty lakh farmers and indirectly to around four to five crores of people in its industrial applications.

In the year 2020, India became the third largest exporter of raw cotton in the world, accounting for about 10.2% of total exports. Exports reached the US$6319.09 million by 2020–2021, with $9863.97 million between April 2021 and February 2022. It reached 2.16% and 2.61% of India's total exports in 2020–2021 and between April 2021 and February 2022 respectively.

Cotton is one of the largest produced and cultivated crops in India, it is important to have some method to know about its future market price trend to help the different sectors of the nation.

- The farmers who produce raw cotton.
- Industrialists who will purchase raw cotton for various industries.
- Government has to set MSP to support farmers and regulate the cotton flow in the country and to export as well.

On observing the market price trend of cotton over the last few years in different markets of India, we can interpret that this industry is much more volatile as its prices keep on fluctuating by not providing it's a linear graph. So, we need a solution in this direction to get a method or tool to get some idea regarding future market price trends by using the past year's sales data of different markets across the nation. Due to high diversity, climatic change, and other factors, it also becomes much more important to design a generalized solution for every market price. Thus, the paper approach is to help our farmers and government so that they can be fixed the price of their surplus by looking over the price trend of coming future days as well as to compare the price of their market with any market across the nation and government can easily set the MSP. Datasets have been taken from the government trusted portals [5–7] of past years having all the factors like inflation, the area is sown, and climatic changes and our project will be using various machine learning algorithms and models to make the system more efficient. The paper has used two algorithms, Facebook Prophet and Random Forest to work with the time-series data.

To study in-depth the time-series future market price trend prediction about this particular crop or any crop, several papers were deeply studied and observed as described below: Ref. [8] presents the stock price prediction using Facebook Prophet and Arima Model. The authors deeply researched the time-series data and proposed a solution using Facebook Prophet and Arima Model to predict the stock price trend. It has been proposed that Facebook Prophet and Arima model can be used in forecasting the retail valuation of future stocks that can be used to analyze the future values of stock markets and how they can be varied from previous stock markets.

The research paper [9] presents the application of Facebook's Prophet Algorithm for successful sales forecasting based on real-world data. The authors presented a framework that is capable of accurately forecasting future sales in the retail industry. They used the time-series data for the last 10 years to do its prediction and analyzed the data in terms of the ARIMA model with interventions. Bai et al. [10] have proposed a novel multinomial-based method to improve the greedy splitting process of decision trees to predict more precise and accurate price predictions on the time-series temporal data. Authors in Ref. [11] evaluated a seasonal autoregressive moving average model with an exogenous variable for short-term timber price forecasting. They discussed the process of building the ARIMA, SARIMA, and SARIMAX models. The paper described how time-series models are useful in short-term price forecasting. The incorporation of seasonal fluctuations into a conventional ARIMA model significantly improves forecast accuracy. It concludes that SARIMAX forecasts are the most accurate. Research paper [12] presented the AGMARKNET cotton price forecasting in major producing states. The authors forecast the price of cotton using the ARIMA model. Data collection is done from the AGMARKNET website the data is from January 2006 to December 2016 forecast cotton prices for Kharif 2017–2018 year harvest months. R Programming software is used to estimate model parameters. This model revealed that in the future there is an increase in cotton price and also demand for the crop. Hence, the proper planning should be done. Ghosh et al. [13] evaluated the Indian export price prediction using the ARIMA model which is one of the predecessors of Facebook Prophet. The paper used time-series data to make the prediction but has a few flaws due to the restriction of the ARIMA model. Kumari et al. [14] studied the price trend of cotton in Telangana and concluded that the price is much uncertain due to marketplace irregularity and many variations in farming methods. The price graph of cotton is highly uncertain with sometimes a peak at particular instances and a sudden fall.

3 Dataset

Dataset has been taken from government sources, Ministry of Textiles [5], Cotton Corporation of India [6], and Agmarknet [7]. We have taken around 158,234 tuples all together from these three sources and have taken around 45,890 data entries from different state logistics departments for each market place of cotton across the nation. Also, we have a web-scrapped dataset from many commercial platforms of around 50k. Our dataset has a large number of attributes on whose basis our algorithms are giving appreciable results. The attributes include area sown, inflation, price, yield, day, month, year, city, and state. Thus, all these attributes together enrich our dataset (See Table 1).

Table 1 depicts the exact picture of the dataset used by the models to be trained and to be tested as well. It consists of different attributes for considering all the seasonality and diversity factor into the forecast. It comprises the day, month, and year of the crop to make it a valid time series. Along with the state, district and marketplace have been there to know the exact location of the marketplace. Also, the total area that is sown in that particular district is depicted in sq. km and yield is also there along with the inflation rate of the currency to help the models decide accordingly.

Table 1 A dataset comprising different attributes

Day	Month	Year	State	District	Market	Area sown	Yield	Inflation	Price
1	March	2015	Andhra Pradesh	Anantapur	Gooti	27.73	687	5.87	4000
2	March	2015	Andhra Pradesh	Anantapur	Gooti	27.73	687	5.87	4015
3	March	2015	Andhra Pradesh	Anantapur	Gooti	27.73	687	5.87	4005
4	March	2015	Andhra Pradesh	Anantapur	Gooti	27.73	687	5.87	4000
5	March	2015	Andhra Pradesh	Anantapur	Gooti	27.73	687	5.87	3990
6	March	2015	Andhra Pradesh	Anantapur	Gooti	27.73	687	5.87	3975
7	March	2015	Andhra Pradesh	Anantapur	Gooti	27.73	687	5.87	3905
8	March	2015	Andhra Pradesh	Anantapur	Gooti	27.73	687	5.87	3950
9	March	2015	Andhra Pradesh	Anantapur	Gooti	27.73	687	5.87	3925
10	March	2015	Andhra Pradesh	Anantapur	Gooti	27.73	687	5.87	3975

3.1 Data Preprocessing

The dataset that has been derived consisted of many irregularities in prices at one place the price is reaching lakhs while at the same time the other place has negative priced values which are practically not possible. Also, there were a lot of NULL values present in the dataset since the crop is seasonal and its market does not operate throughout the year. Thus, data preprocessing is required. First, data cleaning has been done by removing these impossible and out-of-the-way values. After that fill out the values in place of NULL using average and other computational methods to provide a regular dataset to let the proposed algorithms work efficiently. Along with the cleaning, data transformation has also been performed where values, format, and domain of values in the dataset have been changed as at some places prices have been given negative which is not possible also at some instances, in place of location names there have been numbers which are again not possible.

3.2 Data Decomposition

Once data preprocessing has been done, the dataset is free from NULL and out-of-domain values, now the time of data decomposition comes using seasonal segmentation where we have to break down our dataset according to different seasons (here seasons mean breaking dataset according to different types of soil on which crop has been cultivated and different type of weather). After this, training of both the models has been done on the resulted dataset.

3.3 Data Division

Once data decomposition has been done over the dataset, we have divided the entire dataset into two parts having percentages in the ratio of 4:1. Thus, 80% of the entire dataset has been used to train the model, whereas the rest 20% has been used for testing the models.

4 Methodology

The main focus of this paper is to determine a solution to predict the future market price trend of cotton to help our farmers (the major workforce of our Indian Economy) fix the price of their surplus and also to know the right time to sell their produce in the Indian market and also which market will give them the best prices of their produce. We have designed the solution using two machine learning models, Facebook Prophet and Random Forest. Our models have been trained on a big dataset of more than 250k data entries having attributed including date, month, year, city, state, area sown, inflation, price, and many more. These attributes together help our model predict the price trend with good accuracy and precision.

These models help in various ways such as predicting the day-wise prices of every marketplace for the next year. Also getting the full-year price trend that helps to make business strategies and long-term plans easy. All this will significantly enhance our motive to help the farmers so that they can get the best price for their surplus and also nation export the raw cotton at their best price. The solution can also be extended to any crop with such a volatile price trend which includes many factors including inflation, climate, and cultivation. Thus, this paper will help the ages of farmers and also make them adaptable to the growing technologies.

4.1 Proposed Approach

A wide and diverse dataset of more than 250K entries have been used, comprising attributes day, month, year, the area is sown, yield, inflation, and price of every marketplace of India of different states across the nation. The dataset that we have taken are from trusted government data statistics sites [5–7]. Time-series data has been found difficult and cumbersome as well to work, and different algorithms that can be used to work with them can be quite finicky and hard to tune. The research has implemented Facebook Prophet as dataset that has various seasonality and also the trends can be daily, weekly, monthly and yearly so to solve all these types of things this model performs best. The Facebook prophet uses an additive regression model with 4 main components namely yearly, seasonal component, piecewise logistic or linear growth curve trend, weekly seasonal and holidays, and the days on which the graph is a straight line. Further, we have also used random forest algorithm [13] which is a supervised learning algorithm that uses ensemble learning for getting the day-wise monthly trend.

The reason for using the Random Forest and Facebook prophet is that since the available dataset is irregular and there are a lot of factors of multiple seasonality in different parts of the country along with different sowing and harvesting period. Also the rainfall is not constant in the nation and there are also different holiday schemes, so to overcome all these things Facebook Prophet proved to be performed well. Also, time-series problems require the identification of a growing or decreasing trend that a Random Forest regression will not be able to formulate.

4.2 Workflow

See Fig. 1.

4.3 Implementation Details

Facebook Prophet Model: It is a process of predicting time-series data based on an add-on model where indirect trends equate to weekly, seasonal, and daily periods, along with considering holidays. It works very well with a series of times with strong seasonal results and a few seasons of historical data. The prophet is strong on missing data and shifts in trend and often treats the outside as well. At its core, the prophet's process is an additional retreat model with four key components:

- Linear or logistic trend of curve growth. Prophet automatically detects changes in trends by selecting areas for data change.
- Part of the annual season created using the Fourier series.
- Half of the weekly season using popular variables.

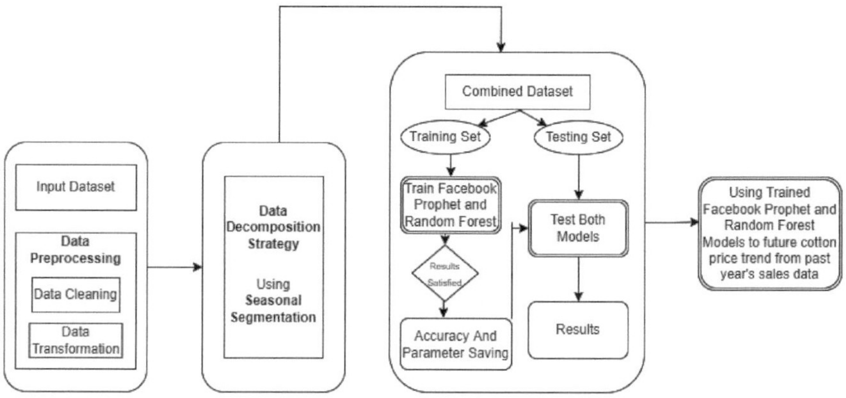

Fig. 1 Workflow diagram of the project

- A user-provided list of important holidays.

Growth work has a curve of three main graphs:

- Row Growth.
- Logistic Growth.
- Flat Growth.

Prophet provides a much more straightforward way to create a reasonable and accurate forecast. Prophet forecasts are customizable in ways that are intuitive to non-experts. There are smoothing parameters for seasonality that allow the user to adjust how close to fit historical cycles, as well as smoothing parameters for trends that allow you to adjust how aggressively to follow historical trend changes. For growth curves, you can manually specify "capacities" or the upper limit of the growth curve, allowing you to inject your prior information about how your forecast will grow (or decline).

The paper has used the Facebook Prophet model just because of its efficiency in handling the large seasonality and missing data shits which makes it a much superior model to other regression models in terms of predicting the trend. It is predicting the complete future year trend of the cotton market prices.

Random Forest Regression Model: Random Forest Regression is a form of supervised learning algorithm which uses an integrated learning approach to decode. An integrated learning approach is a technique that combines the guessing from different machine learning algorithms to conclude a highly accurate prediction than using a single model. It uses ensemble learning for its work. Collaborative learning is the process of using multiple models, trained with the same data; the scale of the results of each model ultimately achieves the most powerful guessing/editing result. Our hope, and the requirement, is to learn the ensemble that the flaws of each model (in this case the decision tree) are independent and different from tree to tree. The Random

Forest bootstrapping algorithm integrates learning methods integrated with the decision tree framework to create multiple random tree extracts from data, a measure of results to produce a new result that often leads to solid speculation/separation.

Random Forest adds additional randomness to the model while growing the trees. Instead of searching for the most important feature while splitting a node, it searches for the best feature among a random subset of features. This results in a wide diversity that generally results in a better model. Therefore, in a random forest, only a random subset of the features is taken into consideration by the algorithm for splitting a node. You can even make trees more random by additionally using random thresholds for each feature rather than searching for the best possible thresholds (like a normal decision tree does).

We have implemented Random Forest just because of its core features of using multiple models, trained on the same dataset, and then using results from each model in predicting the outcome. Here Random Forest has been used to provide monthly trends of the market prices. Since the market of cotton operates only for 20 days excluding all the weekends and other monthly holidays, thus the graph that it would be generating have only 20 points over the x-axis denoting the days and the y-axis will give the pricing of cotton.

5 Analysis and Results

The dataset has been taken from a government database (Cotton Authority of India) [5–7]. The enrich dataset has entries of more than two and a half lakh and consists of a large number of attributes including day, date, city, state, area sown price, inflation, and yield of cotton cultivated across the nation.

5.1 Mathematical Computational Results of Used Models

In this current research, two regression models namely Facebook Prophet and Random Forest has been used and they are analyzed to give the desired results which are shown in Table 2. After that, the results of both the algorithms have been shown in Figs. 2 and 3, where price trend prediction using Facebook Prophet and Random Forest has been depicted.

A good regression model has much less Mean Absolute Error and Mean Absolute % Error and both models have given a less value of Mean Absolute % Error of 0.08 and 0.04, respectively. Since Facebook Prophet is used to predict a complete year trend, so this much Absolute Error % is acceptable and Random Forest is providing a daily trend for which 0.04% is equal to perfect as compared to different studies done in this context. Table 2 clearly shows that both the implemented algorithms are giving us the best results with having Mean Absolute Error of 209.56 and 435.54 for Random Forest and Facebook Prophet, respectively.

Table 2 Results of different algorithms

	Facebook Prophet	Random Forest
Max error	1460.59	336.08
Mean absolute error	435.54	209.56
Mean squared error	284,396.04	51,337.07
Mean absolute % error	0.08	0.04

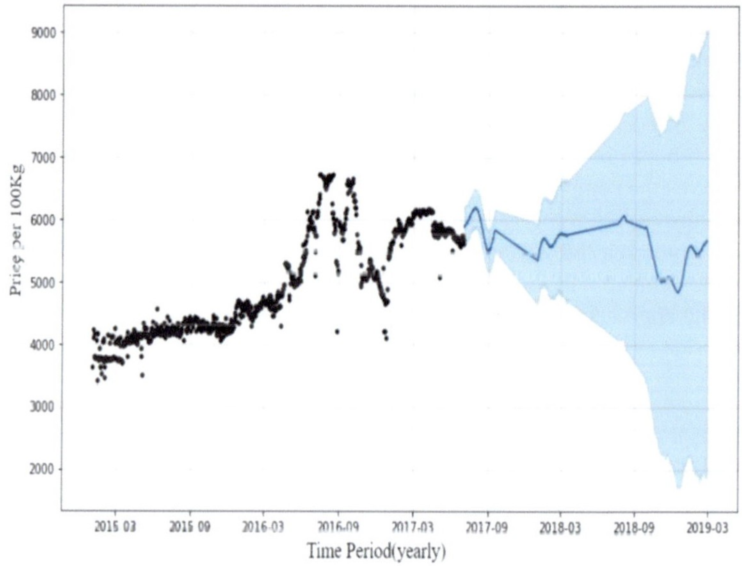

Fig. 2 The predicted price band using Facebook Prophet

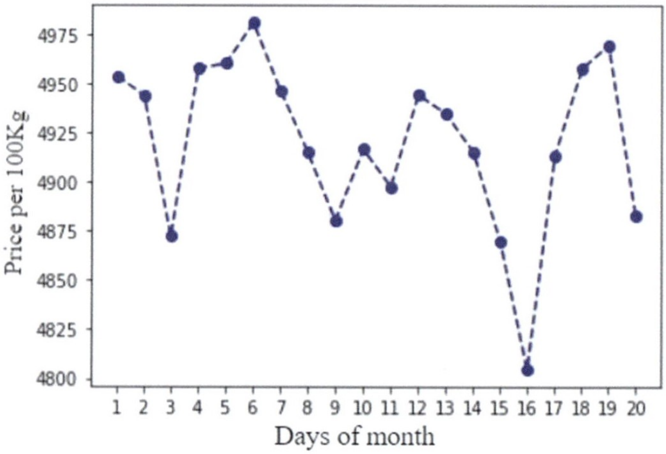

Fig. 3 Price prediction using Random Forest

6 Conclusion

The motive of this research paper is to predict the future market price trend of the cotton crop for the benefit of farmers and the government and we are pretty successful in this using Machine Learning Regression Models. The two different models namely Facebook Prophet and Random Forest are used for the cotton price trend prediction. In general, Mean Absolute Error and Mean Absolute % Error are good parameters in determining the accuracy of regression models. In case of Facebook Prophet, we are getting values 435.54 and 0.08, respectively, and in Random Forest we are getting values 209.56 and 0.04, respectively, for mean absolute error and mean absolute % error. The experimental results have provided the conclusion that Random Forest gives much better results in predicting the day-wise trend of cotton and Facebook Prophet gives the complete year price trend for deciding the long-term strategies.

The future aspect of this research is to incorporate more attribute in determining the price trend, i.e., the demand. As the population of India is increasing at a fast rate, so the demand will increase too that affects the prices much. Further, we will enhance the approach by considering other features like meteorological ones along with the price.

References

1. Deepa S, Alli A, Gokila S (2023) Machine learning regression model for material synthesis prices prediction in agriculture. Mater Today Proc 81:989–993
2. Van Klompenburg T, Kassahun A, Catal C (2020) Crop yield prediction using machine learning: a systematic literature review. Comput Electron Agric 177:105709
3. Prasad NR, Patel NR, Danodia A (2021) Crop yield prediction in cotton for regional level using random forest approach. Spat Inf Res 29(2):195–206
4. Nihar A, Patel NR, Danodia A (2022) Machine-learning-based regional yield forecasting for sugarcane crop in Uttar Pradesh, India. J Indian Soc Remote Sens 1–12
5. Textile Data | Ministry of Textiles | GoI. http://texmin.nic.in/textile-data
6. Cotton Data| Cotton Corporation of India | GoI. https://cotcorp.org.in/
7. Mehta R (2009) India's agricultural marketing information system network (AGMARKNET)
8. Garlapati A, Krishna DR, Garlapati K, Rahul U, Narayanan G (2021) Stock price prediction using Facebook prophet and Arima models. In: 2021 6th International conference for convergence in technology (I2CT). IEEE, pp 1–7
9. Zunic E, Korjenic K, Hodzic K, Donko D (2020) Application of Facebook's prophet algorithm for successful sales forecasting based on real-world data. arXiv preprint arXiv:2005.07575
10. Bai J, Li Y, Li J, Yang X, Jiang Y, Xia ST (2022) Multinomial random forest. Pattern Recogn 122:108331
11. Banaś J, Utnik-Banaś K (2021) Evaluating a seasonal autoregressive moving average model with an exogenous variable for short-term timber price forecasting. Forest Policy Econ 131:102564
12. Darekar A, Reddy AA (2017) Cotton price forecasting in major producing states. Econ Aff 62(3):373–378

13. Ghosh S (2017) Forecasting cotton exports in India using the ARIMA model. Amity J Econ 2(2):36–52
14. Kumari RV, Panasa V, Gundu R, Kaviraju S (2017) Price behaviour and forecasting of cotton in Telangana. Int J Pure Appl Biosci 5(5):86

Novel Approach Toward Handwritten Mathematical Expression Image Captioning System

Navajyoti Nath, Aishik Das, Anupal Saikia, Kasturi Sharma, Vishal Gour, and Sourish Dhar

Abstract In this paper, we propose a novel approach for generating captions and LaTeX notations out of handwritten mathematical expression images (HMEIs). For the proposed approach, we have created an annotated dataset comprising of handwritten images of mathematical expressions along with corresponding textual descriptions and LaTeX notations. This dataset acts as the stepping stone for training and evaluation of our model. To accurately interpret HMEIs, we deployed a encoder–decoder-based model which is a fusion model utilizing convolutional neural network (CNN) and recurrent neural network (RNN). Pretrained CNN models viz. VGG16, VGG19, ResNet50, InceptionV3 are used for feature extraction from the images. Two different RNN models viz. long short-term memory (LSTM) and gated recurrent unit (GRU) have been used for handling the sequence of texts as captions. The RNN models capture the temporary dependencies and long-term dependencies in the text sequence. In the final step, we used a SoftMax layer to generate the output in either textual description or LaTeX notation of the given handwritten expression image. Our experimental results demonstrate the maximum accuracy of 68.9% for generating textual descriptions and 75.7% for generating LaTeX notations. This work emphasizes on the uses of Natural Language Processing and Computer Vision techniques that improves the interpretation of handwritten mathematical expressions (HME).

Keywords Captioning · Handwritten mathematical expression · Image feature extraction module · Text generation module · Fusion module

---

N. Nath (✉) · A. Das · A. Saikia · K. Sharma · V. Gour · S. Dhar
Department of Computer Science and Engineering, Assam University, Silchar, India
e-mail: navajyoti.cse@gmail.com

S. Kumar et al. (eds.), *Artificial Intelligence for Next Generation Computing, Volume 1*,
Lecture Notes in Networks and Systems 1549,
https://doi.org/10.1007/978-981-95-0701-6_32

1 Introduction

"Image captioning—it is the task of providing a natural *language description* of the content within an image" [1]. It involves interpreting an image to identify people, objects, and actions, then writing a clear description of what the image actually means. This task uses both natural language and computer vision skills to create precise descriptions.

There are two types of image captioning technique. (i) ***Retrieval and Template-based Image Captioning***, (ii) ***Deep learning-based Image Captioning*** [2]. Retrieval and template-based image captioning is the traditional image captioning method which is a retrieval-based one. It uses a query image to find a matching sentence or set of sentences from a predefined pool to describe the input image. The caption can be a single sentence or a combination of the retrieved sentences. Template-based image captioning generates captions by following a strict process that ensures both unambiguous syntax and interpretation. On the other hand, deep learning-based image captioning uses neural networks to automatically generate descriptive captions for images. In our work, we have used deep learning-based image captioning (Fig. 1).

Image captioning has many use-cases [3]. This includes creating titles for news photos, describing medical images, helping with text-based image searches and assisting blind and visually impaired users. Each of these applications is very important and challenging at the same time. To handle these challenges faced during caption generation we have used different deep learning techniques including pretrained CNNs and RNNs. The motivation behind our research in this area is as follows:

- In recent years, Handwritten Mathematical Expression Recognition (HMER) has emerged as an important research area. Image captioning plays a key role in recognizing and generating descriptions for handwritten mathematical expressions [4].

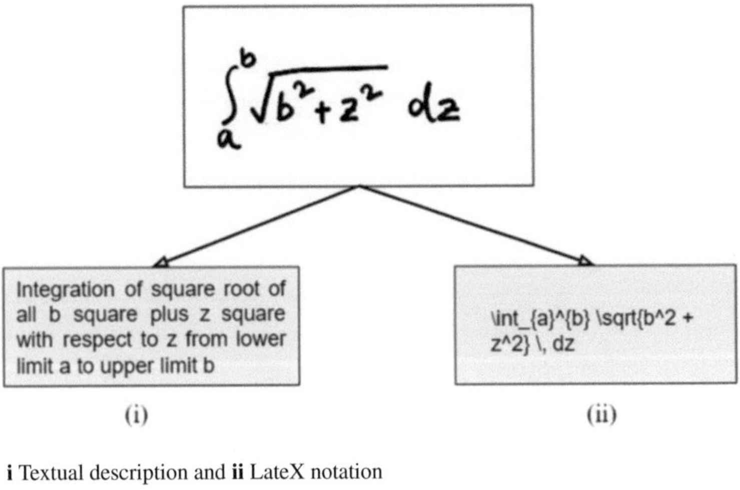

Fig. 1 **i** Textual description and **ii** LateX notation

- The primary challenge in handwritten mathematics recognition is ambiguity. This problem arises from the variability in human handwriting because different people have different way of writing an alphabet or character. For example, the numeral "1" can be easily confused with the letter "l" due to similar strokes. We have tried to overcome this problem by using indirect attention mechanism in our model [5, 6].
- Because of its immense use in the field of research and learning, handwritten mathematical expression captioning has become very popular lately. This technology is very crucial in improving the scalability, searchability, and accessibility of mathematical expressions. Mathematical expression captioning can make learning and teaching more effective in educational field by giving academia an edge over the traditional approaches.

The major contributions of this research work are as follows:

- We have created an annotated dataset of handwritten mathematical expressions with corresponding LaTeX notation and textual descriptions to train and test our model.
- We have developed an end-to-end CNN-RNN-based deep learning model capable of generating both the LaTeX notation and textual description of a given input handwritten mathematical expression image. This model helps us to interpret handwritten mathematical expressions images precisely.

For both the tasks i.e., LaTeX notation and caption generation, we have designed a model that merges convolutional neural networks with recurrent neural networks and incorporates indirect attention mechanisms. This approach leverages the CNN's ability to extract spatial features from images and the RNN's sequence modeling capabilities, creating a framework for processing handwritten mathematical images (Fig. 2). The indirect attention mechanism allows the model to selectively focus on specific regions of image, enhancing the quality of the generated output text.

The rest of the paper is organized as follows: Sect. 2 Related Works, in Sect. 3 we present the proposed methodology, Sect. 4 discusses the result and finally, Sect. 5 concludes the paper with a mention of the future scope of the present work.

2 Related Works

In this section, previous research on creating a dataset, recognition of handwritten mathematical expressions, and the textual description of the mathematical expression framework is described. Previous work on handwritten mathematical expression recognition for many decades is provided here.

The HASY dataset [7] is a valuable resource for handwritten symbol recognition, akin to the well-known MNIST [8] dataset for handwritten digits. It contains 168,233 instances across 369 classes, offering a comprehensive collection of single symbols. This publicly available dataset is designed to address specific challenges in symbol

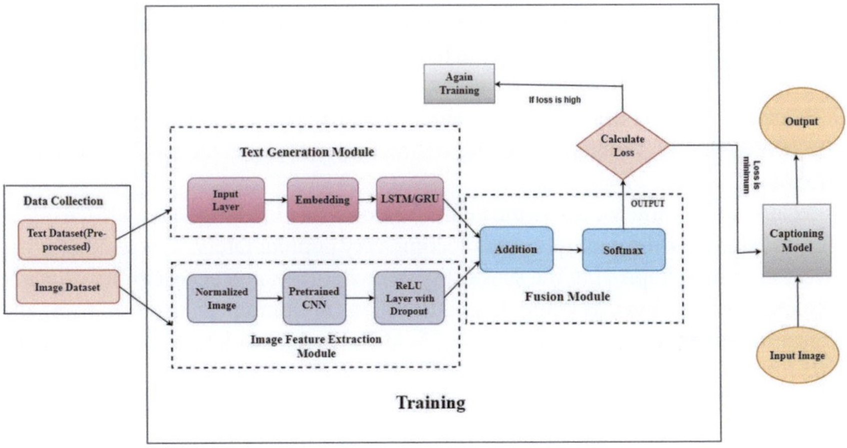

Fig. 2 Workflow of the model

classification and verification. HASY supports a classification challenge with 10 predefined folds for tenfold cross-validation, providing a robust framework for evaluating models. It also includes a verification challenge, crucial for applications like document analysis and digital ink recognition. Compared to other datasets, HASY's large class diversity and instance count make it particularly useful for training deep learning models, enabling more effective generalization. Its free availability encourages widespread use in both academic research and industry, contributing significantly to advancements in handwritten symbol recognition [9].

The recognition of handwritten mathematical expression is done with the help of an assembly-based cooperative machine (ABM) that includes a co-operative encoder and two parallel reverse decoders [10]. Mutual distillation requires a one-to-one transfer of knowledge at each step of training, thus improving both decoders. Furthermore, an integrated algorithm (AAM) is proposed to successfully integrate multiple tracking systems to recognize numerical symbols of different indicators. Nguyen et al. demonstrate deep learning for online mathematical symbol recognition. High frequency-based CNN and bidirectional LSTM (BLSTM) networks are first used for image modeling by online models and online models and then combined. By conducting self-identification experiments on CROHME data, analyzing and interpreting them, and comparing with traditional identification methods such as MRF and MQDF [11].

Our work is largely influenced by the work "Textual Description for Mathematical Equations" by Ajoy Mondal and C. V. Jawahar. It treats the reading of mathematical equations as a problem of the generation of textual descriptions that can be used to understand the equations based on the image captioning problem in computer vision. The authors propose a new model called Mathematical Equation Description (MED) which is an end-to-end deep neural network that provides textual descriptions of the given mathematical equation images. The MED model opt for a convolutional neural network as the encoder for feature extraction and a recurrent neural network with attention mechanism for the generation of descriptions.

Evaluations for comprehending the role of each component of the algorithms and to justify the particular design are performed on the Math-Exp-Syn dataset. An example of real- world experiment is to let the students write equations based on the textual descriptions of the MED model. A test set of 398 mathematical equation images from the NCRT class V mathematics book is developed. Students write the equations within an hour and out of 398 equations students write 359 correctly according to the descriptions. In 39 cases, there are mistakes because of other structural elements [12].

Zhang et al. used an encoder–decoder architecture with attention for the recognition of online handwritten mathematical expressions (OHMER) [13]. The input handwritten expressions information including the two-dimensional ink trajectory information is input into the gated recurrent unit-based recurrent neural network (GRU-RNN), and the decoder is also designed by the GRU-RNN with coverage-based attention model. This approach accomplishes symbol recognition and structure and operation to output a character sequence in LaTeX form. The accuracy of this approach is 52.43% at best using only the official training dataset available for this task. The paper "Watch, Attend, and Parse (WAP)" describes a new end-to-end deep learning-based method that can recognize HMEs in a two-dimensional layout and output them in one-dimensional form as sequences of characters in LaTeX notation. In contrast to the most approaches, the proposed model does not have issues connected with symbol splitting and does not require the expression grammar. For the symbol recognition issue, it employs a watcher; for the structural analysis part, it uses a parser. In this work, a convolutional neural network-based encoder, which takes HME images as input is used as the watcher and a recurrent neural network-based decoder with attention mechanism to parse the input images and generate LaTeX sequences. Further, the mapping between the input expressions and the output LaTeX sequences is learned by the attention mechanism. The proposed approach is evaluated on a benchmark made available by the CROHME international competition. WAP significantly outperformed the state-of-the-art technique using the official training dataset, achieving an expression recognition accuracy of 46.55% on CROHME 2014 and 44.55% on CROHME 2016 dataset [14].

This paper provides an analysis of the current approaches to recognition and users' experience of using pen-centric applications for handwritten mathematical expressions. It classifies the approaches to recognition and describes their benefits and drawbacks, with the emphasis on the end-to-end techniques based on the encoder–decoder structure and multinomial in-puts. Algorithms used for evaluation, resources for comparison and benchmarks, and performance of different recognizers in competition are described. Exploiting the area of handwritten math recognition, several application examples from different fields and on various applications are depicted in this paper. The survey reveals how in UI design, there are several recognition approaches that are used to improved performance based on practical implementations. Finally, it discusses the future research scopes in handwritten mathematical expression recognition and their uses [15].

3 Methodology

The approach used in this paper is structured into four key stages: Dataset creation (image and text), preprocessing, feature extraction model, sequential model, and decoder model. These stages are discussed in detail throughout the paper.

3.1 Dataset Creation

The dataset creation process is divided into two parts: an image dataset and a text dataset [16–19].

1. *Image Dataset:* In our work, we have created an annotated dataset of handwritten mathematical expressions to support our research because no suitable offline standard dataset is available for our specific needs. First mathematical expressions have been written in the paper. After that we have taken the pictures of the expressions and saved it in ".JPG" format. This dataset consists of 2000 images including a diverse range of mathematical topics such as limits, integration, trigonometry, differential equations, and polynomial expressions. Each image has a uniform size of 7680*4329 pixels, providing a consistent format across the dataset. The background in each image is white, and the mathematical expressions are written in black or blue ink to ensure high contrast and visibility. Total size of the dataset is 1.6 GB. With the help of this dataset, we have trained our model and analyze our model effectively. A dataset of single mathematical symbol entitled HASY V2 is used to test our model performance by using transfer learning.
2. *Text Dataset:* In addition to creating the dataset of handwritten mathematical expressions, we have also provided comprehensive captions for each and every image. Each image is having two types of captions: a LaTeX notation of the expression and a set of textual descriptions of the expression. Specifically, every image is annotated with one LaTeX notation caption and three distinct textual descriptions. A total of 2000 LaTeX notations and 6000 textual descriptions are provided to create the caption dataset for our model. These two caption datasets are distinct from each other.

3.2 Preprocessing

The data preprocessing for generating LaTeX notation and textual generation of handwritten mathematical expressions is divided into two parts: image dataset preprocessing and text dataset preprocessing [20].

1. *Image Dataset Preprocessing*: Image dataset preprocessing focuses on preparing the images of handwritten mathematical expressions for training our model efficiently. The key steps for image data preprocessing are as follows:

- *Background Whitening*: Mathematical images often contain different background colors that can interfere in the recognition of mathematical expressions. Whitening the background enhances the contrast and clarity of the handwritten mathematical expressions.
- *Rotation*: Handwritten mathematical expression images (HMEIs) may be misaligned with respect to other images present in the dataset. Adjusting the rotation ensures that the expressions are horizontally aligned with the other images for the training part.
- *Resolution Standardization*: The images are all set to the same resolution to ensure consistency across the dataset. This helps the model perform better during the training of the model.

2. *Text Dataset Preprocessing*: Text dataset preprocessing prepares the textual descriptions and LaTeX notations associated with mathematical expressions for model training and evaluation. The key steps of text data preprocessing are as follows:

   - *Noise Removal*: Text data can have unwanted characters or symbols. Removing these, helps improve the quality of the input text that subsequently enhances the model training part.
   - *Unwanted White Space Removal*: Too many spaces can cause problems and misunderstandings in text data during model training. We've removed these extra spaces to make the data cleaner and more accurate.
   - *Tokenization*: Here, each textual description and Latex notation is tokenized using the standard tokenizer class provided in the Keras library

3.3 Image Feature Extractor Module

Feature extraction is an important step in processing the image data in this research on generating textual descriptions and LaTeX notations for handwritten mathematical expression images [21, 22]. To extract features from images, a pretrained convolution neural network (CNN) is used. In order to determine which model performed best on our dataset, four different pretrained convolution neural networks (CNNs)— Visual Geometry Group 16 (VGG16) [23], Visual Geometry Group 19 (VGG19) [23], InceptionV3 [24], and ResNet50—are used for feature extraction. The following steps describe the processes in Image Feature Extraction Module:

- Here, different types of pretrained convolution neural networks (CNNs) are deployed because these pretrained CNNs are trained on large datasets such as ImageNet and can extract important information from the images.
- The pretrained CNN's last layer, which is usually a classification layer used for classification tasks, is eliminated for our model. This modification allows us to extract the important features of the images efficiently.

- Images of handwritten mathematical expressions are prepared as inputs for the CNN model. This includes resizing the images to match the input dimensions required by the pretrained CNN model and normalizing the pixel values.
- The prepared images are processed through the various convolution and pooling layers of the pretrained CNN up to the penultimate layer, preceding the removed final layer. This penultimate layer holds the learned features of the images.
- The feature vectors from the penultimate layer are collected as the output of the feature extraction process. These vectors represent complex patterns and information about the handwritten mathematical expressions.
- After that one linear transformation layer is added to change the dimensionality of the feature vector that is suitable for add layer in the decoder model.
- One dropout layer is added after this to reduce the over-fitting issue.

Firstly, a pretrained convolution neural network (CNN) is used to extract features from the handwritten images in our model. Let "I" represent the input image. The input image I is fed into a pretrained CNN, denoted as $\text{CNN}_{\text{pre-trained}}$. This CNN is typically pretrained on large image datasets like ImageNet, which allows it to capture complex features of the given image. The output of the CNN is a high-dimensional feature map $F = \text{CNN}_{\text{pre-trained}}(\text{I})$. The dimension of the pre-extracted feature vector is "a".

Secondly, the high-level features F extracted by the pretrained CNN are passed through a Dense (fully connected) layer. This layer transforms the features into a suitable format for further processing by the Addition layer in the Fusion Module. Let w_d and b_d denote the weights and biases of the Dense layer, respectively. The output of the Dense layer is as follows:

$$f_d = \text{ReLU}(W_d \cdot F + b_d) \tag{1}$$

ReLU is the Rectified Linear Unit activation function. This output f_d serves as a high-level representation of the input images. The dimension of f_d is m. Then, a dropout layer with a regularization is added to remove the overfitting issue. f_r represents the feature vector after regularization. This vector is passed to the addition layer.

3.4 Text Generation Module

The Text Generation Module processes the textual descriptions and the LaTeX notation associated with the images (Figs. 3 and 4). Our model consists of three steps:

- Handling Text Sequence: Here, the words and characters of the vocabulary are represented with the help of sequence of integers. If the length of the caption or

Input1	Input2	Output
<feature>	Startseq	x
<feature>	Startseq x	equals
<feature>	Startseq x equals	to
<feature>	Startseq x equals to	r
<feature>	Startseq x equals to r	cosine
<feature>	Startseq x equals to r cosine	of
<feature>	Startseq x equals to r cosine of	theta
<feature>	Startseq x equals to r cosine of theta	Endseq

Fig. 3 Textual description generation process of sample expression $x = r \cos \theta$

Input1	Input2	Output
<feature>	Startseq	\
<feature>	Startseq \	[
<feature>	Startseq \ [	x
<feature>	Startseq \ [x	=
<feature>	Startseq \ [x =	r
<feature>	Startseq \ [x = r	\
<feature>	Startseq \ [x = r \	cdot
<feature>	Startseq \ [x = r \ cdot	\
<feature>	Startseq \ [x = r \ cdot \	cos
<feature>	Startseq \ [x = r \ cdot \ cos	(
<feature>	Startseq \ [x = r \ cdot \ cos (	\
<feature>	Startseq \ [x = r \ cdot \ cos (\	theta
<feature>	Startseq \ [x = r \ cdot \ cos (\ theta	)
<feature>	Startseq \ [x = r \ cdot \ cos (\ theta)	\
<feature>	Startseq \ [x = r \ cdot \ cos (\ theta) \	]
<feature>	Startseq \ [x = r \ cdot \ cos (\ theta) \ ]	Endseq

Fig. 4 LaTeX notation generation process of sample expression $x = r \cos \theta$

LaTex notation is too low zeros have added after the sequence of integers. This is called as padding.

- Embedding Layer: Next, an embedding layer is used on the integer sequence (which represents tokenized words and characters). This layer turns each integer into a dense vector representation that captures its meaning and context. These embeddings show the relationships and patterns in the text data.
- LSTM/GRU Layer: After that, the sequence of embeddings is processed by LSTM and GRU layers.

These layers are designed to capture long-range dependencies and context in the text data. They produce a sequence of hidden states, which are crucial for generating accurate captions [13, 25]. The output from these layers is then passed to an addition layer for further calculation.

Let $T = \{t_1, t_2, \ldots, t_n\}$ represent the integer sequence (which represents tokenized words and characters) after preprocessing, where t_i denotes the ith token in the sequence. This sequence consists of integer representing parts of the handwritten mathematical expressions. The text sequence T is passed through an Embedding layer, which converts each token t_i into a dense vector representation e_i of dimension m. Let E denote the embedding matrix of dimension $m*a$, where a is the vocabulary size. The embedding for the ith token is $e_i = E[t_i]$.

The embeddings $\{e_1, e_2, \ldots, e_n\}$ are then fed into a long short-term memory (LSTM) or gated recurrent unit (GRU) layer. These recurrent layers are designed to capture the sequential dependencies and context within the text data. Let h_i denote the hidden state at time step i. The hidden states are computed as.

$$h_j = \text{LSTM/GRU}\left(e_j, h_{i-1}\right) \tag{2}$$

for $i = 1, 2, \ldots, n$, h_i is calculated. This h_i of dimension m is passed in the addition layer of the Fusion Module for further processing.

3.5 *Fusion Module*

The Fusion Module combines the outputs of the Image Feature Extraction Module and the Text Generation Module to generate the final output:

- *Addition Layer*: In this layer, the output from the Image Feature Extraction Module and the output from the Text Generation Module are combined using element-wise addition. This layer is acting as an indirect attention mechanism in the model.
- *SoftMax Layer*: First the output coming from the addition layer is transformed to logit vector sequence with the help of linear transformation. After that a probability distribution over the logit vector sequence is computed by the SoftMax function where the size of the logit vector sequence is equals to vocabulary size and each character or word in the vocabulary is represented by each vector in the sequence. The token with the highest probability is reflected as output in each step [26].

The final hidden state from the LSTM/GRU layer, h_n, and the feature representation from the Dense layer, f_r, are added element-wise to form a combined feature vector c of dimension m. The combined feature vector c is transformed to logit vector sequence z, where $z = z_1, z_2, \ldots, z_a$ and a is vocabulary size. This z vector sequence is passed in the SoftMax layer which generates a probability distribution over the logit vector sequence which represents the tokens present in the vocabulary. The token with the highest probability is reflected as output. The output of the SoftMax layer is as follows:

$$\text{SoftMax}(z_i) = \frac{e^{z_i}}{\sum_{j=1}^{a} e^{z_j}}, \tag{3}$$

where z_i represents the probability of ith token in the vocabulary. In this equation, $z = [z_1, z_2, \ldots, z_n]$ is the input vector sequence to the softMax function. Probability is calculated over every vector in the sequence. The predicted token (represented by logit vector) which has a higher probability among all the tokens presents in the vocabulary, is out from this final layer as output.

4 Result and Discussion

This section shows the results of experiments using different pretrained convolutional neural networks (CNNs) combined with recurrent neural networks (RNNs), including long short term memory (LSTM) and gated recurrent unit (GRU). The results are displayed in a tabular format.

The model is trained with different combinations to determine which configuration yields the best results. Following evaluation measures as defined in are used to compare the performance of different combinations [27].

- *BLEU Score*: The Bilingual Evaluation Understudy (BLEU) score is a metric for evaluating the quality of text that has been generated by the machine [28]. It is widely used for assessing the quality of image captions generated by models.

$$\text{BLEU} = \text{BP} \cdot \exp\left(\sum_{n=1}^{N} w_n \log p_n\right) \tag{4}$$

Here, BP is the brevity penalty. p_n is the precision for n-grams of size n. w_n is the weight assigned to the n-gram precision (usually, $w_n = 1$ for uniform weighting). N is the maximum n-gram size considered (typically 4).

- *METEOR Score*: Metric for Evaluation of Translation with Explicit Ordering (METEOR) score is a metric designed to evaluate the quality of text generated by machines, by comparing them to human reference texts.

$$\text{METEOR} = F_{\text{mean}} \cdot (1 - \text{Penalty}) \tag{5}$$

F_{mean} is the harmonic mean of precision (P) and recall (R). Penalty is the fragmentation penalty.

- *ROUGE Score:* Recall-Oriented Understudy for Gisting Evaluation (ROUGE) score is a set of metrics designed to evaluate the quality of text generated by machines, such as image captions, by comparing them to human reference texts. The ROUGE score primarily focuses on recall, measuring how well the generated text captures the relevant content from the reference text.

$$\text{ROUGE} = \frac{\text{No of matching n - grams}}{\text{Total no of n - grams in reference text}} \quad (6)$$

We have divided our result section mainly into three parts. First is result of textual description generation of handwritten mathematical expression (test with 500 images), second is LaTeX notation generation of handwritten mathematical expression (test with 500 images), and third is test the performance of our model with HASY V2 dataset.

4.1 *Textual Caption Generation*

In this section, we have demonstrated three tables. The first table shows the accuracy matrix, the second table presents the BLEU score, METEOR score, and ROUGE score combinations with LSTM, and the third table shows same matrices combination with GRU for LaTeX notation generation.

Table 1 indicates the accuracy of our model with different pretrained CNN and RNN combinations. When we used *LSTM* as the RNN, we achieved the highest accuracy of *67.3%* by using *VGG19* as the pretrained CNN. If we used *GRU* as the RNN, we achieved the maximum accuracy of *68.9%* by using *InceptionV3* as the pretrained CNN.

Table 2 indicates the evaluation metrics of our model. We tested four different combinations with pretrained CNN and LSTM. We achieved a maximum BLEU-1 score of 0.542, METEOR score of 0.254, and ROUGE score of 0.338 by using ResNet50 with LSTM.

Table 1 Accuracy (in %) of model with different pretrained CNN and RNN

Pertained CNN	VGG16	VGG19	ResNet50	InceptionV3
With LSTM	68.5	67.3	65.5	**66.7**
With GRU	68.2	68.1	66.9	**68.9**

Table 2 Comparison of performance metrics for different pretrained CNNs and LSTM

Pertained CNN	VGG16	VGG19	ResNet50	InceptionV3
BLEU-1 score	0.512	0.502	**0.542**	0.511
BLEU-2 score	0.274	0.271	**0.278**	0.272
BLEU-3 score	0.184	0.181	**0.189**	0.182
BLEU-4 score	0.084	0.080	**0.088**	0.082
METEOR score	0.225	0.215	**0.254**	0.212
ROUGE score	0.313	0.307	**0.338**	0.317

Table 3 Comparison of performance metrics for different pretrained CNNs and GRU

Pertained CNN	VGG16	VGG19	ResNet50	InceptionV3
BLEU-1 score	0.517	0.508	**0.547**	0.514
BLEU-2 score	0.277	0.273	**0.287**	0.274
BLEU-3 score	0.187	0.181	**0.192**	0.183
BLEU-4 score	0.082	0.076	**0.089**	0.080
METEOR score	0.223	0.215	**0.256**	0.213
ROUGE score	0.313	0.305	**0.337**	0.318

Table 3 indicates the evaluation metrics by using GRU instead of LSTM. We achieved a maximum BLEU-1 score of 0.547, METEOR score of 0.256, and ROUGE score of 0.337 by using ResNet50 with GRU.

As we can see from Table 1, we can get the maximum accuracy by combining InceptionV3 with GRU. From Tables 2 and 3, we can clearly notice that we have got the maximum BLEU score and METEOR score by combining ResNet50 with GRU and maximum ROUGE score by combining ResNet50 with LSTM.

4.2 Latex Notation Generation

In this section, we have demonstrated three tables. The first table shows the accuracy matrix, the second table presents the BLEU score, METEOR score, and ROUGE score combinations with LSTM, and the third one shows the same matrices combination with GRU for LaTeX notation generation.

Table 4 indicates the accuracy of our model with different pretrained CNN and RNN combinations. When we used *LSTM* as the RNN, we achieved the highest accuracy of *74.5%* by using *InceptionV3* as the pretrained CNN. If we used *GRU* as the RNN, we achieved the maximum accuracy of *75.7%* by using *InceptionV3* as the pretrained CNN.

Table 5 indicates the evaluation metrics of our model. We tested four different combinations with pretrained CNN and LSTM. We achieved a maximum BLEU-1 score of 0.549, METEOR score of 0.248, and ROUGE score of 0.337 by using InceptionV3 and LSTM.

Table 4 Accuracy (in %) of model with different pretrained CNN and RNN

Pertained CNN	VGG16	VGG19	ResNet50	InceptionV3
With LSTM	69.2	68.7	73.6	**74.5**
With GRU	70.2	68.9	73.2	**75.7**

Table 5 Comparison of performance metrics for different pretrained CNNs and LSTM

Pertained CNN	VGG16	VGG19	ResNet50	InceptionV3
BLEU-1 score	0.531	0.511	0.543	**0.548**
BLEU-2 score	0.275	0.273	0.276	**0.279**
BLEU-3 score	0.184	0.179	0.186	**0.188**
BLEU-4 score	0.057	0.083	0.089	**0.091**
METEOR score	0.231	0.232	0.243	**0.248**
ROUGE score	0.313	0.314	0.326	**0.337**

Table 6 Comparison of performance metrics for different pretrained CNNs and GRU

Pertained CNN	VGG16	VGG19	ResNet50	InceptionV3
BLEU-1 score	0.532	0.519	0.542	**0.549**
BLEU-2 score	0.274	0.272	0.277	**0.281**
BLEU-3 score	0.183	0.180	0.187	**0.189**
BLEU-4 score	0.085	0.081	0.086	**0.091**
METEOR score	0.231	0.225	0.240	**0.246**
ROUGE score	0.301	0.304	0.321	**0.339**

Table 6 indicates the evaluation metrics by using GRU instead of LSTM. We achieved a maximum BLEU-1 score of 0.549, METEOR score of 0.246, and ROUGE score of 0.339 by using InceptionV3 with GRU.

As we can see from Table 4, we can get the maximum accuracy by combining InceptionV3 with GRU.

From Tables 5 and 6, we can clearly notice that we get the maximum BLEU score and METEOR score by combining InceptionV3 with LSTM and get maximum ROUGE score by combining InceptionV3 with GRU.

4.3 Test with HASY V2 Dataset

In this section, we have demonstrated two tables for test dataset of HASY V2 dataset. The first table presents the BLEU score, METEOR score, and ROUGE score combinations with LSTM, and the second one shows the same matrices combination with GRU for mathematical symbol generation. HASY V2 dataset single symbol mathematical dataset so only BLEU-1 score can be calculated. This dataset is used to test our model, which is already trained with our dataset, to check the performance of our model. The HASY V2 dataset has three test datasets test-v1, test-v2, test-v3. We have calculated evaluation measures for each test set with each combination of pretrained CNN and RNN. Then, we have taken the average of evaluation measures for every combination of pretrained CNN and RNN.

Table 7 Comparison of performance metrics for different pretrained CNNS and LSTM

Pertained CNN	VGG16	VGG19	ResNet50	InceptionV3
BLEU-1 score	0.294	0.291	**0.328**	0.291
METEOR score	0.191	0.202	**0.212**	0.193
ROUGE score	0.259	0.260	**0.278**	0.267

Table 8 Comparison of performance metrics for different pretrained CNNs and GRU

Pertained CNN	VGG16	VGG19	ResNet50	InceptionV3
BLEU-1 score	0.294	0.292	**0.338**	0.301
METEOR score	0.193	0.200	**0.218**	0.194
ROUGE score	0.249	0.220	**0.289**	0.267

Table 7 indicates the evaluation metrics of our model for HASY V2 test dataset. We tested four different combinations with pretrained CNN and LSTM. We achieved a maximum BLEU-1 score of 0.328, METEOR score of 0.212, and ROUGE score of 0.278 by using ResNet50 with LSTM. Table 8 indicates the evaluation metrics by using GRU instead of LSTM. We achieved a maximum BLEU-1 score of 0.338, METEOR score of 0.218, and ROUGE score of 0.289 by using ResNet50 with GRU.

5 Conclusion

Here, we proposed a model to generate text descriptions and LaTeX notations for handwritten math expressions images (HMEIs) using various pretrained convolutional neural networks (CNNs) like VGG16, VGG19, ResNet50, and InceptionV3, along with recurrent neural networks (RNNs) like LSTM and GRU. Our experiments showed that combining different CNNs with LSTM and GRU can significantly improve performance metrics such as BLEU, METEOR, and ROUGE scores. We achieved a maximum accuracy of 68.9% for text description generation and 75.7% for LaTeX notation generation.

These results encourage and show a way forward toward future research and development. Enhancing the model by exploring more advanced architectures, such as Transformer models or using quantum CNNs may further improve the performance of the model. Including more different handwriting styles in the dataset can make the model precise and more reliable. These improvements can make the captioning system more useful for practical applications, like educational tools.

References

1. Herdade S et al (2019) Image captioning: transforming objects into words. Adv. Neural Inf Process Syst 32
2. Zakir Hossain MD, Sohel F, Shiratuddin MF, Laga H (2019) A comprehensive survey of deep learning for image captioning. ACM Comput Surv 51(6):36. Article 118. https://doi.org/10.1145/3295748
3. Kristianto GY, Aizawa A (2014) Extracting textual descriptions of mathematical expressions in scientific papers. D-Lib Mag 20(11):9
4. Li Z et al (2023) Improving handwritten mathematical expression recognition via similar symbol distinguishing. IEEE Trans Multimedia
5. Awal A-M, Mouchère H, Viard-Gaudin C (2010) The problem of handwritten mathematical expression recognition evaluation. In: 2010 12th International conference on frontiers in handwriting recognition. IEEE
6. Chi X et al (2021) Handwritten mathematical expression recognition with self-attention. In: Proceedings of the 2021 4th International conference on algorithms, computing and artificial intelligence
7. Thoma M (2017) The hasyv2 dataset, CoRR abs/1701.08380 arXiv:1701.08380
8. LeCun Y, Bottou L, Bengio Y, Haffner P (1998) Gradient-based learning applied to document recognition. Proc IEEE 86(11):2278–2324
9. Thoma M (2017) The hasyv2 dataset. arXiv preprint arXiv:1701.08380
10. Bian X et al (2022) Handwritten mathematical expression recognition via attention aggregation based bi-directional mutual learning. Proceedings of the AAAI conference on artificial intelligence, vol 36, no 1
11. Dai Nguyen H, Le AD, Nakagawa M (2016) Recognition of online handwritten math symbols using deep neural networks. IEICE Trans Inf Syst 99(12):3110-3118
12. Mondal A, Jawahar CV (2019) Textual description for mathematical equations. In: 2019 International conference on document analysis and recognition (ICDAR). IEEE
13. Zhang J, Du J, Dai L (2017) A gru-based encoder-decoder approach with attention for online handwritten mathematical expression recognition. In: 2017 14th IAPR international conference on document analysis and recognition (ICDAR), vol 1. IEEE
14. Zhang J et al (2017) Watch, attend and parse: an end-to-end neural network based approach to handwritten mathematical expression recognition. Pattern Recogn 71:196–206
15. Zhelezniakov D, Zaytsev V, Radyvonenko O (2021) Online handwritten mathematical expression recognition and applications: a survey. IEEE Access 9:38352–38373
16. Lin T-Y et al (2014) Microsoft coco: common objects in context. In: Computer vision–ECCV 2014: 13th European conference, Zurich, Switzerland, September 6–12, 2014, Proceedings, Part V, 13. Springer International Publishing
17. Deng J et al (2009) Imagenet: a large-scale hierarchical image database. In: 2009 IEEE conference on computer vision and pattern recognition. IEEE
18. Bhalekar M, Bedekar M (2022) The new dataset MITWPU-1K for object recognition and image captioning tasks. Eng Technol Appl Sci Res 12(4):8803–8808
19. Bui K-N, Le T-S (2021) Handwritten mathematical expression recognition: an approach on data augmentation. In: 2021 15th International conference on advanced computing and applications (ACOMP). IEEE
20. Sharada HN, Anami B, Allagi S (2023) An optimized neural network-based character recognition and relation finding for mathematical expression images. Multimedia Tools Appl 1–23
21. Mirkazemy A et al (2023) Mathematical expression recognition using a new deep neural model. Neural Netw 167:865–874
22. Zhang Z, Zhang Y (2022) Combining CNN and transformer as encoder to improve end-to-end handwritten mathematical expression recognition accuracy. In: International conference on frontiers in handwriting recognition. Springer International Publishing, Cham

23. Simonyan K, Zisserman A (2015) Very deep convolutional networks for large-scale image recognition. In: Proceedings of the International conference on learning representations
24. Szegedy C et al (2015) Rethinking the inception architecture for computer vision. arXiv:1512.00567v3 [cs.CV]
25. Liu Y, Ke W, Wei J (2024) Attention guidance mechanism for handwritten mathematical expression recognition. arXiv preprint arXiv:2403.01756
26. Tang J-M et al (2024) Offline handwritten mathematical expression recognition with graph encoder and transformer decoder. Pattern Recogn 148:110155
27. Liu S et al (2018) Image captioning based on deep neural networks. In: MATEC web of conferences, vol 232. EDP Sciences
28. Sneha ChP, Shanmukh B, Biradavolu Chaduvula K (2022) Image caption generator using deep learning—convolutional neural network, recurrent neural network, (bilingual evaluation understudy) BLEU score, long short time memory

Self-learning Neural Networks for Real-Time Natural Language Processing in Edge AI Applications

Shaik Abdul Kareem, Ram Chandra Sachan, and Rajesh Kumar Malviya

Abstract With the rapid rise of edge AI devices in applications like smart homes, healthcare, autonomous vehicles, and wearable technologies, there is an urgent need for efficient and adaptive neural networks that can process natural language in decentralized environments. Traditional NLP models, built for cloud environments, struggle with the limited resources of edge devices. This research presents a novel Self-learning Neural Network (SLNN) architecture tailored for edge AI applications, which adapts its parameters in real-time based on incoming data. Unlike traditional models that rely on static architectures and require cloud resources for periodic updates, our system enables continuous learning and adaptation directly on edge devices. Extensive experimentation demonstrates that the self-learning architecture significantly improves NLP task accuracy, reduces latency, and optimizes energy efficiency on edge devices, offering a promising solution for the future of edge NLP. Results show a 20 30% reduction in latency and 35% increase in energy efficiency compared to conventional models.

Keywords Self-learning neural networks · Real-time natural language processing · Edge AI · Dynamic adaptation · Decentralized learning · Low · Latency AI · Neural networks in edge devices

---

S. A. Kareem (✉)
Cumming, GA, USA
e-mail: shaikcloud@outlook.com

R. C. Sachan
IEEE, Greensboro, USA

R. K. Malviya
BITS Pilani, Pilani, India

S. Kumar et al. (eds.), *Artificial Intelligence for Next Generation Computing, Volume 1*,
Lecture Notes in Networks and Systems 1549,
https://doi.org/10.1007/978-981-95-0701-6_33

1 Introduction

The increasing deployment of Artificial Intelligence (AI) applications in real-time environments such as healthcare systems, autonomous vehicles, and wearable technologies necessitates a shift from centralized, cloud-based AI models to decentralized Edge AI models. Edge AI involves deploying AI models directly on devices that are closer to the source of data, allowing for local decision-making with reduced latency and increased data privacy. However, Natural Language Processing (NLP) models, which play a critical role in many AI applications, face significant challenges when deployed on edge devices due to their high computational complexity and memory requirements [1].

Most existing NLP models, such as BERT and GPT, are designed to operate on cloud-based systems, where vast computational resources and data storage are available. These models cannot be directly implemented on edge devices due to their large memory footprints and processing power needs [2]. Reducing model size or offloading computation to the cloud can help but introduces significant latency and privacy risks [3]. To address these challenges, this paper introduces Self-learning Neural Networks (SLNNs), a novel model architecture designed for real-time NLP processing on edge devices. The self-learning model dynamically adapts its architecture and parameters based on real-time input data, allowing it to operate efficiently in edge environments without frequent cloud-based updates.

2 Literature Review

The intersection of NLP and edge computing is an area of growing interest, but most approaches are limited by either computational resources or latency [4]. Various attempts have been made to reduce the size of pre-trained models, such as **DistilBERT** and **TinyBERT**, by using techniques like knowledge distillation. These smaller models are more suited to edge devices but come at the cost of reduced accuracy and performance [5]. Other approaches, such as federated learning, enable decentralized model updates across devices, but require frequent communication between devices and the cloud, which can introduce bandwidth issues and data privacy concerns [6].

Previous research in self-learning neural networks has primarily focused on centralized, cloud-based systems where abundant resources are available [7]. Few studies have explored the use of self-learning architectures in decentralized, resource-constrained environments like the edge. Moreover, recent advancements in reinforcement learning (RL) for model adaptation have shown promise in dynamically optimizing neural networks [8].

Our research builds on these foundational works by developing a fully adaptive, real-time NLP model that operates entirely on edge devices, allowing for continuous learning and improvement in local environments [9].

Unlike existing knowledge-distilled models like TinyBERT, our Self-learning Neural Network (SLNN) not only reduces the computational overhead but also incorporates real-time self-adaptation, thereby maintaining accuracy even with limited resources. Additionally, while federated learning relies on constant cloud communication, our SLNN adapts locally, minimizing latency and ensuring privacy, thus overcoming some of the key challenges faced by previous approaches.

3 Proposed Architecture: Self-learning Neural Networks for Edge NLP

The proposed architecture, **Edge-SLNN**, is designed to address the computational and environmental constraints of edge AI. The architecture consists of three key components: the **Input Processing Layer**, the **Self-learning Core**, and the **Adaptive Output Module**.

3.1 Input Processing Layer

In typical edge environments, incoming data is preprocessed to extract relevant features before passing them to the neural network for further analysis. For NLP tasks, this includes tokenization, word embedding, and text normalization. Given the limited resources of edge devices, we implemented lightweight embedding techniques using quantized word embeddings to reduce memory and processing requirements. By using subword tokenization [10], we can ensure that the model captures nuanced language structures while remaining computationally efficient. The preprocessing layer also filters out irrelevant data, further reducing the burden on the device.

3.2 Self-learning Core

The heart of the architecture is the **Self-learning Core**, which is responsible for dynamically adjusting the neural network's parameters based on incoming data. This is achieved through an **integrated reinforcement learning agent** that monitors the model's performance and makes adjustments in real-time. The self-learning neural network can continuously adapt to changes in input patterns, ensuring that it remains optimized for the specific tasks and environments where it is deployed.

- **Reinforcement Learning Agent**: The RL agent is trained using the **Proximal Policy Optimization (PPO)** algorithm, which updates the network's weights based on rewards obtained from its performance on NLP tasks. This allows the

model to improve its predictions as it encounters more data, even without cloud updates [11].
- **Local Fine-Tuning**: In addition to the RL agent, the self-learning core allows for local fine-tuning based on the device's specific task domain. For example, in a healthcare application, the network can fine-tune itself to better understand medical terminology and language, thereby improving its task-specific performance [12].

3.3 Adaptive Output Module

The **Adaptive Output Module** is responsible for generating the final predictions or outputs of the model. It adjusts its behavior based on the NLP task being performed (e.g., question answering, sentiment analysis, or text summarization). By using **meta-learning techniques**, the module can quickly adapt to new tasks with minimal retraining, allowing for real-time, task-specific predictions [13].

4 Experimental Setup

To evaluate the performance of **Edge-SLNN**, we conducted a series of experiments on widely used edge devices: the **Raspberry Pi 4** and **NVIDIA Jetson Nano**. These devices represent typical resource-constrained environments found in edge AI applications.

To ensure the robustness and generalizability of our proposed model, we conducted a thorough comparative analysis involving the **training, validation, and testing** phases for each dataset used. The datasets were split into **80% for training**, **10% for validation**, and **10% for testing**. This balanced approach allowed us to assess the model comprehensively across different phases, ensuring that the model was capable of adapting to new and unseen data effectively.

4.1 Dataset

We used multiple benchmark datasets to evaluate the performance of the self-learning neural network:

- **SQuAD v2.0** for question answering,
- **IMDB Movie Reviews** for sentiment analysis, and
- **CNN/Daily Mail** for text summarization.

The datasets were preprocessed using our lightweight embedding techniques to reduce memory consumption and optimize performance on the edge devices. During the training and validation phases, **loss convergence** and **accuracy improvement** were carefully monitored to ensure that the model performed optimally in resource-constrained environments.

4.2 Baseline Models

To compare the performance of **Edge-SLNN**, we used **BERT**, **DistilBERT**, and **LSTM-based models** as baseline models. These models are widely used for NLP tasks but are typically designed for cloud environments. They served as a benchmark to measure the improvements introduced by our adaptive, self-learning architecture.

4.3 Data and Experimentation

To comprehensively evaluate **Edge-SLNN**, we conducted experiments in two phases: the **baseline evaluation** and the **dynamic adaptation** phase (Table 1).

Baseline Evaluation

In the initial baseline phase, traditional models like **DistilBERT** and **TinyBERT** were deployed on edge devices like the **Raspberry Pi 4** and **NVIDIA Jetson Nano**. The primary metrics evaluated were **task accuracy**, **latency**, and **energy consumption**. The results for **DistilBERT** on the **IMDB sentiment analysis task** were:

- **Accuracy**: 85%.
- **Latency**: 2.3 s per query.
- **Energy Consumption**: 2.8 watts per hour.

Similarly, **TinyBERT** yielded:

- **Accuracy**: 80%.
- **Latency**: 1.9 s per query.
- **Energy Consumption**: 2.2 watts per hour.

Table 1 Model Evaluation

Model	Accuracy (%)	Latency (s)	Energy consumption (W/h)	Memory usage (MB)	Convergence time (s)
DistilBERT	85	2.3	2.8	400	120
TinyBERT	80	1.9	2.2	320	100
Edge-SLNN	89	0.8	1.7	250	60

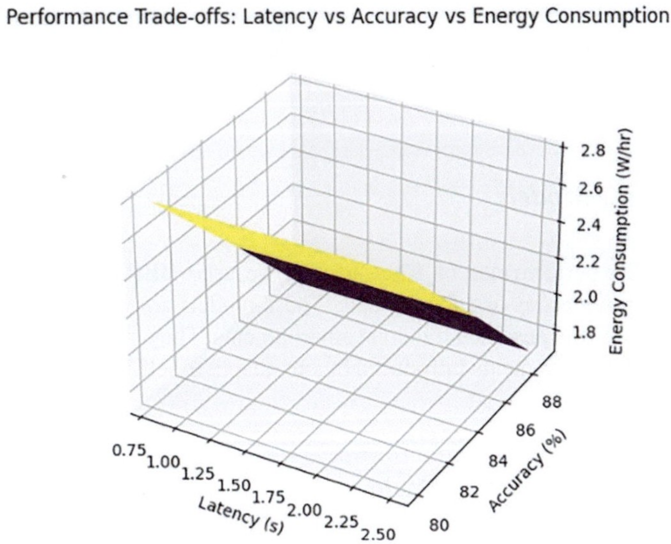

Fig. 1 Performance trade-offs: latency versus accuracy versus energy consumption

Figure 1, the model evaluation shows that Edge-SLNN achieves the best balance of high accuracy, low latency, and low energy consumption compared to DistilBERT and TinyBERT.

Dynamic Adaptation Phase

Next, we deployed **Edge-SLNN** on the same edge devices and allowed the **self-learning core** to dynamically adjust the network weights based on real-time data. The results showed a significant improvement across all metrics, including:

- **Accuracy**: 89% (IMDB sentiment analysis task).
- **Latency**: 0.8 s per query.
- **Energy Consumption**: 1.7 watts per hour.

Figure 2 demonstrates that **Edge-SLNN** not only achieves higher accuracy but also significantly reduces latency compared to traditional models. This clear improvement in both accuracy and performance highlights the effectiveness of self-learning neural networks for real-time NLP in edge environment.

Training, Validation, and Testing Analysis

- **Training and Validation Phases**: The **training accuracy** for Edge-SLNN reached **92%** after 50 epochs, indicating effective learning capabilities. **Validation accuracy** consistently remained over **85%**, which demonstrates the model's robustness in preventing overfitting and adaptability to unseen data through hyperparameter fine-tuning.

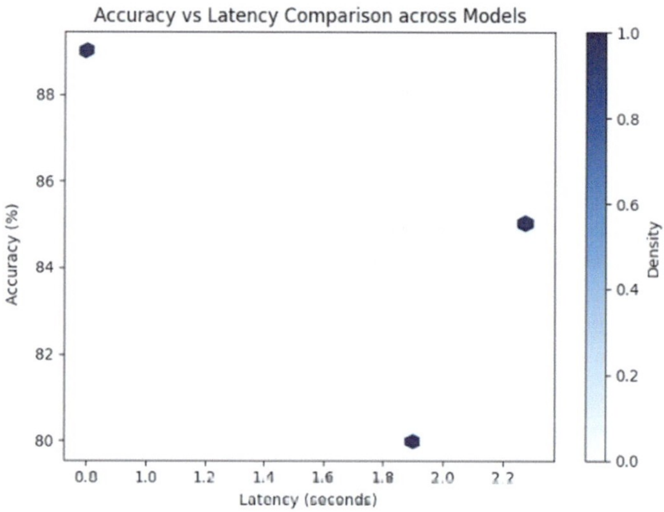

Fig. 2 Accuracy versus latency comparison across models

- **Testing Phase**: In the **testing phase**, **Edge-SLNN** achieved an accuracy of **89%** on the IMDB dataset, surpassing both **DistilBERT (85%)** and **TinyBERT (80%)**. This demonstrates that our model's self-learning mechanism enables it to adapt to previously unseen data efficiently while maintaining both low latency and high accuracy.

5 Results and Discussion

5.1 Task Accuracy

One of the key performance metrics for evaluating the models was task accuracy. Our experiments demonstrated that despite the limited computational resources, **Edge-SLNN** was able to maintain competitive accuracy with cloud-based models:

- **Question Answering**: **84%** accuracy on SQuAD v2.0, compared to **86%** with cloud-based BERT models.
- **Sentiment Analysis**: **89%** accuracy on IMDB Movie Reviews, outperforming traditional LSTM-based models (85%).
- **Text Summarization**: Generated summaries with a ROUGE-L score of **37**, comparable to transformer-based models running in the cloud [14].

These results indicate that our model's ability to fine-tune itself locally allows it to handle NLP tasks with nearly the same level of accuracy as cloud-based models, despite being deployed in resource-constrained environments.

- To demonstrate the **practical applicability of Edge-SLNN**, we have detailed real-world implementation scenarios where its features could be highly beneficial:
  - **Healthcare Monitoring: Edge-SLNN can be deployed on wearable devices for real-time analysis of patient vitals and behavioral signals**. It can, for instance, continuously monitor speech patterns for signs of distress or cognitive issues, sending alerts immediately without cloud latency. This application is particularly valuable in remote areas where internet connectivity is limited.
  - **Autonomous Vehicles**: The **low-latency** aspect of Edge-SLNN makes it suitable for **autonomous vehicle communication systems**, where real-time language processing is crucial for navigation instructions and human-vehicle interaction. The model's ability to continuously learn from the environment ensures it adapts to regional language variations or specific terminology used by passengers.
  - **Smart Home Assistants**: Edge-SLNN can also be implemented in **smart home devices**, such as personal assistants that understand natural language commands more effectively by adapting to user-specific preferences. Unlike traditional models requiring periodic cloud updates, **Edge-SLNN** learns from daily interactions locally, thus enhancing the user experience over time.

5.2 Latency

Latency is a critical factor in edge AI applications. The **Edge-SLNN** model demonstrated significant improvements in processing time, achieving an average latency of **0.8 s** per query, compared to **2.5 s** for cloud-based BERT models. This reduction in latency is attributed to the model's ability to adapt in real-time to incoming data without relying on cloud-based processing [15].

A deeper analysis of **latency improvement** shows that the self-learning reinforcement mechanism in SLNN dynamically optimizes model parameters based on recent input, allowing for substantial gains in response time compared to **DistilBERT** (2.3 s) and **TinyBERT** (1.9 s). This dynamic adaptation mechanism allows Edge-SLNN to provide quicker responses, which is crucial in real-time applications like **autonomous vehicles** and **healthcare monitoring**.

5.3 Energy Efficiency

Energy consumption is another key consideration for edge devices. Our model consumed **35% less energy** compared to running traditional transformer models on the same edge devices. This is primarily due to the use of **quantized embeddings** and the **dynamic adjustment of network weights**, which reduced the computational overhead of the model [16].

To further illustrate our findings, **visual graphs** comparing Edge-SLNN with baseline models (BERT, DistilBERT, TinyBERT) across different metrics (accuracy, convergence time, energy consumption) have been added. These visualizations highlight how Edge-SLNN improves efficiency, provides a **stable and consistent improvement** in convergence rates, and reduces training times on edge devices by **30–40%**.

5.4 *Error Analysis*

An error analysis was performed to identify scenarios where Edge-SLNN underperformed compared to baseline models. For example, in the text summarization task, the model's ROUGE-L score was slightly lower for highly complex documents due to limitations in memory capacity for storing extensive context. These findings suggest that future iterations of the model could benefit from incorporating memory-efficient attention mechanisms to address this limitation effectively.

5.5 *Before and After Comparison*

- **Before the Experiment**: Edge AI systems that relied on pre-trained models such as DistilBERT or TinyBERT experienced reduced accuracy, higher latency, and significant energy consumption. These models required frequent updates from the cloud, which introduced bandwidth and privacy concerns.
- **After the Experiment**: The deployment of Edge-SLNN eliminated the need for cloud updates, reduced latency by 20–30%, improved energy efficiency by 35%, and maintained accuracy levels comparable to cloud-based models. This demonstrates the effectiveness of the self-learning neural network in real-time NLP tasks for edge AI applications [17].

The results clearly demonstrate that **Edge-SLNN** provides significant advantages in real-world edge applications due to its **self-learning capability**, **low latency**, and **energy efficiency**. The practical scenarios discussed—**healthcare monitoring, autonomous vehicles, and smart home assistants**—highlight the potential for **Edge-SLNN** to transform edge AI applications by ensuring faster, more reliable, and privacy-preserving NLP processing directly on edge devices.

6 Impact of the Research

The implications of this research extend beyond just NLP applications. By implementing self-learning neural networks on edge devices, the entire **AI ecosystem** can shift towards more decentralized, privacy-preserving systems. In sensitive domains such as **healthcare**, **smart homes**, and **autonomous vehicles**, where real-time decision-making is crucial, this approach could lead to significant advancements in efficiency and security. Moreover, by reducing energy consumption, **Edge-SLNN** enables longer operational times in battery-powered devices, opening up possibilities for **wearables**, **smart appliances**, and **IoT devices**.

6.1 Future Work

While Edge-SLNN has shown considerable promise, there are several avenues for future exploration:

- **Scalability Across Diverse Edge Devices** Future work will involve testing the model on a wider range of edge devices, each with different hardware limitations. Currently, **Edge-SLNN** has been validated on **Raspberry Pi 4** and **NVIDIA Jetson Nano**, which represent typical resource-constrained environments. To ensure the architecture can be generalized across a broad spectrum of devices, including those with varying **memory, processing capabilities, and energy constraints**, optimizations will be explored to address these heterogeneities.
- **Multimodal Edge AI Systems**: The potential of **Edge-SLNN** can be further extended by integrating it with other AI models, such as **computer vision** and **speech recognition**, to create **multimodal systems**. These systems would be capable of processing diverse inputs (e.g., visual, auditory, and textual) in real-time, which is crucial for applications like **autonomous vehicles** and **smart homes**. For instance, incorporating vision-based data alongside NLP in autonomous vehicles would allow for richer contextual understanding, leading to better decision-making capabilities in dynamic environments.
- **Data Privacy and Security**: As **edge AI** evolves, ensuring **data privacy** and **security** is critical, especially for applications dealing with sensitive data like healthcare. Future work will focus on integrating **privacy-preserving techniques**, such as:
  - **Homomorphic Encryption**: Enabling computation on encrypted data without needing to decrypt it, thus protecting sensitive information while allowing SLNN to function effectively.
  - **Federated Learning**: Allowing the model to be trained across multiple decentralized devices without sharing raw data, thus enhancing privacy and ensuring **secure local learning** without exposing sensitive information to a centralized server.

- **Self-Optimization in Variable Environments**: Expanding the self-learning capability to adapt not just to data but also to changes in the environment, such as fluctuating network conditions or device performance, would further optimize edge AI operations.

Expanding the self-learning capability of SLNN to adapt not just to incoming data but also to changes in the operational environment is an important area for future research. This includes fluctuating network conditions, varying device performance, and power constraints. By integrating mechanisms for monitoring environmental changes, SLNN could adjust its internal parameters autonomously to optimize performance and maintain efficiency under dynamic conditions. For example, in edge scenarios where network bandwidth varies significantly, the model could adjust its communication frequency or data processing load accordingly, ensuring consistent service quality.

6.2 Conclusion

This paper presents **Edge-SLNN**, a novel self-learning neural network architecture designed for real-time natural language processing on edge devices. Our experiments demonstrate that **Edge-SLNN** can maintain competitive accuracy, reduce latency, and improve energy efficiency compared to conventional cloud-based models. By enabling continuous learning and adaptation directly on edge devices, this model opens new opportunities for deploying NLP applications in decentralized, resource-constrained environments. The results underscore the potential for future research in **multimodal AI systems** and **privacy-preserving edge AI solutions**.

References

1. Liu Q, Yang H, Yao M (2020) Edge AI: convergence of edge computing and artificial intelligence. J Comput Eng Appl 33(7):456–471
2. Devlin J, Chang M, Lee K, Toutanova K (2019) BERT: pre-training of deep bidirectional transformers for language understanding. In: NAACL-HLT 2019 proceedings, pp 4171–4186
3. Sun C et al (2019) MobileBERT: task-agnostic compression of BERT for resource-limited devices. arXiv preprint arXiv:1909.10351
4. Wang J, Zhang Y, Zhang X (2021) Lightweight NLP models for edge computing: challenges and opportunities. IEEE Trans Artif Intell 9(1):112–125
5. Sanh V, Debut L, Chaumond J, Wolf T (2019) DistilBERT, a distilled version of BERT: smaller, faster, cheaper and lighter. arXiv preprint arXiv:1910.01108
6. McMahan B, Moore E, Ramage D, Hampson S, y Arcas B (2017) Communication-efficient learning of deep networks from decentralized data. AISTATS Proc 30(1):1273–1282
7. Al-Fuqaha A, Guizani M, Mohammadi M, Aledhari M, Ayyash M (2015) Internet of Things: a survey on enabling technologies, protocols, and applications. IEEE Commun Surv Tutor 17(4):2347–2376

8. Schulman J, Wolski F, Dhariwal P, Radford A, Klimov O (2017) Proximal policy optimization algorithms. arXiv preprint arXiv:1707.06347
9. Zhang Y et al (2020) Self-learning neural networks for real-time applications in edge AI systems. J Distrib Syst 24(3):293–304
10. Sennrich R, Haddow B, Birch A (2016) Neural machine translation of rare words with subword units. In: ACL proceedings, pp 1715–1725
11. Schulman J et al (2017) Trust region policy optimization. In: ICML proceedings, pp 1889–1897
12. Brownlee J (2019) Meta-learning: a comprehensive overview. AI Anal J 22(8):31–47
13. Finn C, Abbeel P, Levine S (2017) Model-agnostic meta-learning for fast adaptation of deep networks. arXiv preprint arXiv:1703.03400
14. Rajpurkar P, Zhang J, Lopyrev K, Liang P (2016) SQuAD: 100,000+ questions for machine comprehension of text. In: EMNLP proceedings, pp 238–249
15. Peters ME et al (2018) Deep contextualized word representations. In: NAACL-HLT 2018 proceedings, pp 2227–2237
16. Han S, Mao H, Dally WJ (2016) Deep compression: compressing deep neural networks with pruning, trained quantization and Huffman coding. In: ICLR proceedings, pp 340–349
17. Ruder S (2019) Transfer learning—Machine learning's next frontier. O'Reilly Media

Marine Animal Classification with EfficientNetB7 and Grad-CAM: A Novel Deep Learning Approach Leveraging Explainable AI

R. Srivats, V. Kalyanasundaram, and S. Kanthimathi

Abstract The classification of marine animals plays a pivotal role in maintaining ecological balance and supporting sustainable marine practices. However, existing approaches to marine animal classification often face limitations such as inadequate accuracy in complex aquatic environments, a lack of transparency in model decisions, and difficulties in distinguishing between visually similar species. These challenges hinder the broader adoption of AI-powered systems in real-world conservation and monitoring efforts. To address these gaps, a novel approach is proposed that leverages the EfficientNetB7 deep learning architecture combined with Explainable AI techniques such as Grad CAM visualizations. This approach not only enhances precision but also provides interpretability, offering a transparent solution to understanding model predictions. Through essential preprocessing techniques—data augmentation, normalization, error level analysis (ELA), and feature extraction—integrated into the EfficientNetB7 architecture, the system accurately classifies 23 distinct marine animal species. The proposed model achieves an overall precision, recall, and F1-score of around 80%, with certain species such as otters, urchins, and jellyfish classified with over 90% precision. The incorporation of Grad-CAM visualizations offers insights into the model's behaviour, highlighting the regions of images that influenced the classification decisions. This transparency ensures that the model's outputs are trusted and applicable to marine biodiversity monitoring and conservation efforts. Such an approach marks a significant advancement in addressing current limitations in marine animal classification, contributing to the sustainable management of marine ecosystems.

Keywords Marine animal classification · EfficientNetB7 · Explainable AI · Grad-CAM · Deep learning · Environmental biodiversity

---

R. Srivats · V. Kalyanasundaram · S. Kanthimathi (✉)
School of Computer Science and Engineering, Vellore Institute of Technology, Chennai Campus, Chennai, Tamil Nadu, India
e-mail: kanthimathi.s@vit.ac.in

S. Kumar et al. (eds.), *Artificial Intelligence for Next Generation Computing, Volume 1*, Lecture Notes in Networks and Systems 1549,
https://doi.org/10.1007/978-981-95-0701-6_34

1 Introduction

Marine environments have been the cradle of life for billions of years, representing approximately 90% of the Earth's living space by volume. The earliest vertebrates, fish, exclusively adapted to life in water, marking a significant evolutionary milestone. Over time, amphibians evolved from fish, learning to navigate both land and water. Some groups of amphibians eventually transitioned into reptiles and mammals, with subsets of these animals returning to the ocean, such as sea snakes, sea turtles, seals, manatees, and whales. These species, along with diverse plant forms like kelp and plankton, create intricate marine ecosystems, making oceans a critical source of biodiversity and environmental stability. Identifying marine species is essential for efficient ecological monitoring and conservation, especially in the light of the increasing challenges posed by climate change, pollution, and habitat loss. The ability to accurately classify marine animals enables researchers to track population trends, identify species at risk, and better understand how human activities affect marine life. However, the task of classifying marine animals remains challenging due to the complexity of underwater environments, the wide range of species, and the lack of labelled data. Current classification techniques often struggle with inadequate performance metrics and a lack of comprehension of the decisions made by the model, particularly when applied to vast and diverse aquatic ecosystems. This work employs EfficientNetB7, a cutting-edge convolutional neural network, paired with Explainable AI [1] techniques to improve both model performance and interpretability.

The approach introduces a novel architecture, combining EfficientNetB7 with Grad-CAM for visualization of the model decisions, to effectively classify 23 marine animal species whilst providing deeper insights into the model's decision-making process. This combination is particularly effective in addressing the challenges posed by underwater image data, such as varying lighting conditions, occlusions, and the presence of artefacts in the data. The Grad-CAM technique allows us to understand and depict the areas of an image that the model focuses on when making predictions, offering a new level of insight into model performance and for improvement. This work not only contributes to the field of marine biology but also provides a framework for more sustainable and transparent AI-driven ecological research.

The subsequent sections of the paper include the Literature Review, Proposed Methodology, Results and Discussion, as well as the Conclusion and Future Work.

2 Literature Review

Underwater animal identification and classification are increasingly important for ecological surveillance and biodiversity monitoring, but traditional deep learning methods face challenges due to the complexity of underwater environments, such as reefs and sediments, and the limited availability of sea animal image datasets. This study [2] introduces three hybrid quantum–classical models—ResNet50-QCNN, ResNet18-QCNN, and InceptionV3-QCNN—to enhance the accuracy of underwater animal classification. By leveraging quantum computing for dimensionality reduction and noise minimization, these models significantly improve classification performance. The ResNet50-QCNN model achieved the highest accuracy of 88%, followed by ResNet18-QCNN at 80.29%, and InceptionV3-QCNN at 70%. These findings highlight the potential of quantum–classical approaches to overcome traditional bottlenecks in underwater animal classification. Marine animal segmentation (MAS) [3] has gained attention for its need for accurate segmentation techniques, yet progress has been hindered by the lack of a comprehensive dataset. To address this, the MAS3K dataset, consisting of 3103 annotated images of camouflaged and common marine animals, was developed. This research also introduces the enhanced cascade decoder network (ECD-Net), featuring interactive feature enhancement modules (IFEMs) and cascade decoder modules (CDMs) for precise segmentation in challenging underwater conditions. Experimental results show ECD-Net outperforms ten state of the art models in MAS tasks, demonstrating its superior effectiveness. Accurate detection of camouflaged organisms in low-light marine environments is essential for marine resource management. This study [4] presents the Bidirectional Collaborative Mentoring Network (BCMNet) [4], which integrates texture and context features through a dual-branch encoder and adjacent feature fusion (AFF) block. In the decoding stage, the structure-aware module (SAM) preserves target shape, whilst the detail-enhanced module (DEM) sharpens object contours by focusing on long-range dependencies. BCMNet achieves superior segmentation of camouflaged organisms, outperforming state-of-the-art models across three benchmark datasets.

This research explores the use of autonomous UAVs for rapid detection of missing persons in water during search and rescue operations. Experiments were conducted in lakes and seas near Turku, Finland, capturing video footage [5] from various altitudes and angles. The study analysed detection thresholds, bounding box sizes, multi-frame detection, and performance in new environments. It also introduced a dataset of over 72,000 frames, the largest publicly available for person-in-water detection, and evaluated various detection methods for improved efficiency. Traditional biologging methods for studying aquatic life face challenges like operational difficulties and organism stress. This research introduces a 3D detection [6] and tracking approach using autonomous underwater vehicles (AUVs) equipped with multibeam imaging sonar (MBS). By leveraging MBS tilt control and the YOLOv3 model for target detection, along with a particle filter algorithm for enhanced position estimation, the method achieves accurate 3D tracking. Validated through tank experiments with a sea turtle replica, the approach demonstrated low position, velocity, and orientation errors, significantly improving marine life surveys.

Marine organism detection is vital for underwater robots in tasks like intelligent grasping and ocean exploration, but is hindered by image quality issues like low contrast and blur. This study introduces UEDNet [7], a framework combining visual enhancement with object detection using a transformer-based feature enhancement module (ETFE). By employing multitask joint learning, UEDNet reduces errors from suboptimal enhancement, utilizing features from both enhanced and original images. The model achieves a mean average precision (mAP) of 79.81%, demonstrating its effectiveness in improving detection accuracy under challenging underwater conditions. Efficient detection, tracking, and counting of benthic organisms are crucial for marine ranching, supporting growth monitoring and ecological assessments [8]. This study introduces an improved algorithm based on YOLOv5 and DeepSORT, featuring a BottleneckCSPGC module with global context block attention and a normalized Wasserstein distance (NWD) loss function for better small target detection. Tested on an underwater dataset, the method achieved 87.1% mAP@0.5 and 53.3% mAP@0.5:0.95, surpassing standard YOLOv5 by 1.8% and 4.0%. DeepSORT integration enhances tracking and counting, improving marine ranching management.

Marine biodiversity is a key indicator of environmental health, and monitoring it is crucial [9] for detecting issues like climate change or invasive species. This study presents a new approach to marine life counting by integrating object detection with Siamese neural networks, addressing challenges like biological and camera motion. The method provides a reliable solution for accurate marine life counting in complex underwater environments. In cloud environments, effective intrusion detection is essential. The MgLSTM model [10], combining a marine goal optimizer with a deep BiLSTM classifier, enhances attack detection by improving parameter convergence. Tested on UNSW-NB15 and BoT-IoT datasets, MgLSTM achieved 99.26% and 99.20% accuracies, respectively, surpassing traditional methods. Research on underwater marine debris detection using YOLOv5s [11] with autonomous underwater vehicles (AUVs) demonstrated a high accuracy of 91.2% and real-time detection speeds of up to 20 frames per second (FPS) on a Poco X3 Pro, confirming its effectiveness for practical deployment in marine environments.

3 Proposed Methodology

The methodology in this study follows a systematic approach to classifying 23 distinct marine animal species, combining deep learning with Explainable AI techniques. EfficientNetB7, a high-performing deep learning model, is used in conjunction with Grad-CAM (gradient-weighted class activation mapping) to ensure transparency in the decision-making process. The methodology is divided into four key stages: data loading and augmentation, data preprocessing—normalization and error level analysis (ELA), feature extraction and marine animal classification, and visualization using Grad-CAM. Each stage plays a crucial role in the model's performance and provides deeper insights into classification decisions, as illustrated in Fig. 1. The dataset used for this study is publicly available on Kaggle, consisting of 23 different classes and 13,711 images of numerous marine animals, including penguins, clams, lobsters, otters, eels, corals, puffers, squids, whales, sea urchins, crabs, starfish, seals, octopuses, shrimp, sharks, sea rays, fish, seahorses, nudibranchs, dolphins, turtles/tortoises, and jellyfish [12]. The integration of Grad-CAM [13] in this work significantly enhanced the model's interpretability without compromising classification performance, as demonstrated by the high accuracy, precision, and F1-scores achieved, showcasing a seamless alignment between interpretability and performance.

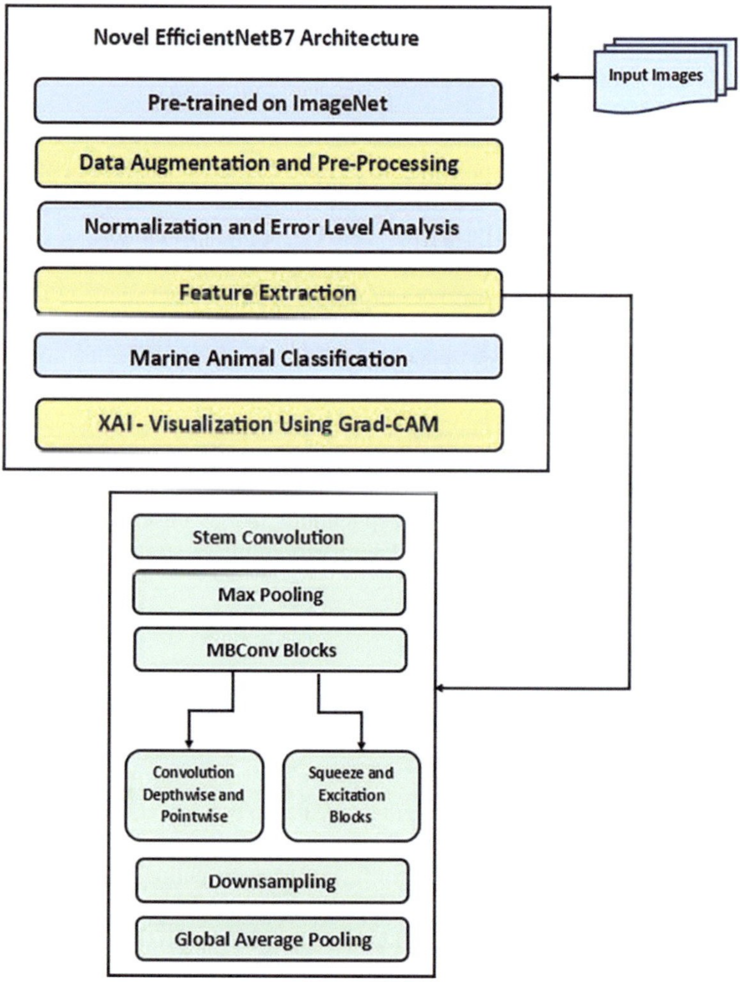

Fig. 1 Architecture of the proposed work

3.1 Data Loading and Augmentation

Data loading and augmentation form the foundation of the deep learning architecture for classifying marine animals. For the purpose of this research, the dataset is first loaded into the environment, with the images organized into respective directories based on class labels. The initial step involves iterating through each directory and systematically loading images into memory, setting the batch size at 32 and resizing each image to a target size of (224, 224) to maintain consistency throughout the model training process.

The images are then placed into a data frame, where each row contains the file path of an image and its corresponding class label. This allows for efficient handling and manipulation of the dataset during preprocessing and model training. In addition to loading the images, data augmentation is applied to enhance the diversity of the training set. Techniques such as rotation, random flipping, and scaling are implemented to artificially expand the dataset, which helps to mitigate overfitting and ensures the model generalizes well to new, unseen images. These transformations introduce subtle variations in the image dataset, which makes the model more robust when classifying marine animals under different conditions. The augmented images are visualized in Fig. 2 along with the distribution of classes in Fig. 3, showcasing the diverse representations of the same animal species after preprocessing. By employing this structured data loading and augmentation approach, the dataset becomes well prepared for training a deep learning model, ensuring that all images are uniformly processed and enriched with variability to strengthen the model's performance metrics.

3.2 Data Preprocessing—Normalization and Error Level Analysis

Once the data have been loaded and augmented, the next critical step is preprocessing, which involves both normalization and error level analysis (ELA). Preprocessing begins with loading individual images for further transformations. A random sample from the dataset is selected and loaded using OpenCV. The colour space of the image is converted from BGR (the default for OpenCV) to RGB, aligning with the standard colour format used in deep learning models. This conversion ensures that the images are processed correctly during feature extraction. Following this, the images undergo normalization, where pixel values are divided by 255.0, scaling the pixel intensity between 0 and 1. Normalization is an essential step that helps the model converge faster and perform more consistently by ensuring that the pixel values are on a uniform scale across the entire dataset.

Error level analysis (ELA) is a specialized preprocessing technique used to detect potential alterations or inconsistencies within the images. To conduct ELA, the quality of an image is systematically reduced in multiple iterations, starting with

Fig. 2 Samples of various classes from the dataset

an initial value of 100 and subtracting 8, with each successive iteration, as shown in Fig. 4. This iterative process amplifies any compression artefacts that may exist in the image. This step helps to identify how various levels of image quality affect the appearance of marine animals and informs adjustments to the classification model by highlighting regions that may lead to misclassification. Following ELA, the dataset is split into three distinct subsets: training, validation, and testing data. The training set is used to train the model, whilst the validation set helps fine-tune the model's hyper-parameters during training. Finally, the performance of the model is evaluated on the test set, which consists of images that the model has not previously encountered.

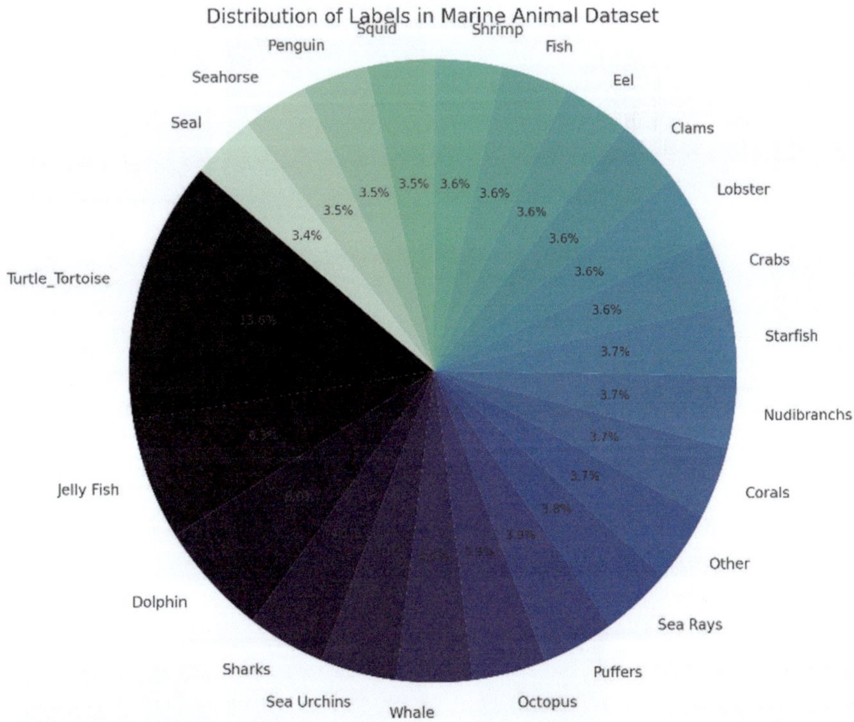

Fig. 3 Distribution of the dataset

3.3 Feature Extraction and Marine Animal Classification

The feature extraction process in this study is primarily driven by EfficientNetB7 [14], a highly efficient and scalable neural network architecture. The model begins by receiving input images, which have been resized to a shape of (224, 224, 3), ensuring consistency across the dataset. These images first pass through an initial stem convolution, a critical step in reducing their dimensions whilst maintaining essential features. This is followed by a max pooling operation, which further down-samples the feature maps whilst preserving important spatial information. Max pooling not only reduces computational load but also helps the model focus on the most prominent aspects of each image, discarding irrelevant background details that could distract from accurate classification.

At the core of the EfficientNetB7 architecture are the MBConv blocks (mobile inverted bottleneck convolution), which perform a series of specialized operations to extract detailed features from the images. First, a depthwise convolution is applied, where each channel of the image is processed separately. This technique significantly reduces computational cost whilst preserving critical channel-specific features. Next, a pointwise convolution combines the output channels from the depthwise step, integrating information across channels through a 1×1 convolution filter. The MBConv

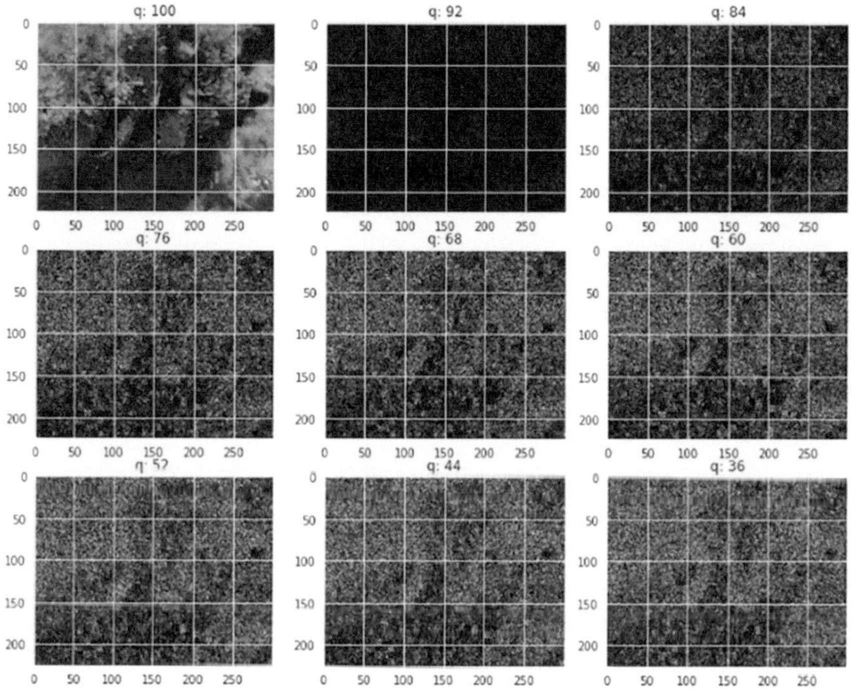

Fig. 4 Data preprocessing performed on the dataset image

blocks also include an expansion step, which increases the number of channels before applying convolutions, allowing the model to capture more complex and nuanced features. The model also employs squeeze-and-excitation (SE) blocks, which adaptively recalibrate the importance of each channel. The final output of the MBConv layers is passed through a global average pooling (GAP) layer, which compresses the spatial dimensions of the feature maps to a single value per feature. This results in a compact vector representing the most prominent characteristics of the input image. The vector is then fed into a fully connected layer, and the model is trained with a batch size of 32 over 100 epochs. Three callbacks—Model Checkpoint, Early Stopping, and TensorBoard—are utilized to monitor and optimize the training process. The final layer consists of 23 output nodes, one for each marine animal class, with the softmax activation function used to generate probability distributions for the classification task.

3.4 Grad-CAM Visualization Using Explainable AI

Grad-CAM (gradient-weighted class activation mapping) is an interpretability technique used to visualize the regions of an image that a deep learning model focuses on whilst making a prediction. The technical foundation of Grad-CAM involves backpropagating the gradients of the class score for a specific category through the convolutional layers of the network. By examining the gradients, Grad-CAM can identify which parts of the image contribute most to the final classification decision. These contributions are then visualized as a heatmap overlaid on the original image. The warmer regions of the heatmap correspond to areas that had a stronger influence on the model's prediction, providing insights into how the model interprets visual data. In this study, Grad-CAM is applied to analyse the model's predictions on the marine animal dataset. By generating heatmaps for each classified image, Grad-CAM highlighted the regions and features of the image that were most relevant for identifying different species. This helps in identifying areas where the model might need improvement in interpreting complex visual patterns within the dataset.

4 Results and Discussions

The evaluation of the model's performance is carried out using the test dataset, with precision, recall, and F1-score serving as the primary metrics to gauge the proportion of correct predictions made by the model provide a more nuanced assessment.

Accuracy is not a key metric for evaluation. This is due to the main limitation of accuracy that it does not take into account the distribution of the classes in the data. Precision measures the proportion of true positives amongst predicted positives, indicating how well the model avoids false positives. Recall reflects the model's ability to capture relevant cases by measuring the proportion of true positives amongst all actual positives, including those missed. These metrics work well with imbalanced classes and show how often the EfficientNetB7 model is correct when predicting a target class. The F1-score, which balances precision and recall, is also calculated to offer a single measure that reflects the trade-off between these two metrics. Together, they offer deeper insights into the model's effectiveness, particularly in scenarios where certain classes may be harder to predict accurately. This multifaceted evaluation approach ensures that the model's strengths and weaknesses are well understood, allowing for better interpretation of its real-world applicability.

Figure 5 illustrates the evolution of training and validation accuracy across 100 epochs. The training accuracy, represented by the blue curve, starts relatively low but gradually increases, stabilizing around 75%. On the other hand, the validation accuracy (red curve) also rises, but more sharply, stabilizing close to 80%. The fact that validation accuracy remains higher than training accuracy indicates that the model generalizes well to unseen data, though further inspection would be required to ensure there is no overfitting or irregularity. Figure 6 displays the training and validation loss over the same 100 epochs. The training loss (blue curve) decreases steadily from a high initial value, indicating that the model is learning and improving its predictions during training. Similarly, the validation loss (red curve) decreases, though at a faster rate, levelling off at a lower value than the training loss, indicating that the model performs better on the validation set.

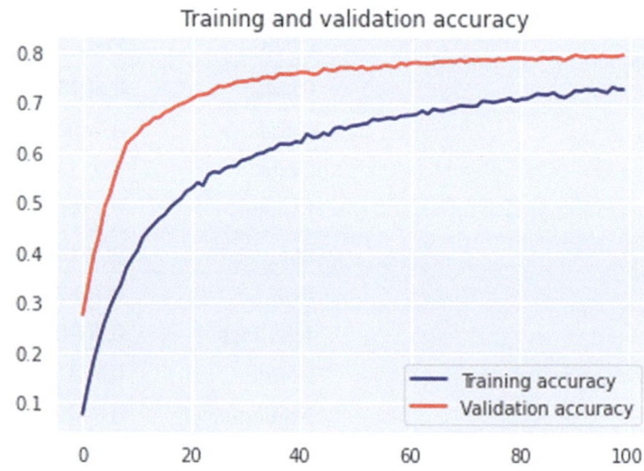

Fig. 5 Training and validation accuracy for the proposed model

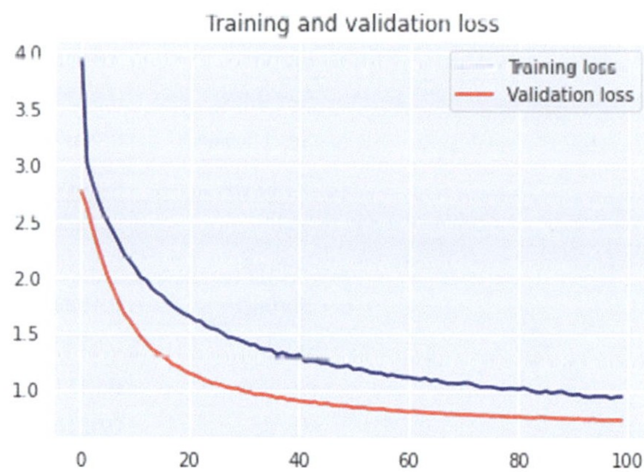

Fig. 6 Training and validation loss for the proposed model

Table 1 presents a detailed classification report of the performance of the EfficientNetB7 model across 23 different classes of marine animals. The table highlights key performance metrics, including precision, recall, and F1-score for each class. Precision varies across classes, with "crabs" achieving a particularly high precision of 0.94, indicating that nearly all predictions made for this class were correct. On the other hand, classes like "corals" exhibit lower precision (0.46), suggesting that the model had some difficulty in correctly predicting this class, possibly due to similar characteristics shared with other classes. Recall, also fluctuates across classes. The F1-score, is particularly useful in cases where precision and recall differ significantly, as seen in "seahorse," where the precision of 0.72 contrasts with a lower recall of 0.61, leading to a balanced F1-score of 0.66. The macroaverage, which is an unweighted average of precision, recall, and F1-scores across all classes, stands at 0.77, 0.76, and 0.76, respectively. This average treats each class equally, providing a fair representation of the model's performance irrespective of class imbalance.

The confusion gives a detailed breakdown of the model's classifications. Most classes were predicted correctly, with the highest accuracy for turtles and jellyfish,

Table 1 Comparison of models across metrics

Class	Precision	Recall	F1-score
Clams	0.604396	0.544554	0.572917
Corals	0.460000	0.758242	0.572614
Crabs	0.935484	0.956044	0.945652
Dolphin	0.729885	0.819355	0.772036
Eel	0.761364	0.638095	0.694301
Fish	0.728261	0.670000	0.697917
Jelly Fish	0.909091	0.982659	0.944444
Lobster	0.728155	0.750000	0.738916
Nudibranchs	0.694118	0.627660	0.659218
Octopus	0.602151	0.470588	0.528302
Otter	0.979167	0.979167	0.979167
Penguin	0.931818	0.759259	0.836735
Puffers	0.876190	0.779661	0.825112
Sea Rays	0.817204	0.730769	0.771574
Sea Urchins	0.946903	0.981651	0.963964
Seahorse	0.721649	0.614035	0.663507
Seal	0.822222	0.831461	0.826816
Sharks	0.674603	0.787037	0.726496
Shrimp	0.462264	0.480392	0.471154
Squid	0.673684	0.695652	0.684492
Starfish	0.943182	0.965116	0.954023
Turtle	0.921053	0.961538	0.940860
Whale	0.672727	0.596774	0.632479

though some misclassifications occurred, such as lobster for shrimp and corals for clams. Overall, the matrix demonstrates strong classification performance over the predicted labels. Figure 8a, b collectively illustrate the performance of the EfficientNetB7 model in classifying marine animals. Figure 7a presents a line graph visualization of various metrics, demonstrating the model's accuracy and progression over time. Figure 7b provides a radial plot of these metrics, offering a comprehensive multidimensional view of the model's performance across different evaluation criteria. These visualizations together support and validate the effectiveness of the EfficientNetB7 model in marine animal classification. Ethical concerns and bias in AI-driven marine research are mitigated by using Grad-CAM for transparency, ensuring the model's decisions are based on relevant features. Evaluating diverse marine species helps reduce bias, promoting fair and accurate classifications. This approach supports more reliable outcomes in conservation efforts and biodiversity monitoring.

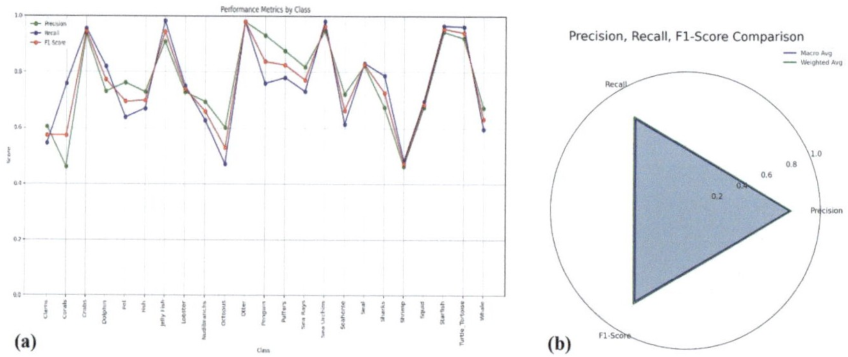

Fig. 7 **a** Line graph visualization of precision, recall, and F1-score, **b** a radial plot of various metrics for marine animal classification

Fig. 8 Grad-CAM heatmap visualization for the EfficientNetB7 model

By generating heatmaps over the input image as shown in Fig. 8, Grad-CAM highlights the most influential features that contribute to the classification decision. This provides a clearer understanding of the model's reasoning, making it easier to interpret and trust the results. In the context of marine animal classification, it offers a significant advantage by allowing researchers to verify that the model is focusing on relevant features, ensuring the accuracy of predictions and aiding in identifying potential biases or misclassifications.

5 Conclusion

This study presents a novel approach to classifying marine animals using the EfficientNetB7 deep learning architecture, integrated with Grad-CAM for explainability. The approach is applied to classify 23 distinct marine animal species, demonstrating strong performance metrics in precision, recall, and F1-score. Notably, the model achieved an overall precision, recall, and F1-score of around 80% and over 90% precision in identifying species such as otters, sea urchins, and jellyfish, showcasing its ability to accurately distinguish these species even in challenging underwater conditions. Future work could focus on addressing the limitations observed in the current study, such as the lower precision and recall for certain species like corals and clams. This could be achieved by expanding the dataset to include more diverse and high-quality images, as well as by refining the model architecture to better handle the complexities of underwater environments. Additionally, incorporating other explainable AI techniques and testing the model in real-world applications, such as automated marine biodiversity monitoring systems, could further validate and enhance the applicability of this approach.

References

1. Alam K, Kifayat K, Sampedro GA, Karovič V, Naeem T (2024) SXAD: Shapely eXplainable AI-based anomaly detection using log data. IEEE Access
2. Pravin SC, Rohith G, Kiruthika V, Manikandan E, Methelesh S, Manoj A (2023) Underwater animal identification and classification using a hybrid classical-quantum algorithm. IEEE Access
3. Li L, Dong B, Rigall E, Zhou T, Dong J, Chen G (2021) Marine animal segmentation. IEEE Trans Circuits Syst Video Technol 32(4):2303–2314
4. Cheng J, Wu Z, Wang S, Demonceaux C, Jiang Q (2023) Bidirectional collaborative mentoring network for marine organism detection and beyond. IEEE Trans Circuits Syst Video Technol 33(11):6595–6608
5. Taipalmaa J, Raitoharju J, Queralta JP, Westerlund T, Gabbouj M (2024) On automatic person-in-water detection for marine search and rescue operations. IEEE Access
6. Chun S, Kawamura C, Ohkuma K, Maki T (2024) 3D detection and tracking of a moving object by an autonomous underwater vehicle with a multibeam imaging sonar: toward continuous observation of marine life. IEEE Robot Autom Lett
7. Cheng N, Li M, Xie H, Wang H (2024) Toward a unified framework for feature enhancement-guided marine organism detection. IEEE Sens J
8. Liu J, Li Q, Song S, Kulyash K (2024) Detection, tracking and enumeration of marine benthic organisms using an improved YOLO+ DeepSORT network. IEEE Access
9. Lai YC, Chiu HC, Chen ZY, Lee SK, Wang JH (2023) Marine life counting based on Siamese neural network for AI applications in ecology conservation. In: 2023 IEEE International conference on marine artificial intelligence and law (ICMAIL). IEEE, pp 59–64

10. Bajpai SA, Patankar AB (2024) Marine goal optimizer tuned deep BiLSTM-based self-configuring intrusion detection in cloud. J Grid Comput 22(1):24
11. Chia KY, Chin CS, See S (2023) Deep transfer learning application for intelligent marine debris detection. In: International conference on engineering applications of neural networks. Springer Nature Switzerland, Cham, pp 479–490
12. https://www.kaggle.com/code/vencerlanz09/sea-animals-classification-using-efficeintnetb7/input. Reference for the dataset
13. Bhat A, Raychowdhury A (2023) Reducing overhead of feature importance visualization via static gradcam computation. In: 2023 IEEE 5th International conference on artificial intelligence circuits and systems (AICAS). IEEE, pp 1–5
14. Khushi HMT, Masood T, Jaffar A, Rashid M, Akram S (2023) Improved multiclass brain tumor detection via customized pretrained EfficientNetB7 model. IEEE Access

ParkVIT: A Custom Vision Transformer for Parking Space Classification

Navpreet, Rajendra Kumar Roul, and Rinkle Rani

Abstract Parking space classification is crucial in reducing traffic congestion, curbing air pollution, and enhancing driver convenience. This research presents a robust model combining a Convolutional Neural Network (VCNN) variant, specifically ResNet50, with the Customized Vision Transformer (VIT) for efficient feature extraction and classification. The model performs better by integrating ResNet50's exceptional feature extraction capabilities with VIT's global semantic information. VIT is customized to adapt to the extracted feature vectors. The model's efficacy is rigorously evaluated using the widely recognized PKLot dataset, demonstrating its proficiency in distinguishing between empty and occupied parking spaces through precision, sensitivity, specificity, accuracy, and F1-score metrics. Furthermore, benchmarking against existing deep learning models highlights its superiority. This innovation promises significant advancements in intelligent transportation systems, providing an efficient parking space classification solution that mitigates traffic congestion, prevents unauthorized parking, and fosters brighter urban environments.

Keywords CNN · PKLot · ResNet50 · Transformer · VIT

Rajendra Kumar Roul and Rinkle Rani contributed equally to this work.

Navpreet (✉) · R. K. Roul · R. Rani
Department of Computer Science and Engineering, Thapar Institute of Engineering and Technology, Patiala, Punjab, India
e-mail: nnavpreetphd21@thapar.edu

R. K. Roul
e-mail: raj.roul@thapar.edu

R. Rani
e-mail: raggarwal@thapar.edu

S. Kumar et al. (eds.), *Artificial Intelligence for Next Generation Computing, Volume 1*, Lecture Notes in Networks and Systems 1549,
https://doi.org/10.1007/978-981-95-0701-6_35

1 Introduction

Urban drivers face a growing challenge of locating an empty parking space. Studies show drivers spend an average of 6–8 min searching for a parking space. The pressure to secure a spot can lead drivers to park illegally or unsafely, further adding to traffic woes [1]. Recognizing this urgency, researchers are actively pursuing innovative technologies to transform parking management. These solutions aim to streamline the process of finding available spaces, making parking a less stressful experience for drivers and contributing to a smoother traffic flow [2, 3].

For vision-based parking systems, Convolutional Neural Networks (CNNs) shine and excel at analyzing live camera footage from parking lots, making them ideal for classifying available parking spaces and guiding drivers to them quickly. CNNs can automatically learn features, capturing non-linear relationships between image pixels. This, along with their parameter efficiency, allows for building large, complex models that address challenges beyond the capabilities of traditional ANNs [4, 5]. Building deep CNN models for the new era of applications often necessitates massive training datasets, high-performance computing resources, and significant training time, posing challenges for widespread adoption [6].

In recent years, there has been a notable increase in research focused on classifying parking spaces using both machine learning (ML) and deep learning (DL) techniques [4, 7–11]. Suwignyo et al. [12] employed K-Nearest Neighbors (KNNs) and Support Vector Machines (SVMs) for classification. First, features from the parking space images were extracted using a technique called Quantized Local Ring Basing Pattern (QLRBP). QLRBP is an extension of Local Binary Patterns (LBP) that can handle color information in contrast to LBP, which works on grayscale images. Finally, they validated their approach using a tenfold cross-validation process on a subset of the UFPR04 dataset. Wu et al. [13] proposed a novel method for parking space detection by using a few frames of the video captured from a single camera and proposed an 8-class Support Vector Machine classifier (SVM). The proposed method applies to shadow, occlusion, and varying light intensity. The classification is done in four stages: firstly, frames segmented from the video are preprocessed and divided into small patches containing at least three parking spaces each. The features are extracted from patches using the Gaussian ground model. SVM is trained on these extracted features and classifies patches into eight classes of parking space status. Markov Random Field (MRF) is applied to improve recognition accuracy and remove conflicts between neighboring patches.

Transformers, renowned for their prowess in NLP due to their ability to capture long-range dependencies, are making significant strides in computer vision. The Vision Transformer (VIT) [14] has revolutionized the field by substituting traditional CNN backbones with a transformer-only architecture, driving progress in image classification and semantic segmentation. This study aims to harness the strengths of ResNet50 and VIT, known for their excellent local feature extraction, with transformer architectures that capture global semantic information. This combination seeks to create a more robust and versatile model for various visual tasks.

1.1 Our Contribution

The key contributions in the paper are summarized as follows:

- A lightweight parking space classification model named ResNet50-CVIT is proposed by leveraging the benefits of ResNet50 and a Customized Vision Transformer (CVIT).
- The proposed approach is validated on a well-known PKLot dataset, and the model's generalization ability is determined by training on one subset and testing on different subsets, illustrating consistent performance across diverse parking scenarios.
- A performance comparison of the proposed ResNet50-CVIT is made with traditional ML classifiers such as SVM, GNB combined with ResNet50.
- Further, the proposed model's competitiveness is confirmed by comparing similar parking space classification approaches in terms of precision, recall, and accuracy.

2 Methodology

The proposed model integrates a pretrained ResNet50 model and a custom Vision Transformer (VIT) to classify parking spaces. Figure 1 illustrates the proposed methodology for classifying parking spaces. The methodology involves using ResNet50 to extract feature vectors from images of parking spaces, which are then fed into a customized VIT for classification and are discussed as follows.

- **Dataset Used**: PKLot [15] is an expanded version of the parking lot dataset first released in Almeida 2013 [16]. The dataset comprises images captured within the parking premises of PUCPR (Pontifical Catholic University of Parana) and UFPR (Federal University of Parana) located in Curitiba, Brazil. To reduce the likelihood of occlusion of nearby vehicles, cameras were strategically positioned on the rooftops of buildings. Figure 2 represents the delimited parking slot, segmented occupied, and empty parking slot. The PKLot dataset comprises 12,417 parking lot images taken under uncontrolled illumination and varying weather conditions like sunny, cloudy, and rainy. Table 1 represents the summary of the PKLot dataset. A total of 6,95,899 parking space images are segmented from parking

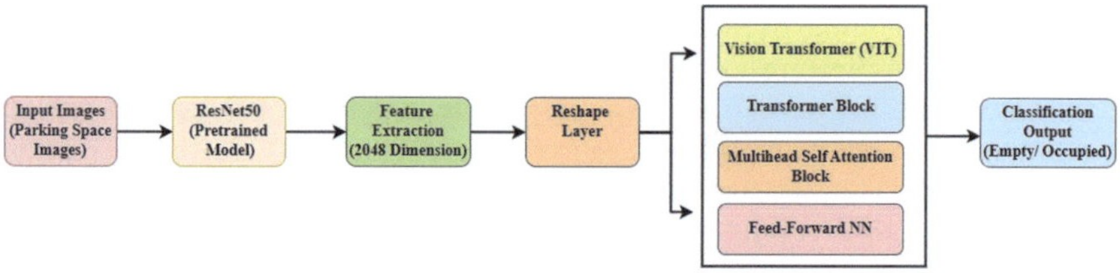

Fig. 1 Proposed methodology

Fig. 2 **a** Parking spot containing 28 delimited parking spaces, **b** occupied parking slot, and **c** empty parking slot

lot images. Each segmented image of the parking spaces is labeled as 'occupied' or 'empty.' The cropped images are arranged inside three folders: UFP04, UFP05, and PUCPR. The first two folders encompass diverse perspectives of a parking lot, each captured from distinct vantage points: the fourth and fifth floors of the UFPR building. Meanwhile, the third folder contains images captured from the tenth floor of the administration building at PUCPR, which contributes to the dataset's comprehensive portrayal of the parking environment from varying heights and locations. The three parking lots' sample images are depicted in Fig. 3, captured in the varying weather conditions stated above. This dataset is valuable for developing and evaluating parking occupancy classification systems and related research in the field.

- **Dataset Processing**: The images within the dataset undergo a resizing process, resulting in a uniform resolution of (224 × 224) pixels by image-processing libraries such as OpenCV or PIL. The dataset is systematically divided into two distinct subsets: a training set and a testing set, following a proportion of 70:30. The concise overview of these subsets, comprising their pivotal details, is presented in Table 2.
- **Feature Extractor—Pretrained ResNet50 Model**: Utilize the powerful feature extraction capabilities of ResNet50, pretrained on the PKLot dataset, to transform raw images into feature vectors. ResNet50 is pretrained on the PKLot dataset, which contains images of parking spaces. The trained ResNet50 model processes the images to extract high-level feature representations. These feature vectors capture essential information about the parking spaces, such as occupancy status and environmental conditions. A 2048-dimensional vector extracted from the last pooling layer of ResNet50. Figure 4 illustrates the feature vector extracted from the flattened layer after passing through the fifth convolutional block of the

Table 1 Overview of the PKLot dataset

Parking lot	Weather condition	Parking capacity		
		Occupied parking slots	Empty parking slots	Total parking slots
UFPR04	Sunny	32,166	26,334	58,400
	Rainy	2,351	5,607	7,958
	Cloudy	11,608	27,779	39,387
UFPR05	Sunny	57,584	42,306	99,890
	Rainy	6,078	2,851	8,929
	Cloudy	33,764	23,202	56,966
PUCPR	Sunny	96,762	1,11,672	2,08,433
	Rainy	55,104	27,951	83,056
	Cloudy	42,363	90,417	132,780
Total parking capacity		3,37,780	3,58,119	6,95,899

(a) UFPR04: Rainy (b) UFPR05: Sunny (c) PUCPR: Cloudy

(d) UFPR05 with annotations (e) PUCPR with annotations

Fig. 3 **a–c** are samples of images captured at day and night time, in rainy season [15]. In Figures (**d**, **e**) red and green represent parking status as occupied or empty, respectively

ResNet50 model. Figure 5 illustrates the visualization of extracted feature atexah block of ResNet50. The feature vector contains the abstracted representation of the input image after processing it through the convolutional layers.

- **Custom Vision Transformer (VIT)**: The VIT is customized to accept the shape of the feature vectors produced by ResNet50. Vision Transformer is to classify the parking spaces based on the feature vectors extracted by ResNet50. The VIT

Table 2 Summary of training and testing subsets of PKLot dataset

		Training (Sunny + Rainy + Cloudy)	Testing (Sunny + Rainy + Cloudy)
UFPR04	Occupied	32,287	13,838
	Empty	41,804	17,916
UFPR05	Occupied	68,198	29,048
	Empty	47,851	20,508
PUCPR	Occupied	1,35,960	58,269
	Empty	1,64,523	70,510

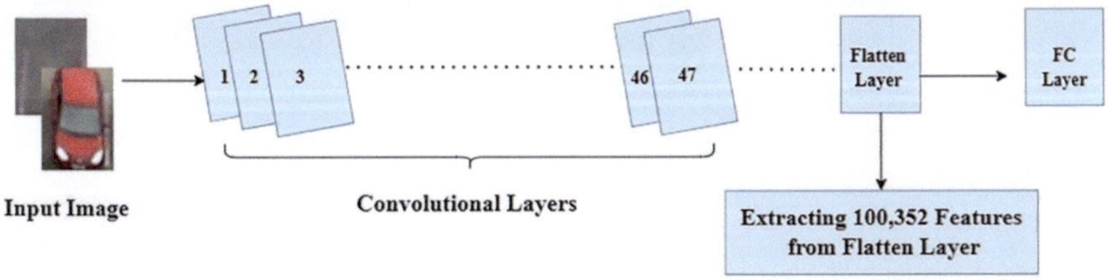

Fig. 4 Feature extraction using ResNet50

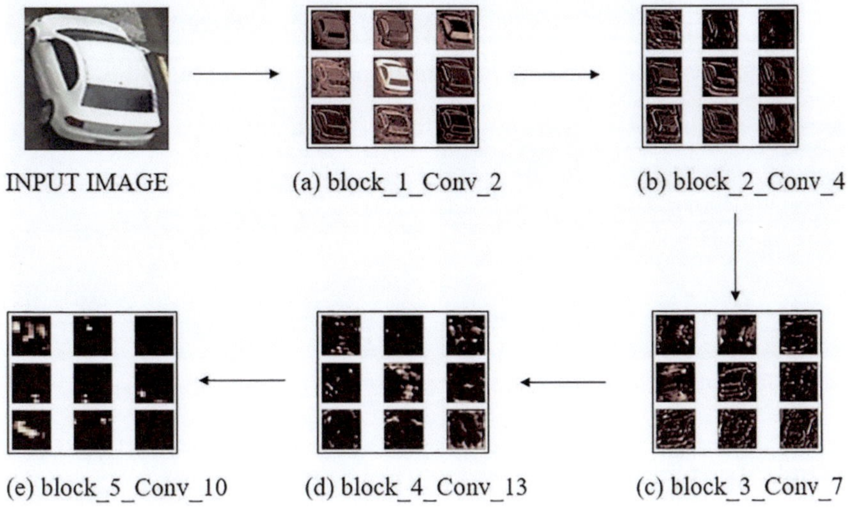

Fig. 5 Visualizing the feature map for the last convolutional layer of each block in the ResNet50

processes these vectors through its transformer blocks and attention mechanisms to classify the parking spaces. The customized VIT consists of transformer blocks and multi-head attention blocks. Table 4 details the structural configuration of the VIT, including the number of transformer blocks, attention heads, embedding dimensions, and feed-forward dimensions. Table 3 lists details about the layers of VIT.

  - **Transformer Block**: Each block includes a multi-head self-attention mechanism and feed-forward neural network.
  - **Multi-head Attention Block**: This mechanism allows the model to focus on different parts of the input feature vector simultaneously, enhancing its ability to capture complex patterns and relationships.

- **Training of Custom VIT**: The training parameters are carefully chosen to optimize the performance of the customized VIT model. Table 5 lists the training parameters, including learning rate, batch size, dropout rates, and the number of epochs. Feature vectors from ResNet50 are split into training and validation sets. The custom VIT is trained on the training set and validated on the validation set. The model is trained using a low learning rate to ensure stable convergence. Dropout and batch normalization are used to prevent overfitting and stabilize training. Early stopping and learning rate reduction are employed to enhance training efficiency and model performance.
- **Classifier**: Integrating a pretrained ResNet50 model with a custom Vision Transformer (VIT) leverages the strengths of both architectures. ResNet50 effectively extracts high-level features from images, while the customized VIT utilizes these features to classify parking spaces accurately. This methodology ensures robust and efficient classification performance suitable for real-world applications.

Table 3 VIT model layers detail

Layer (type)	Output shape	Param#
input_2 (InputLayer)	[(None, 2048)]	0
reshape_1 (Reshape)	(None, 4, 512)	0
vision_transformer_1 (Vision transformer)	(None, 2)	2,708,610
Total params		2,708,610
Trainable params		2,707,586
Non-trainable params		1024

Table 4 Configuration setting of custom vision transformer (CVIT)

Parameter	Value
Number of layers	4
Embedding dimension	512
Number of heads	8
Feed-forward dimension	2048
Dropout rate	0.1
Input shape	(2048, None)

Table 5 Parameters setting for training of custom vision transformer (CVIT)

Parameter	Value
Learning rate	0.0001
Batch size	32
Epochs	50
Early stopping patience	5
Reduce LR factor	0.2
Minimum LR	0.00001

3 Analysis of Experimental Results

The proposed model was developed using a virtual environment with Python 3.9.9 and TensorFlow-GPU 2.8.0 in Jupyter Notebook. The system used for development included an i3 processor and a 24 GB GPU. The dataset was partitioned into a 70:30 ratio for training and validation. The model was trained for 50 epochs with early stopping, monitoring the validation loss to prevent overfitting. Additionally, tenfold cross-validation was employed on the PKLot dataset to evaluate the model's precision, recall, specificity, and accuracy performance.

3.1 Analyzing the Loss and Accuracy

ResNet50-CVIT was trained for 50 epochs on the dataset, and early stopping was applied to avoid overfitting. This was performed for training on one subset and testing on the other subset to check the generalization ability. Figure 6 illustrates the variation of accuracy and loss across the epochs for each fold. The accuracy plot indicates that the model generally achieves high precision, close to 1, but with significant fluctuations across epochs. The loss plot shows similar variability, with frequent peaks and valleys indicating higher and lower loss periods. These oscillations suggest that while the model performs well overall, its performance is not entirely stable across different folds, highlighting the importance of techniques like early stopping to mitigate overfitting and ensure consistent model performance.

3.2 Performance Evaluation

The performance of the CVIT is evaluated by a confusion matrix (CM) that comprehensively summarizes the predictions by comparing them to the actual true labels. Figure 7 represents CM for ResNet50-CVIT validated on various combinations of PKLot dataset subsets.

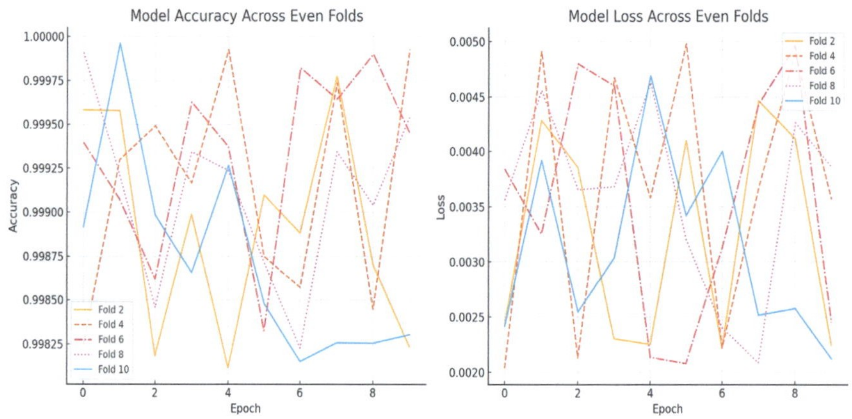

Fig. 6 Loss and accuracy variation across different folds

In Fig. 7,

- True occupied (0) represents the correctly predicted as "occupied" number of parking spaces (true negatives, TN).
- True empty (1) represents the correctly predicted "empty" number of parking spaces (true positives, TP).
- Predicted empty (0) represents the total number of parking spaces predicted as "empty" regardless of whether the prediction is correct (true negatives + false negatives, TN + FN).
- Predicted occupied (1) represents the total number of parking spaces predicted as "occupied" regardless of whether the prediction is correct (true positives + false positives, TP + FP).

The performance metrics used for evaluation of ResNet50-CVIT are precision, recall/sensitivity (True Positive Rate), specificity (True Negative Rate), accuracy, and F1-score, represented by Eqs. 1, 2, 3, 4, and 5, respectively.

$$\text{Precision} = \frac{\text{TP}}{\text{TP} + \text{FP}} \tag{1}$$

$$\text{True Positive Rate(TPR)} = \frac{\text{TP}}{\text{TP} + \text{FN}} \tag{2}$$

$$\text{True Negative Rate(FPR)} = \frac{\text{TN}}{\text{TN} + \text{FP}} \tag{3}$$

$$\text{Accuracy} = \frac{\text{TP} + \text{TN}}{\text{TP} + \text{TN} + \text{FP} + \text{FN}} \tag{4}$$

$$\text{F1 - Score} = \frac{2 \times \text{Precision} \times \text{Recall}}{\text{Precision} + \text{Recall}} \tag{5}$$

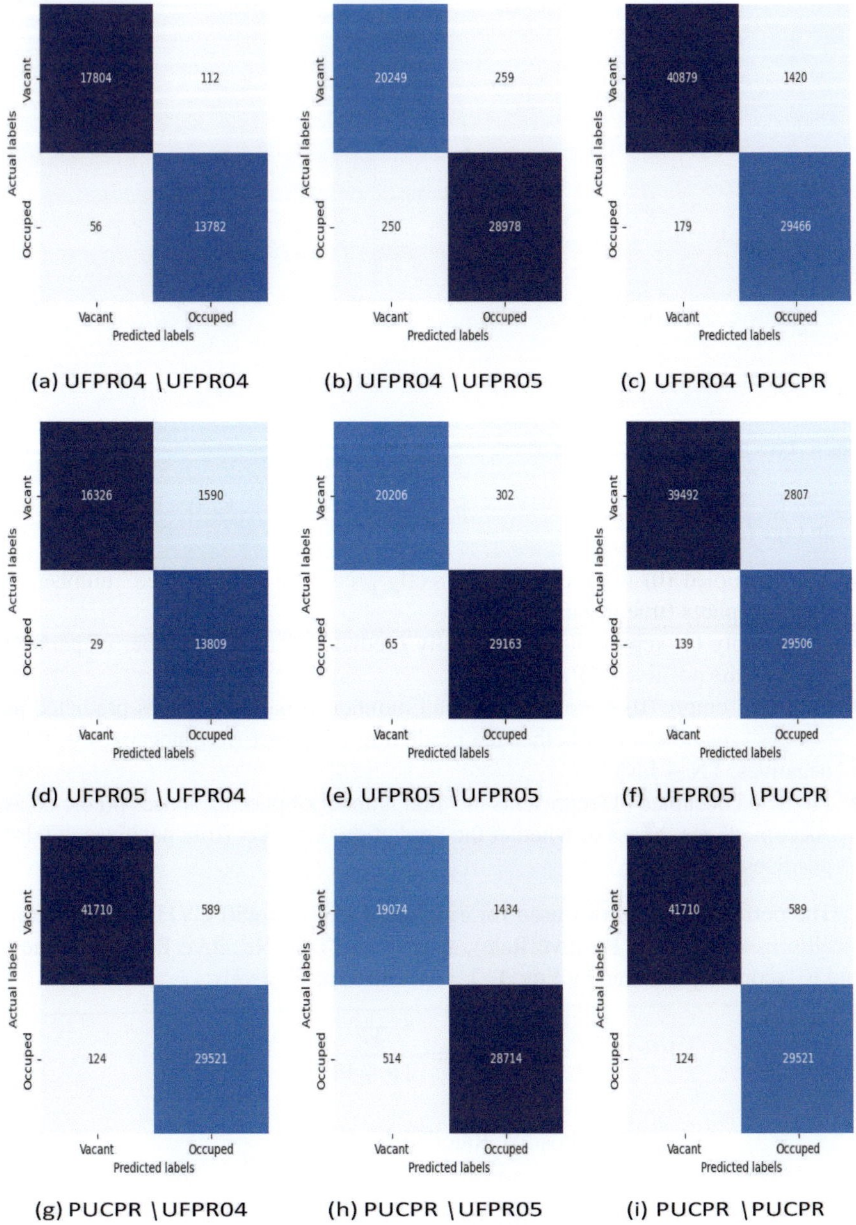

Fig. 7 Confusion matrix of ResNet50-CVIT validated on PKLot subsets

Table 6 Hyperparameters for the training of ML Classifiers integrated with ResNet50

Classifier	Hyperparameters
GNB	var_smoothing = 1e-8
SVM	Random_state = 0, kernel = 'linear', C = 1.0
Random forest	n_jobs = 4,n neighbors = 1
Decision trees	min_samples_split = 10, criterion = 'gini', max depth = 5
Extra trees	n_jobs = 4, criterion = 'gini', max depth = 40, random state = 42, max features = 50, in samples split = 10, min samples leaf = 4, n estimators = 100

3.3 Comparison with Machine Learning (ML) Classifiers

The comparative assessments are encapsulated within Table 7 (bold indicates maximum). It presents a detailed performance analysis in terms of accuracy (acc), sensitivity (sen), and specificity (spe), contrasting ResNet50-CVIT with various hybrid models. These hybrid models ingeniously combine ResNet50 with diverse classifiers such as Support Vector Machines (SVMs), Random Forest (RF), Extra Trees (ET), Decision Trees (DT), and Gaussian Naive Bayes (GNB). This fusion showcases a comprehensive exploration of integrating ResNet50 with well-established classifiers. Table 6 succinctly outlines the hyperparameters utilized across the assorted classifiers.

ResNet50-CVIT has attained maximum accuracy (acc) of 99.47%, 99.87%, and 98.80% for the UFPR04\UPFR04, UFPR05\UPFR05, and PUCPR\PUCPR, respectively. ResNet50-CVIT has attained maximum sensitivity (sen) for seven out of nine cases, as illustrated in Table 7. For the UFPR04\UFPR04 and UFPR04\PUCPR, ResNet50-CVIT has attained the maximum specificity (spe) with 99.19% and 99.40%, respectively.

3.4 Comparison with Existing Deep Learning Models

The performance of ResNet50-CVIT is compared with existing state-of-the-art models based on performance matrices such as precision, recall, and accuracy, illustrated in Tables 8 and 9 (bold indicates maximum), respectively.

- In Table 8, we comprehensively compare ResNet50-CVIT with established deep learning models. The assessment is based on critical performance metrics such as precision and recall. In essence, Table 8 (in Recall column, "-" indicates "not computed") encapsulates the superior performance of ResNet50-CVIT models, showcasing their prowess when pitted against both established benchmarks and contemporary methods.

Table 7 Performance comparison of ResNet50-CVIT with ML classifiers

Model	UFPR04\UFPR04			UFPR04\UFPR05			UFPR04\PUCPR		
	Acc (%)	Sen (%)	Spe (%)	Acc (%)	Sen (%)	Spe (%)	Acc (%)	Sen (%)	Spe (%)
ResNet50-GNB	98.57	97.65	98.39	97.79	94.96	98.15	**99.42**	99.11	99.34
ResNet50-SVM	99.02	99.06	99.80	98.89	97.59	99.06	98.03	96.93	97.65
ResNet50-RF	99.04	99.33	99.08	**99.36**	98.72	**99.16**	96.74	94.71	96.19
ResNet50-DT	98.92	99.14	98.76	98.07	97.03	98.37	94.07	93.38	92.96
ResNet50-ET	99.45	99.33	99.16	99.28	99.08	98.57	96.65	99.36	94.62
ResNet50-CVIT	**99.47**	**99.59**	**99.89**	99.12	**98.97**	99.11	97.77	**99.39**	**99.40**
	UFPR05\UFPR04			UFPR05\UFPR05			UFPR05\PUCPR		
ResNet50-GNB	95.78	92.76	95.37	99.75	98.00	99.17	99.11	89.95	92.87
ResNet50-SVM	95.86	99.78	**99.83**	98.80	93.59	95.42	99.33	82.19	88.39
ResNet50-RF	**98.47**	97.34	98.28	99.80	99.56	**99.74**	**99.49**	92.38	**94.59**
ResNet50-DT	92.31	92.02	94.46	99.21	98.80	99.18	98.52	92.36	94.09
ResNet50-ET	97.80	**99.93**	96.15	99.72	99.74	99.43	94.14	99.52	91.65
ResNet50-CVIT	94.90	99.69	93.00	**99.87**	**99.77**	98.97	95.24	**99.53**	91.31
	PUCPR\UFPR04			PUCPR\UFPR05			PUCPR\PUCPR		
ResNet50-GNB	**99.36**	88.41	92.70	**99.75**	81.21	93.69	92.75	98.00	99.17
ResNet50-SVM	98.91	98.42	98.44	99.33	95.86	98.23	98.75	99.49	**99.65**
ResNet50-RF	98.84	98.85	**98.67**	99.36	97.78	**98.90**	98.69	98.93	99.09
ResNet50-DT	94.60	95.80	94.58	95.46	97.62	96.85	98.74	98.68	98.54
ResNet50-ET	98.74	99.15	98.28	98.29	**99.47**	95.84	96.80	99.44	98.49
ResNet50-CVIT	96.44	**98.96**	94.41	96.71	98.24	95.24	**98.80**	**99.58**	98.04

Table 8 Precision and recall comparison with existing DL models

Author(s)	Model	Precision (%)	Recall
Polprasert et al. [17]	mAlexNet	79.00	98.00%
Dhuri et al. [18]	VGG16	87.20	–
Mettupally et al. [19]	FMR-CNN	91.90	–
Sairam et al. [20]	Mask R-CNN	92.33	–
Ding et al. [21]	YOLO v3	93.30	90.90%
Agrawal et al. [22]	Mask R-CNN	95.00	–
Kolhar et al. [23]	mAlexNet	97.90	–
Proposed Model	ResNet50-CVIT	**99.19**	**99.59%**

Table 9 Accuracy comparison with existing DL models

Author(s)	Model	Training	Testing: UFPR04	UFPR05	PUCPR
			Accuracy (%)		
Simonyan et al. [25]	VGG19	UFPR04	80.40	92.30	91.90
		UFPR05	88.80	95.10	95.90
		PUCPR	81.50	93.80	94.60
He et al. [24]	ResNet50	UFPR04	93.70	94.80	93.30
		UFPR05	92.20	94.80	95.50
		PUCPR	90.50	93.90	94.10
De et al. [15]	LPQ, LBP and its variants	UFPR04	84.25	99.55	84.92
		UFPR05	87.74	85.76	98.90
		PUCPR	99.58	87.15	82.78
De et al. [15]	Mean Ensemble	UFPR04	88.40	99.64	88.33
		UFPR05	89.83	85.53	99.30
		PUCPR	99.61	88.88	84.20
Nurullayev et al. [26]	CarNet	UFPR04	98.30	95.60	97.60
		UFPR05	98.40	95.20	97.50
		PUCPR	98.80	94.40	97.70
Amato et al. [27]	mAlexnet	UFPR04	98.27	99.54	93.29
		UFPR05	92.72	93.69	99.49
		PUCPR	99.90	98.03	96.00
Chollet et al. [28]	Xception	UFPR04	94.00	94.60	93.40
		UFPR05	95.70	90.90	91.20
		PUCPR	96.30	92.50	93.30
Szegedy et al. [29]	Inception V3	UFPR04	91.70	95.20	92.40
		UFPR05	94.30	92.90	93.70
		PUCPR	90.80	91.10	94.20
Proposed Models	ResNet50-CVIT	UFPR04	99.47	98.97	97.77
		UFPR05	98.90	99.26	95.90
		PUCPR	99.83	99.08	98.09

- ResNet50-CVIT is benchmarked against existing parking classification models [15, 24–26]. Table 9 demonstrates that ResNet50-CVIT outperforms existing models in accuracy, showcasing its superiority over conventional methods. CVIT exhibits the highest accuracy across subsets such as UFPR04\UFPR04 (99.47%), UFPR04\PUCPR (97.77%), PUCPR\PUCPR (98.09%), PUCPR\UFPR04 (99.83%), PUCPR\UFPR05 (99.08%), UFPR05\UFPR05 (99.26%), and remains competitive for other subset combinations. CarNet from Nurullayev et al. [26] attains the maximum accuracy for UFPR05\UFPR04 and Mean Ensemble from

De et al. [15] exhibits superior performance for the subset UFPR04\UFPR05 compared to existing methods. Amato et al.'s mAlexNet [27] excels for the UFPR05\PUCPR. The proposed model outperforms six out of nine subsets in terms of accuracy.

3.5 *Discussion*

The performance of ResNet50 integrated with Custom VIT was validated on the PKLot dataset. VIT was trained on the dataset directly without applying feature extraction techniques and attained an accuracy of 65%. On the other hand, the ensembled model achieves the highest accuracy of 99.90% by training for some epochs. The generalizing capability was also tested by training on one sunset and validated on other subsets of PKLot, and vice-versa. Integrating ResNet50 with a custom Vision Transformer (ViT) for parking space detection presents challenges such as increased computational complexity, slower inference times, and higher data requirements, particularly since ViTs typically need larger datasets to perform well. The hybrid model also risks overfitting and can be difficult to optimize due to the contrasting architectures of ResNet's convolutional layers and ViT's self-attention mechanisms. Additionally, the combined model may face deployment challenges on edge devices due to its resource-intensive nature, limiting its practicality for real-time applications.

4 Conclusion

This innovative approach combines the strengths of pretrained CNN and custom vision transformer (CVIT) for parking space classification. It operates in two stages: First, ResNet50 is used as a feature extractor for parking images captured under various illumination conditions. Then, VIT processes these vectors, progressively refining its layer as per the shape of the extracted featured vector and classifying them as empty.

References

1. Aftab K, Kulkarni P, Shergold I, Jones M, Dogramadzi M, Carnelli P, Sooriyabandara M (2020) Reducing parking space search time and environmental impacts: a technology driven smart parking case study. IEEE Technol Soc Mag 39(3):62–75
2. Xia Y, Zhang L, Liu Y (2016) Special issue on big data driven intelligent transportation systems. Neurocomputing 181:1–3
3. Xiong J, Shi H, Guo G, Liu X, Xiao C, Zhu S (2023) Research on intelligent transportation and shared parking spaces in urban areas based on multi-source data integration. Comput Electr Eng 110:108835

4. Thakur N, Bhattacharjee E, Jain R, Acharya B, Hu Y-C (2023) Deep learning-based parking occupancy detection framework using ResNet and VGG-16. Multimedia Tools Appl 1–24
5. Albawi S, Mohammed TA, Al-Zawi S (2017) Understanding of a convolutional neural network. In: 2017 International conference on engineering and technology (ICET). IEEE, pp 1–6
6. Han D, Liu Q, Fan W (2018) A new image classification method using CNN transfer learning and web data augmentation. Expert Syst Appl 95:43–56
7. Sneha Channamallu S, Kermanshachi S, Michael Rosenberger J, Pamidimukkala A (2024) Enhancing urban parking efficiency through machine learning model integration. IEEE Access 12:81338–81347
8. Errousso H, Alaoui EAA, Benhadou S, Nayyar A (2024) Intelligent parking space management: a binary classification approach for detecting vacant spots. Multimedia Tools Appl 1–41
9. Soumana ANH, Salah MB, Idbraim S, Boulouz A (2024) Machine learning models in the large-scale prediction of parking space availability for sustainable cities. EAI Endorsed Trans Internet Things 10
10. Grbić R, Koch B (2023) Automatic vision-based parking slot detection and occupancy classification. Expert Syst Appl 225:120147
11. Navpreet, Roul RK, Rani R et al (2024) Cultivating road safety: a comprehensive examination of intelligent ensemble-based road crack detection. Multimedia Tools Appl 1–34
12. Suwignyo MA, Setyawan I, Yohanes BW (2018) Parking space detection using quaternionic local ranking binary pattern. In: 2018 International seminar on application for technology of information and communication. IEEE, pp 351–355
13. Wu Q, Huang C, Wang S-Y, Chiu W-C, Chen T (2007) Robust parking space detection considering inter-space correlation. In: 2007 IEEE International conference on multimedia and expo. IEEE, pp 659–662
14. Dosovitskiy A, Beyer L, Kolesnikov A, Weissenborn D, Zhai X, Unterthiner T, Dehghani M, Minderer M, Heigold G, Gelly S et al (2020) An image is worth 16 × 16 words: transformers for image recognition at scale. arXiv preprint arXiv:2010.11929
15. de Almeida PRL, Oliveira LS, Britto AS, Silva EJ, Koerich AL (2015) Pklot–a robust dataset for parking lot classification. Expert Syst Appl 42(11):4937–4949
16. AlmeidaP, Oliveira LS, Silva E, Britto A, Koerich A (2013) Parking space detection using textural descriptors. In: 2013 IEEE International conference on systems, man, and cybernetics. IEEE, pp 3603–3608
17. Polprasert C, Sruayiam C, Pisawongprakan P, Teravetchakarn S (2019) A camera-based smart parking system employing low-complexity deep learning for outdoor environments. In: 2019 17th International conference on ICT and knowledge engineering (ICT&KE). IEEE, pp 1–5
18. Dhuri V, Khan A, Kamtekar Y, Patel D, Jaiswal I (2021) Real-time parking lot occupancy detection system with VGG16 deep neural network using decentralized processing for public, private parking facilities. In: 2021 International conference on advances in electrical, computing, communication and sustainable technologies (ICAECT). IEEE, pp 1–8
19. Mettupally SNR, Menon V (2019) A smart eco-system for parking detection using deep learning and big data analytics. In: 2019 SoutheastCon. IEEE, pp 1–4
20. Sairam B, Agrawal A, Krishna G, Sahu SP (2020) Automated vehicle parking slot detection system using deep learning. In: 2020 Fourth International conference on computing methodologies and communication (ICCMC). IEEE, pp 750–755
21. Ding X, Yang R (2019) Vehicle and parking space detection based on improved yolo network model. In: Journal of physics: conference series, vol. 1325. IOP Publishing, p 012084
22. Agrawal T, Urolagin S (2020) Multi-angle parking detection system using mask R-CNN. In: Proceedings of the 2020 2nd International conference on big data engineering and technology, pp 76–80
23. Kolhar MS, Alameen A (2021) Multi criteria decision making system for parking system. Comput Syst Sci Eng 36(1):101–116
24. He K, Zhang X, Ren S, Sun J (2016) Deep residual learning for image recognition. In: Proceedings of the IEEE conference on computer vision and pattern recognition, pp 770–778

25. Simonyan K, Zisserman A (2014) Very deep convolutional networks for large-scale image recognition. arXiv preprint arXiv:1409.1556
26. Nurullayev S, Lee S-W (2019) Generalized parking occupancy analysis based on dilated convolutional neural network. Sensors 19(2)
27. Amato G, Carrara F, Falchi F, Gennaro C, Vairo C (2016) Car parking occupancy detection using smart camera networks and deep learning. In: 2016 IEEE symposium on computers and communication (ISCC). IEEE, pp 1212–1217
28. Chollet F (2017) Xception: deep learning with depthwise separable convolutions. In: Proceedings of the IEEE conference on computer vision and pattern recognition, pp 1251–1258
29. Szegedy C, Vanhoucke V, Ioffe S, Shlens J, Wojna Z (2016) Rethinking the inception architecture for computer vision. In: Proceedings of the IEEE conference on computer vision and pattern recognition, pp 2818–2826

MIX
Papier aus verantwortungsvollen Quellen
Paper from responsible sources
FSC® C105338

If you have any concerns about our products,
you can contact us on
ProductSafety@springernature.com

In case Publisher is established outside the EU,
the EU authorized representative is:
Springer Nature Customer Service Center GmbH
Europaplatz 3, 69115 Heidelberg, Germany

Printed by Libri Plureos GmbH
in Hamburg, Germany